ENVIRONMENTAL MICROBIOLOGY

Environmental Microbiology

Editor:

Sarika Garg, Ph.D.

ArclerPress

www.arclerpress.com

Environmental Microbiology
Editor: Sarika Garg, Ph.D.

© 2017 Arcler Press LLC
708 3rd Avenue, 6th Floor
New York
NY 10017
United States of America
www.arclerpress.com

ISBN: 978-1-68094-478-5

Library of Congress Control Number: 2016951342

About the Editor

Sarika Garg

Ph.D. Montreal, Canada

Sarika Garg procured her doctoral degree at the Max Planck Unit For Structural Molecular Biology, University of Hamburg, Germany in 2011. This was followed by Postdoctoral Scientist position at University of Saskatchewan, Canada. Later, she moved to CRCHUM, University of Montreal for her second Postdoc. She has diverse research interests and has contributed immensely in many research fields viz. Enzyme Technology, Environmental Science, and Neuroscience. She has numerous research papers and review articles to her credit. She has coauthored two books 'Enzymes & Enzyme Technology'; and 'Biochemical Tests: Principles and Protocols'. She has also contributed several chapters in various books including a chapter on Regulation of Gene Expression in Encyclopedia of Molecular Cell Biology and Molecular Medicine. She is a recipient of 2012 SHRF Postdoctoral Fellowship Award and 2012 Laura E. Chapman Research Award. She is currently working as a Career Counselor in her very own venture 'higherstudiescounseling.com', Montreal, Canada and as a freelancer science writer.

Contents

List of Contributors

Dominika Drzewiecka
Department of General Microbiology, Institute of Microbiology, Biotechnology and Immunology, University of Łódź, 90-237 Łódź, Poland

S. Schulz
Helmholtz Zentrum Munchen, German Research Center for Environmental Health, Research Unit Environmental Genomics, ¨ Ingolstadter Landstr. 1, 85764 Neuherberg, Germany

R. Brankatschk
ETH Zurich, Environmental Microbiology, Institute of Biogeochemistry and Pollutant Dynamics, Universitatsstr. 16, 8092 ¨ Zurich, Switzerland

A. Dumig
Technische Universitat M¨ unchen, Lehrstuhl f ¨ ur Bodenkunde, 85350 Freising-Weihenstephan, Germany

I. Kogel-Knabner
Technische Universitat M¨ unchen, Lehrstuhl f ¨ ur Bodenkunde, 85350 Freising-Weihenstephan, Germany
Technische Universitat M¨ unchen, Institute for Advanced Study, Lichtenbergstrasse 2a, 85748 Garching, Germany

M. Schloter
Helmholtz Zentrum Munchen, German Research Center for Environmental Health, Research Unit Environmental Genomics, ¨ Ingolstadter Landstr. 1, 85764 Neuherberg, Germany

J. Zeyer
ETH Zurich, Environmental Microbiology, Institute of Biogeochemistry and Pollutant Dynamics, Universitatsstr. 16, 8092 ¨ Zurich, Switzerland

Yuguang Zhang
Institute of Forestry Ecology, Environment and Protection, and the Key Laboratory of Forest Ecology and Environment of State Forestry Administration, Chinese Academy of Forestry, Beijing, China

Jing Cong
Institute of Forestry Ecology, Environment and Protection, and the Key Laboratory of Forest Ecology and Environment of State Forestry Administration, Chinese Academy of Forestry, Beijing, China
School of Mineral Processing and Bioengineering, Central South University, Changsha, China

Hui Lu
Institute of Forestry Ecology, Environment and Protection, and the Key Laboratory of Forest Ecology and Environment of State Forestry Administration, Chinese Academy of Forestry, Beijing, China

Caiyun Yang
Institute for Environmental Genomics and Department of Microbiology and Plant Biology, University of Oklahoma, Norman, Oklahoma, United States of America

Yunfeng Yang
State Key Joint Laboratory of Environment Simulation and Pollution Control, School of Environment, Tsinghua University, Beijing, China

Jizhong Zhou
Institute for Environmental Genomics and Department of Microbiology and Plant Biology, University of Oklahoma, Norman, Oklahoma, United States of America
State Key Joint Laboratory of Environment Simulation and Pollution Control, School of Environment, Tsinghua University, Beijing, China
Earth Sciences Division, Lawrence Berkeley National Laboratory, Berkeley, California, United States of America

Diqiang Li
Institute of Forestry Ecology, Environment and Protection, and the Key Laboratory of Forest Ecology and Environment of State Forestry Administration, Chinese Academy of Forestry, Beijing, China

Ireneusz Całkosiński
Laboratory of Neurotoxicology and Environmental Diagnosis, Faculty of Health Science, Wroclaw Medical University, 51-618 Wroclaw, Poland

Katarzyna Płoneczka-Janeczko
Division of Infectious Diseases and Veterinary Administration, Department of Epizootiology and Clinic of Birds and Exotic Animals, Faculty of Veterinary Medicine, Wroclaw University of Environmental and Life Sciences, 50-366 Wroclaw, Poland

Magda Ostapska
Laboratory of Neurotoxicology and Environmental Diagnosis, Faculty of Health Science, Wroclaw Medical University, 51-618 Wroclaw, Poland

Krzysztof Dudek
Department of Logistics and Transport Systems, Faculty of Mechanical Engineering, Wroclaw University of Technology, 50-371 Wroclaw, Poland

Andrzej Gamian
Department of Immunology of Infectious Diseases, Institute of Immunology and Experimental Therapy, 53-114 Wroclaw, Poland

Krzysztof Rypuła
Division of Infectious Diseases and Veterinary Administration, Department of Epizootiology and Clinic of Birds and Exotic Animals, Faculty of Veterinary Medicine, Wroclaw University of Environmental and Life Sciences, 50-366 Wroclaw, Poland

João P. S. Cabral
Center for Interdisciplinary Marine and Environmental Research (C. I. I. M. A. R.), Faculty of Sciences, Oporto University, Rua do Campo Alegre, 4169-007 Oporto, Portugal

Zahar Marziah
Malaysia-Japan International Institute of Technology, Universiti Teknologi Malaysia, Jalan Sultan Yahya Petra, Kuala Lumpur 54100, Malaysia

Akbariah Mahdzir
Malaysia-Japan International Institute of Technology, Universiti Teknologi Malaysia, Jalan Sultan Yahya Petra, Kuala Lumpur 54100, Malaysia
Ocean Thermal Energy Centre (OTEC-KL), Universiti Teknologi Malaysia, Jalan Sultan Yahya Petra, Kuala Lumpur 54100, Malaysia

Md. Nor Musa
Ocean Thermal Energy Centre (OTEC-KL), Universiti Teknologi Malaysia, Jalan Sultan Yahya Petra, Kuala Lumpur 54100, Malaysia

Abu Bakar Jaafar
Perdana School of Science, Technology and Innovation Policy, Universiti Teknologi Malaysia, Jalan Sultan Yahya Petra, Kuala Lumpur 54100, Malaysia

Azran Azhim
Kuliyyah of Science, International Islamic University Malaysia, Kuantan 25200, Malaysia

Hirofumi Hara
Malaysia-Japan International Institute of Technology, Universiti Teknologi Malaysia, Jalan Sultan Yahya Petra, Kuala Lumpur 54100, Malaysia

Vessoni Thereza Christina Penna
Department of Biochemical and Pharmaceutical Technology, School of Pharmaceutical Sciences, University of São Paulo, São Paulo, Brazil

Silva Alzira Maria Martins
Department of Biochemical and Pharmaceutical Technology, School of Pharmaceutical Sciences, University of São Paulo, São Paulo, Brazil

Priscila Gava Mazzola
Department of Biochemical and Pharmaceutical Technology, School of Pharmaceutical Sciences, University of São Paulo, São Paulo, Brazil

Shibu Yooseph
Informatics, J. Craig Venter Institute, San Diego, California, United States of America

Cynthia Andrews-Pfannkoch
Synthetic Biology and Bioenergy, J. Craig Venter Institute, Rockville, Maryland, United States of America

Aaron Tenney
Informatics, J. Craig Venter Institute, San Diego, California, United States of America

Jeff McQuaid
Microbial and Environmental Genomics, J. Craig Venter Institute, San Diego, California, United States of America

Shannon Williamson
Microbial and Environmental Genomics, J. Craig Venter Institute, San Diego, California, United States of America

Mathangi Thiagarajan
Informatics, J. Craig Venter Institute, Rockville, Maryland, United States of America

Daniel Brami
Informatics, J. Craig Venter Institute, San Diego, California, United States of America

Lisa Zeigler-Allen
Microbial and Environmental Genomics, J. Craig Venter Institute, San Diego, California, United States of America

Jeff Hoffman
Microbial and Environmental Genomics, J. Craig Venter Institute, San Diego, California, United States of America

Johannes B. Goll
Informatics, J. Craig Venter Institute, Rockville, Maryland, United States of America

Douglas Fadrosh
Microbial and Environmental Genomics, J. Craig Venter Institute, San Diego, California, United States of America

John Glass
Synthetic Biology and Bioenergy, J. Craig Venter Institute, Rockville, Maryland, United States of America

Mark D. Adams
Microbial and Environmental Genomics, J. Craig Venter Institute, San Diego, California, United States of America

Robert Friedman
Microbial and Environmental Genomics, J. Craig Venter Institute, San Diego, California, United States of America

J. Craig Venter
Microbial and Environmental Genomics, J. Craig Venter Institute, San Diego, California, United States of America

Marcus Adonai Castro da Silva
Centro de Ciências Tecnológicas da Terra e do Mar, Universidade do Vale do Itajaí (UNIVALI), Rua Uruguai, 458, Itajaí, SC CEP 88302202, Brazil

Angélica Cavalett
Centro de Ciências Tecnológicas da Terra e do Mar, Universidade do Vale do Itajaí (UNIVALI), Rua Uruguai, 458, Itajaí, SC CEP 88302202, Brazil

Ananda Spinner
Centro de Ciências Tecnológicas da Terra e do Mar, Universidade do Vale do Itajaí (UNIVALI), Rua Uruguai, 458, Itajaí, SC CEP 88302202, Brazil

Daniele Cristina Rosa
Centro de Ciências Tecnológicas da Terra e do Mar, Universidade do Vale do Itajaí (UNIVALI), Rua Uruguai, 458, Itajaí, SC CEP 88302202, Brazil

Regina Beltrame Jasper
Centro de Ciências Tecnológicas da Terra e do Mar, Universidade do Vale do Itajaí (UNIVALI), Rua Uruguai, 458, Itajaí, SC CEP 88302202, Brazil

Maria Carolina Quecine
Departamento de Genética, Universidade de São Paulo, Escola Superior de Agricultura "Luiz de Queiroz", Av. Padua Dias, 11, Piracicaba, SP CEP 13418-260, Brasil

Maria Letícia Bonatelli
Departamento de Genética, Universidade de São Paulo, Escola Superior de Agricultura "Luiz de Queiroz", Av. Padua Dias, 11, Piracicaba, SP CEP 13418-260, Brasil

Aline Pizzirani-Kleiner
Departamento de Genética, Universidade de São Paulo, Escola Superior de Agricultura "Luiz de Queiroz", Av. Padua Dias, 11, Piracicaba, SP CEP 13418-260, Brasil

Gertrudes Corção
Departamento de Microbiologia, Imunologia e Parasitologia, Instituto de Ciências Básicas da Saúde, Universidade Federal do Rio Grande do Sul (UFRGS), Rua Sarmento Leite 500, Cidade Baixa CEP 90050-170Porto AlegreRS, Brazil

André Oliveira de Souza Lima
Centro de Ciências Tecnológicas da Terra e do Mar, Universidade do Vale do Itajaí (UNIVALI), Rua Uruguai, 458, Itajaí, SC CEP 88302202, Brazil

Abhrajyoti Ghosh
Department of Biochemistry and Department of Biotechnology, University of Calcutta, 35, Ballygunge Circular Road, Kolkata- 700019, West Bengal, India
Current address: Max Planck Research Group "Molecular Biology of Archaea", Max Planck Institute for Terrestrial Microbiology, Karlvon-Frisch-Strasse, D-35043 Marburg, Germany

Nirmalya Dey
Department of Biochemistry and Department of Biotechnology, University of Calcutta, 35, Ballygunge Circular Road, Kolkata- 700019, West Bengal, India
Current address: Division of Nephrology Department of Medicine, 7703 Floyd Curl Drive, University of Texas Health Science Center at San Antonio, San Antonio, Texas 78229 3900, USA

Amit Bera
Department of Biochemistry and Department of Biotechnology, University of Calcutta, 35, Ballygunge Circular Road, Kolkata- 700019, West Bengal, India

Amit Tiwari
Department of Biochemistry and Department of Biotechnology, University of Calcutta, 35, Ballygunge Circular Road, Kolkata- 700019, West Bengal, India
Current address: Institute of Molecular Cancer Research, University of Zurich, 17 K 28, Winterthurerstrasse 190, Zurich-8057, Switzerland

KB Sathyaniranjan
Department of Biochemistry and Department of Biotechnology, University of Calcutta, 35, Ballygunge Circular Road, Kolkata- 700019, West Bengal, India

Kalyan Chakrabarti
Department of Agricultural Chemistry & Soil Science, Institute of Agricultural Science, University of Calcutta, 35, Ballygunge Circular Road. Kolkata- 700019, West Bengal, India

Dhrubajyoti Chattopadhyay
Department of Biochemistry and Department of Biotechnology, University of Calcutta, 35, Ballygunge Circular Road, Kolkata- 700019, West Bengal, India
Anish Bhattacharyya
Department of Biochemistry, University of Calcutta, 35 Ballygunge Circular Road, Kolkata, West Bengal 700019, India

Niladri Shekhar Majumder
Roche Diagnostics India Pvt. Ltd., Block 4C, Akash Tower, Near Ruby Hospital, 781 Anandapur, Kolkata 700107, India

Pijush Basak
Department of Biochemistry, University of Calcutta, 35 Ballygunge Circular Road, Kolkata, West Bengal 700019, India

Shayantan Mukherji
Department of Biochemistry, Bose Institute, P1/12, C. I. T. Road, Scheme VIIM, Kolkata, West Bengal 700054, India

Debojyoti Roy
Department of Biochemistry, University of Calcutta, 35 Ballygunge Circular Road, Kolkata, West Bengal 700019, India

Sudip Nag
Department of Biochemistry, University of Calcutta, 35 Ballygunge Circular Road, Kolkata, West Bengal 700019, India

Anwesha Haldar
Department of Geography, University of Calcutta, 35 Ballygunge Circular Road, Kolkata, West Bengal 700019, India

Dhrubajyoti Chattopadhyay
Department of Biochemistry, University of Calcutta, 35 Ballygunge Circular Road, Kolkata, West Bengal 700019, India

Suparna Mitra
Norwich Medical School, University of East Anglia and Institute of Food Research, Norwich Research Park, Norwich, Norfolk NR4 7UA, UK

Maitree Bhattacharyya
Department of Biochemistry, University of Calcutta, 35 Ballygunge Circular Road, Kolkata, West Bengal 700019, India

Abhrajyoti Ghosh
Department of Biochemistry, Bose Institute, P1/12, C. I. T. Road, Scheme VIIM, Kolkata, West Bengal 700054, India

Annette Bollmann
Department of Microbiology, Miami University, Oxford, Ohio, United States of America

George S. Bullerjahn
Department of Biological Sciences, Bowling Green State University, Bowling Green, Ohio, United States of America

Robert Michael McKay
Department of Biological Sciences, Bowling Green State University, Bowling Green, Ohio, United States of America

Krushna Chandra Das
National Research Centre on Mithun, Indian Council of Agricultural Research, Medziphema, India
Biorefining Research Institute, Department of Biology, Lakehead University, Thunder Bay, Canada

Wensheng Qin
Biorefining Research Institute, Department of Biology, Lakehead University, Thunder Bay, Canada

Nada Ouhaibi-Ben Abdeljalil
Higher Agronomic Institute of Chott-Mariem, Sousse University, 4042-Chott Mariem, Tunisia
UR13AGR09-Integrated Horticultural Production in the Tunisian Centre East, Region-

al Centre of Research on Horticulture and Organic Agriculture, University of Sousse, 4042, Chott-Mariem, Tunisia

Jessica Vallance
INRA, UMR1065 Santé et Agroécologie du Vignoble (SAVE), ISVV, F-33140 Villenave d'Ornon, France
Université de Bordeaux, Bordeaux Sciences Agro, ISVV, UMR1065 SAVE, F-33140 Villenave d'Ornon, France

Jonathan Gerbore
BIOVITIS, 15400 Saint Etienne de Chomeil, France

Emilie Bruez
INRA, UMR1065 Santé et Agroécologie du Vignoble (SAVE), ISVV, F-33140 Villenave d'Ornon, France
Université de Bordeaux, Bordeaux Sciences Agro, ISVV, UMR1065 SAVE, F-33140 Villenave d'Ornon, France

Guilherme Martins
USC Oenologie-INRA, Université Bordeaux Segalen, Bordeaux Sciences Agro, ISVV, Villenave d'Ornon, France

Patrice Rey
INRA, UMR1065 Santé et Agroécologie du Vignoble (SAVE), ISVV, F-33140 Villenave d'Ornon, France
Université de Bordeaux, Bordeaux Sciences Agro, ISVV, UMR1065 SAVE, F-33140 Villenave d'Ornon, France

Mejda Daami-Remadi
UR13AGR09-Integrated Horticultural Production in the Tunisian Centre East, Regional Centre of Research on Horticulture and Organic Agriculture, University of Sousse, 4042, Chott-Mariem, Tunisia

Alejandro Gonzalez-Martinez
Department of Civil Engineering, University of Granada, Campus de Fuentenueva, s/n, 18071 Granada, Spain

Alejandro Rodriguez-Sanchez
Institute of Water Research, University of Granada, C/Ramón y Cajal 4, 18071 Granada, Spain

Belén Rodelas[2] Ben A. Abbas
Department of Biotechnology, Technical University of Delft, Julianalaan 67, 2628 BC Delft, Netherlands

Maria Victoria Martinez-Toledo
Institute of Water Research, University of Granada, C/Ramón y Cajal 4, 18071 Granada, Spain

Mark C. M. van Loosdrecht
Department of Biotechnology, Technical University of Delft, Julianalaan 67, 2628 BC Delft, Netherlands

F. Osorio
Department of Civil Engineering, University of Granada, Campus de Fuentenueva, s/n, 18071 Granada, Spain

Jesus Gonzalez-Lopez
Institute of Water Research, University of Granada, C/Ramón y Cajal 4, 18071 Granada, Spain

Yu Zhang
[1] State Key Laboratory of Microbial Metabolism and School of Life Science & Biotechnology, State Key Laboratory of Ocean Engineering, Shanghai Jiao Tong University, Dongchuan Rd. 800, Shanghai, 200240, P. R. China
[2] Laboratory of Microbial Ecology and Technology (LabMET), Ghent University, Coupure Links 653, Gent, 9000, Belgium

Loïs Maignien
Laboratory of Microbial Ecology and Technology (LabMET), Ghent University, Coupure Links 653, Gent, 9000, Belgium

Xianxian Zhao
[1] State Key Laboratory of Microbial Metabolism and School of Life Science & Biotechnology, State Key Laboratory of Ocean Engineering, Shanghai Jiao Tong University, Dongchuan Rd. 800, Shanghai, 200240, P. R. China

Fengping Wang
State Key Laboratory of Microbial Metabolism and School of Life Science & Biotechnology, State Key Laboratory of Ocean Engineering, Shanghai Jiao Tong University, Dongchuan Rd. 800, Shanghai, 200240, P. R. China

Nico Boon
Laboratory of Microbial Ecology and Technology (LabMET), Ghent University, Coupure Links 653, Gent, 9000, Belgium

Chioma Blaise Chikere
Department of Microbiology, University of Port Harcourt, Rivers State, Nigeria

Chinedu Christopher Obieze
Department of Microbiology, University of Port Harcourt, Rivers State, Nigeria

Phillip Okerentugba
Department of Microbiology, University of Port Harcourt, Rivers State, Nigeria

Preface

Microorganisms are one of the first living organisms that populated Earth. Microbial life is exceptionally diverse and microorganisms have inhabited various environments such as soil, water, air, sediments, and also include the plants and animals. Till recent past, conventional approaches were taken into consideration to study the isolation and characterization of selected microbes in defined communities. However, molecular biology approach has transformed the field. It has ameliorated our knowledge of the microbial communities and this modern approach can be deployed to analyse microorganisms in the environment without much hassle. The present molecular biology approach incorporates analysis of DNA, RNA and proteins obtained from environmental samples. At present, the identity and functions of microbes is determined by applying 'omics' approach.

Microorganisms have exceptional influence on the whole ecosystem. Chemosynthetic bacteria help other organisms in light devoid zones by providing them carbon and energy. Few microbes have the potential to recycle the nutrients. Microorganisms are of prime importance for on-site remediation of agricultural and industrial wastes and subsurface pollution. Each microorganism is capable of degrading toxic waste depending on the nature of each contaminant. Special effects of microbe symbiotic relationship (either good or bad) on the biosphere make it significant.

The initiatives to eradicate the environmental problems highlight the need for collating significant published research work of different scientists across the world to help young researchers in accomplishing their ambitious goals. The significance and role of bacteria in natural environments is described in Chapter 1. Chapters 2-4 focus on the analyses of microorganisms in soil. Study of microbial community in marine environment is discussed in Chapters 5-7. A study of airborne microbial communities is reported in Chapter 8. The analysis of bacterial communities in sediment is described in Chapters 9-12. Isolation and characterization of bacteria of cattle is reported in Chapter 13. Chapter 14 describes the characterization of tomato-associated Rhizobacteria. The last three Chapters of the book describe the assessment of microbial species in an artificial environment.

I hope that this book will serve as assistance to students, researchers, teachers, environmental organizations and industries.

Editor

Sarika Garg, Ph.D.

Montreal, Canada

CHAPTER 1

SIGNIFICANCE AND ROLES OF PROTEUS SPP.BACTERIA IN NATURAL ENVIRONMENTS

Dominika Drzewiecka[1]

[1]Department of General Microbiology, Institute of Microbiology, Biotechnology and Immunology, University of Łódź, 90-237 Łódź, Poland

ABSTRACT

Proteus spp. bacteria were first described in 1885 by Gustav Hauser, who had revealed their feature of intensive swarming growth. Currently, the genus is divided into Proteus mirabilis, Proteus vulgaris, Proteus penneri, Proteus hauseri, and three unnamed genomospecies 4, 5, and 6 and consists of 80 O-antigenic serogroups. The bacteria are known to be human opportunistic pathogens, isolated from urine, wounds, and other clinical sources. It is postulated that intestines are a reservoir of these proteolytic organisms. Many wild and domestic animals may be hosts of Proteus spp. bacteria, which are commonly known to play a role of parasites or commensals. However, interesting examples of their symbiotic relationships with higher organisms have also been described.Proteus spp. bacteria present in soil or water habitats are often regarded as indicators of fecal pollution, posing a threat of poisoning when the contaminated water or seafood is consumed. The health risk may also be connected with drug-resistant strains sourcing from intestines. Positive aspects of the bacteria presence in water and soil are connected with exceptional features displayed by autochthonic Proteus spp. strains detected in these environments. These rods acquire various metabolic abilities allowing their adaptation to different environmental conditions, such as high concentrations of heavy metals or toxic substances, which may be exploited as sources of energy and nutrition by the bacteria. TheProteus spp. abilities to tolerate or utilize polluting

Citation: Dominika Drzewiecka, Significance and Roles of Proteus spp. Bacteria in Natural Environments, DOI 10.1007/s00248-015-0720-6.

compounds as well as promote plant growth provide a possibility of employing these microorganisms in bioremediation and environmental protection.

INTRODUCTION

Proteus Like in Homer's Poem...

Microorganisms belonging to the genus *Proteus* were first described in 1885 by a German microbiologist Gustav Hauser, who had revealed their ability to swarm on solid surfaces. The name *Proteus* came from Homer's "Odyssey" and its character Proteus, who could change his shape and had an ability of endless transformation. Hauser described two species of the genus: *Proteus vulgaris* and *Proteus mirabilis* [81]. The swarming ability, connected with a change of short swimmer cells into long, poli-nucleous and hyper-flagellated swarmer cells, is especially visible in the second species. This is a possible source of the name of *P. mirabilis*, which in Latin means amazing, marvelous, splendid. Hauser might have considered *P. vulgaris* to be more common and ordinary; therefore, he gave it the Latin name meaning widespread, usual. The ability of swarming growth is used in a simple and effective Dienes test to differentiate *Proteus* spp. strains with discriminatory power comparable to ribotyping [114]. The phenomenon described by Dienes in 1946 consists in forming boundaries between the swarming growth of different strains, while isogenic strains merge with each other (Fig. 1). However, the background of the expression of the territorial competition between two swarming non-isogenic strains still remains unclear [8, 48]. The formation of the boundaries (Dienes lines) may depend on different profiles of produced proticines (*Proteus* bacteriocines) and different profiles of strain sensitivity [129]. Budding et al. [19] suggested that one of the meeting strains dominated over the other killing it in the cell–cell contact because round dead cells were observed near the border in the swarm of the dominated strain. The Dienes reaction is connected with at least three recognized gene clusters enabling the self-recognition of *P. mirabilis* bacteria. Two of them (*idr* and *tss*) encode cytotoxins and a type VI secretion system, respectively [151], while the third one (*ids*) encodes the Ids proteins responsible for self-identification [21, 49]. However, the mutation in the *ids* genes does not provoke the killing of parent or mutant strain cells, although the Dienes line is visible between their swarms [48].

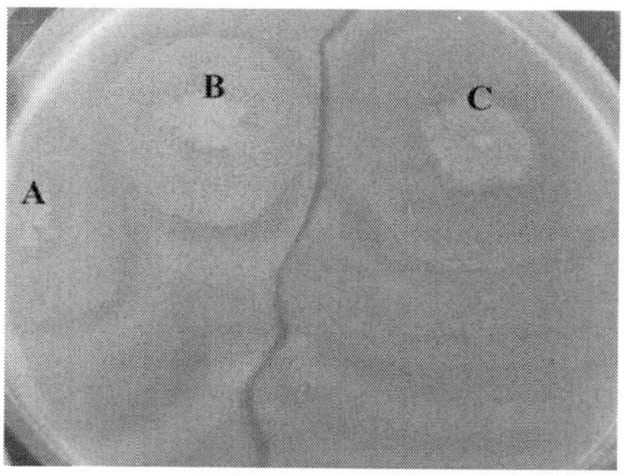

Figure 1: *P. mirabilis* strains swarming on the surface of an agar plate: isogenic A and B (no line of demarcation is visible) versus unrelated C (clear Dienes line of demarcation).

The genus *Proteus* includes Gram-negative, facultative anaerobic, heterotrophic, and proteolytic rods being human opportunistic pathogens. The taxonomic classification of these bacteria has changed several times. Lately, the only *Proteus* species with no clinical significance, *Proteus myxofaciens*, has been postulated to be moved from the genus*Proteus* to a new genus *Cosenzaea* [47]. Among other changes, the exclusion from the genus *Proteus* of several species that created new genera *Providencia* and *Morganella* is worth mentioning. These three closely related genera have formed the tribe *Proteeae* in the family *Enterobacteriaceae* [81]. Currently, the genus *Proteus* consists of *P. mirabilis*,*P. vulgaris*, *Proteus penneri*, *Proteus hauseri*, and three genomospecies 4, 5, and 6. *P. hauseri* as well as the genomospecies were separated from *P. vulgaris* on the grounds of molecular studies and form *P. vulgaris* group. The genomospecies 4, 5, and 6 are marked only with numbers as there have been no metabolic properties indicated to allow their full differentiation [104]. The main biochemical features typical of the genus and distinguishing *Proteus* species are summarized in Table 1.

Table 1: The main metabolic characteristics of *Proteus* spp. [103, 104]: "+"positive in 100 % strains; "–"negative in 100 % strains; in the other cases, the percentage of strains exhibiting positive reaction is given in parentheses

Feature	***P. mirabilis***	***P. penneri***	***P. vulgaris***	***P. hauseri***	***P.*genomospecies 4, 5, and 6**
Typical of the genus and distinguishing it from the other *Enterobacteriaceae*					
Phenylalanine deaminase	+ (98 %)	+ (99 %)	+	+	+
Lysine decarboxylation	–	–	–	–	–
Arginine hydrolysis	–	–	–	–	–
Urease production	+ (98 %)	+	+ (86 %)	+	+
Glucose fermentation	+	+	+	+	+
Mannose fermentation	–	–	–	–	–
Lactose fermentation	– (2 %)	– (1 %)	–	–	–
Methyl-red	+ (97 %)	+	+ (86 %)	+	+
Growth in KCN	+ (98 %)	+ (99 %)	+ (99 %)	+	+ (97 %)
Tyrosine clearing	+ (91 %)	+	+	+	+
Typical of the species and allowing their differentiation					
Ornithine decarboxylation	+ (99 %)	–	–	–	–
Indole production	– (2 %)	–	+	+	+
Lipase production	+ (92 %)	– (35 %)	– (14 %)	–	+ (97 %)
DNase production	+/– (50 %)	– (12 %)	+	–	+ (85 %)
Maltose fermentation	–	+ (96 %)	+	+	+
Sucrose fermentation	– (15 %)	+	+	+	+
Salicin fermentation	–	–	+	–	– (3 %)
Esculin hydrolysis	–	–	+	–	– (3 %)
Citrate utilization	+ (65 %)	–	– (29 %)	–	–

Bacteria from the genus *Proteus* can also be differentiated on the basis of their O-antigen variability, although serotyping is not included in the routine diagnostics of these rods. So far, there have been established 80 O-antigenic serogroups in the genus, some of them divided into subgroups [7, 67, 135], and many new O serotypes are still being discovered (D. Drzewiecka, unpublished data). The chemical structure of the sugar part of the lipopolysaccharide may play an important role in the adaptation of *Proteus* spp. bacteria to environmental conditions and enhancing their pathogenicity, as some O serotypes are more prevalent and more frequently isolated from clinical sources than the others (Table2) [7, 35, 37, 72, 111, 133].

Table 2: Pathogenicity of *Proteus* spp. rods to humans [7, 8, 35, 37, 72, 81, 111, 125, 133, 148]

Infections	Virulence factors	Prevalent O-serotypes
• Urinary tract infections (cystitis, prostatitis, pyelonephritis, kidney stone formation) • Wounds and burns infections, abscesses • Respiratory tract infections • Bacteremia • Meningitis • Intestine colonization (diarrhea) • Nosocomial infections • Rheumatoid arthritis (?) – autoaggressive antibodies may arise due to the molecular mimicry between-*Proteus* hemolysin and urease, and human white blood cells	• Fimbriae – adhesion • Flagella – swimming, swarming growth • Urease – urine pH elevation, kidney stones formation • LPS – endotoxin • O and capsular polysaccharides – swarming facilitation, kidney stones and biofilm formation, serospecificity • Biofilm – protection • Invasiveness – internalization into host cells • Haemolysins – cytotoxicity • *Proteus* toxic agglutinin – cell–cell aggregation, cytotoxicity • Proteases – antibodies degradation • Deaminases – α-keto-acid siderophores production and iron acquisition • Zinc and phosphate transport systems – utilization	• O3, O6, O10, O11, O13, O23, O24, O27, O28, O29, O30 – including*P. mirabilis* and*P. vulgaris* strains • O17, O61, O64, O65 – including *P. penneri*strains mainly • O78 – including *P. mirabilis* strains (data only from Poland) • O79 including *P.*genomospecies strains (data only from Poland)

Proteus spp. bacteria have been isolated from different human and non-human environments and their presence in higher organisms, soil, and water is well documented.

However, their specific features and roles played in their natural habitats have not been summarized so far.

HUMAN BEINGS AND *PROTEUS* BACTERIA—COMMENSALS AND PARASITES

Proteus spp. bacteria are mostly known as opportunistic human pathogens. Their roles in the pathogenesis of human beings as well as their virulence factors enabling the bacteria to reach different niches of the host organism and survive have been extensively studied and reviewed [8, 36, 81, 89, 103, 124, 125] and are summarized in Table 2. The bacteria cause infections mainly in people with an impaired immunity system, and most all of them may be a source of complicated urinary tract and wound infections as well as nosocomial infections. Urinary tract infections are most frequently ascending, often connected with the presence of urinary catheters. It should be noticed that *Proteus* spp. are the most prevalent bacteria, isolated from bladder and kidney stones (70 % of cases). *P. mirabilis* is the main cause of all *Proteus* spp. infections accounting for 80–90 % of them.

It is postulated that human intestines are a reservoir of *Proteus* bacteria, especially those belonging to prevailing *P. mirabilis* species, and they are members of natural fecal microflora of several percent of human population [124]. Lately, Porres-Osante et al. [118] reported the presence of *Proteus* spp. bacteria (one *P. mirabilis* and one *P. vulgaris*strain) in fecal samples from 4 % of healthy Spanish volunteers. Zilberstein et al. [160], using cultivation methods to study the microbiota in particular parts of the digestive tract, stated that *Proteus* spp. bacteria were absent from the mouth and the esophagus of healthy volunteers. However, the bacteria were present in comparatively high mean concentrations in the stomach of 8.1 % of volunteers (10^5 colony forming units(CFU)/mL), the duodenum (45.5 %, 10^2), the proximal jejunum (45.5 %, 10^4), the proximal ileum (20 %, 10^6), the distal ileum (19 %, $10^{3.5}$), the cecum (12.5 %, 10^4), the colon (ascending, 33.3 %, 10^5; transverse, 37.5 %, 10^5; descending, 25 %, 10^5; and sigmoid, 34.8 %, 10^7), and in the rectum (30 %, 10^7). The other members of the family*Enterobacteriaceae—Escherichia coli*, *Enterobacter* spp., and *Klebsiella* spp.—were detected in the same places in the lower digestive tract, but they were more prevalent.

Proteus spp. are regarded to be an undesired element of intestinal microflora, as the bacteria may also become a causative agent of diarrhea. Although Ikeobi et al. [59] did not notice significant differences in the presence of *Proteeae* members in the intestines of healthy individuals and diarrheal patients, Müller [91, 92] isolated *P. mirabilis* strains statistically more frequently from feces of patients suffering from diarrhea than from healthy individuals. The author suggested that the bacteria may become opportunists when the illness is caused by other intestinal pathogens or they may lead to intestinal disorders independently. Also, *P. penneri* strains were more frequently isolated from sick than from healthy people, while *P. vulgaris* were present in almost the same number of individuals in each group.

Thus, the presence of *Proteus* spp. bacteria in the gastrointestinal tract may also be treated as a carrier state, because in some conditions, it may lead to cross-infections and autoinfections especially in the urinary tract, which was observed many years ago by de Louvois [73] using serotyping as well as the Dienes test and confirmed in other studies [8,48]. Chow et al. [26] emphasized the role of the intestinal tract as a reservoir of *P. mirabilis* bacteria leading to person-to-person transmitted nosocomial infection. Drzewiecka et al. [34, 35], by the use of serological and molecular methods as well as the Dienes test, proved that *P. mirabilis* strains isolated from feces and urine of patients were in fact the same spreading clone causing autoinfection and nosocomial infection. Wang et al. [148] reported the case of food poisoning in a restaurant in Beijing, China, caused by *P. mirabilis*. The same clone, identified by genotyping and the Dienes method, was detected in the consumed food and in stools of consumers as well as the cook handler and the waiter who, due to the lack of hygiene, most probably had transported the bacteria to the meal. The fact that dirty hands may be an important step in feces-to-hand-to-mouth spread of *Proteus* spp. bacteria was also confirmed by other researchers. Smith et al. [136] found *P. vulgaris* among the bacteria and yeasts isolated from roadside telephone receivers in Lagos, Nigeria. Padaruth and Biranjia-Hurdoyal [110] detected *Proteus* spp. on hands of primary school pupils in Mauritius while Shojaei et al. [132] on hands of food handlers in Iran (but simple washing resolved the problem). Also, Qadripur et al. [119] found *P. mirabilis* to colonize hand skin between nail plate and nail fold in motor mechanics.

Peerbooms et al. [113] compared *P. mirabilis* strains isolated from urine and feces, and all the isolates seemed to display similar virulence potential, which confirmed that the strains attacking the urinary tract may come from the intestinal reservoir. Also, Senior and Leslie [130] concluded that frequent isolation of *P. mirabilis* from feces and rare occurrence of *P. vulgaris* in the intestines of healthy individuals and those suffering from gastroenteritis may explain the fact that it is the first species and not the second one that is strongly connected with urinary tract infections.

The presence of *P. mirabilis* in the intestines may also be connected with obesity. Lecomte et al. [74], conducting research on an animal model, have recently reported on the changes in gut microflora depending on diet. Studying the intestinal microbiota of rats fed a high-fat diet (43 or 51 % of fat), the authors noticed higher numbers of *P. mirabilis* as compared to the control group (12 % fat diet). Moreover, a significant positive correlation was found between the abundance of *P. mirabilis* (as well as *Phascolarctobacterium* and*Veillonellaceae*) and all ten analyzed metabolic parameters associated with obesity.

PROTEUS SPP. IN ANIMALS—ADVERSE AND FRIENDLY

Not only rats but also many wild and domestic animals (mammals, birds, reptiles, amphibians, insects, and "seafood") are the hosts of *Proteus* spp. bacteria. The relation of the bacteria to their host organism is sometimes not determined; in some cases, it may be symbiotic or change from neutral/commensal to parasitic (Table 3). The microorganisms are an element of animal pathogenic or physiological microflora, especially in the intestines—a reservoir of the bacteria.

Table 3: Different kinds of relationship between *Proteus* spp. and other organisms (details in the text): "+" positive (beneficial), "?" neutral/commensal/not determined, "–" negative (antagonistic, pathogenic)

organisms	*Proteus* sp.	*P. mirabilis*	*P. vulgaris*group	*P. penneri*	selected references
Humans		? / –	? / –	? / –	8, 36, 81, 103, 124, 125, 148
Gorillas		?	?		15
Dogs	–	? / –	?	?	38, 44, 69, 99, 144
Cats, feral cats	–	–			54, 69, 99
Pigs		?	?		68, 75, 147
Horses	? / –				84, 152
Donkeys	–				83
Cow, cattle, calf		?	? / –	?	2, 56, 76, 123, 138
Raccoon dog			?		66
Flying fox		+	+		6
Rats		?			73
Birds, poultry	– (eggs)	?/–	? / –	? / –	9, 43, 46, 61, 62, 68, 71, 123, 154, 155, 161
Snakes		?	?	?	16, 53, 131
Alligator	–				101
Turtles	–	?	? / –	? / –	5, 10, 42, 53, 102, 108, 127
Amphibians			?		53
Fishes	+ / ?	+	+ / ? /–	+	17, 66, 70, 97, 107, 141
Oysters		?			39
Shrimps	+ / ?	+ / ?	+ / ?	+ / ? / –	20, 82, 97, 98
Lobsters	+	+	+	+	97
Blue crab				?	109
Sponges		+/?	?		50, 64
Millipede		+			4

Lepidopteran			–		85
Cockroaches		?	?		142, 145, 149
Honey bees	?				137
Flies	?	? / +	?		18, 51, 79, 80, 85, 87, 94, 134, 144, 150
Mites			+		2
Nematodes				–	76
Leguminous plants			+		12, 14, 120
Wild grass		+			121
Tea			+		12
Cabbage			+		156
Maize		+			60
Mould fungi	–		–		11, 12
C. albicans		–			50
B. bacteriovorus				–	20

P. mirabilis and *P. vulgaris* were found in fecal samples of western lowland gorillas (*Gorilla gorilla gorilla*), collected at two locations in south-central Cameroon, proving to inhabit the intestines of these great wild apes [15]. Like in humans, the presence of*Proteus* spp. in animal intestines may pose a threat of autoinfection and cross-infection. The example of such autoinfection was described by Gaastra et al. [44], who in the Netherlands isolated *P. mirabilis* strains from feces and urine of dogs suffering from recurrent urinary tract infections. It is probable that the intestine was a reservoir of the bacteria infecting the urinary tract of the dogs, because *P. mirabilis* strains were not isolated from feces of healthy controls. Also, Kroemer et al. [69] reported *P. mirabilis*strains as a reason for urinary tract infections in dogs and cats in European countries. Normand et al. [99] indicated *Proteus* spp. isolated from different materials (mostly feces, urine, skin swabs, upper respiratory tracks) as an important cause of illness in dogs and cats examined in Glasgow, UK. Moreover, 26 % of the strains were identified as multiple drug resistant. *P. mirabilis* was isolated from the rectum, the vagina, the mouth, the nose, and wound/abscess of feral cats analyzed in Grenada, West Indies [54]. However, all the isolates were sensitive to most of the used antibiotics, so it was concluded that feral cats did not pose a risk to humans or other cats. Kitamikado and Lee [66] isolated from feces of raccoon dog *Nyctereutes procyonoides*, a *P. vulgaris* strain producing chondroitinase, which may be regarded as one of the virulence factors produced by the microorganism because chondroitin sulfate is distributed in animal connective tissues.

On the other hand, there was no connection found between *Proteus* spp. and ulcerative keratitis [146] or transmissible venereal tumor [38] in dogs, as *P.*

penneri and *P. vulgaris*strains were isolated from the eyes of healthy dogs in Beijing, China, and, respectively, *P. mirabilis* from the vagina of a healthy dog in Nigeria, but not from sick animals.

Proteus spp. bacteria may also be the members of natural microflora of the esophagus [84] and the skin of horses, although they were also isolated from wounds [152]. However, the wound isolates displayed strong adherence and significantly stronger attachment than skin isolates proving enhanced virulence of pathogenic strains, as compared to normal skin isolates. The bacteria are virulent also to donkeys causing urinary tract infections. They were found to account for 6.7 % of donkey uropathogens isolated in Ethiopia, among which *Streptococcus* spp. (43.3 % isolates) and *E. coli* (20 %) were dominating [83].

P. mirabilis and *P. vulgaris* were also detected as the elements of the microbial community in pigs. Lowe et al. [75] revealed that the family *Enterobacteriaceae* was a minor but significant component of pig tonsil microflora in which *P. mirabilis* or *P. vulgaris* clones dominated. Wang et al. [147] reported a *P. vulgaris* strain isolated from nasal swab of a pig from a food producing animal farm (sic!) that was carrying a chromosomally located staphylococcal multiresistance *cfr* gene encoding the resistance to linezolid, but also to other chemically unrelated classes of antimicrobial agents. The gene was for the first time reported in a naturally occurring Gram-negative bacterium. Still Kobashi et al. [68] isolated *P. mirabilis* strains from pig feces detecting the efflux genes responsible for their resistance to tetracycline (*tetH* and *tetJ*).

All *Proteus* species were detected in the cow. Hawkey et al. [56] revealed that *P. vulgaris*(including the strains currently numbered among genomospecies) as well as *P. mirabilis*species were, respectively, the first and the third species among the tribe *Proteeae* most commonly isolated from beddings contaminated with feces and urine in two calf farms in South West England. The authors concluded that the high similarity of the O-serotype profile of isolated strains (e.g., serotypes O23 and O30) to those reported for human infections (Table 2) suggests that food animals may be a source of *Proteeae* strains carried in human gut. Lu et al. [76] reported on the isolation from a cow dung in China of *P. penneri* strains displaying strong nematicidal activity. *P. vulgaris* seems to belong to normal skin microflora of cattle, but the involvement of these bacteria in the damage to tissues around the skin lesions caused by *Demodex bovis* mite and their synergistic influence was also revealed [2]. The symbiosis between the parasites relies on the fact that the mite introducing the bacteria into the skin on the exoskeleton or in the gut receives suitable microclimate for the establishment and replication due to many virulence factors and enzymes produced by the cooperating microorganisms.

Rogers [123] suggested a possibility of the transmission of potentially pathogenic bacteria, including *Proteus* spp., between wild birds and cattle. The author stated the presence of *P. vulgaris* in feces of 13 % dairy cattle in five studied farms, and in fecal and cloacal samples from 7.8 % analyzed birds (sparrows, blackbirds, cowbirds, but not starlings).

Other authors also emphasized the role of wild birds in the transmission and spread of pathogenic bacteria to domestic poultry, cattle, or humans, resulting in the change of their status from bird fecal microflora members to the etiological agent of pathogenesis. Yong et al. [155] found *P. mirabilis* to be prevalent in feces of large-billed crows (*Corvus* spp.) inhabiting the surroundings of a minimarket in Bangsar, Malaysia, and becoming a health hazard due to their large numbers. However, no bacteria (with the exception of *Klebsiella*spp.) were present in the air samples from the market place. Winsor et al. [154] conducted studies on fecal microflora of apparently healthy wild turkey vultures (*Cathartes aura*) in Texas, USA, because the diet of these birds, which are carrion-feeding animals, must include animals that have died of infectious diseases. The content of the studied bird intestines was dominated by *E. coli* but, in fact, *P. mirabilis* was the second predominant species detected in 50 % birds and *P. vulgaris* was isolated from one bird. Jahantigh [61,62] stated that *Proteus* spp. strains in 2010 accounted for 5 % and in 2012 for 12.5 % isolates from eggs of the ostrich (*Struthio camelus*). The authors indicated fecal contamination that may lead to the penetration of the bacteria into the egg interior and the infection which may be a reason for a relatively high ratio of embryonic mortality in ostrich eggs in Iran. Awad-Alla et al. [9] suggested a possible role of white ibises (*Nipponia nippon*) in Egypt in the transmission of some pathogens to poultry as in the internal organs of apparently healthy birds they found several strains classified as *P. mirabilis* (although the indicated ability of indole production suggests that the strains should be identified as *P. vulgaris*; see Table 1). Next, Foti et al. [43] conducted studies on the health status of birds belonging to several orders migrating to Africa through the Ustica Island (Italy). From fecal swabs and internal organs of birds found dead, they isolated only one *P. mirabilis* strain, so the probability of dissemination of these bacteria was very low. Kwiecińska-Piróg et al. [71] revealed that *P. mirabilis* naturally occurring in the crop and the cloaca of white stork (*Ciconia ciconia*) healthy nestlings in Poland were more susceptible to antibiotics than clinical strains isolated from human beings and no ESBL production was detected.

The above data indicate that the presence of *Proteus* spp. in birds does not pose a real risk of dissemination; moreover, *Proteus* spp. bacteria seem to be a member of normal poultry microflora. Among the bacteria inhabiting the beak cavity and the cloaca of reproductive goose flocks (190–800 birds) from 17 farms in Poland, *Proteus* spp. colonized approximately 10–25 % of birds [161], while *E. coli*, *Enterococcus* spp., and *Streptococcus*spp. were found in approximately 60–70 % of birds and coagulase-negative*Staphylococcus* was isolated from 80–90 % birds, independently of the sampling site. Other bacteria were isolated sporadically. Also, Kobashi et al. [68] isolated *P. mirabilis*and *Proteus* sp. strains from poultry feces, detecting the *tetM* gene, encoding their resistance to tetracycline. However, Giacopello et al. [46] recognized antibiotic-resistant *P. mirabilis*, *P. vulgaris*, and *P. penneri* present in feces of sick domestic canaries as pathogenic microflora.

Proteus spp. colonize also amphibians and reptiles, e.g., *P. vulgaris* is reported as a common inhabitant both in the oral cavity and in the cloaca of water amphibians (*Lissotriton vulgaris* newts and *Pelophylax ridibundus* frogs) and reptiles (*Mauremys rivulata* turtles and *Natrix natrix* grass snakes) inhabiting the Kavak Delta, Turkey [53]. In the oral bacterial flora of Chinese cobra, *Naja atra*, and bamboo pit vipers,*Thimeresurus albolabris*, two snake species common in Hong Kong [131], *P. penneri*, *P. vulgaris*, and *P. mirabilis* strains were isolated among many other potentially pathogenic bacteria. Blaylock [16] reported on the common isolation of *Proteus* sp., *P. mirabilis*, and*P. vulgaris* from oral swabs of four different house snake species from southern Africa. There is also a report on pathogenic *Proteus* sp. as a cause (together with *Morganella morganii*) of the septicemia and death of a captive alligator held in the Savannah River Ecology Laboratory in Aiken, SC, USA [101].

There are several reports available concerning the presence of *Proteus* spp. bacteria in water turtles, sometimes connected with illness. They were the fourth genus most frequently isolated from lesions of sea turtles from the Canary Islands, Spain, and were considered as one of the causes of their diseases and mortality [108]. Al-Bahry et al. [5] interpreted the presence of *Proteus* spp. and *P. vulgaris* (resistant to ampicillin, streptomycin, and tetracycline) in oviductal fluids of nesting sea green turtles (*Chelonia mydas*) as an indicator of pollution in the surrounding areas between the Gulf of Oman and the Arabian Sea (see the next paragraph). In contrast, in nesting green turtles from Costa Rica, *P. mirabilis* and *P. vulgaris* bacteria were recognized as a non-pathogenic constituent of microflora. They were isolated from nasal and cloacal swabs from apparently healthy turtle females with no signs of disease. There was no correlation observed between the turtles and seawater bacteria composition, so it was suggested that*Proteus* bacteria should be treated as the physiological microflora of green turtles [127]. *P. mirabilis* was also commonly found as inhabiting the cloaca of European pond turtles (*Emys orbicularis*), especially captive ones, from Poland [102]. *P. vulgaris* was the predominant species, and some *P. mirabilis* strains were also identified in cloacal and oral swabs of loggerhead sea turtles (*Caretta caretta*) inhabiting the Sicilian Channel, the South Tyrrhenian Sea, and the Ionian Sea [42]. The loggerhead turtle is included in the Red List of the world conservation union, and the microbial contamination of the turtle eggs is suspected to play a role in embryonic death and a low loggerhead hatch success rate in Georgia, USA. Indeed, among other Gram-negative isolates from unhatched eggs of the turtle on Jekyll Island, Georgia, *P. penneri* and *P. vulgaris* strains were identified [10].

Although *Proteus* spp. bacteria are commonly known as opportunistic pathogens, there can be found interesting examples of positive relations between the microorganisms and the host animal (Table 3). *P. mirabilis* and *P. vulgaris* strains isolated from the intestine of the Indian flying fox (*Pteropus giganteus*) were recognized as the members of symbiotic physiological microflora of this big fruit bat [6]. Although this is unusual for the two species (Table 4), the isolates were able to produce cellulolytic and xylanolytic enzymes just as three other isolated species, *Citrobacter freundii*, *Serratia liquefaciens*, and*Klebsiella oxytoca*, and contrary to the other six gut isolates unable to digest cellulose. Fruits and leaves, which are the animal's main food, are built of up to 50 % of cellulose, hemicellulose (xylan), lignocellulose, and pectin cellulose, and because mammals do not possess proper enzymes to degrade the polymers, the symbiotic enterobacteria play an essential role in their nourishment and digestion. Another example of a cellulolytic and xylanolytic *P. mirabilis* strain isolated as a gut symbiont of millipede (*Xenobolus carnifex*) was reported by Alagesan et al. [4].

Table 4: Unusual physiological features displayed by *Proteus* spp. strains isolated from different habitats (details in the text)

Feature	***Proteus* sp.**	***P. mirabilis***	***P. vulgaris***	***P. hauseri***
Heterotrophic nitrification		Coastal seawater		
Cellulose digestion		Flying fox, millipede	Flying fox	
Lipase production/ hydrocarbons utilization (including aromatic ones)	Contaminated soil; waste sludge	Contaminated soil	Contaminated soil; contaminated fish	
Phenol utilization		Contaminated soil		
Methyl tert-butyl ether (MTBE) degradation		Contaminated soil		
ε-Caprolactam utilization		Contaminated soil		
Hexachlorocyclohexane (HCH) pesticide utilization	Contaminated soil			
Phorate pesticide utilization	Contaminated soil			
Chlorpyrifos, methyl parathion, and p-nitrophenol pesticides degradation	Contaminated soil (in consortium)	Contaminated soil (in consortium)		
DDT reduction			Mouse	
Azo dyes decolorizing	Waste site	Contaminated soil; wastewater sludge		Hot spring
Phosphate solubilization	Phorate contaminated soil	Wild grass rhizosphere		
Copper tolerance		Wild grass rhizosphere; wastewater	Soil	Hot spring
Chromium,cobalt, cadmium, zinc,mercury, nickel, lead, arsenic tolerance		Wild grass rhizosphere; wastewater; contaminated soil		
Silver tolerance		Wastewater		
Chromate tolerance	Contaminated seawater			

Thermotolerance				Hot spring
Halotolerance	Salt lake	Oysters	Halophyte glasswort rhizosphere	
Acidotolerance	Soil contaminated by hydrocarbons		Acidic soil	Hot spring

Also, a symbiotic role has been suggested for *P. mirabilis* and *P. vulgaris* isolated from hematophagous sand fly *Phlebotomus papatasi* being a vector of *Leishmania* parasite, as the bacteria were the most prevalent in larvae, pupae, and mature male and female insects gut [80]. Another very interesting hypothesis indicating a close relationship between *P. mirabilis* and its blowfly host *Lucilia sericata* was formed by Ma et al. [79]. These proteolytic microorganisms are able to produce volatile components, for example, putrescine (from ornithine) and ammonia (see Table 1), important for their swarming ability and, at the same time, attracting flies to the carcass. The authors stated that putrescine, which is an extracellular signal required for the swarming phenomenon and used by *P. mirabilis* in *quorum sensing* [89], may be an interkingdom signal sensed by both the insects and the bacteria. The same role was suggested for ammonia. Both compounds as well as four other attractants (NaOH, KOH, phenol, and lactic acid) restored the swarming motility in different swarming-deficient mutants of the *P. mirabilis*strain isolated from maggot salivary glands where they dominated, so the researchers linked fly attraction with *Proteus* swarming. The bacteria were detected also in adult *L. sericata* flies and in *Lucilia cuprina* sister species [134]. The symbiosis between blowflies and *Proteus* bacteria is also indicated by the fact that *P. mirabilis* biofilm, contrary to biofilms constructed by *Staphylococcus aureus* and *Enterobacter cloacae*, is not disrupted or is even stimulated by *L. sericata* maggots used in the debridement therapy. Simultaneously, *P. mirabilis* is antagonistic to some bacteria eliminated by maggots from wounds, protecting the larvae from harmful microorganisms [18]. Complete sterility of *L. cuprina* maggots for wound therapy was achieved in all cases except for *P. mirabilis* [87]. Additionally, Wei et al. [150] revealed that both sensitive and antibiotic-resistant *P. mirabilis* strains could persist for several days among the gut microflora of the green bottle fly (*L. sericata*) or the housefly (*Musca domestica* L.) when introduced by feeding. It has been speculated that the mechanisms stimulated by the fly host may contribute to the maintaining of antibiotic-resistant strains in particular, and in that way, their transmission is imminent. It was also suggested [144] that flies, as the vectors of bacteria to raw meat devoted to dog breeding, were responsible for a high percentage of morbidity and mortality due to intestinal infections among pups in greyhound dog kennels in Kansas, USA. A high percentage of bacterial contamination among blowflies (different species) and domestic flies (*M. domestica*), stable flies (*Stomoxys calcitrans* L.), flesh flies, and others was observed, while *Proteus* spp. were proved to be the most common bacteria among Gram-negative and lactose-negative ones isolated from flies, followed by*Providencia* spp., *Pseudomonas* spp., and *Salmonella* spp. Nazni et al. [94] identified*Proteus* as the second (after *Enterobacter*) dominating genus on the external body surface, while in the fly gut, *Proteus* spp. dominated among the bacteria isolated from the housefly (*M. domestica*) on a poultry farm in Malaysia. *P. mirabilis* with *Providencia* spp. was also prevalent in gut microflora of flies captured in different public places in India [51].

Not only flies but also cockroaches are regarded as common vectors of different microorganisms, including *P. mirabilis* and *P. vulgaris* strains, carrying them on their bodies and posing a threat of their dissemination, food contamination, and spoiling as well as infection of humans. The bacteria were found in Nigerian cockroaches in Ekpoma, a village characterized by poor sanitary conditions [142], and in Iranian brown-banded cockroaches collected in kitchens of Ahvaz houses [145]. Wannigama et al. [149] reported on the isolation of *P. mirabilis* from 8.9 % cockroaches found in households and food-handling establishments in Varanasi, India.

P. vulgaris was isolated from tissues of an American fly *Drosophila paulistorum* [85], although it was not determined if the microorganisms were parasitic, mutualistic, or symbiotic to their host. However, they displayed strong pathogenicity toward lepidopterans like *Heliothis virescens*. Other insects which may carry *Proteus* spp. bacteria in the intestines are bees [137]. It is suggested that the source of the bacteria is pollen consumption and bees in the colony are infected one by one during food exchange. Honey pollution by *Proteus* spp. may pose a threat to consumers.

A similar situation may be observed in water animals, so called "seafood," where the presence of *Proteus* spp. may result in food spoiling and poisoning, e.g., scombroid poisoning of fish meat as a result of histidine decarboxylation leading to a rise in the level of toxic histamine [100]. A source of bacteria, including antibiotic-resistant ones, may be fecally contaminated water (see the next paragraph), because *P. vulgaris* and *P. mirabilis*rods were found as absorbed on body structures of the commercially important sponge*Spongia officinalis*, an animal inhabiting the Aegean Sea and feeding by the seawater filtration [64]. Also, Graça et al. [50] reported on the isolation of several *Proteus* sp. strains (closely related to *P. mirabilis*) from the marine sponge (*Erylus deficiens*) collected 150 km off the southwest coast of Portugal. The authors revealed that *Proteus* sp. strains (together with *Pseudoalteromonas* and *Microbacterium* spp. isolates) presented the strongest bioactivity against pathogenic bacteria and *Candida albicans*, thus protecting the host animal and assisting its survival, also due to the fact that the absorbed bacteria constitute 50–60 % of the sponge biomass.

However, the occurrence of these opportunistic human pathogens in oysters poses a health risk if the shellfish are consumed raw. Fernandez-Delgado et al. [39] found halophilic (growth in saline concentrations from 3 to 8 %) *P. mirabilis* as prevailing in the bodies of two oyster species in Venezuela and resistant to several tested antibiotics (mainly tetracycline, ampicillin and penicillin, and cefoxitin and cefazolin). Also, Matyar et al. [82] isolated several antibiotic-resistant *P. vulgaris* and *P. penneri* strains from the intestines of shrimps inhabiting the Iskenderun Bay, Turkey, which was most probably due to the Iskedrun Bay contamination by industrial and domestic wastes, including hospital ones. It is worth noting that *Proteus* spp. strains were absent from seawater and sediments, although high amounts of fecal coliforms indicated strong fecal contamination of the Bay. Nimrat et al. [98] identified *Proteus* spp. and *P. mirabilis* among other bacteria detected in spermatophores from black tiger shrimps (*Penaeus monodon*) collected from the Andaman Sea, Thailand. Cao et al. [20] recognized *P. penneri* as an agent causing red body disease of commercial white shrimps *Penaeus vannamei*, cultivated in Xiaoshan, Zheijang, China. Interestingly, the mortality of shrimps was successfully inhibited by the predatory activity of *Bdellovibrio bacteriovorus* against the *P. penneri* pathogen; thus, the authors propose this bacterial predator as a potential biocontrol agent. *P. penneri* strain inter alia was also detected on the body surface of the wild blue crab (*Callinectes sapidus*) in the Akyatan Lagoon (the south of the Mediterranean Sea) [109].

Proteus spp. rods are found in marine fishes. In Atlantic mackerel (*Scomber scombrus*), *P. vulgaris* and *Proteus* sp. were detected in gills, skin, and gut as the only members of the family *Enterobacteriaceae* [141]. In *Scomber japonicus* mackerel (the intestine and the stomach) or in *Limanda herzensteini* flat fish (the gills), *P. vulgaris* was stated as producing a big yield of intracellular and extracellular chondroitinase, which may be directed against the host connective tissues, containing mucopolysaccharide chondroitin [66]. Also, freshwater Nile tilapias (*Oreochromis niloticus*) from experimental freshwater aquaculture in Brazil [17] and tilapias from Lake Victoria, Kenya [107], were sporadically colonized by *P. vulgaris* or *Proteus* sp. Kumar et al. [70] reported on *P. hauseri* as a causative agent of hemorrhage and mortality in a koi carp (*Cyprinus carpio*) farm in India. This is the first animal habitat reported for this poorly recognized species, while the sources of isolation of two previously described *P. hauseri* strains (one human) remain unknown [104] and two other strains were isolated from human urine [63].

The presence of *Proteus* spp. bacteria was also revealed to have a surprisingly positive influence on water animals. Many *P. mirabilis*, *P. vulgaris*, *P. penneri*, *P.* genomospecies 4, and unidentified *Proteus* sp. strains isolated from the intestines of black tiger shrimp, cobia marine fish, snubnose pompano marine fish, and ornate spiny lobster in Vietnam demonstrated probiotic properties linked to bacteriocin production and their antagonistic activity towards many pathogenic bacteria [97].

PROTEUS AS AN INDICATOR OF FECAL POLLUTION

The presence of *Proteus* spp. bacteria in water and soil may indicate the fecal pollution of the environments where these proteolytic bacteria are treated as allochthonic. Human and animal feces are probably an important source of these rods in natural environments. Water animals may absorb pollutant microorganisms from water. There is a potential risk of their spread in the marine food chain as well as dissemination during the food processing and transmission to humans after consumption. For instance, bacteria (in that number *Proteus* spp.) associated with sponges (*S. officinalis*), which are the main filter feeders in marine environment, were proposed to be treated as indicators of fecal contamination of marine ecosystems [64]. Other marine animals that are reported to accumulate *Proteus* spp. from water environment are oysters [39], loggerhead turtles (*C. caretta*) [42], and green turtles (*C. mydas*) [5], although Santoro et al. [127] did not find the correlation between green turtle microflora and seawater microorganisms.

Also, antibiotic-resistant *Proteus* spp. strains may be released from the intestines which are a source of drug-resistance genes [126]. Some authors [5, 39, 82] emphasize the fact that marine environment seems to be a reservoir of genes responsible for the antibiotic resistance of polluting bacteria, as many antibiotic-resistant strains, including *Proteus*spp., are isolated from water and sea animals (see the previous paragraph). A good example is also the Jiaozhou Bay on the western coast of the Yellow Sea, China, which is highly contaminated due to intensive industrial development and urbanization. Effluents from hospitals and wastewater may be the sources of drug-resistant bacteria, allowing the resistance genes transfer to environmental microflora. Many bacterial strains resistant to tetracycline or chloramphenicol, including *P. mirabilis*, were isolated from the seawater by Dang et al. [28, 29]. Also, in Cameroon studies of water samples collected for a period of 8 months from Douala Lagoon, contaminated by industrial and domestic wastes, resulted in the isolation of *P. vulgaris* strains among other fecal bacteria in each sample, which poses a serious health problem [3].

The fact of *Proteus* spp. survival in marine environments indicates the ability of these microorganisms to adapt to higher salinity conditions, similarly to the strains from oysters, mentioned earlier. Indeed, several halotolerant *Proteus* spp. strains were detected in water samples from El Golea Salt Lake, Algerian Sahara, proving that domestic wastes from the El-Goela oasis could be a source of the fecal microflora surviving in these conditions [52].

Proteus spp. can also be found in freshwater, as an indicator of its contamination by feces, even in unexpected habitats. Microbiological investigation of water and sediments from the Vrelo Cave, the Republic of Macedonia [33], revealed mostly the presence of *Bacillus*spp. strains (83 %), but, additionally, allochthonic strains were detected, including *P. penneri*, in a water sample taken from a place located 400 m from the cave entrance and 100 m deep. Moreover, although physical, chemical, and biochemical parameters indicated a high water quality, big numbers of total and fecal coliform bacteria were observed both in water and in sediment samples, suggesting their contamination from an animal source.

The detection of bacteria belonging to the genus *Proteus* in drinking water disqualifies its suitability for consumption due to its fecal pollution, which would threaten with waterborne infections. The problem occurs in India. *P. vulgaris* was claimed to be the main microbial pollutant of drinking water in Rajasthan [140]. Then, in bore well waters in Mysore City, *P. mirabilis* and *P. vulgaris* strains dominated over the other H_2S-producing strains (considered as associated to fecal coliforms in drinking water) [93]. Poonia et al. [117] reported on the alarming presence of multidrug resistant *P. mirabilis*and *P. vulgaris* strains among other bacteria in drinking water from springs and streams in the rural areas of Sikkim. Also, in Nigeria, *Proteus* spp. bacteria were detected in two of five studied well waters, treated as a source of drinking water [1].

P. mirabilis, *P. vulgaris*, and *P. penneri* strains, in addition to other members of*Enterobacteriaceae*, were recognized as bacteria connected with an anthropic influence, indicating the fecal contamination of rice field water from irrigation channels and rice field plots in Rio Grande do Sul, Brazil [122].

Also, in soil, the presence of *Proteus* spp. bacteria is regarded as the evidence for its fecal contamination. Srinivasan et al. [138] reported on the isolation of *P.*

mirabilis strains from the soil samples from a dairy farm area that was regularly treated with cow manure, but not from the control forest soil samples with no history of agriculture. Trawińska et al.[143] detected both *Proteus* spp. and *E. coli* strains in soil samples, collected near or 300 m from the reproductive layer farm in Poland, although the grounds were not treated with manure or exploited. Thus, it was suggested that the bacteria associated with fecal contamination of soil may have come from other animals, because the soil samples also contained *Toxocara* spp. eggs, which are not found in birds.

PROTEUS IN BIOREMEDIATION AND PLANT GROWTH PROMOTION

Proteus spp. bacteria, as an element of intestinal microflora of humans and animals, are often treated as zymogens in soil or water. In fact, these proteolytic microorganisms come to these environments with feces or waste, and after digestion of approachable organic matter, they decay due to the lack of nutrients.

However, these microorganisms are also found in such habitats as autochthones, well adapted to the environmental conditions, exhibiting unusual and exceptional metabolic features (Table 4), although this aspect of the genus *Proteus* lifestyle is less known. An interesting example is a *P. mirabilis* strain isolated from coastal seawater in China, which was characterized as a heterotrophic nitrifier [158]. The strain effectively removed ammonia (NH_4^+) ions by oxidation. Moreover, only trace amounts of NO_2^- and NO_3^- were detectable after cultivation as the ions were simultaneously reduced by the bacterium to gaseous nitrogen (N_2) in the denitrification processes. The processes and the cell growth were inhibited in the absence of any organic source of carbon. The strain is suggested to be used for toxic NH_4^+-N removal, although its activity in oligotrophic water was limited probably due to the lack of carbon sources. Similar metabolic activity was displayed by the*Proteus* sp. strain related to *P. mirabilis*, isolated from effluents from a fish processing plant in India [86]. Simultaneous aerobic nitrification and denitrification leading to the efficient removal of nitrogen by this heterotrophic bacterium are suggested to be applied in fish waste treatment.

Proteus spp. bacteria, though well known as opportunistic pathogens, in natural environments show more positive aspects of their existence. They may play a role of effective and specialized plant-growth-promoting rhizobacteria (PGRP) or bioremediators of hydrocarbons, pesticides, herbicides, aromatic compounds, azo dyes, and heavy metals in contaminated environments.

Lipase production is not typical of all *Proteus* species (Table 1), but the rods in natural environments are able to effectively degrade hydrocarbons, including oils, and to remove these hazardous substances efficiently and inexpensively. Kim et al. [65] reported on a *P. vulgaris* strain producing extracellular alkaline lipase isolated from soil samples collected near a sewage disposal plant in South Korea. The lipase was stable from pH 5 to 11 and had a maximum activity at pH 10. Lu et al. [77] isolated from soil in China a *P. vulgaris* strain which produced alkaline non-position-specific lipase, whereas Whangsuk et al. [153] found in a waste-sludge from beer factory in

Taiwan a *Proteus* sp. strain actively giving off organic-solvent-tolerant lipase named LipA, which effectively conversed palm oil into biodiesel.

The first report on a *Proteus* sp. strain which could be useful as a sole degrader of oil hydrocarbons in soil was presented by Hernandez-Rivera et al. [57]. The strain may be presumed to belong to *P. vulgaris* group due to indole production and maltose fermentation abilities observed by the authors (see Table 1). It was found in tropical soil containing total petroleum hydrocarbons (TPHs) in the Tabasco region, Mexico, highly contaminated by oil spills during 20 years of pollution and was able to remove the superficial hydrocarbon layer (the only carbon source) in the culture medium forming a stable emulsion, most probably due to the production of biosurfactants by the strain itself. The results were better than in the medium containing urea (actively utilized by the bacteria; see Table 1). Then, Ibrahim et al. [58] studying soil samples collected from the rhizosphere of legumes planted on crude-oil-contaminated soil in Kaduna, Nigeria, found*P. mirabilis* and *P. vulgaris* strains as belonging to the most active crude oil degraders among the several isolated species, although not displaying biosurfactant production. Earlier, another hydrocarbon degrading *P. vulgaris* strain had been isolated from newly killed fish samples that were collected near the point of spill in the Niger Delta, where most of the crude oil in Nigeria is found. The isolated strain was able to utilize Bonny light crude oil, diesel, and kerosene, generating organic acids. The finding is promising for this region, where the oil spills are a source of significant air, soil, and water pollution, destroying biodiversity in the ecosystem [105]. Lutz et al. [78] described a wild-type bacterial cocktail marketed as Superbugs® and composed of bacteria belonging to genera*Proteus*, *Bacillus*, *Pseudomonas*, *Citrobacter*, and *Enterobacter*, as effective in the biodegradation of ethyl biodiesel made from palm triacylglycerols in Costa Rica.

There are several reports on *Proteus* spp. strains able to biodegrade polyaromatic hydrocarbons (PAHs). The genus *Proteus* dominated after *E. coli* in the group of 60 different bacterial strains able to degrade hydrocarbons, classified as hazardous waste in soil (benzene, toluene, octane, heptanes, biphenyl, naphthalene, camphor, and phenanthrene), isolated from soil samples near different petrol pumps of Karachi City, Pakistan [139]. The genes responsible for the hydrocarbon degradation ability were located either chromosomally or extra-chromosomally. The plasmid location is promising for bioremediation processes since the genes can be conjugated to other microorganisms in polluted environments. Ceyhan [22] reported on a *P. vulgaris* strain, isolated from biofilm in wastewater of the petrochemical industry in Turkey, which was effective in the degradation of pyrene (four-ring PAH) as a sole source of carbon and energy. Moreover, the degradation of this highly toxic and carcinogenic hydrocarbon resulted in non-toxic and non-accumulating metabolites, proving a big biodegradation potential of the strain.

Gasoline-contaminated soil from a gas station in Chihuahua, Mexico, was the habitat of a*P. mirabilis* strain degrading with medium efficiency methyl tert-butyl ether (MTBE), which is a toxic synthetic compound added to gasoline in a high concentration of 15 %, as a blending component [90]. Next, a *P. mirabilis* strain was found, which was able to efficiently degrade phenol as a sole carbon and energy source in an oil-contaminated soil sample collected in India [88]. Sanuth et al. [128] identified a *P. mirabilis* strain which was the most effective in ε-caprolactam degradation among the bacteria isolated from soils collected from the major solid waste dumpsites in Lagos State, Nigeria. ε-Caprolactam is a monomer for nylon-6 production, found in wastewater effluents from nylon-producing factories, toxic to plants, animals, and humans. The strain seems to be a potential candidate for its bioremediation.

Bacteria from the genus *Proteus* solely or in consortia also display an ability to neutralize different toxic herbicides and pesticides that may cause heavy pollution in terrestrial and aquatic ecosystems, especially when inappropriately used. *P. vulgaris* from a mouse intestine was one of the first reported bacteria solely able to reduce dichloro-diphenyl-trichloroethane (DDT) pesticide to dichloro-diphenyl-dichloroethane (DDD) [41]. Although DDD is also a toxic pesticide, its production is the first step during the degradation and mineralization of DDT, and it is utilized in further transformations.

Correa and Steen [27] found a *P. mirabilis* strain to be the fastest degrader of the propanil (a commonly used herbicide) among the natural microflora inhabiting a pristine lake in northeast Georgia, USA, followed by the bacteria belonging to the genera *Aeromonas*,*Aerobacter*, and *Acinetobacter*.

Proteus spp. bacteria isolated from the rhizosphere of rice in West Bengal, India, used hexachlorocyclohexane (HCH) pesticide [30] or phorate insecticide [31, 32] as a source of carbon and energy, and the addition of these chemicals to soil intensified the growth of bacteria. It is worth mentioning that *Proteus* spp. were the only members of the family*Enterobacteriaceae* isolated from these soil samples besides typical soil bacteria, actinomycetes, and fungi. Another example of autochthonous *Proteus* sp. strain, isolated from phorate-contaminated soil in India and able to degrade this insecticide, was described by Bano and Musarrat [11]. The strain exhibited also siderophore production, phosphate solubilizing capacities, and strong antifungal effect on phytopathogen*Fusarium oxysporum*, thus indicating the potential possibilities of its exploitation as PGPR as well.

The presence of *P. vulgaris* and *Proteus* sp. strains in microbial consortia capable of pesticide degradation was detected in highly contaminated soils in Colombia [116]. From soil samples exposed to different kinds of waste (including hazardous ones), a microbial consortium was isolated composed of ten strains, including *P. vulgaris* and *Proteus* sp. The consortium was able to degrade two widely used organophosphatic pesticides, chlorpyrifos and methyl parathion, both in culture medium and in soil. A similar bacterial consortium was found in highly contaminated soil samples from Moravia, Medellin, an area that was used as a garbage dump from 1974 to 1982 [115]. The consortium was able to degrade methyl parathion and p-nitrophenol as the only source of carbon and to decrease their toxicity in the medium and in soil at different depths.

Azo dyes frequently used for textile dyeing and paper printing and their metabolites present in effluents may be toxic, mutagenic, or carcinogenic. Their physicochemical neutralization is costly, while biological degradation is cost-effective and friendly to the environment, but difficult. For that reason, identification of the effective *Proteus* spp. degraders is a promising perspective for the environmental protection. Patil et al. [112] have isolated a *Proteus* sp. strain (related to *P. vulgaris*) from a waste site of the textile industry in India, which was capable of degrading seven textile azo dyes, but the most effective decolorization and detoxification of Navy Blue Rx were observed due to the induction of laccase and other enzymes. Chen et al. [25] isolated from sludge, obtained from a dyeing wastewater treatment plant in Taiwan, a *P. mirabilis* strain exhibiting a great capacity of efficient enzymatic reduction and decolorization associated with the biosorption of deep red RED RBN as well as deep black BK-5 azo dyes. Furthermore, Olukanni et al. [106] isolated from municipal dump site soil near Lagos, Nigeria, a *P. mirabilis* strain able to degrade a Reactive Blue 13 azo dye to phyto-non-toxic products. However, its laccase (the azo-dye-degrading oxidoreductase) activity was 70-fold lower than that of another very effective strain, *P. hauseri*, isolated from the Chiao-His hot spring in Taiwan [95, 157]. It effectively decolorized several mono- and di-azo dyes [23,24] and was copper-tolerant as its laccase was a copper-induced enzyme with thermophilic and acidophilic properties [159]. However, the presence of copper in the environment reduced the swarming activity of the bacterium [96]. The isolate is the first reported non-pathogenic strain in the species *P. hauseri*, coming from natural water habitat.

There are more reports on the high tolerance of *Proteus* spp. bacteria to copper and other heavy metals. Ge et al. [45], from the East Sea (China) contaminated by chromate, isolated a *Proteus* sp. strain (related to *P. penneri* and *P. hauseri*) which was highly tolerant to Cr (VI) and could significantly reduce the amounts of this toxic metal in seawater. Hassen et al. [55] in Tunisia isolated from wastewater many Gram-negative bacteria with dominating *Pseudomonas* spp. and *P. mirabilis* strains, highly tolerant especially to copper and also to chromium, cobalt, cadmium, zinc, and mercury. Also, a highly resistant to several heavy metals (including mercury, copper, zinc, cadmium, cobalt, silver, and others) *P. mirabilis* strain was identified in wastewater samples from Casablanca City, Moroccco, also exhibiting resistance to naphthalene, anthracene, and antibiotics [40]. Another strongly copper resistant strain identified as *P. vulgaris* and isolated from soil samples collected near the Panki power plant in Kampur, India, was reported by Rani et al.[120]. The strain used to inoculate pigeon pea (*Cajanus cajan*) seeds protected the pea against the inhibitory effect of copper, reducing its amounts both in soil and in the plant. Moreover, the bacterium was able to produce siderophores providing pigeon pea with iron and preventing chlorosis.

There are more examples of soil *Proteus* spp. strains beneficial to plants (Table 3). Islam et al. [60] reported on a *P. mirabilis* strain highly resistant to zinc, found in agricultural field soil, irrigated with industrial effluents in Faisalabad City, Pakistan. The strain was able to block zinc absorption by maize (*Zea mays*) roots, reduce oxidative stress in the plant, and increase its tolerance to zinc, promoting maize growth in the presence of the heavy metal. Simultaneously, the bacterium enhanced phytoremediation processes conducted by this copper-resistant plant. Rau et al. [121] characterized a *P. mirabilis* strain isolated from rhizosphere of wild grass *Saccharum ravennae* colonizing fly ash dumps in Delhi, India. The strain displayed strong resistance to arsenic and was medium resistant to copper, chromium, cobalt, cadmium, zinc, mercury, nickel, and lead. Additionally, the strain produced siderophores with great capacity and was an active phosphate solubilizer, thus enhancing the growth of grass. Similarly, a *Proteus* sp. strain from a rice field in Pakistan inoculated on wheat seeds effectively solubilized rock phosphate composted with poultry litter, stimulating the plant growth [14]. Yet, rhizobacterial *P. vulgaris* strains, identified in acidic soil from a tea (*Camellia sinensis*) plantation in India, expressed siderophore production and antifungal properties both in acidic and in neutral pH conditions [12]. The strains successfully protected three tested species of leguminous plants from the pathogenic fungus *Fusarium moniliformae*; when the seeds were bacterized during the sowing, they were able to colonize roots and promote the growth of host plants. Lu et al.[76] found in cow dung in China *P. penneri* strains which produced volatile organic compounds killing within 1 h 98–100 % of tested nematodes: soil *Caenorhabditis elegans*and plant-pathogenic *Meloidogyne incognita*, thus expressing the possibility of being used as a protecting plant biocontrol agent. Yu and Lee [156] stated that indole production by the *P. vulgaris* strain (Table 1) isolated from the rhizosphere of halophyte glasswort (*Salicornia herbacea* L.) effectively promoted the growth of Chinese cabbage seedling (*Brassica campestris* ssp. *pekinensis*), while Bhattacharyya et al. [13] reported that the same strain stimulated the growth of *Arabidopsis thaliana* (a small flowering plant from the family *Brassicacae*) by indole production influencing the auxin, cytokinin, and brassinosteroid signaling pathways.

These promising results open wider possibilities of using specialized *Proteus* spp. strains as bioremediators and PGPR.

CONCLUSION

Bacteria from the genus *Proteus* are well known as human opportunistic pathogens and intestinal microorganisms indicating fecal pollution of water or soil. Their presence in drinking water poses a threat of infection. On the other hand, the bacilli are believed to play an important role in removing organic pollutants of animal origin, especially fecal ones, by decomposing the dead organic matter in water or soil environments. Their proteolytic and ureolytic properties may place the bacteria among the most efficient saprobes taking part in the enrichment of manured soils in ammonia salts and therefore participating in the nitrogen cycle. However, the significance of these microorganisms in the mineralization processes of nitrogen-containing organic compounds is not obvious. This aspect of their metabolism requires more research. It would be reasonable to establish their exact contribution as destruents to the circulation of matter in nature.

Proteus spp. strains are sometimes found to co-operate with higher organisms, playing a role of animal and plant symbionts. Their antagonistic or commensal relationships with numerous animals have also been documented. However, environments other than the human body are less recognized as habitats of *Proteus* spp. rods. In the future, more attention should be paid to the interdependences between *Proteus* spp. and animals or plants coexisting with the bacteria in the environment. Relatively little information is available also about the mutual relations between *Proteus* spp. bacilli and other bacteria in natural habitats. Establishing what kind of relationships these are and which organisms benefit from this system is the basis for determining the reasons why the bacilli occur in a given ecosystem and the consequences of their activity.

Nowadays, molecular techniques provide proper tools to investigate the microorganisms in their natural habitats. The differences in the biochemical features do not exclude the strain from a genus as long as the similarities on the genome level are big enough. It is important to employ molecular techniques in the identification of *Proteus* spp. strains, especially those belonging to the new genomospecies, the biochemical differentiation of which is impossible. Metagenomic analysis of soil and water environments as well as of different niches in human, animal, and plant organisms using proper primer sequences may reveal the presence of *Proteus* spp. strains in these habitats even if they exhibit atypical activities and play unusual roles.

Proteus spp. bacteria often display exceptional metabolic features, allowing them to adapt to various conditions and reach distant niches in natural environments where they may be recognized as autochthons. Although the genus is usually not mentioned among PGPR or bioremediators, *Proteus* spp. environmental strains possess the abilities of immobilization of heavy metals, utilization of toxic pollutants, and plant growth stimulation, as well as nematicidal and fungicidal properties. Numerous examples suggest that the application of these microorganisms may be an effective way of protecting plants and groundwater. The biological treatment of polluted natural environments is more efficient and cost-effective than physicochemical methods and can be used in large areas due to simple application and an ability to remove pollution completely. Bioremediation could involve not only the stimulation of the natural autochthonic *Proteus* spp. strains but also introducing these microorganisms to the polluted environment.

However, it is important to remember that the bacteria belonging to the genus *Proteus* are opportunistic human and animal pathogens. It would be useful and interesting to compare the features (especially virulence factors, serotypes, metabolic apparatus, and the antibiotic resistance) expressed by *Proteus* spp. clinical isolates and by the strains existing in animal, soil, and water ecosystems. Do environmental *Proteus* spp. strains differ significantly from clinical ones? Do they possess genes coding the factors of pathogenicity identified in the strains isolated from patients? If so, do they express these factors and how do they use them in non-pathogenic conditions? What is the risk of transformation of a strain which could be used in bioremediation and environmental protection into a virulent one posing a threat to the animals or humans inhabiting the area? Do the environmental strains carry antibiotic resistance genes, which might lead to the transfer of these genes into other microorganisms in the environment? Answers to these questions could explain which factors contribute to and in what way they are important in different lifestyles presented by *Proteus* spp. bacteria in various environments.

ACKNOWLEDGMENTS

I thank Prof. Antoni Różalski for critical reading of the manuscript.

REFERENCES

1. Aboh EA, Giwa FJ, Giwa A (2015) Microbiological assessment of well waters in Samaru, Zaria, Kaduna, State, Nigeria. Ann Afr Med 14:32–38
2. Abu-Samra MT, Shuaib YA (2014) A study on the nature of association between

Demodex mites and bacteria involved in skin and meibomian gland lesions of demodectic mange in cattle. Vet Med Int. doi:10.1155/2014/413719

3. Akoachere J-FTK, Oben PM, Mbivnjo BS, Ndip LM, Nkwelang G, Ndip RN (2008) Bacterial indicators of pollution of the Douala lagoon, Cameroon: public health implications. Afr Health Sci 8(2):85–89
4. Alagesan P, Ashok KB, Muthukrishnan J, Gunasekaran P (2003) Isolation and characterization of gut bacteria of millipede, Xenobolus carnifex (Fabricius). Indian J Microbiol 43:111–113
5. Al-Bahry SN, Al-Zadjali MA, Mahmoud IY, Elshafie AE (2012) Biomonitoring marine habitats in reference to antibiotic resistant bacteria and ampicillin resistance determinants from oviductal fluid of the nesting green sea turtle, Chelonia mydas. Chemosphere 87:1308–1315
6. Anand AAP, Sripathi K (2004) Digestion of cellulose and xylan by symbiotic bacteria in the intestine of the Indian flying fox (Pteropus giganteus). Comp Biochem Physiol 139:65–69
7. Arbatsky NP, Drzewiecka D, Palusiak A, Shashkov AS, Zabłotni A, Siwińska M, Knirel YA (2013) Structure of a Kdo-containing O polysaccharide representing Proteus O79, a newly described serogroup for some clinical Proteus genomospecies isolates from Poland. Carbohydr Res 379:100–105
8. Armbruster CE, Mobley HLT (2012) Merging mythology and morphology: the multifaceted lifestyle of Proteus mirabilis. Nat Rev Microbiol 10(11):743–754
9. Awad-Alla ME, Abdien HMF, Dessouki AA (2010) Prevalence of bacteria and parasites in white ibis in Egypt. Vet Ital 46(3):277–286
10. Awong-Taylor J, Craven KS, Griffiths L, Bass C, Muscarella M (2008) Comparison of biochemical and molecular methods for the identification of bacterial isolates associated with failed loggerhead sea turtle eggs. J Appl Microbiol 104:1244–1251
11. Bano N, Musarrat J (2003) Isolation and characterization of phorate degrading soil bacteria of environmental and agronomic significance. Lett Appl Microbiol 36:349–353
12. Barthakur M, Bezbaruah B (1999) Plant beneficial effect of two strains of Proteus vulgaris isolated from tea plantations. Indial J Exp Biol 37:919–924
13. Bhattacharyya D, Garladinne M, Lee Y-H (2015) Volatile indole produced by rhizobacterium Proteus vulgaris JBLS202 stimulates growth of Arabidopsis thaliana through auxin, cytokinin, and brassinosteroid pathways. J Plant Growth Regul 34:158–168
14. Billah M, Bano A (2015) Role of plant growth promoting rhizobacteria in modulating the efficiency of poultry litter composting with rock phosphate and its effect on growth and yield of wheat. Waste Manag Res 33(1):63–72
15. Bittar F, Keita MB, Lagier J-C, Peeters M, Delaporte E, Raoult D (2014) Gorilla gorilla gorilla gut: a potential reservoir of pathogenic bacteria as revealed using culturomics and molecular tools. Sci Rep. doi:10.1038/srep07174

16. Blaylock RS (2001) Normal oral bacterial flora from some southern African snakes. Onderstepoort J Vet Res 68(3):175–182

17. Boari CA, Pereira GI, Valeriano C, Silva BC, de Morais VM, Figueiredo HCP, Piccoli RH (2008) Bacterial ecology of tilapia fresh fillets and some factors that can influence their microbial quality. Cienc Tecnol Aliment 28(4):863–867

18. Bohova J, Majtan J, Majtan V, Takac P (2014) Selective antibiofilm effects of Lucilia sericata larvae secretions/excretions against wound pathogens. Evid Based Complement Alternat Med. doi:10.1155/2014/857360

19. Budding AE, Ingham CJ, Bitter W, Vandenbroucke-Grauls CM, Schneeberger PM (2009) The Dienes phenomenon: competition and territoriality in swarming Proteus mirabilis. J Bacteriol 191:3892–3900

20. Cao H, He S, Lu L, Yang X, Chen B (2014) Identification of a Proteus penneri isolate as the causal agent of red body disease of the cultured white shrimp Penaeus vannamei and its control with Bdellovibrio bacteriovorus. Antonie Van Leeuwenhoek 105:423–430

21. Cardarelli L, Saak C, Gibbs KA (2015) Two proteins form a heteromeric bacterial self-recognition complex in which variable subdomains determine allele-restricted binding. mBio 6:e00251–15

22. Ceyhan N (2012) Biodegradation of pyrene by a newly isolated Proteus vulgaris. Sci Res Essays 7(1):66–77

23. Chen B-Y, Wang Y-M, Ng I-S (2011) Understanding interactive characteristics of bioelectricity generation and reductive decolorization using Proteus hauseri. Bioresour Technol 102:1159–1165

24. Chen B-Y, Zhang M-M, Chang C-T, Ding Y, Lin K-L, Chiou C-S, Hsueh C-C, Xu H (2010) Assessment upon azo dye decolorization and bioelectricity generation by Proteus hauseri. Bioresour Technol 101(12):4737–41

25. Chen K-C, Huang W-T, Wu J-Y, Houng J-Y (1999) Microbial decolorization of azo dyes by Proteus mirabilis. J Ind Microbiol Biotechnol 23:686–690

26. Chow AW, Taylor PR, Yoshikawa TT, Guze LB (1979) A nosocomial outbreak of infections due to multiply resistant Proteus mirabilis: role of intestinal colonization as a major reservoir. J Infect Dis 139(6):621–627

27. Correa IE, Steen WC (1995) Degradation of propanil by bacterial isolates and mixed populations from a pristine lake. Chemosphere 30(1):103–116

28. Dang H, Ren J, Song L, Sun S, An L (2008) Diverse tetracycline resistant bacteria and resistance genes from coastal waters of Jiaozhou Bay. Microb Ecol 55:237–246

29. Dang H, Ren J, Song L, Sun S, An L (2008) Dominant chloramphenicol-resistant bacteria and resistance genes in coastal marine waters of Jiaozhou Bay, China. World J Microbiol Biotechnol 24:209–217

30. Das AC, Chakravarty A, Sukul P, Mukherjee D (1995) Insecticides: their effect on microorganisms and persistence in rice soil. Microbiol Res 150:187–194

31. Das AC, Chakravarty A, Sukul P, Mukherjee D (2003) Influence and persistence of phorate and carbofuran insecticides on microorganisms in rice field. Chemosphere 53:1033–1037

32. Das AC, Mukherjee D (2000) Soil application of insecticides influences microorganisms and plant nutrients. Appl Soil Ecol 14:55–62

33. Davalieva K, Kungulovski D, Atanasova-Pancevska N, Bojkovska R, Stafilov T, Efremov GD (2011) Microbiological and chemical characteristics of water and sediment from Vrelo Cave, Republic of Macedonia. Prilozi 32(2):169–186

34. Drzewiecka D, Arbatsky NP, Shashkov AS, Stączek P, Knirel YA, Sidorczyk Z (2008) Structure and serological properties of the O-antigen of two clinical Proteus mirabilis strains classified into a new Proteus O77 serogroup. FEMS Immunol Med Microbiol 54:185–194

35. Drzewiecka D, Arbatsky NP, Stączek P, Shashkov AS, Knirel YA, Sidorczyk Z (2010) Structural and serological studies of the O-polysaccharide of strains from a newly created Proteus O78 serogroup prevalent in Polish patients. FEMS Immunol Med Microbiol 58:269–276

36. Drzewiecka D, Sidorczyk Z (2005) Characterization of Proteus penneri species – human opportunistic pathogens. Post Mikrobiol 44:113–126 (in Polish)

37. Drzewiecka D, Zych K, Sidorczyk Z (2004) Characterization and serological classification of a collection of Proteus penneri clinical strains. Arch Immunol Ther Exp 52:121–128

38. Eze CA, Ugwu J, Eze JI, Nnaji TO, Nweze NE, Ngene AA (2014) Characterisation of the bacterial isolates from transmissible venereal tumor lesions of dogs and their antibiogram resistance in Nsukka area, Southeastern Nigeria. Comp Clin Pathol 23:641–645

39. Fernandez-Delgado M, Contreras M, Garcia-Amado A, Gueneau P, Suarez P (2007) Occurrence of Proteus mirabilis associated with two species of Venezuelan oysters. Rev Inst Med Trop S Paulo 49(6):355–359

40. Filali BK, Taoufik J, Zeroual Y, Dzairi FZ, Talbi M, Blaghen M (2000) Waste water bacterial isolates resistant to heavy metals and antibiotics. Curr Microbiol 41:151–156

41. Foght J, April T, Biggar K, Aislabie J (2001) Bioremediation of DDT-contaminated soils: a review. Bioremediat J 5(3):225–246

42. Foti M, Giacopello C, Bottari T, Fisichella V, Rinaldo D, Mammina C (2009) Antibiotic resistance of Gram negatives isolates from loggerhead sea turtles (Caretta caretta) in the central Mediterranean Sea. Mar Pollut Bull 58:1363–1366

43. Foti M, Rinaldo D, Guercio A, Giacopello C, Aleo A, De Leo F, Fisichella V, Mammina C (2011) Pathogenic microorganisms carried by migratory birds passing through the territory of the island of Ustica, Sicily (Italy). Avian Pathol 40(4):405–409

44. Gaastra W, van Oosterom RAA, Pieters EWJ, Bergmans HEN, van Dijk L, Agnes A, ter Huurne HM (1996) Isolation and characterization of dog uropathogenic Proteus mirabilis strains. Vet Microbiol 48:57–71

45. Ge S, Dong X, Zhou J, Ge S (2013) Comparative evaluations on bio-treatment of hexavalent chromate by resting cells of Pseudochrobactrum sp. and Proteus sp. in wastewater. J Environ Manag 126:7–12

46. Giacopello C, Foti M, Fisichella V, Lo Piccolo F (2015) Antibiotic-resistance patterns of Gram-negative bacterial isolates from breeder canaries (Serinus canaria domestica) with clinical disease. J Exot Pet Med 24:84–91

47. Giammanco GM, Grimont PA, Grimont F, Lefevre M, Giammanco G, Pignato S (2011) Phylogenetic analysis of the genera Proteus, Morganella, and Providencia by comparison of rpoB gene sequence of type and clinical strains suggests the reclassification of Proteus myxofaciens in a new genus, Cosenzaea gen. nov., as Cosenzaea myxofaciens comb. nov. Int J Syst Evol Microbiol 61:1638–1644

48. Gibbs KA, Greenberg EP (2011) Territoriality in Proteus: advertisement and aggression. Chem Rev 111(1):188–194

49. Gibbs KA, Wenren LM, Greenberg EP (2011) Identity gene expression in Proteus mirabilis. J Bacteriol 193:3286–3292

50. Graça AP, Viana F, Bondoso J, Correia MI, Gomes L, Humanes M, Reis A, Xavier JR, Gaspar H, Lage OM (2015) The antimicrobial activity of heterotrophic bacteria isolated from the marine sponge Erylus deficiens (Astrophorida, Geodiidae). Front Microbiol 6:389. doi:10.3389/fmicb.2015.00389

51. Gupta AK, Nayduch D, Verma P, Shah B, Ghate HV, Patole MS, Shouche YS (2012) Phylogenetic characterization of bacteria in the gut of house flies (Musca domestica L.). FEMS Microbiol Ecol 79:581–593

52. Hačene H, Rafa F, Chebhouni N, Boutaiba S, Bhatnagar T, Baratti JC, Ollivier B (2004) Biodiversity of prokaryotic microflora In El Golea Salt lake, Algerian Sahara. J Arid Environ 58:273–284

53. Hacioglu N, Tosunoglu M (2014) Determination of antimicrobial and heavy metal resistance profiles of some bacteria isolated from aquatic amphibian and reptile species. Environ Monit Assess 186:407–413

54. Hariharan H, Matthew V, Fountain J, Snell A, Doherty D, King B, Shemer E, Oliveira S, Sharma RN (2011) Aerobic bacteria from mucous membranes, ear canals, and skin wounds of feral cats in Grenada, and the antimicrobial drug susceptibility of major isolates. Comp Immunol Microbiol Infect Dis 34(2):29–34

55. Hassen A, Saidi N, Cherif M, Boudabous A (1998) Resistance of environmental bacteria to heavy metals. Bioresour Technol 64:7–15

56. Hawkey PM, Penner JL, Linton AH, Hawkey CA, Crisp LJ, Hinton M (1986) Speciation, serotyping, antimicrobial sensitivity and plasmid content of Proteeae from the environment of calf-rearing units in South West England. J Hyg Camb 97:405–417

57. Hernandez-Rivera MA, Ojeda-Morales ME, Martinez-Vazquez JG, Villegas-Cornelio VM, Cordova-Bautista Y (2011) Optimal parameters for in vitro development of the hydrocarbonoclastic microorganism Proteus sp. J Soil Sci Plant Natur 11(1):29–43

58. Ibrahim ML, Ijah UJJ, Manga SB, Bilbis LS, Umar S (2013) Production and partial characterization of biosurfactant produced by crude oil degrading bacteria. Int Biodeterior Biodegrad 81:28–34

59. Ikeobi CC, Ogunsanya TO, Rotimi VO (1996) Prevalence of pathogenic role of Morganella-Proteus-Providencia-group of bacteria in human faeces. Afr J Med Med Sci 25(1):7–12

60. Islam F, Yasmeen T, Riaz M, Arif MS, Ali S, Raza SH (2014) Proteus mirabilis alleviates zinc toxicity by preventing oxidative stress in maize (Zea mays) plants. Ecotoxicol Environ Saf 110:143–152

61. Jahantigh M (2010) Bacteriological study of dead-in-shell embryos of ostrich. Iran J Vet Res 11(1):88–90

62. Jahantigh M (2012) A study of bacteriologic status of infertile ostrich (Struthio camelus) eggs. Comp Clin Pathol 21:1049–1051

63. Janda JM, Abbott SL, Khashe S, Probert W (2001) Biochemical identification and characterization of DNA groups within the Proteus vulgaris complex. J Clin Microbiol 39(4):1231–1234

64. Kefalas E, Castritsi-Catharios J, Miliou H (2003) Bacteria associated with the sponge Spongia officinalis as indicators of contamination. Ecol Indicators 2:339–343

65. Kim H-K, Lee J-K, Kim H, Oh T-K (1996) Characterization of an alkaline lipase from Proteus vulgaris K80 and the DNA sequence of the encoding gene. FEMS Microbiol Lett 135:117–121

66. Kitamikado M, Lee Y-Z (1975) Chondroitinase-producing bacteria in natural habitats. Appl Microbiol 29(3):414–421

67. Knirel YA, Perepelov AV, Kondakova AN, Senchenkova SN, Sidorczyk Z, Różalski A, Kaca W (2011) Structure and serology of O-antigens as the basis for classification of Proteus strains. Innate Immun 17:70–96

68. Kobashi Y, Hasebe A, Nishio M, Uchiyama H (2007) Diversity of tetracycline resistance genes in bacteria isolated from various agricultural environments. Microbes Environ 22(1):44–51

69. Kroemer S, Garh FE, Galland D, Petit J-L, Woehrle F, Boulouis H-J (2014) Antibiotic susceptibility of bacteria isolated from infections in cats and dogs throughout Europe (2002-2009). Comp Immunol Microbiol Infect Dis 37:97–108

70. Kumar R, Swaminathan TR, Kumar RG, Dharmaratnam A, Basheer VS, Jena JK (2015) Mass mortality in ornamental fish, Cyprinus carpio koi caused by a bacterial pathogen, Proteus hauseri. Acta Trop 149:128–134

71. Kwiecińska-Piróg J, Bogiel T, Gospodarek E, Kasprzak M, Jerzak L, Kamiński P (2011) The susceptibility of Proteus mirabilis strains isolated from white stork (Ciconia ciconia). Med Dosw Mikrobiol 63:139–144 (in Polish)
72. Larsson P (1984) Serology of Proteus mirabilis and Proteus vulgaris. Meth Microbiol 14:187–214
73. Lecomte V, Kaakoush NO, Maloney CA, Raipuria M, Huinao KD, Mitchell HM, Morris MJ (2015) Changes in gut microbiota in rats fed a high fat diet correlate with obesity-associated metabolic parameters. PloS One 10(5), doi:10.1371/journal.pone.0126931
74. de Louvois J (1969) Serotyping and the Dienes reaction on Proteus mirabilis from hospital infections. J Clin Pathol 22:263–268
75. Lowe BA, Marsh TL, Isaacs-Cosgrove N, Kirkwood RN, Kiupel M, Mulks MH (2011) Microbial communities in the tonsils of healthy pigs. Vet Microbiol 147:346–357
76. Lu H, Wang X, Zhang K, Lu Y, Zhou L, Li G (2014) Identification and nematicidal activity of bacteria isolated from cow dung. Ann Microbiol 64:407–411
77. Lu Y, Lin Q, Wang J, Wu Y, Bao W, Lv F, Lu Z (2010) Overexpression and characterization in Bacillus subtilis of a positionally nonspecific lipase from Proteus vulgaris. J Ind Microbiol Biotechnol 37:919–925
78. Lutz G, Chavarria M, Arias ML, Mata-Segreda JF (2006) Microbial degradation of palm (Elaeis giuneensis) biodiesel. Rev Biol Trop 54(1):59–63
79. Ma Q, Fonseca A, Liu W, Fields AT, Pimsler ML, Spindola AF, Tarone AM, Crippen TL, Tomberlin JK, Wood TK (2012) Proteus mirabilis interkingdom swarming signals attract blow flies. ISME J 6:1356–1366
80. Maleki-Ravasan N, Oshaghi MA, Hajikhani S, Saeidi S, Akhavan AA, Gerami-Shoar M, Shirazi MH, Yakhchali B, Rassi Y, Afshar D (2014) Aerobic microbial community of insectary population of Phlebotomus papatasi. J Arthropod-Borne Dis 8(1):69–81
81. Manos J, Belas R (2006) The genera Proteus, Providencia, and Morganella. In: Dworkin M, Falkow S, Rosenberg E, Schleifer K-H, Stackebrandt E (eds) The Prokaryotes (3rd edn.) Springer pp. 245-269, DOI: 10.1007/0-387-30746-x_12
82. Matyar F, Kaya A, Dinçer S (2008) Antibacterial agents and heavy metal resistance in Gram-negative bacteria isolated from seawater, shrimp and sediment in Iskenderun Bay, Turkey. Sci Total Environ 407:279–285
83. Mekibib B, Abrham E, Kassaye A, Tesfaye D, Amenu K (2015) Clinicopathologic findings associated with urinary tract problems in working donkeys, Southern Ethiopia: a case study. Comp Clin Pathol 24:417–422
84. Meyer W, Kacza J, Schnapper A, Verspohl J, Hornickel I, Seeger J (2010) A first report on the microbial colonization of the equine oesophagus. Ann Anat 192(1):42–51

85. Miller SG, Campbell BC, Becnell J, Ehrman L (1995) Bacterial entomopathogens from the Drosophila paulistorum semispecies complex. J Invertebr Pathol 65:125–131

86. Mishra SS, Markande AR, Keluskar RP, Karunasagar I, Nayak BB (2015) Simultaneous nitrification and denitrification by novel heterotrophs in remediation of fish processing effluents. J Basic Microbiol 55:772–779

87. Mohd Masri S, Nazni WA, Lee HL, Tengku RTAR, Subramaniam S (2005) Sterilization of Lucilia cuprina (Wiedemann) maggots used in therapy of intractable wounds. Trop Biomed 22(2):185–189

88. Mohite BV, Pawar SP, Morankar A (2011) Isolation, selection and biodegradation profile of phenol degrading bacteria from oil contaminated soil. Bull Environ Contam Toxicol 87(2):143–146

89. Morgenstein RM, Szostek B, Rather PN (2010) Regulation of gene expression during swarmer cell differentiation in Proteus mirabilis. FEMS Microbiol Rev 34:753–763

90. Muňoz-Castellanos LN, Torres-Muňoz JV, Keer-Rendon A, Manzanares-Papayanopoulos LI, Nevarez-Moorillon GV (2006) Aerobic biodegradation of methyl tert-butyl ether (MTBE) by pure bacterial cultures isolated from contaminated soil. World J Microb Biot 22:851–855

91. Müller H (1986) Occurrence and pathogenic role of Morganella-Proteus-Providencia group bacteria in human feces. J Clin Microbiol 23(2):404–405

92. Müller H (1989) The role of Proteeae in diarrhea. Zbl Bakt 272(1):30–35

93. Nagaraju D, Sastri JCV (1999) Confirmed faecal pollution to bore well waters of Mysore city. Environ Geol 38(4):322–326

94. Nazni WA, Seleena B, Lee HL, Jeffery J, Rogayah TAR, Sofian MA (2005) Bacteria fauna from the house fly, Musca domestica (L.). Trop Biomed 22(2):225–231

95. Ng I-S, Zheng X, Chen B-Y, Chi X, Chang C-S (2013) Proteomics approach to decipher novel genes and enzymes characterization of a bioelectricity-generating and dye-decolorizing bacterium Proteus hauseri ZMd44. Biotechnol Bioproc Eng 18:8–17

96. Ng I-S, Zheng X, Wang N, Chen B-Y, Zhang X, Lu Y (2014) Copper response of Proteus hauseri based on proteomic and genetic expression and cell morphology analysis. Appl Biochem Biotechnol 173:1057–1072

97. Nguyen VD, Pham TT, Nguyen THT, Nguyen THX, Hoj L (2014) Screening of marine bacteria with bacteriocin-like activities and probiotic potential for ornate spiny lobster (Panulirus ornatus) juveniles. Fish Shellfish Immunol 40:49–60

98. Nimrat S, Bart AN, Keatsaksit A, Vuthiphandchai V (2008) Microbial flora of sphermatophores from black tiger shrimp (Penaeus monodon) declines over long-term cryostorage. Aquaculture 274:247–253

99. Normand EH, Gibson NR, Reid SWJ, Carmichael S, Taylor DJ (2000) Antimicrobial-resistance trends in bacterial isolates from companion-animal community practice in the UK. Prev Vet Med 46:267–278

100. Novak SM (1998) Foodborne illness – chemical fish and shellfish poisoning. Clin Microbiol Newsl 20(3):17–21

101. Novak SS, Seigl RA (1986) Gram-negative septicemia in American alligators (Alligator mississippiensis). J Wildl Dis 22(4):486–487

102. Nowakiewicz A, Ziółkowska G, Zięba P, Dziedzic BM, Gnat S, Wójcik M, Dziedzic R, Kostruba A (2015) Aerobic bacterial microbiota isolated from the cloaca of the European pond turtle (Emys orbicularis) in Poland. J Wildl Dis 51(1):255–9

103. O'Hara CM, Brenner FW, Miller JM (2000) Classification, identification, and clinical significance of Proteus, Providencia, and Morganella. Clin Microbiol Rev 13:534–546

104. O'Hara CM, Brenner FW, Steigerwalt AG, Hill BC, Holmes B, Grimont PAD, Hawkey PM, Penner JL, Miller JM, Brenner D (2000) Classification of Proteus vulgaris biogroup 3 with recognition of Proteus hauseri sp. nov., nom. rev. and unnamed Proteus genomospecies 4, 5 and 6. Int J Syst Evol Microbiol 50:1869–1875

105. Olajide PO, Ogbeifun LB (2010) Hydrocarbon biodegrading potentials of a Proteus vulgaris strain isolated from fish samples. Am J Applied Sci 7(7):922–928

106. Olukanni OD, Osuntoki AA, Kalyani DC, Gbenle GO, Govindw SP (2010) Decolorization and biodegradation of Reactive Blue 13 by Proteus mirabilis LAG. J Hazard Mater 184(1-3):290–298

107. Onyango MD, Wandili S, Kakai R, Waindi EN (2009) Isolation of Salmonella and Shigella from fish harvested from the Winam Gulf of Lake Victoria, Kenya. J Infect Dev Ctries 3(2):99–104

108. Oros J, Torrent A, Calabuig P, Deniz S (2005) Diseases and causes of mortiality among sea turtles stranded in the Canary Islands, Spain (1998 – 2001). Dis Aquat Org 63:13–24

109. Özoğul F, Küley E, Özoğul Y (2010) Usefulness of API test strips for identification of bacterial flora in blue crab (Callinectes sapidus) caught from Akyatan Lagoon (Adana – Turkey). J Fisheries Sciencescom 4(1):1–7

110. Padaruth SK, Biranjia-Hurdoyal SD (2014) Hygiene practices and faecal contamination of the hands of children attending primary school in Mauritius. Int Health. doi:10.1093/inthealth/ihu080

111. Palusiak A (2013) Immunochemical properties of Proteus penneri lipopolysaccharides – one of the major Proteus sp. virulence factors. Carbohydr Res 380:16–22

112. Patil PS, Phugare SS, Kalyani DC, Surwase SN, Jadhav JP (2012) Bioremediation perspective of Navy Blue Rx-containing effluent by bacterial isolate. Bioremediat J 16(4):185–194

113. Peerbooms PG, Verweij AM, Oe PL, MacLaren DM (1986) Urinary pathogenicity of Proteus mirabilis strains isolated from faeces or urine. Antonie Van Leeuwenhoek 52(1):53–62

114. Pfaller MA, Mujeeb I, Hollis RJ, Jones RN, Doern GV (2000) Evaluation of the discriminatory powers of the Dienes test and ribotyping as typing methods for Proteus mirabilis. J Clin Microbiol 38:1077–1080

115. Pino NJ, Dominguez MC, Peñuela GA (2011) Isolation of a selected microbial consortium capable of degrading methyl parathion and p-nitrophenol from a contaminated soil site. J Environ Sci Health B 46(2):173–80

116. Pino N, Peñuela G (2011) Simultaneous degradation of the pesticides methyl parathion and chlorpyrifos by an isolated bacterial consortium from a contaminated site. Int Biodeterior Biodegrad 65:827–831

117. Poonia S, Singh TS, Tsering DC (2014) Antibiotic susceptibility profile of bacteria isolated from natural sources of water from rural areas of East Sikkim. Indian J Community Med 39(3):156–160

118. Porres-Osante N, Saenz Y, Somalo S, Torres C (2015) Characterization of beta-lactamases in faecal Enterobacteriaceae recovered from healthy humans in Spain: focusing on AmpC polymorphisms. Microb Ecol 70:132–140

119. Qadripur SA, Schauder S, Schwartz P (2001) Black nails from Proteus mirabilis colonization. Hautarzt 52:658–661 (in German)

120. Rani A, Shouche YS, Goel R (2008) Declination of copper toxicity in pigeon pea and soil system by growth-promoting Proteus vulgaris KNP3 strain. Curr Microbiol 57:78–82

121. Rau N, Mishra V, Sharma M, Das MK, Ahaluwalia K, Sharma RS (2009) Evaluation of functional diversity in Rhizobacterial taxa of a wild grass (Saccharum ravennae) colonizing abandoned fly ash dumps in Delhi urban ecosystem. Soil Biol Biochem 41:813–821

122. Reche MHLR, Fiuza LM (2005) Bacterial diversity in rice-field water in Rio Grande do Sul. Braz J Microbiol 36:253–257

123. Rogers KH (2006) Prevalence of pathogenic enteric bacteria in wild birds associated with agriculture in Humboldt County, California. Dissertation, Humboldt State University, Arcata, California, USA

124. Różalski A, Stączek P (2010) Proteus. In: Liu D (ed) Molecular detection of human bacterial pathogens. Taylor & Francis Group, LLC, CRC Press, Boca Raton, pp 981–996

125. Różalski A, Torzewska A, Moryl M, Kwil I, Maszewska A, Ostrowska K, Drzewiecka D, Zabłotni A, Palusiak A, Siwińska M, Stączek P (2012) Proteus sp. – an opportunistic bacterial pathogen – classification, swarming growth, clinical significance and virulence factors. Folia Biol Oecol 8:1–17

126. Salyers AA, Gupta A, Wang Y (2004) Human intestinal bacteria as reservoirs for antibiotic resistance genes. Trends Microbiol 12(9):412–416

127. Santoro M, Hernandez G, Caballero M, Garcia F (2006) Aerobic bacterial flora of nesting Green turtles (Chelonia mydas) from Tortuguero National Park, Costa Rica. J Zoo Wildl Med 37(4):549–552

128. Sanuth HA, Yadav A, Fagade OE, Shouche Y (2013) ε-Caprolactam utilization by Proteus sp. and Bordetella sp. isolated from solid waste dumpsites in Lagos State, Nigeria, first report. Indian J Microbiol 53(2):221–226

129. Senior BW (1977) The Dienes phenomenon: identification of the determinants of compatibility. J Gen Microbiol 102:235–244

130. Senior BW, Leslie DL (1986) Rare occurrence of Proteus vulgaris in faeces: a reason for its rare association with urinary tract infections. J Med Microbiol 21(2):139–144

131. Shek KC, Tsui KL, Lam KK, Crow P, Ng KHL, Ades G, Yip KT, Grioni A, Tan KS, Lung DC, Lam TSK, Fung HT, Que TL, Kam CW (2009) Oral bacterial flora of the Chinese cobra (Naja atra) and bamboo pit viper (Trimeresurus albolabris) in Hong Kong SAR, China. Hong Kong Med J 15:183–190

132. Shojaei H, Shooshtaripoor J, Amiri M (2006) Efficacy of simple hand-washing in reduction of microbial hand contamination of Iranian food handlers. Food Res Int 39:525–529

133. Sidorczyk Z, Toukach FV, Zych K, Arbatsky NP, Drzewiecka D, Ziółkowski A, Shashkov AS, Knirel YA (2002) Structural and serological characterization of the lipopolysaccharide from Proteus penneri 20 and classification of the cross-reacting Proteus penneri strains 10, 16, 18, 20, 32 and 45 in Proteus serogroup O17. Arch Immunol Ther Exp 50:345–350

134. Singh B, Crippen TL, Zheng L, Fields AT, Yu Z, Ma Q, Wood TK, Dowd SE, Flores M, Tomberlin JK, Tarone AM (2014) A metagenomic assessment of the bacteria associated with Lucilia sericata and Lucilia cuprina (Diptera: Calliphoridae). Appl Microbiol Biotechnol. doi:10.1007/s00253-014-6115-7

135. Siwińska M, Levina EA, Ovchinnikova OG, Drzewiecka D, Shashkov AS, Różalski A, Knirel YA (2015) Classification of a Proteus penneri clinical isolate with a unique O-antigen structure to a new Proteus serogroup, O80. Carbohydr Res 407:131–136

136. Smith SI, Opere B, Goodluck HT, Akindolire OT, Folaranmi A, Odekeye OM, Omonigbehin EA (2009) Antibiotic susceptibility pattern of Staphylococcus species isolated from telephone receivers. Singapore Med J 50(2):208–211

137. Snowdon JA, Cliver DO (1996) Microorganisms in honey. Int J Food Microbiol 31:1–26

138. Srinivasan V, Nam H-M, Sawant AA, Headrick SI, Nguyen LT, Oliver SP (2008) Distribution of tetracycline and streptomycin resistance genes and class 1 integron in Enterobacteriaceae isolated from dairy and nondairy farm soils. Microb Ecol 55:184–193

139. Survery S, Ahmad S, Subhan SA, Ajaz M, Rasool SA (2004) Hydrocarbon degrading bacteria from Pakistani soil: isolation, identification, screening and genetical studies. Pak J Biol Sci 7(9):1518–1522

140. Suthar S, Chhimpa V, Singh S (2009) Bacterial contamination in drinking water: a case study in rural areas of northern Rajasthan, India. Environ Monit Assess 159:43–50

141. Svanevik CS, Lunestad BT (2011) Characterization of the microbiota of Atlantic mackerel (Scomber scombrus). Int J Food Microbiol 151:164–170

142. Tatfeng YM, Usuanlele MU, Orukpe A, Digban AK, Okodua M, Oviasogie F, Turay AA (2005) Mechanical transmission of pathogenic organisms: the role of cockroaches. J Vect Borne Dis 42:129–134

143. Trawińska B, Polonis A, Tymczyna L, Popiołek-Pyrz M, Bombik T, Saba L (2006) Bacteriological and parasitologic pollution of the environment and birth health state around the reproductive layer farm. Annales Universitatis Mariae Curie-Sklodowska, EE, XXIV, 50:371-376

144. Urban JE, Broce A (1998) Flies and their bacterial loads in greyhound dog kennels in Kansas. Curr Microbiol 36:164–170

145. Vazirianzadeh B, Dehghani R, Mehdinejad M, Sharififard M, Nasirabadi N (2014) The first report of drug resistant bacteria isolated from the brown-banded cockroach, Supella longipalpa, in Ahvaz, south-western Iran. J Arthropod-Borne Dis 8(1):53–59

146. Wang L, Pan Q, Zhang L, Xue Q, Cui J, Qi C (2008) Investigation of bacterial microorganisms in the conjunctival sac of clinically normal dogs and dogs with ulcerative keratitis in Beijing, China. Vet Ophthalmol 11:145–149

147. Wang Y, Wang Y, Wu C-M, Schwarz S, Shen Z, Zhang W, Zhang Q, Shen J-Z (2011) Detection of the staphylococcal multiresistance gene cfr in Proteus vulgaris of food animal origin. J Antimicrob Chemother 66:2521–2526

148. Wang Y, Zhang S, Yu J, Zhang H, Yuan Z, Sun Y, Zhang L, Zhu Y, Song H (2010) An outbreak of Proteus mirabilis food poisoning associated with eating stewed pork balls in brown sauce, Beijing. Food Control 21:302–305

149. Wannigama DL, Dwivedi R, Zahraei-Ramazani A (2014) Prevalence and antibiotic resistance of Gram-negative pathogenic bacteria species isolated from Periplaneta americana and Blattella germanica in Varanasi, India. J Arthropod-Borne Dis 8(1):10–20

150. Wei T, Miyanaga K, Tanji Y (2014) Persistence of antibiotic-resistant and -sensitive Proteus mirabilis strains in the digestive tract of the housefly (Musca domestica) and green bottle flies (Calliphoridae). Appl Microbiol Biotechnol 98:8357–8566

151. Wenren LM, Sullivan NL, Cardarelli L, Septer AN, Gibbs KA (2013) Two independent pathways for self-recognition in Proteus mirabilis are linked by type VI-dependent export. mBio 4(4):e00374–13

152. Westgate SJ, Percival SL, Knottenbelt DC, Clegg PD, Cochrane CA (2011) Microbiology of equine wounds and evidence of bacterial biofilms. Vet Microbiol 150:152–159

153. Whangsuk W, Sungkeeree P, Thiengmag S, Kerdwong J, Sallabhan R, Mongkolsuk S, Loprasert S (2013) Gene cloning and characterization of a novel highly organic solvent tolerant lipase from Proteus sp. SW1 and its application for biodiesel production. Mol Biotechnol 53:55–62

154. Winsor DK, Bloebaum AP, Mathewson JJ (1981) Gram-negative, aerobic, enteric pathogens among intestinal microflora of wild turkey vultures (Cathartes aura) in West Central Texas. Appl Environ Microbiol 42(6):1123–1124

155. Yong LH, Ambu S, Devi S, Maung M (2008) Detection of protozoan and bacterial pathogens of public health importance in faeces of Corvus spp. (large-billed crow). Trop Biomed 25(2):134–139

156. Yu S-M, Lee YH (2013) Plant growth promoting rhizobacterium Proteus vulgaris JBLS202 stimulates the seedling growth of Chinese cabbage through indole emission. Plant Soil 370:485–495

157. Zhang M-M, Chen W-M, Chen B-Y, Chang C-T, Hsueh C-C, Ding Y, Lin K-L, Xu H (2010) Comparative study on characteristics of azo dye decolorization by indigenous decolorizers. Bioresour Technol 101:2651–2656

158. Zhang WW, Andong ZY, Zhang M, Wang QN, Wei YQ, Chen LX (2014) Isolation and characterization of a heterotrophic nitrifier Proteus mirabilis strain V7 and its potential application in NH4 +-N removal. Ann Microbiol 64:1231–1238

159. Zheng X, Ng I-S, Ye C, Chen B-Y, Lu Y (2013) Copper ion-stimulated McoA-laccase production and enzyme characterization in Proteus hauseri ZMd44. J Biosci Bioeng 115(4):388–93

160. Zilberstein B, Quintanilha AG, Santos MAA, Pajecki D, Moura EG, Alves PRA, Filho FM, de Souza JAU, Gama-Rodriques J (2007) Digestive tract microbiota in healthy volunteers. Clinics 62(1):47–54

161. Ziółkowska G, Tokarzewski S, Nowakiewicz A, Kosturba A (2006) Bacteriological flora isolated from geese reproductive flocks. Annales Universitatis Mariae Curie-Sklodowska, DD, LXI, 9:5-86

2

CHAPTER

THE ROLE OF MICROORGANISMS AT DIFFERENT STAGES OF ECOSYSTEM DEVELOPMENT FOR SOIL FORMATION

S. Schulz[1], R. Brankatschk[2], A. Dumig ¨ [3], I. Kogel-Knabner[3,4], M. Schloter[1], and J. Zeyer[2]

[1]Helmholtz Zentrum Munchen, German Research Center for Environmental Health, Research Unit Environmental Genomics, ¨ Ingolstadter Landstr. 1, 85764 Neuherberg, Germany

[2]ETH Zurich, Environmental Microbiology, Institute of Biogeochemistry and Pollutant Dynamics, Universitatsstr. 16, 8092 ¨ Zurich, Switzerland

[3]Technische Universitat M¨ unchen, Lehrstuhl f ¨ ur Bodenkunde, 85350 Freising-Weihenstephan, Germany

[4]Technische Universitat M¨ unchen, Institute for Advanced Study, Lichtenbergstrasse 2a, 85748 Garching, Germany

ABSTRACT

Soil formation is the result of a complex network of biological as well as chemical and physical processes. The role of soil microbes is of high interest, since they are responsible for most biological transformations and drive the development of stable and labile pools of carbon (C), nitrogen (N) and other nutrients, which facilitate the subsequent establishment of plant communities. Forefields of receding glaciers provide unique chronosequences of different soil development stages and are ideal ecosystems to study the interaction of bacteria, fungi and archaea with their abiotic environment. In this review we give insights into the role of microbes for soil development. The results presented are based on studies performed within the Collaborative Research Program DFG SFB/TRR 38 (http://www.tu-cottbus.de/ecosystem) and are supplemented by data from other studies. The review focusses on the microbiology of major steps of soil formation. Special attention is given to the development of nutrient cycles on the formation of biological soil crusts (BSCs) and on the establishment of plant–microbe interactions.

Citation: S. Schulz, R. Brankatschk, A. Dumig, I. Kogel-Knabner, M. Schloter, and J. Zeyer, The role of microorganisms at different stages of ecosystem development for soil formation, doi:10.5194/bg-10-3983-2013

INTRODUCTION

Microbial communities can be considered as architects of soils (Rajendhran and Gunasekaran, 2008) and many ecosystem services that are linked to terrestrial ecosystems, including plant production, safeguarding of drinking water or C sequestration, are closely linked to microbial activities and their functional traits (Torsvik and Øvreas, 2002). Vice versa, ° the soil matrix as well as chemical and physical properties of soils, like quality and amount of soil organic matter, pH, and redox conditions, have a pronounced influence on the dynamics of the microbial community structure and function in soils (Lombard et al., 2011). This close interplay between abiotic conditions and the soil biosphere is one of the most fascinating issues as far as earth sciences are concerned, with huge implications on environmental as well as human health (van Elsas et al., 2008). Due to the complex interactions, it is not surprising that the formation of soils with a high level of fertility is a result of more than hundreds of years of soil "evolution" (Harrison and Strahm, 2008). As a result of global change in general and the loss of soil quality in particular, many soils are threatened. Thus, there is a huge need to develop strategies for a sustainable protection of soils for future generations. In this respect the knowledge gained from soil chronosequences might help to improve our understanding about the development of biotic–abiotic interplays and to identify factors that drive the formation of soils (Doran, 2002).

Studies on the development of abiotic and biotic interactions are very complex. They require both different spatial and temporal scales (Ollivier et al., 2011). Microbes act on a scale of μm3 and form biogeochemical interfaces with the soil matrix, shaping their own environment (Totsche et al., 2010). It remains largely unknown how many interfaces are connected and how many interfaces are needed for the stability of soil. Furthermore, microbes can change their phenotype within minutes depending on the present environmental conditions at those interfaces by gene induction or repression (Sharma et al., 2012). The corresponding transcripts often have half times in the range between seconds and minutes. Putting this in the frame of soil formation which may take centuries is a highly challenging issue.

In addition to that, the diversity of soil microbes is huge and can still be considered as a black box (Simon and Daniel, 2011); consequently nobody is so far able to give exact numbers on the species or, even more important, on the ecotype richness in one unit of soil. Microbes are also able to easily exchange genetic information, which induces a very fast and ongoing diversification of organisms in natural environments, and the genetic flexibility of the whole soil microbiome can be considered enormous (Monier et al., 2011). Finally, most functional traits, for example the degradation of plant litter or the development of food web structures and closed nutrient cycles, are not a result of a single organism but of microbial communities which closely interact which each other (Aneja et al., 2006). Even the development of symbiotic interactions between plants and microbes in soil (e.g. mycorrhization of plants or legume–rhizobia interactions) are much more complex than described in text books, including the involvement of a diverse number of "helper organisms" during the infection phase (Frey-Klett et al., 2007). Timescales for community development and stable microbiomes are therefore still a highly challenging topic of research, and in many cases concepts do not even exist for the formation of microbial communities on the basis of single organisms being present at a certain point in time of ecosystem genesis.

Forefields of receding glaciers are ideal field sites to study the initial steps of soil formation, as in a close area of some square kilometres a chronosequence of soils of different development stages can be found. As time is substituted by space, a simultaneous comparison of the formation of organismic interactions and of abiotic–biotic interfaces at different development stages is possible. Since the end of the Little Ice Age around 150 yr ago, most alpine glaciers have been receding at an increasing rate (Paul et al., 2004). A recent survey on 97 Swiss glaciers revealed that today most glaciers show an annual recession of dozens of metres (Paul et al., 2007). A similar trend can be observed in alpine zones dominated by permafrost. Permanent permafrost is more and more occurring in deeper soil horizons (Paul et al., 2007).

Detailed studies on glacier recession and soil formation have been performed at the Damma Glacier in Central Switzerland (Kobierska et al., 2011). The length of this glacier has been monitored since 1921, and the rate of recession is currently about 10 m per year (http://glaciology. ethz.ch/swiss-glaciers/). The forefield has a north-eastern exposition, an inclination of about 21 %, and is located at 2050 m a.s.l. (http://map.geo. admin.ch/). As depicted in Fig. 1, the glacier forefield is flanked by two lateral moraines, which emerged during the Little Ice Age around 1850. Two brief advancements of the glacier in 1928 and 1992 resulted in two further moraines, which divided the forefield in three parts: initial (6–13 yr), intermediate (50–80 yr) and developed sites (110–150 yr).

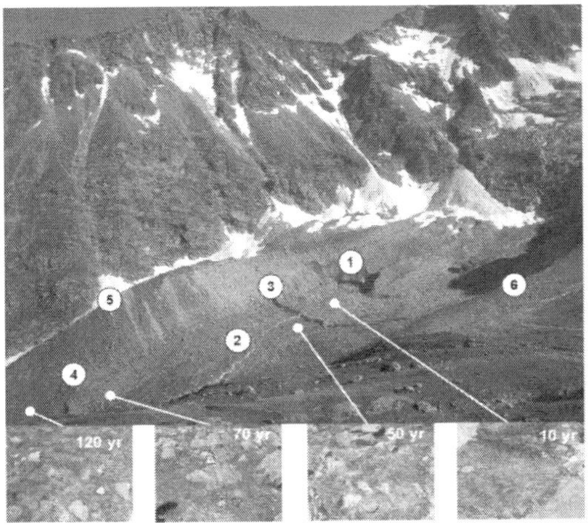

Figure 1: The forefield of the Damma Glacier (Switzerland) as it developed in response to the continuous retreat of the glacier. The numbers mark important corner points of the forefield: (1) the glacier terminus, (2) the glacier stream, (3) moraine from 1992, (4) moraine from 1928, (5) the south flanking moraine and (6) the north flanking moraine, which both date back to the end of the Little Ice Age in 1850. The small pictures are closeups from 10, 50, 70 and 120 yr ice-free soils, where the 10 yr site is situated in the initial part of the glacier forefield (6–13 yr), 50 and 70 yr sites in the intermediately developed sites (50–80 yr) and the 120 yr site in the most developed part of the forefield (110–160 yr). Photos: A. Zumsteg.

In this review we summarized our knowledge about the role of microorganisms in soil development by using current data from the chronosequence of the forefield of the Damma Glacier and how this improves our view of soils as the most important bioreactor on earth.

SETTING THE STAGE FOR MICROBIAL ACTIVITY: THE ROLE OF WEATHERING

The formation of fertile soils from inorganic bedrock implies a complex interaction of physical, chemical and biological processes. The rate of the soil development is dominated by variables such as climate, bedrock type, topography, time, microorganisms and plants (Paul and Clark, 1996; Egli et al., 2011). Usually hundreds of years are required to convert inorganic precursors to humus and eventually distinct soil horizons.

The climate in alpine zones is characterized by high precipitation and pronounced fluctuations of temperatures. For example, in the cities of Basel, Bern and Zurich (Switzerland, all at an altitude of < 550 m a.s.l.) the annual precipitation is in the range of 140–160 cm yr^{-1} (http://www.meteosuisse. admin.ch/web/en.html). In contrast, in the Alps, at an altitude of > 2500 m a.s.l., the precipitation is usually > 250 cm yr^{-1} . Nevertheless, microbes can be subject to dry stress in parts of the glacier forefield, which are dominated by rocks and sand with a low water holding capacity. In addition, solar radiation (UV and visible range) increases with altitude due to decreasing optical air masses (Blumenthaler et al., 1997). For UVB, the increase is about 9 % per 1000 m and in highaltitude glacier forefields; severe effects of UVB on microorganisms cannot be excluded. UVB damage results from a direct absorption of radiation by target molecules such as DNA and proteins, and usually microorganisms produce secondary metabolites as a photoprotective mechanism.

In the mountains not only the radiation during daytime but also the back radiation during night-time is higher than in the lowlands. As a consequence, the temperature fluctuations on the rock and soil surfaces are very pronounced and can easily go beyond 40 °C within 12 hours (unpublished data). It is remarkable that the snow cover in winter and spring provides a good insulation and temperatures on snow covered soils usually do not fall below 0 °C. Exceptions can be time periods in fall when the air temperatures drop significantly below 0 °C and the snow cover only amounts to a few centimetres (Korner et al., 1999)

The forefield of a receding glacier consists of very heterogeneous and distinct morphotypes such as moraines, rock fields, floodplains, sand hills, erosion channels and mudslides. From a geomorphological point of view, these features are formed by complex interactions of glacial, periglacial, fluvial and gravitational forces. In order to allow a meaningful analysis of experimental data along the chronosequence, these morphotypes have to be considered, and in addition to that geostatistical methods have to be employed. In the past, the mapping of the morphotypes required laboursome field work using triangulation, while today it is usually done by lidar, radar interferometry and photogrammetry (Harris et al., 2009). The weathering of rocks depends on the composition of the bedrock as well as on the environmental conditions. For example, calcareous rocks are mainly subject to chemical weathering (acidic dissolution of the calcium carbonate) whereas siliceous rocks are mainly fractured as a consequence of freezing–thawing cycles. At the Damma Glacier for example, the bedrock consists of Aare granite composed of quartz, plagioclase, potassium feldspar, biotite and muscovite (Kobierska et al., 2011). The mineralogy of the sand, silt and clay fractions in the glacier forefield reflects the composition of the bedrock.

Besides chemical and physical processes at this stage, bacteria and fungi can also contribute substantially to the weathering of mineral surfaces. Several biogeochemical processes have been described which catalyse the decay of minerals and the mobilization of nutrients. The mechanisms include enzymatically catalysed reactions, the local reduction of the pH, and the production of complexing agents such as cyanide, oxalate and gluconic acid and the excretion of transport vehicles such as siderophores (Mavris et al., 2010; Styriakova et al., 2012). However, many reports on the role of bacteria and fungi for weathering mainly cover phenomenological experiments in the laboratory (e.g. using fluorescence staining of biofilms on mineral surfaces or scanning electron microscopy of weathered minerals) and do not necessarily allow the prediction of the rates of biological weathering under field conditions. Direct field observations are rare and include for example a study on the role of microorganisms in phosphorus (P) cycling in the forefield of the Damma Glacier (Tamburini et al., 2010), which revealed a shift from substrate derived P at initial sites to internal P turnover at more developed sites.

For the establishment of microbial life in a glacier fore- field, both the chemical composition of the bedrock as well as the physical structure of the weathered fractions are important. Siliceous rocks contain a number of minerals which contain essential elements (e.g. apatite is a source of phosphorous) and thus favour microbial life. In contrast, the weathering of calcareous rocks releases very few elements which facilitate the growth of microorganisms. Some rocks such as serpentinites even release toxic elements (e.g. Nickel, Cadmium) which prevent plant life (Bratteler et al., 2006) and which may also inhibit microbial activities. Soil aggregation is of upmost importance in controlling microbial structures, functions and plant life. Ideally, the weathered fractions in a glacier forefield include sand (fraction 2 mm to 63 μm), silt (63 to 2 μm) and clay (< 2 μm), and thus allow a good diffusion of gases and bacterial motility as well as high ion exchange and water holding capacities. Particularly clay is often found in glacier forefields (Kobierska et al., 2011; Mavris et al., 2011), and this is essential for aggregate formation and the stabilization of soil organic matter (Paul and Clark, 1996). In the forefield of the Damma Glacier, there is little change in the phyllosilicate clay mineralogy, whereas the amount of poorly crystalline Fe oxides and Al phases increased with soil development, reflecting a growing potential for soil organic C stabilization (Dumig et al., 2011). In contrast to ¨ strongly increasing quantities, only small changes in the composition of the Fe and Al pools were detected during initial pedogenesis. Fe oxides and inorganic Al phases mainly remained poorly crystalline.

DEVELOPMENT OF INITIAL C AND N CYCLES

At the Damma Glacier important macronutrients, such as phosphorus and sulfur (S), are part of the mineral composition. Therefore, microbial activity might accelerate the release of those elements from the bedrock to supply living organisms with P and S. In contrast, C and N are not part of the mineral composition and are scarce in the initial soils, stressing the importance of studying these geochemical cycles from the view of microorganisms. The concentrations of the most abundant nutrients are summarized in Table 1.

In the initial soils, total C content fluctuates around 700 μg C g^{-1} . Microbial C reached up to 50 μg C g^{-1} , which is a value also found in semi-arid grassland soils (Dijkstra et al., 2006), indicating a high turnover rate of C (Table 1). This is in agreement with soil respiration rates, which are in the range of 130 μg C g^{-1} a^{-1} (Gulland et al., 2013a). The source ¨ of organic C in the initial soils, however, is a matter of controversy. Three different sources potentially contribute C to the initial soil. First, the deposition of allochthonous organic matter, such as plant litter, insects, and soot particles, contributes considerable amounts of C (Hodkinson et al., 2003). Measurements of allochthonous C deposition range from 7.5 kg C ha^{-1} a^{-1} at the Damma Glacier (Brankatschk et al., 2011) to 34 kg C ha^{-1} a^{-1} at Toolik Lake, Alaska (Fahnestock et al., 1998). Second, C inputs from close-by cyanobacterial and algal communities, such as cryoconite holes or patches of snow algae, might contribute C to the forefield (Kastovsk ˇ a et al., 2007; Sawstrom et al., 2002; Stibal et al., ´ 2008). Third, ancient C might be present in the forefield of the Damma Glacier. During the Holocene, glacier basins were vegetated. Dating the sub-fossil remains of trees and peat, warm periods, e.g. around the years 2000, 3900 and 4900 BP, were identified (Joerin et al., 2006). At this time glacier valleys were covered by peat bogs with birch and willow trees. Similarly, at the Damma Glacier it appears possible that C originating from ancient vegetation is mixed into the initial soil. This hypothesis is supported by ^{14}C measurements on the carbon dioxide released from the initial soils by respiration. The δ ^{14}C value of −68.1 indicates the degradation of ancient C as the main C source in the initial soils (Gulland et al., 2013b). In addition, the C content at 5-10 ¨ cm depth (400–500 μg C g^{-1}) was similar to the C content in the top soil of the most recently deglaciated soils (700– 1100 μg C g^{-1} , Bernasconi et al., 2011), indicating that organic material has been blended into the soil.

Esperschutz et al. (2011) used ¨ 13C labelled litter to study the microbial food web in the initial soils. C flow through the food web was traced using phospholipid fatty acids and phospholipid ether lipids. In the initial soils, archaea, fungi and protozoa were enriched in 13C. The community pattern changed only slightly towards the developed soils, where actinomycetes were involved in litter degradation. After twelve weeks of incubation, the litter degradation in the initial soils was comparable to that in the developed soils, again highlighting the activity of heterotrophic microorganisms in the initial soil.

Gulland et al. (2013a) studied C losses from the ini- ¨ tial soils at the Damma Glacier that were ice-free for 10 yr. Within the study period of three summer months, 33 g C m^{-2} were released via respiration and 2 g C m^{-2} leached from the soil. Taking into consideration the total C stocks of 90 g C m^{-2} at the Damma Glacier, these data indicate a highly active microbial community degrading the soil organic matter. Similarly, Bardgett and Walker (2004) described a heterotrophic stage of C decomposition at Odenwinkelkees Glacier, Austria.

In contrast to small changes in clay mineralogy, pronounced shifts of soil organic matter quality with increasing age of the clay fractions were found at the Damma Glacier (Dumig et al., 2012). Clay-bound organic matter ¨ from the 15-yr-old soils was mainly inherited organic C rich in aromatic compounds and rich in compounds carrying carboxyl groups. With increasing age of the clay fractions (75 and 120 yr), the formation of organo-mineral associations started with the sorption of proteinaceous compounds and microbial-derived carbohydrates on mineral surfaces. In the acidic soils, ferrihydrite (determined as oxalate-soluble iron) was the main provider of mineral surfaces and thus important for the stabilization of organic matter. We assume that sorption is not the only protective mechanism as poorly crystalline Fe phases also interact with organic matter by coprecipitation or micro-aggregation. These results show that organo-mineral associations already evolve in early stages of soil development, whereby mineral weathering and organic matter accumulation proceed in different timescales (Dumig ¨ et al., 2012).

Different pools of N are present in the initial soils. As summarized in Table 1, the content of total N ranges around 70 $\mu g\ N\ g^{-1}$, while the contents of microbial N (6 $\mu g\ N\ g^{-1}$), nitrate (0.1 $\mu g\ N\ g^{-1}$), and ammonium (0.03 $\mu g\ N\ g^{-1}$) are considerably lower. As for C, different N sources contribute N to the glacier forefield: (i) N fixation by microorganisms and (ii) N deposition. As shown in Fig. 2, N fixation is very low in the initial soils at the Damma Glacier. Duc et al. (2009) detected N fixation activity in the range of 2 pmol C2H4 $g^{-1}\ h^{-1}$, using the acetylene reduction assay. Another study found N fixation rates below 0.2 pmol $N\ h^{-1}\ g^{-1}$ in the initial soils, using the stable isotope incorporation method (Brankatschk et al., 2011). The lowest N fixation activity was accompanied by the lowest abundance of the N fixation marker gene nifH (2×10^6 copies g^{-1}) at initial sites, confirming the presence of few microorganisms capable of the N fixation process. In contrast to the N fixation, the deposition of N is several orders of magnitude higher (Brankatschk et al., 2011). Estimations for wet deposition of nitrate and ammonium ranged between 7 and 11 $kg\ N\ ha^{-1}\ a^{-1}$. The total N deposition is estimated to be 10 to 15 $kg\ N\ ha^{-1}\ a^{-1}$. This indicates the importance of the N deposition as the primary N source in the fore- field, and is supported by stable isotope measurements. The N in the initial soils exhibits $\delta\ ^{15}N$ values of −4 to −2, which appear to be typical for initial soils in cold climates and can be explained by the negative $\delta\ ^{15}N$ of atmospheric N (Smittenberg et al., 2012). The quantification of organic detritus on snow revealed that approximately 0.6 $kg\ N\ ha^{-1}\ a^{-1}$ is deposited as particulate organic matter. Decomposition and mineralization of this organic matter was suggested to be the dominant N transformation process in the initial soils at the Damma Glacier (Brankatschk et al., 2011), which is declared as initial phase in Fig. 2. Also, the marker genes for the breakdown of organic matter, chitinase (chiA) and protease (aprA) were detected in the initial soils. The abundance of chiA gene was 7×10^5 copies g^{-1} , and the aprA gene was quantified with 6×10^6 copies g^{-1} . Activities of other N turnover processes such as nitrification and denitrification were low (< 2 $nmol\ N\ g^{-1}\ h^{-1}$) in the initial soils (Brankatschk et al., 2011).

Table 1: Plant and soil parameters from initial (6–13 yr), transient (60-80 yr) and developed (110-150 yr) sites and a reference site (> 2000 yr) outside the glacier forefield. All nutrient values and the microbial biomass C are given in μg g^{-1} . Data were summarized from Bernasconi et al. (2011), Brankatschk et al. (2011), Goransson et al. (2011), H ¨ ammerli et al. (2007) and Noll and Wellinger (2008)

	Approximate soil age			
	6–13 yr	60–80 yr	110–150 yr	2000 yr
Plant coverage	< 10 %	> 70 %	100 %	100 %
Dominant vegetation	*Agrostis gigantea, Rumex scutatus, Cerastium uniflorum, Oxyria digyna*	*Agrostis gigantea, Salix* spp., *Deschampsia cespitosa, Athyrium alpestre*	*Rhododendron ferrugieum, Salix* spp., *Agrostis gigantea, Festuca rubra*	*Agrostis gigantea*
pH_{CaCl2}	5.5	4.7	4.6	3.6
Maximum water holding capacity [%]	3.5	2.4	4.8	6.7
Microbial biomass C	58	122	241	902
Microbial biomass N	6	19	29	115
Total C	700	4700	12 000	39 600
Total N	70	250	730	2670
Dissolved organic C	7	18	42	78
Dissolved organic N	0.3	0.9	3	12
NO_3^-	0.1	0.2	1.3	0.8
NH_4^+	0.03	0.2	7	13
$NaHO_3$ extractable P	6.7	9.7	7.8	n.d.
S	0.25	1.25	870.7	1300

The abundance of the marker gene for nitrification amoA of ammonia oxidizing bacteria (AOB) was two orders of magnitude higher (2×10^6 copies g^{-1}) than for ammonia oxidizing archaea (AOA) (Brankatschk et al., 2011), which stands in contrast to many others studies (Leininger et al., 2006; Schauss et al., 2009). It might be that the conditions in the initial soils are more ideal for AOB. On the one hand ammonium is supplied from atmospheric deposition and mineralization of organic matter, while competition about ammonium with plants is low. On the other hand the low pH of the soil is more favourable for AOB than AOA (Gubry-Rangin et al., 2011). However, at the same time potential nitrification measurements were low (Brankatschk et al., 2011). This can be explained by two scenarios: (i) The AOB community at the initial sites of the glacier forefield is inactive per se or (ii) the AOB community is adapted to the harsh conditions at the initial sites and is not able to adapt to laboratory conditions, thus turnover rates during potential nitrification measurements are low.

From the analyzed marker genes for denitrification, the nirK gene was the most abundant one with 1.5×10^8 copies g^{-1} . Since potential denitrification activity was low, the high nirK gene abundance might indicate the presence of facultative anaerobic bacteria in the initial soils, and would support the hypothesis of the dominance of mineralizing microorganisms that are adapted to temporarily waterlogged conditions, e.g. during heavy rains or snow melt. The established C and N cycle in the glacier forefield leads to the accumulation of protein-rich organic matter in the very early stages of soil development, which includes a stronger accumulation of carbohydrate-rich material in the course of time, both most probably of microbial origin (Dumig et al., 2012). ¨ P, next to N, is frequently found to be the limiting nutrient in terrestrial ecosystems. Therefore, the release of P from the minerals in the bedrock was at the centre of studies at the Damma Glacier. The granite bedrock contains approximately

400–600 μg P g^{-1} soil that is bound to apatite. However, recent field studies in alpine habitats suggest that apatite has a limited impact on the microbial structures and function (Ragot et al., 2013). The amount of organic P and deposited P in the initial soil is low, accounting for 50–60 μg P g^{-1} (Tamburini et al., 2010; Goransson ¨ et al., 2011) and 0.1–1.2 kg P ha^{-1} a^{-1} (Binder et al., 2009). An even smaller proportion of the P, however, is freely available as phosphate (0.8 μg P g^{-1}) as determined by resin bag experiments (Bernasconi et al., 2011). Nevertheless, the heterotrophic microorganisms as well as plants from the initial soils were not P limited (Goransson et al., 2011). However, ¨ the isotopic analysis of the plants P source showed that the minerals are the major P source and deposition plays a minor role (Tamburini et al., 2010). This might underline the microbial importance of mineral dissolution during the initial soil formation process. The dissolution of P and other nutrients from the minerals might be accelerated by microorganisms releasing organic acids or chelators. This was investigated in mineral dissolution experiments using bacterial isolates from the Damma forefield (Lapanje et al., 2012). Selected isolates were screened for high mineral dissolution potential; however, the abiotic controls using citric and hydrochloric acid released elements at significantly higher rates than the bacterial isolates did (Lapanje et al., 2012). Future studies need to investigate the P release from mineral bedrock in the initial soils, as the mechanisms have not been studied in detail, but are crucial for the development of the ecosystem.

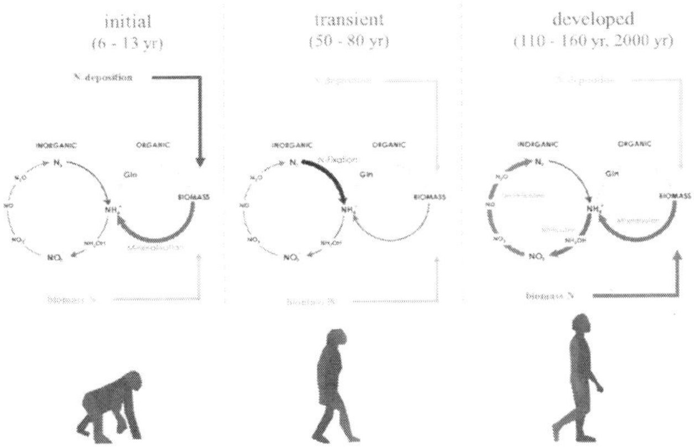

Figure 2: The development of the microbial N cycle during soil formation. To exclude the strong influence of the increasing biomass along the glacier forefield, underlying data were related to ng DNA to be able to compare results from different development stages. Figure modified after Ollivier et al. (2011).

Significant amounts of S are required to maintain the high ecosystem productivity that was measured at the forefield of the Damma Glacier (Smittenberg et al., 2012). In the bedrock a total of approximately 5 μg S g^{-1} is present (Lazzaro et al., 2009; Bernasconi et al., 2011), and concentrations of available sulfate in the initial soils range from 0.25 μg S g^{-1} (Table 1). However, in the developed soils the S content is in the range of 300 µg S g^{-1} (Bernasconi et al., 2011). Borin et al. (2010) pointed out the importance of S-oxidising bacteria as first colonizers in microbial communities at the Midtre Lovenbreen Glacier, Svalbard. If similar processes prevailed ´ at the Damma Glacier, this would indicate that sulfate is quickly released from the minerals of the initial soil and might therefore be available for plants early on. A study on desulfonating bacteria, i.e. bacteria that release sulfate from organic matter, found very high diversity at the Damma fore- field (Schmalenberger and Noll, 2010). Therefore, it was hypothesized S might be a limiting nutrient. As indicated above, the S content in the soil is low in the initial soils and the S stocks in the developed soil cannot be explained by mineral dissolution only. Therefore, the S deposition might signifi- cantly contribute to the S budget of the forefield. The deposition of S is estimated to be 200 to 350 mg SOX m^{-2} a $^{-1}$ (Nyiri et al., 2009); however, a detailed analysis on S sources in the Damma forefield, e.g. deposition measurements or isotope analysis, is lacking.

BIOFILMS AND SOIL CRUSTS AS HOTSPOTS OF NUTRIENT TURNOVER

Initial sites of glacier forefields are characterized by less vegetation and low nutrient contents. In 2005 the vegetation cover at the Damma Glacier was below 20 % at a distance of 80 m to the glacier front, which corresponds to approximately 13 yr of soil development (Hammerli et al., 2007). ¨ Therefore, it is obvious that the initial processes of soil formation and input of nutrients rely on the activity of microorganisms. In principal two main functions can be assigned to the microbes: (i) the biological weathering of the bedrock material and (ii) the formation of interfaces for nutrient turnover at vegetation free sites.

Regarding biological weathering, Frey et al. (2010) demonstrated that isolates from the granitic sand in front of the Damma Glacier were able to effectively dissolve the siliceous bedrock material. The main underlying mechanism for the dissolution is the formation of biofilms on the mineral surface. Organisms being organized in such biofilms exude organic acids like oxalic acid, which on the one hand lead to a ligand-promoted dissolution and on the other hand to a proton-promoted dissolution because of the decreasing pH. The released elements are then captured in the polysaccharide matrix of the biofilm and display therefore a nutrient hotspot in the bare substrate. In addition to that mechanism, Budel et al. (2004) attributed an important role in biologi- ¨ cal weathering to cryptoendolithic Cyanobacteria, which enhance weathering by alkalization of the substrate during photosynthesis.

It was ascertained by several researchers that the trophic base at initial sites of glacier forefields is established by first colonizers like Cyanobacteria, green algae, lichens, mosses and fungi, which often conglomerate and form biological soil crusts (BSCs) (Belnap et al., 2001a). The formation of BSCs is strongly linked to the environmental conditions present, as well as to the parental material. As shown in Fig. 3, BSC development at the Damma Glacier is very heterogeneous and strongly depends on the right equilibrium of water availability and water holding capacity of the substrate. These BSCs fulfil different important roles in the ecosystem development. Most BSCs forming organisms are able to perform photosynthesis and/or N fixation, and thus enhance C and N content of the soils. In this regard, researchers stated that under optimal conditions the performance per unit ground surface of BSC is similar or even higher than that of vascular plants (Yoshitake et al., 2009; Pointing and Belnap, 2012). Moreover, Dickson et al. (2000) showed that the N_2-fixation activity of BSCs was already measurable at 3 °C. Thus, the "vegetation period" of BSCs starts much earlier than that of vascular plants and nutrient input is prolonged, which is especially advantageous at glacier forefields. Nutrient acquisition is further supported by the excretion of exopolysaccharides by several Cyanobacteria, which are often coated with clay particles. The negatively charged clay particles are associated with positively charged nutrients in turn, and prevent those from leaching (Belnap et al., 2001a). At the Damma Glacier the highest numbers of Cyanobacteria have been found at the initial sites, including Lecanoromycetes, to which a lot of lichen-forming species belong (Zumsteg et al., 2012). A second study, which focused on nifH-carrying microbes, revealed a cyanobacterial community comparable to mature BSC from the Colorado plateau (Duc et al., 2009; Yeager et al., 2004). The dominant species were Nostoc sp. and Scytonema sp. Both are able to produce pigments, which enable them to withstand high solar radiation. This property is a big advantage at glacier forefields because radiation strongly increases during summertime and day temperatures can reach up to 40 °C at the soil surface (Miniaci et al., 2007), while the thermal radiation at night hampers the storage of heat energy from the day (Landolt, 1992). Thus, poikilohydric organisms, which form BSCs, have a big advantage as they are able to withstand drought and high solar radiation, and as soon as they are rewetted the activity strongly increases. Although crust forming bacteria have been detected at initial sites of the Damma Glacier, the formation of bacterial dominated crusts at the sampling site was not observed (Duc et al., 2009). This might be mainly attributed to the exposed position of some parts of the initial sites to the glacier tongue, which leads to regular disturbances of the surface by the glacial stream. However, as soon as sites are more protected against erosion, either due to moraines or their location on hydrologic islands, moss and lichen dominated crusts develop (Bernasconi et al., 2011, Fig. 3), which is in accordance with observations in the Negev Desert (Israel) (Zaady et al., 2000). Interestingly, these types of crusts are often associated with vegetation patches (Duc et al., 2009). It is very likely that crusts were there before plants established, as they are known to pave the way for vascular plants, because on the one hand BSCs improve soil fertility and on the other hand they create advantageous micro-environments for plant germination and growth (Belnap et al., 2001b).

Regarding the soil fertility, data from different BSCs indicate that crusts have 200 % higher N content than uncrusted soils from the same site (Harper and Belnap, 2001; Rogers and Burns, 1994; Pointing and Belnap, 2012). However due to increased microbial activities and leaching of N to deeper soil layers (Johnson et al., 2007), nitrogen is still one of the limiting factors in BSCs. Brankatschk et al. (2011) demonstrated that deposited N and C are important drivers for the ecosystem development at initial sites; the ability of Cyanobacteria to trap nutrient-rich deposits via their exopolysaccharide sheath even facilitates that effect (Reynolds et al., 2001). Thus, also decomposing microbes are stimulated in BSCs, which in turn supply plants in the vicinity with a lot of different micronutrients and trace elements. As arctic and alpine plant species often have a shallow root system due to the thin soil layers, they can easily acquire the nutrient sources below BSCs (Billings, 1987).

Figure 3: BSCs at initial sites of the forefield of the Damma Glacier. The left picture shows that BSCs do not develop when water has destructive properties (1) or when the water holding capacity of the substrate is too low (2). However, under optimal hydrological conditions and in areas protected against erosion, BSCs develop (3). The right picture illustrates the frequently observed proximity of BSC and plants. Photos: J. Zeyer.

BSCs did not only enhance the soil nutrient status, but also improved the physical conditions for plant establishment. The rough surface structure and the dark colour of BSCs create beneficial conditions for seedling settlement and germination, because they are sheltered against external disturbances and the temperature is higher compared to bare surrounding soils (Escudero et al., 2007; Breen and Levesque, ´ 2008; Gold, 1998). Moreover, the stress for pioneering plants is generally reduced – on the one hand because of the already mentioned nutrient conditions, and on the other hand because of the volumetric water content being higher in BSCs (Breen and Levesque, 2008). Altogether, BSCs are able to ´ buffer the diurnal changes in climatic conditions at glacier forefields and therefore reduce stress conditions, which otherwise might hamper the plant development.

Although a lot of effort is still necessary to directly characterize the BSCs at the forefield of the Damma Glacier, a lot of data indirectly underline their pivotal role in the ecosystem development. In summary, one can assign several functions to BSCs and biofilms at the forefield of the Damma Glacier. At the beginning of soil formation, microorganisms organized in biofilms contribute to the weathering of bedrock material and the dissolution of trace elements. Afterwards, their main function is the input of nutrients to the bare soil and the stabilization of the surface. That result then is the basis for the settlement and growth of plants, which will finally completely replace BSCs.

ROLE OF PLANTS

During the succession at the forefield of the Damma Glacier, the coverage of plants increased with soil age from 20 % to 70 % (Hammerli et al., 2007). Plants play an important role ¨ in the stabilization of the slope (Korner, 2004) and the soil ¨

development, as their root exudates and decaying litter mate rial are the main sources of organic matter at initial sites of the glacier forefield (Duc et al., 2009). Although the microbial biomass is much lower at initial sites, the decomposition of plant material is as fast as it is at developed sites over a time period of twelve weeks (Esperschutz et al., 2011). Inter- ¨ estingly, initial sites of the Damma Glacier and other alpine glaciers are dominated by forbs and grasses like Leucanthemopsis alpina, Agrostis gigantea or Cerastium uniflorum, whereas legumes like Lotus alpinus or Trifolium pallescens appear later in succession although their symbiosis with rhizobia might be an advantage at N-poor initial sites (Miniaci et al., 2007; Goransson et al., 2011; Tscherko et al., 2003). ¨ There are two reasons explaining the absence of legumes from the initial sites. First, many legumes like Lotus alpinus form heavier seeds than L. alpina or C. uniflorum (Pluess et al., 2005; Tackenberg and Stocklin, 2008), and thus dis- ¨ persal via wind is more difficult. Second, the establishment and maintenance of a rhizobia–legume symbiosis is a very energy consuming process (Merbach et al., 1999), and therefore energy consumption might be too high under these harsh conditions. However, in contrast to alpine glacier forefields, at Glacier Bay (Alaska) symbiotic N fixers like Dryas drummondii and a single species of alder are already dominant at initial sites. That difference might be mainly attributed to climatic conditions. While at Glacier Bay a mild and maritime climate with small annual and diurnal temperature changes predominates, the forefield of the Damma Glacier is characterized by strong temperature variations and irregularly distributed rainfall during the year, both being unfavourable for effective plant establishment (Bernasconi, 2008; Landolt, 1992; Miniaci et al., 2007).

Several studies at the Damma Glacier showed that pioneering plants display a nutrient hotspot at initial sites (Towe et ¨ al., 2010; Duc et al., 2009; Miniaci et al., 2007). Thus, much higher abundances and activities of microbes were detected in the rhizosphere of pioneering plants. In initial soils this effect is even expanded up to a distance of 20 cm (Miniaci et al., 2007). The phenomenon of enhanced microbial activity and abundance in the rhizosphere of plants is known as the "rhizosphere effect" (Butler et al., 2003; Hartmann et al., 2008). Regarding the Damma Glacier the rhizosphere effect is generally more pronounced at initial sites compared to developed ones (Towe et al., 2010; Edwards et al., 2006). This ¨ observation seems to be a general effect independent of plant species, bedrock material or climatic conditions. For example, Deiglmayr et al. (2006) detected only a significant difference in the nitrate reductase activity at the initial sites of the Rotmoosferner Glacier. This activity was 23 times higher in the rhizosphere of Poa alpina compared to the bulk soil. The enhancement of microbial activity and abundance in the rhizosphere is mainly attributed to uncontrolled leakage or controlled exudation of organic substances like malate, citrate or oxalate. Thus, it is unquestionable that plants display a nutrient hotspot in terms of C, as they provide up to 40 % of the photosynthetically fixed CO2 to the microbes (Paterson and Sim, 2000). In return, microbes supply the plant with N, phosphate or other nutrients and additionally protect them against herbivores or parasites.

Interestingly, the highest abundances of N fixers in the bulk soil in connection with highest N fixation activity in the rhizosphere of L. alpina have been detected at intermediate development stages (transient phase in Fig. 2), where the N content is still low but plant coverage has already strongly increased, which hints at a competition between microbes and plants for N (Brankatschk et al., 2011; Duc et al., 2009). This theory is further corroborated by results from Towe et ¨ al. (2010), where the abundance of nifH carrying microbes was highest in the rhizosphere of L. alpina planted in a 10 yr soil, which was connected with the lowest C / N ratios of plant biomass. However, during incubation the N content of L. alpina grown in the 10 yr soil strongly increased while plant and root biomass stayed stable, as shown in Fig. 4, which fits with the assumption that microbes are able to win the competition over a short timescale because of their higher volume–surface ratio, higher growth rate and substrate affinity (Hodge et al., 2000). On the contrary, plants are more effective over a long time period because of their longer lifespan and the ability to retain the assimilated N (Hodge et al., 2000; Nordin et al., 2004). In addition to the enrichment of N fixing microbes, heterotrophic mineralizers like chiA-containing microbes are also enhanced (Tscherko et al., 2004; Towe et al., 2010). The degradation of chitin and pro- ¨ teins has the advantage that low-weight organic compounds consisting of C and N are released. The ability of plants to assimilate amino acids, amino sugar or small peptides seems to be ubiquitously distributed among different ecosystems and plays a particular role in cold and wet habitats (Nasholm et ¨ al., 2009; Lipson and Monson, 1998; Schimel and Bennett, 2004). In parallel, functional groups leading to N losses like nitrifying or denitrifying microbes were reduced in the rhizosphere, which is in line with the assumption that plants are able to actively influence their rhizosphere community (Singh et al., 2004) by changing their exudation pattern or by actively excreting substances like tannins, polyphenolic substances or monoterpenes (Briones et al., 2003; Kowalchuk and Stephen, 2001; Ward et al., 1997; Cocking, 2003). In this regard, Edwards et al. (2006) detected changes in the exudation pattern of L. alpina along the forefield of the Damma Glacier mainly attributed to a strong reduction of oxalic and citric acid concentrations. In general, it seems that indeed microbial activity and abundance is much higher in the rhizosphere of pioneering plants compared to the bulk soil, but the community composition is strongly driven by the bulk soil community, whereas at developed sites the plant determines the microbial community in the rhizosphere (Duc et al., 2009; Miniaci et al., 2007). The strength of that phenomenon has been proven for functional groups like nifHcontaining microbes as well as for overall bacterial diversity from different sites and by different methods like PLFA analysis (Tscherko et al., 2004; Bardgett and Walker, 2004), clone libraries (Duc et al., 2009), pyrosequencing (Knelman et al., 2012) and fingerprinting methods (Deiglmayr et al., 2006; Miniaci et al., 2007).

With ongoing succession a shift from a competition for N to a competition for phosphate takes place (Vitousek and Farrington, 1997). While phosphate is not limited at the beginning of succession as it is released during weathering of the siliceous bedrock material (Ragot et al., 2013), the concentration of bioavailable phosphate steadily decreases along the glacier chronosequence, thus favouring ecto- and ericoid mycorrhizal associations (Bernasconi et al., 2011). A similar trend was found at Glacier Bay (Alaska), where a symbiosis with legumes has been found already in very young soils, but the first mycorrhiza-forming plants appear later in succession (Chapin et al., 1994).

Anyhow, as soon as vegetation cover is nearly closed, the ecosystem development speeds up so that ecosystem properties are similar to mature ecosystems. This includes increasing amounts of organic C and N, the formation of soil horizons, increasing microbial biomass and enzyme activities even in the bulk soil (Dumig et al., 2011; Brankatschk ¨ et al., 2011; Duc et al., 2009; Sigler and Zeyer, 2002). Once a stable plant community has developed, a positive feedback loop establishes. The plants still provide fixed CO2 via rhizodeposition to the microbial community. Additionally, the high input of dead plant material provides a broad nutrient source for decomposing microbes. Thus, in contrast to sparsely vegetated sites, sufficient amounts of N are released during mineralization of high molecular compounds, which is then again available for plants. Moreover, climatic conditions like water content, temperature and radiation are more stable below a closed plant cover, leading to a reduction of environmental stressors, which are strongly pronounced in initially poorly vegetated sites.

Finally, one can summarize that plants play a crucial role in ecosystem development, but their function changes during succession. At initial stages their pivotal role in stabilizing the slope and providing C is unquestionable, but similarly they compete for scarce nutrients like N with the microbes. The succession of plant establishment strongly depends on the external conditions. Thus, at alpine glacier forefields higher plants establish later compared to glacier forefields with maritime climate, where shrubs and small trees can already be found at very young sites (Bardgett and Walker, 2004).

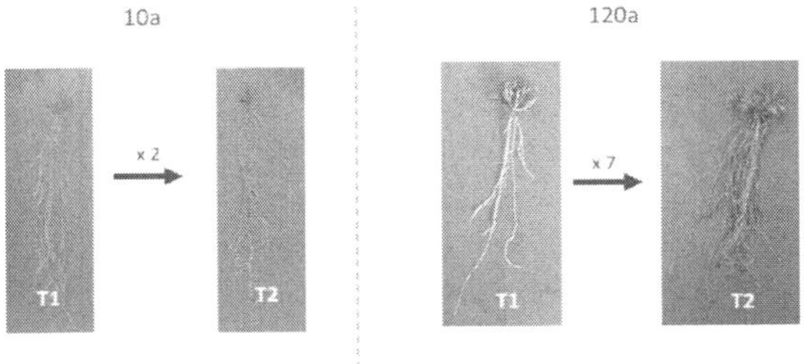

Figure 4: Development of root biomass of L. alpina grown in soils obtained from the Damma chronosequence that have been ice free for 10 and 120 yr, respectively, at two different points in time of plant development (2 weeks and 8 weeks of growth) in a green house experiment.

LESSONS LEARNED FROM THE DAMMA GLACIER CHRONOSEQUENCE FOR ECOSYSTEM DEVELOPMENT

1. Initial ecosystems can emerge as a consequence of different events. These systems include, for example, fore- fields of retreating glaciers, chronosequences related to volcanic eruptions, post-mining areas and permanent mechanical disturbance at coastal and inland dunes. For initial ecosystems the following phases of development can be assumed: (i) the system properties are driven by physical structures and hydrological processes, (ii) the importance of chemical processes increases and (iii) biological communities drive ecosystem properties and development (Schaaf et al., 2011). However, the individual time of development

strongly depends on the external conditions like bedrock material, altitude, meteorological properties, initial C, or the degree of mechanical disturbance.

2. The knowledge gained from the Damma Glacier showed that not only the presence of biota but also the interaction of organisms is the key for ecosystem development. Regarding the microbial N cycle, development takes place in three phases. Initially, allochthonous N sources are mineralized. As soon as plant biomass and with that competition on the rare N sources increase, N fixation becomes more important. Finally, when the surface is completely covered with plants the N cycling community becomes more complex, and N is mostly derived from biomass. It is obvious that biological "hotspots" like BSCs can accelerate the development of initial ecosystems. It has been shown that the development of the nitrogen cycle in the Damma chronosequence and that in the BSCs of inland dunes are characterized by similar phases.
3. The presented review indicates that the development of initial ecosystems cannot be accelerated by "simple" management strategies like reforestation, irrigation or fertilization. Biotic–abiotic interactions and appropriate soil structures must be established first to pave the way for higher plants. For example, the growth of higher plants in an initial phase can be limited because applied fertilizers will leach into the groundwater and essential mycorrhizal communities are lacking. Therefore, it will be a future challenge to find and establish adequate strategies to accelerate the development of initial ecosystems.

ACKNOWLEDGEMENTS

This study is part of the Transregional Collaborative Research Centre 38 (SFB/TRR 38), which is financially supported by the Deutsche Forschungsgemeinschaft (DFG, Bonn) and the Brandenburg Ministry of Science, Research and Culture (MWFK, Potsdam). Moreover, we thank Anita Zumsteg for providing Fig. 1. Finally, we thank the anonymous reviewers for their helpful comments to improve the manuscript.

REFERENCES

1. Aneja, M., Sharma, S., Schloter, M., and Munch, J. C.: Microbial degradation of beech litter – Influence of soil type and litter quality on the structure and function of microbial populations involved in the turnover process, Microb. Ecol., 52, 127–135, 2006.
2. Bardgett, R. D. and Walker, L. R.: Impact of coloniser plant species on the development of decomposer microbial communities following deglaciation, Soil Biol. Biochem., 36, 555–559, 2004.
3. Belnap, J., Budel, B., and Lange, O. L.: Biological Soil Crusts: ¨ Characteristics and Distribution, in: Biological Soil Crusts: Structure, Function, and Management, edited by: Belnap, J. and Lange, O. L., Springer, Berlin, Heidelberg, 2001a.
4. Belnap, J., Prasse, R., and Harper, K. T.: Influence of biological soil crusts on soil environments and vascular plants, in: Biological Soil Crusts: Structure,

Function, and Management, edited by: Belnap, J. and Lange, O. L., Springer, Berlin, Heidelberg, 281– 300, 2001b.

5. Bernasconi, S. M.: Weathering, soil formation and initial ecosystem evolution on a glacier forefield: a case study from the Damma Glacier, Switzerland, Mineral Mag, 72, 19–22, 2008.
6. Bernasconi, S. M., Bauder, A., Bourdon, B., Brunner, I., Bunemann, ¨ E., Chris, I., Derungs, N., Edwards, P., Farinotti, D., Frey, B., Frossard, E., Furrer, G., Gierga, M., Goransson, H., G ¨ ulland, K., ¨ Hagedorn, F., Hajdas, I., Hindshaw, R., Ivy-Ochs, S., Jansa, J., Jonas, T., Kiczka, M., Kretzschmar, R., Lemarchand, E., Luster, J., Magnusson, J., Mitchell, E. A. D., Venterink, H. O., Plotze, ¨ M., Reynolds, B., Smittenberg, R. H., Stahli, M., Tamburini, F., ¨ Tipper, E. T., Wacker, L., Welc, M., Wiederhold, J. G., Zeyer, J., Zimmermann, S., and Zumsteg, A.: Chemical And Biological Gradients Along The Damma Glacier Soil Chronosequence, Switzerland, Vadzone, 10, 867–883, 2011.
7. Billings, W. D.: Constraints to Plant Growth, Reproduction, and Establishment in Arctic Environments, Arctic Alpine Res., 19, 357–365, 1987.
8. Binder, C. R., de Baan, L., and Wittmer, D.: Phosphorflusse der ¨ Schweiz, Bundesamt fur Umwelt (BAFU), 28, 1–163, 2009.
9. Blumenthaler, M., Ambach, W., and Ellinger, R.: Increase in solar UV radiation with altitude, J. Photochem. Photobiol., 39, 130– 134, 1997.
10. Borin, S., Ventura, S., Tambone, F., Mapelli, F., Schubotz, F., Brusetti, L., Scaglia, B., Acqui, L., Solheim, B., Turicchia, S., Marasco, R., Hinrichs, U., Baldi, F., Adani, F., and Daffonchio, D.: Rock weathering creates oases of life in a High Arctic desert, Environ. Microbiol., 12, 293–303, 2010.
11. Brankatschk, R., Towe, S., Kleineidam, K., Schloter, M., and Zeyer, ¨ J.: Abundances and potential activities of nitrogen cycling microbial communities along a chronosequence of a glacier forefield, ISME J., 5, 1025–1037, 2011.
12. Bratteler, M., Lexer, C., and Widmer, A.: Genetic architecture of traits associated with serpentine adaptation of Silene vulgaris, J Evol. Biol., 19, 1149–1156, 2006.
13. Breen, K. and Levesque, E.: The Influence of Biological Soil Crusts ´ on Soil Characteristics along a High Arctic Glacier Foreland, Nunavut, Canada, Arct. Antarct. Alp. Res., 40, 287–297, 2008.
14. Briones, A. M., Okabe, S., Umemiya, Y., Ramsing, N.-B., Reichardt, W., and Okuyama, H.: Ammonia-oxidizing bacteria on root biofilms and their possible contribution to N use efficiency of different rice cultivars, Plant Soil, 250, 335–348, 2003.
15. Budel, B., Weber, B., K ¨ uhl, M., Pfanz, H., S ¨ ultemeyer, D., and Wes- ¨ sels, D.: Reshaping of sandstone surfaces by cryptoendolithic cyanobacteria: bioalkalization causes chemical weathering in arid landscapes, Geobiology, 2, 261–268, 2004.
16. Butler, J. L., Williams, M. A., Bottomley, P. J., and Myrold, D. D.: Microbial community dynamics associated with rhizosphere carbon flow, Appl. Environ. Microbiol., 69, 6793–6800, 2003.

17. Chapin, F. S., Walker, L. R., Fastie, C. L., and Sharman, L. C.: Mechanisms of primary succession following deglaciation at Glacier Bay, Alaska, Ecol. Monogr., 64, 149–175, doi:10.2307/2937039, 1994.

18. Cocking, E. C.: Endophytic colonization of plant roots by nitrogen- fixing bacteria, Plant Soil, 252, 169–175, 2003.

19. Deiglmayr, K., Philippot, L., Tscherko, D., and Kandeler, E.: Microbial succession of nitrate-reducing bacteria in the rhizosphere of Poa alpina across a glacier foreland in the Central Alps, Environ. Microbiol., 8, 1600–1612, 2006.

20. Dickson, L. G.: Constraints to nitrogen fixation by cryptogamic crusts in a polar desert ecosystem, Devon Island, NWT, Canada, Arct. Antarct. Alp. Res., 32, 40–45, 2000.

21. Dijkstra, P., Ishizu, A., Doucett, R., Hart, S. C., Schwartz, E., Menyailo, O. V., and Hungate, B. A.: C-13 and N-15 natural abundance of the soil microbial biomass, Soil Biol. Biochem., 38, 3257–3266, 2006.

22. Doran, J.: Soil health and global sustainability: translating science into practice, Agricult. Ecosyst. Environ., 24, 119–127, 2002.

23. Duc, L., Noll, M., Meier, B., Burgmann, H., and Zeyer, J.: High di- ¨ versity of diazotrophs in the forefield of a receding alpine glacier, Microb. Ecol., 57, 179–190, 2009.

24. Dumig, A., Smittenberg, R., and K ¨ ogel-Knabner, I.: Concurrent ¨ evolution of organic and mineral components during initial soil development after retreat of the Damma glacier, Switzerland, Geoderma, 163, 83–94, 2011.

25. Dumig, A., H ¨ ausler, W., Steffens, M., and K ¨ ogel-Knabner, I.: Clay ¨ fractions from a soil chronosequence after glacier retreat reveal the initial evolution of organo-mineral associations, Geochim. Cosmochim. Ac., 85, 1–18, 2012.

26. Edwards, I. P., Burgmann, H., Miniaci, C., and Zeyer, J.: Variation in microbial community composition and culturability in the rhizosphere of Leucanthemopsis alpina (L.) heywood and adjacent bare soil along an alpine chronosequence, Microb. Ecol., 52, 679–692, 2006.

27. Egli, M., Wernli, M., Burga, C., Kneisel, C., Mavris, C., Valboa, G., Mirabella, A., Ploetze, M., and Haeberli, W.: Fast but spatially scattered smectite-formation in the proglacial area Morteratsch: An evaluation using GIS, Geoderma, 164, 11–21, 2011.

28. Escudero, A., Mart´ınez, I., de la Cruz, A., Otalora, M. A. G., and ´ Maestre, F. T.: Soil lichens have species-specific effects on the seedling emergence of three gypsophile plant species, J. Arid Environ., 70, 18–28, 2007.

29. Esperschutz, J., P ¨ erez-de-Mora, A., Schreiner, K., Welzl, G., Bueg- ´ ger, F., Zeyer, J., Hagedorn, F., Munch, J. C., and Schloter, M.: Microbial food web dynamics along a soil chronosequence of a glacier forefield, Biogeosciences, 8, 3283–3294, doi:10.5194/bg- 8-3283-2011, 2011.

30. Fahnestock, J. T., Jones, M. H., Brooks, P. D., Walker, D. A., and Welker, J. M.: Winter and early spring CO2 flux from tundra communities of northern Alaska, J. Geophys. Res., 102, 925– 929, 1998.

31. Frey, B., Rieder, S. R., Brunner, I., Plotze, M., Koetzsch, S., Lapanje, A., Brandl, H., and Furrer, G.: Weathering-associated bacteria from the Damma glacier forefield: physiological capabilities and impact on granite dissolution, Appl. Environ. Microbiol., 76, 4788–4796, 2010.

32. Frey-Klett, P., Garbaye, J., and Tarkka, M.: The mycorrhiza helper bacteria revisited, New Phytol., 176, 22–36, 2007.

33. Gold, W. G.: The influence of cryptogamic crusts on the thermal environment and temperature relations of plants in a high Arctic polar desert, Devon Island, NWT, Canada, Arctic Alpine Res., 30, 108–120, 1998.

34. Goransson, H., Olde Venterink, H., and B ¨ a˚ath, E.: Soil bacterial ˚ growth and nutrient limitation along a chronosequence from a glacier forefield, Soil Biol. Biochem., 43, 1333–1340, 2011.

35. Gubry-Rangin, C., Hai, B., Quince, C., Engel, M., Thomson, B. C., James, P., Schloter, M., Griffiths, R. I., Prosser, J. I., and Nicol, G. W.: Niche specialization of terrestrial archaeal ammonia oxidizers, P. Natl. Acad. Sci., 108, 21206–21211, 2011.

36. Gulland, K., Espersch ¨ utz, J., Bornhauser, D., Bernasconi, S. M., ¨ Kretzschmar, R., and Hagedorn, F.: Mineralisation and leaching of C from 13C labelled plant litter along an initial soil chronosequence of a glacier forefield, Soil Biol. Biochem., 57, 237–247, 2013a.

37. Gulland, K., Hagedorn, F., Smittenberg, R. H., G ¨ oransson, H., ¨ Bernasconi, S. M., Hajdas, I., and Kretzschmar, R.: Evolution of carbon fluxes during initial soil formation along the forefield of Damma glacier, Switzerland, Biogeochemistry, 113, 545–561, 2013b.

38. Hammerli, A., Waldhuber, S., Miniaci, C., Zeyer, J., and Bunge, ¨ M.: Local expansion and selection of soil bacteria in a glacier forefield, Eur. J. Soil Sci., 58, 1437–1445, 2007.

39. Harper, K. T. and Belnap, J.: The influence of biological soil crusts on mineral uptake by associated vascular plants, J. Arid Environ., 47, 347–357, 2001.

40. Harris, C., Arenson, L. U., Christiansen, H. H., Etzelmuller, B., ¨ Frauenfelder, R., Gruber, S., Haeberli, W., Hauck, C., Holzle, ¨ M., Humlum, O., Isaksen, K., Ka¨ab, A., Kern-L ¨ utschg, M. A., ¨ Lehning, M., Matsuoka, N., Murton, J. B., Notzli, J., Phillips, ¨ M., Ross, N., Seppal¨ a, M., Springman, S. M., and Vonder M ¨ uhll, ¨ D.: Permafrost and climate in Europe: Monitoring and modelling thermal, geomorphological and geotechnical responses, EarthSci. Rev., 92, 117–171, 2009.

41. Harrison, R. B. and Strahm, B.: Soil Formation, Encyclopedia of Ecology 3291–3295, 2008

42. Hartmann, A., Rothballer, M., and Schmid, M.: Lorenz Hiltner, a pioneer in rhizosphere microbial ecology and soil bacteriology research, Plant Soil, 312, 7–14, 2008.

43. Hodge, A., Robinson, D., and Fitter, A.: Are microorganisms more effective than plants at competing for nitrogen?, Trends Plant Sci., 5, 304–308, 2000.

44. Hodkinson I. D., Coulson S. J., and Webb, N. R.: Community assembly along proglacial chronosequences in the High Arctic: vegetation and soil development in north-west Svalbard, J Ecol., 91, 651–663, 2003.

45. Joerin, U., Stocker, T., and Schluchter, Ch.: Multicentury glacier ¨ fluctuations in the Swiss Alps during the Holocene, The Holocene, 16, 697–704, 2006.

46. Johnson, S. L., Neuer, S., and Garcia-Pichel, F.: Export of nitrogenous compounds due to incomplete cycling within biological soil crusts of arid lands, Environ. Microbiol., 9, 680–689, 2007. Kastovsk ˇ a, K., Stibal, M., ´ Saback ˇ a, M., ´ Cern ˇ a, B., ´ Santr ˇ u˚ckov ˇ a,´ H., and Elster, J.: Microbial community structure and ecology of subglacial sediments in two polythermal Svalbard glaciers characterized by epifluorescence microscopy and PLFA, Polar Biol., 30, 277–287, 2007.

47. Knelman, J. E., Legg, T. M., O'Neill, S. P., Washenberger, C. L., Gonzalez, A., Cleveland, C. C., and Nemergut, D. R.: Bacterial ´ community structure and function change in association with colonizer plants during early primary succession in a glacier fore- field, Soil Biol. Biochem., 46, 172–180, 2012.

48. Kobierska, F., Jonas,T., Magnusson, J., Zappa, M., Bavay, M., Bosshard, T., Paul, F., and Bernasconi, S.: Climate change effects on snow melt and discharge of a partly glacierized watershed in Central Switzerland (SoilTrec Critical Zone Observatory), App. Geochem., 26, S60–S62, 2011.

49. Korner, C.: Alpine plant life, Springer, Heidelberg, 1999.

50. Korner, C.: Mountain biodiversity, its causes and function, Ambio, ¨ 13, 11–17, 2004.

51. Kowalchuk, G. A. and Stephen, J. R.: Ammonia-oxidizing bacteria: A model for molecular microbial ecology, Annu. Rev. Microbiol., 55, 485–529, 2001.

52. Landolt, E.: Unsere Alpenflora, Gustav Fischer Verlag, Stuttgart, 1992.

53. Lapanje, A., Wimmersberger, C., Furrer, G., Brunner, I., and Frey, B.: Pattern of elemental release during the granite dissolution can be changed by aerobic heterotrophic bacterial strains isolated from Damma Glacier (central Alps) deglaciated granite sand, Microb. Ecol., 63, 865–882, 2012.

54. Lazzaro, A., Abegg, C., and Zeyer, J.: Bacterial community structure of glacier forefields on siliceous and calcareous bedrock, Europ. J. Soil Sc., 15, 291–301, 2009.

55. Leininger, S., Urich, T., Schloter, M., Schwark, L., Qi, J., Nicol, G. W., Prosser, J. I., Schuster, S. C., and Schleper, C.: Archaea predominate among ammonia-oxidizing prokaryotes in soils, Nature, 442, 806–809, 2006.

56. Lipson, D. A. and Monson, R. K.: Plant-microbe competition for soil amino acids in the alpine tundra: effects of freeze-thaw and dry-rewet events, Oecologia, 113, 406–414, 1998.

57. Lombard, N., Prestat, E., van Elsas, J. D., and Simonet, P.: Soilspecific limitations

for access and analysis of soil microbial communities by metagenomics, FEMS Microbiol. Ecol., 78, 31–49, 2011.

58. Mavris, C., Egli, M., Ploetze, M., Blum, J. D., Mirabella, A., Giaccai, D., and Haeberli, W.: Initial stages of weathering and soil formation in the Morteratsch proglacial area (Upper Engadine, Switzerland), Geoderma, 155, 359–371, 2010.
59. Mavris, C., Ploetze, M., Mirabella, A., Giaccai, D., Valboa, G., and Egli, M.: Clay mineral evolution along a soil chronosequence in an Alpine proglacial area, Geoderma, 165, 106–117, 2011.
60. Merbach, W., Mirus, E., Knof, G., Remus, R., Ruppel, S., Russow, R., Gransee, A., and Schulze, J.: Release of carbon and nitrogen compounds by plant roots and their possible ecological importance, J. Plant Nutr. Soil Sci., 162, 373–383, 1999.
61. Miniaci, C., Bunge, M., Duc, L., Edwards, I., Burgmann, H., and Zeyer, J.: Effects of pioneering plants on microbial structures and functions in a glacier forefield, Biol. Fert. Soils, 44, 289–297, 2007.
62. Monier, J. M., Demaneche, S., Delmont, T. O., Mathieu, A., Vogel, ` T. M., and Simonet, P.: Metagenomic exploration of antibiotic resistance in soil, Curr. Opin. Microbiol., 14, 229–235, 2011.
63. Nasholm, T., Kielland, K., and Ganeteg, U.: Uptake of organic ni- ¨ trogen by plants, New Phytol., 182, 31–48, 2009.
64. Nordin, A., Schmidt, I. K., and Shaver, G. R.: Nitrogen uptake by arctic soil microbes and plants in relation to soil nitrogen supply, Ecology, 85, 955–962, 2004.
65. Noll, M. and Wellinger, M.: Changes of the soil ecosystem along a receding glacier: Testing the correlation between environmental factors and bacterial community structure, Soil Biol. Biochem., 40, 2611–2619, 2008.
66. Nyiri, A., Gauss, M., and Klein, H.: Transboundary data by main pollutants (S, N, O3) and PM, Country reports 2010 MSC-W Data Note 1/2010, ISSN 1890-0003, 2009.
67. Ollivier, J., Towe, S., Bannert, A., Hai, B., Kastl, E. M., Meyer, A., ¨ Su, M.X., Kleineidam, K., and Schloter, M.: Nitrogen turnover in soil and global change, FEMS Microbiol. Ecol., 78, 3–16, 2011.
68. Paterson, E. and Sim, A.: Effect of nitrogen supply and defoliation on loss of organic compounds from roots of Festuca rubra, J. Exp. Bot., 51, 1449–1457, 2000
69. Paul, E. A. and Clark, F.: Soil Microbiology and Biochemistry, 2nd Edition, Academic Press, New York, 1996.
70. Paul, F., Ka¨ab, A., Maisch, M., Kellenberger, T., and Haeberli, W.: ¨ Rapid disintegration of Alpine glaciers observed with satellite data, Geophys. Res. Lett., 31, 1–4, 2004.
71. Paul, F., Ka¨ab, A. and Haeberli, W.: Recent glacier changes in the ¨Alps observed from satellite: Consequences for future monitoring strategies, Global Planet. Change, 56, 111–122, 2007.

72. Pluess, A., Schutz, W., and St ¨ ocklin, J.: Seed weight increases with ¨ altitude in the Swiss Alps (2007) between related species but not among populations of individual species, Oecologia, 144, 55–61, 2005.

73. Pointing, S. B. and Belnap, J.: Microbial colonization and controls in dryland systems, Nat. Rev. Microbiol., 10, 551–562, 2012.

74. Ragot, S., Zeyer, J., Zehnder, L., Reusser, E., Brandl, H., and Lazzaro, A.: Bacterial community structures of an alpine apatite deposit, Geoderma, 202–203, 30–37, 2013.

75. Rajendhran, J. and Gunasekaran, P.: Strategies for accessing soil metagenome for desired applications, Biotechnol. Adv., 26, 576– 590, 2008.

76. Reynolds, R., Belnap, J., Reheis, M., Lamothe, P., and Luiszer, F.: Aeolian dust in Colorado Plateau soils: Nutrient inputs and recent change in source, P. Natl. Acad. Sci., 98, 7123–7127, 2001.

77. Rogers, S. L. and Burns, R. G.: Changes in aggregate stability, nutrient status, indigenous microbial populations, and seedling emergence, following inoculation of soil, Biol. Fert. Soils, 18, 209– 215, 1994.

78. Sawstrom, C., Mumford, P., Marshall, W., Hodson, A., and Laybourn-Parry, J.: The microbial communities and primary productivity of cryoconite holes in an arctic glacier (Svalbard 79◦ N), Polar Biol., 25, 591–596, 2002.

79. Schaaf, W., Bens, O., Fischer, A., Gerke, H. H., Gerwin, W., Grunewald, U., Holl ¨ ander, H. M., K ¨ ogel-Knabner, I., Mutz, M., ¨ Schloter, M., Schulin, R., Veste, M., Winter, S., and Huttl, R. F.: ¨ Patterns and processes of initial terrestrial-ecosystem development, J. Plant Nutr. Soil Sci., 174, 229–239, 2011.

80. Schauss, K., Focks, A., Leininger, S., Kotzerke, A., Heuer, H., Thiele-Bruhn, S., Sharma, S., Wilke, B.-M., Matthies, M., Smalla, K., Munch, J. C., Amelung, W., Kaupenjohann, M., Schloter, M., and Schleper, C.: Dynamics and functional relevance of ammonia-oxidizing archaea in two agricultural soils, Environ. Microbiol., 11, 446–456, 2009.

81. Schimel, J. P. and Bennett, J.: Nitrogen mineralization: Challenges of a changing paradigm, Ecol, 85, 591–602, 2004.

82. Schmalenberger, A. and Noll, M.: Shifts in desulfonating bacterial communities along a soil chronosequence in the forefield of a receding glacier, FEMS Microb. Ecol., 71, 208–217, 2010.

83. Sharma, S., Mehta, R., Gupta, R., and Schloter, M.: Improved protocol for the extraction of bacterial mRNA from soils, J. Microbiol. Methods, 91, 62–64, 2012.

84. Sigler, W. V. and Zeyer, J.: Microbial diversity and activity along the forefields of two receding glaciers, Microb. Ecol., 43, 397–407, 2002.

85. Simon, C. and Daniel, R.: Metagenomic analyses: past and future trends, Appl. Environ. Microbiol., 77, 1153–1161, 2011.

86. Singh, B. K., Millard, P., Whiteley, A. S., and Murrell, J. C.: Unravelling rhizosphere-microbial interactions: opportunities and limitations, Trends Microbiol., 12, 386–393, 2004.

87. Smittenberg, R., Gierga, M., Goransson, H., Christl, I., Farinotti, D., ¨ and Bernasconi, S.: Climate-sensitive ecosystem carbon dynamics along the soil chronosequence of the Damma glacier forefield, Switzerland, Glob. Change Biol., 18, 1941–1955, 2012.

88. Stibal, M., Tranter, M., Benning, L. G., and Reh ˇ ak, J.: Microbial ´ primary production on an Arctic glacier is insignificant in comparison with allochthonous organic carbon input, Environ. Microbiol., 10, 2172–2178, 2008.

89. Styriakova, I., Styriak, I., and Oberhansli, H.: Rock weathering by ¨ indigenous heterotrophic bacteria of Bacillus spp. at different temperature: a laboratory experiment, Mineral. Petrol., 105, 135– 144, 2012.

90. Tackenberg, O. and Stocklin, J.: Wind dispersal of alpine plant ¨ species: A comparison with lowland species, J. Veg. Sci., 19 109–118, 2008.

91. Tamburini, F., Bernasconi, S. M., Angert, A., Weiner, T., and Frossard, E.: A method for the analysis of the d18O of inorganic phosphate in soils extracted with HCl, Eur. J. Soil Sc., 61, 1025– 1032, 2010.

92. Torsvik, V. and Øvreas, L.: Microbial diversity and function in soil: ˚ from genes to ecosystems, Curr. Opin. Microbiol., 5, 240–245, 2002.

93. Totsche, K. U., Rennert, T., Gerzabek, M. H., Kogel-Knabner, I., ¨ Smalla, K., Spiteller, M., and Vogel, H.-J. : Biogeochemical interfaces in soil: The interdisciplinary challenge for soil science, J. Plant Nutr. Soil. Sci., 173, 88–99, 2010.

94. Towe, S., Albert, A., Kleineidam, K., Brankatschk, R., D ¨ umig, A., ¨ Welzl, G., Munch, J., Zeyer, J., and Schloter, M.: Abundance of microbes involved in nitrogen transformation in the rhizosphere of Leucanthemopsis alpina (L.) Heywood grown in soils from different sites of the Damma glacier forefield, Microb. Ecol., 60, 762–770, 2010.

95. Tscherko, D., Rustemeier, J., Richter, A., Wanek, W., and Kandeler, E.: Functional diversity of the soil microflora in primary succession across two glacier forelands in the Central Alps, Eur. J. Soil. Sci., 54, 685–696, 2003.

96. Tscherko, D., Hammesfahr, U., Marx, M. C., and Kandeler, E.: Shifts in rhizosphere microbial communities and enzyme activity of Poa alpina across an alpine chronosequence, Soil Biol. Biochem., 36, 1685–1698, 2004.

97. van Elsas, J. D., Costa, R., Jansson, J., Sjoling, S., Bailey, M., Nalin, ¨ R., Vogel, T. M., and van Overbeek. L.: The metagenomics of disease-suppressive soils – experiences from the METACONTROL project, Trends Biotechnol., 26, 591–601, 2008.

98. Vitousek, P. M. and Farrington, H.: Nutrient limitation and soil development: Experimental test of a biogeochemical theory, Biogeochemistry, 37, 63–75, 1997.

99. Ward, B. B., Courtney, K. J., and Langenheim, J. H.: Inhibition of Nitrosomonas europaea by monoterpenes from coastal redwood (Sequoia sempervirens) in whole-cell studies, J. Chem. Ecol., 23, 2583–2598, 1997.

100. Yeager, C. M., Kornosky, J. L., Housman, D. C., Grote, E. E., Belnap, J., and Kuske, C. R.: Diazotrophic community structure and function in two successional stages of biological soil crusts from the Colorado Plateau and Chihuahuan Desert, Appl. Environ. Microbiol., 70, 973–983, 2004.

101. Yoshitake, S., Uchida, M., Koizumi, H., Kanda, H., and Nakatsubo, T.: Production of biological soil crusts in the early stage of primary succession on a High Arctic glacier foreland, New Phytol., 186, 451–460, 2009.

102. Zaady, E., Kuhn, U., Wilske, B., Sandoval-Soto, L., and Kesselmeier, J.: Patterns of CO2 exchange in biological soil crusts of successional age, Soil Biol. Biochem., 32, 959–966, 2000.

103. Zumsteg, A., Luster, J., Goransson, H., Smittenberg, R., Brunner, ¨ I., Bernasconi, S., Zeyer, J., and Frey, B.: Bacterial, Archaeal and Fungal Succession in the Forefield of a Receding Glacier, Microb. Ecol., 63, 552–564, 2012.

3

CHAPTER

AN INTEGRATED STUDY TO ANALYZE SOIL MICROBIAL COMMUNITY STRUCTURE AND METABOLIC POTENTIAL IN TWO FOREST TYPES

Yuguang Zhang1, Jing Cong1,2, Hui Lu1, Caiyun Yang3, Yunfeng Yang[4], Jizhong Zhou[3,4,5], Diqiang Li[1]

[1] Institute of Forestry Ecology, Environment and Protection, and the Key Laboratory of Forest Ecology and Environment of State Forestry Administration, Chinese Academy of Forestry, Beijing, China

[2] School of Mineral Processing and Bioengineering, Central South University, Changsha, China

[3] Institute for Environmental Genomics and Department of Microbiology and Plant Biology, University of Oklahoma, Norman, Oklahoma, United States of America

[4] State Key Joint Laboratory of Environment Simulation and Pollution Control, School of Environment, Tsinghua University, Beijing, China

[5] Earth Sciences Division, Lawrence Berkeley National Laboratory, Berkeley, California, United States of America

ABSTRACT

Soil microbial metabolic potential and ecosystem function have received little attention owing to difficulties in methodology. In this study, we selected natural mature forest and natural secondary forest and analyzed the soil microbial community and metabolic potential combing the high-throughput sequencing and GeoChip technologies. Phylogenetic analysis based on 16S rRNA sequencing showed that one known archaeal phylum and 15 known bacterial phyla as well as unclassified phylotypes were presented in these forest soils, and Acidobacteria,Protecobacteria, and Actinobacteria were three of most abundant phyla. The detected microbial functional gene groups were related to different biogeochemical processes, including carbon degradation, carbon fixation, methane metabolism, nitrogen cycling, phosphorus utilization, sulfur cycling,

Citation: Zhang Y, Cong J, Lu H, Yang C, Yang Y, Zhou J, et al. (2014) An Integrated Study to Analyze Soil Microbial Community Structure and Metabolic Potential in Two Forest Types. PLoS ONE 9(4): e93773. doi:10.1371/journal.pone.0093773.

etc. The Shannon index for detected functional gene probes was significantly higher (P<0.05) at natural secondary forest site. The regression analysis showed that a strong positive (P<0.05) correlation was existed between the soil microbial functional gene diversity and phylogenetic diversity. Mantel test showed that soil oxidizable organic carbon, soil total nitrogen and cellulose, glucanase, and amylase activities were significantly linked (P<0.05) to the relative abundance of corresponded functional gene groups. Variance partitioning analysis showed that a total of 81.58% of the variation in community structure was explained by soil chemical factors, soil temperature, and plant diversity. Therefore, the positive link of soil microbial structure and composition to functional activity related to ecosystem functioning was existed, and the natural secondary forest soil may occur the high microbial metabolic potential. Although the results can't directly reflect the actual microbial populations and functional activities, this study provides insight into the potential activity of the microbial community and associated feedback responses of the terrestrial ecosystem to environmental changes.

INTRODUCTION

Microorganisms are the most abundant organisms on earth and play key roles in natural ecosystem, including the biogeochemical cycling of carbon, nitrogen, sulfur, phosphorus, and metals, and biodegradation or stabilization of environmental contaminants [1]–[3]. Because of their important roles, changes in soil microbial community may directly affect soil ecosystem function, particularly carbon and nitrogen cycling [4]. A deeply analysis of microbial community structure and their roles in ecological processes would improve our understanding of the biogeochemical elemental cycles affected by microbial communities in natural or man-made environments [5]. However, our understanding of soil microbial communities in terms of structure, composition, and functional activity are still limited, especially for soil microbial metabolic activity and ecosystem function have received little attention to date. Numerous studies have investigated the structure and diversity of microbial communities and their relationships with the surrounding environments [4]–[8]. Previous studies have shown inconsistent or contradictory results, which might be caused by 1) high microbial community complexity and plasticity in natural environments, 2) different methods used to assess microbial diversity, 3) frequency and magnitude of environmental disturbance, and 4) limited knowledge on the temporal and spatial scales of microbial community in ecosystems [4]–[7],[9]. To date, most studies have only described microbial community complexity at the phylogenetic level using 16S rRNA or single functional gene analysis [8]. Further studies of the metabolic potential and physiological traits of microorganisms are necessary to detect specific metabolic processes and functions [4]. Functional gene microarrays, with quantitative and high resolution capabilities, may be a useful tool for this purpose [10], [11]. With the increasing efforts made towards implementing sustainable forest management and conservation strategies since last century [12], natural secondary forest, which form naturally after disturbance of the primary forest by human activities or by extreme natural events, are evident in temperate regions around the world [13]. Shennongjia Mountain is located in the northwestern region of Hubei Province, Central China, and this region is considered to have one of the highest levels of biodiversity in China owing to its unique geographic location and complex terrain [14]. In this study, we selected natural mature forest and a natural secondary forest, which formed in the 1970s following the fall of the primary deciduous broadleaf forest on Shennongjia Mountain and analyzed the soil microbial community and metabolic potential using both GeoChip and high-throughput 16S rRNA sequencing approaches. The aims of this study were to determine (i) the composition and structure of soil microbial community in two deciduous broadleaf forest types, (ii) the soil microbial functional gene diversity and the metabolic potential involved in carbon and nitrogen cycling, and (iii) the major environmental factors governing the soil microbial community structure.

MATERIALS AND METHODS

Site and sampling

The study site was located on Shennongjia Mountain, Hubei Province, China, where deciduous broadleaf forest is a typical and important forest type. Two forest types were selected in this study: natural mature forest (MAT), and natural secondary forest (SEC), which was formed in the 1970s following the felling of the primary deciduous broadleaf forest. The two study sites had similar geography in terms of slope (30°), aspect (NE30°), and elevation (1715–1813 m), and were within 10 km of each other. The soil type was mountain yellow brown soil [15]. The plant survey and soil collected were permitted by the administrative bureau of Shennongjia National Nature Reserve. The detail location for the study site is N 31°29′, E110°21′ for MAT and N 31°25′, E11°20′ for SEC, respectively. At each study site, four plots (20 m×20 m) were selected with about 20 meters between adjacent plots. Ten to fifteen soil cores, at a depth of 0–10 cm, were taken from each plot and combined to obtain about 400 g of soil. Samples were sieved with 2 mm mesh to remove roots and stones, then mixed thoroughly. Soil samples were preserved at −80°C until DNA extraction.

Plant diversity, soil geochemical properties and microbial biomass analyses

Plant diversity was surveyed at each plot, including the plant species, number, height and canopy of each tree or shrub, and diameter at breast (1.3 m) height of trees (DBH>5 cm) and shrubs (DBH>1 cm). Average soil temperature at each plot was measured by placing the Long-Stem Thermomter (SPECTRUM, US) at a depth of 10 cm in relatively open patches. Soil moisture, soil pH, total soil organic carbon and nitrogen, available phosphorous, available water-dissolved organic carbon, and oxidizable organic carbon were measured as previously described [16]. Microbial biomass carbon was determined by the chloroform fumigation-extraction method [17].

DNA extraction, purification, and quantification

Soil microbial genomic DNA was extracted directly from each soil sample (5 g) using a protocol that included liquid nitrogen grinding, freezing and thawing, and treatment with sodium dodecyl sulfate for cell lysis, as previously described [18]. The freshly extracted DNA was purified twice using 0.5% low melting point agarose gel. Final DNA concentrations were quantified by the PicoGreen method [19] using a FLUOstar Optima microplate reader (BMG Labtech, Jena, Germany).

DNA amplification, sequencing, and data analysis

The V4 hypervariable region of the 16S rRNA gene was amplified in each soil sample using PCR primers 515F (5′-GTGCCAGCMGCCGCGGTAA-3′) and 806R (5′-GGACTACH VGGGTWTCTAAT-3′), using a sample tagging approach [20], [21]. The amplification mix contained 10 units of AccuPrime High Fidelity *Taq* polymerase (Invitrogen, Grand Island, NY), 1× AccuPrime PCR buffer II (Invitrogen), 200 μM dNTPs (Amersham, Piscataway, NJ), and 10 μM of each primer in a volume of 25 μl. Genomic DNA (10 ng) was added to the PCR mix. Each sample was amplified under the following conditions: 94°C for 1 min, 30 cycles of 94°C for 20 s, 53°C for 25 s, and 68°C for 45 s, then 10 min at 68°C. The purified PCR amplicons were analyzed using a Miseq Benchtop Sequencer for 2×150 bp paired-end sequencing (Illumina, San Diego, CA).

To minimize the effect of random sequencing errors, only the first 250 bp after the proximal PCR primer of each sequence was analyzed. Sequence that did not perfectly match the PCR primer, had non-assigned tags, or had reads <250 bp was removed. All sequences were aligned using the Ribosomal Database Project Infernal Aligner [22], and the complete linkage clustering method was used to define operational taxonomic units (OTUs) using 97% identity as a cutoff [21]. The number of detected OTUs and sequences at different levels of classification were counted. Details of amplicon preparations, sequencing, and data analysis (e.g. classification, OTU identification) were described in previous reports [20], [21]. To standardize samples, a sub-sample of 20,000 sequences per soil sample was used to compare relative difference among samples. Singletons were removed for downstream analyses. All the 16S rRNA sequences were deposited in GenBank database and the accession number is SRP035449.

GeoChip hybridization and data analysis

GeoChip 4.0 was used for DNA hybridization. This microarray contains >83,000 oligonucleotide probes targeting >150,000 genes in 410 gene categories involved in biogeochemical cycling of carbon, nitrogen, phosphorus, and sulfur, and bioremediation of metal and organic contaminants [24]. To produce consistent hybridizations from all samples, the amplification was used to generate approximately 3.0 μg of DNA from 50 ng of purified DNA using a TempliPhi Kit (GE Healthcare, Piscataway, NJ) following the manufacturer's instructions. Amplified DNA was labeled with Cy5 fluorescent dye (GE Healthcare) using a random priming method [23]. All hybridizations were carried out at 45°C for 10 h with 50% formamide using a TECAN HS4800 (TECAN, US) microarray hybridization station, and arrays were scanned using a ScanArray 5000 analysis system (Perkin-Elmer, Wellesley, MA).

Signal intensities of each spot were measured using ImaGene 6.0 (Biodiscovery, El Segundo, CA), and only the spots automatically scored as positive in the raw data output were used for further data analysis [25]. Spots with a signal-to-noise ratio (SNR=(signal intensity - background intensity)/standard deviation of the background) greater than 2.0 were used for further analysis [23]. The GeoChip data were further analyzed using the following steps: (i) genes detected in only two out of four samples from the same forest type were removed; (ii) the signal intensity of each spot was normalized by dividing by the mean value of eight samples of total signal intensity; and (iii) the data were transformed to the natural logarithmic form.

Statistical analyses

Relative abundance of the 16S rRNA gene was based on the proportional frequencies of those DNA sequences that could be classified at different OTUs. Diversity of the microbial community based on sequencing analysis and diversity of functional gene from GeoChip analysis were calculated using Simpson's reciprocal index (1/D) and the Shannon index (H′) using online software (http://ieg.ou.edu/). Detrended correspondence analysis (DCA) was used to determine the differences in overall microbial community structure and functional gene diversity between the two forest types. To examine the differences in soil microbial community structure and functional diversity between the two study sites, we used the Bray-Curtis similarity index to calculate distance matrices from OTUs and GeoChip hybridization data with the multi-response permutation procedure (MRPP) [26] and Adonis [27]. Mantel tests, canonical correspondence analysis (CCA), and variation partitioning analysis were used to evaluate the linkages between microbial community structure, metabolic potential related to carbon and nitrogen cycles, and environmental factors. The variance inflation factor was used for step-wise removal of redundant variables in CCA modeling [23]. All the analyses were performed by functions in the Vegan package (v. 1.15-1) in R (v. 2.9.1) (http://www.r-project.org/).

RESULTS

Plant diversity and soil geochemical properties

Plant diversity and soil properties in the MAT and SEC forests were analyzed (Table 1). The number of plant species (including trees and shrubs) was 63 and 91 in MAT and SEC, respectively. The dominant trees were *Fagus engleriana*, *Quercus aliena var. acuteserrata*, and*Carpinus viminea* at the MAT forest site, and *Toxicodendron verniciflua*, *Tilia paucicostata*,*Juglans cathayensis*, and *Q. aliena var. acuteserrata* at the SEC site, respectively. The Shannon index of plant was significantly lower for MAT (1.18) than for SEC (2.12).

Table 1: Plant diversity and soil geochemical properties in two forest types

Forest type	MAT	SEC	Forest type	MAT	SEC
non index of tree and shrubs**	1.18±0.27	2.12±0.17	Soil organic carbon (g/kg)**	36.40±3.41	86.07±6.70
Species number of tree*	6.00±0.41	12.50±2.10	Dissolved organic carbon (g/kg)	0.17±0.01	0.26±0.08
Height of tree**	14.14±0.54	8.73±0.34	Labile organic carbon (g/kg)**	2.00±0.48	12.37±1.58
Diameter at breast height of tree**	19.85±1.02	12.06±0.75	Microbial biomass carbon (g/kg)**	0.81±0.05	1.31±0.07
Soil Moisture (%) **	50.12±1.23	27.32±2.45	Total nitrogen (g/kg)**	2.33±0.43	5.83±0.45
Temperature at 10 cm depth (°C)	15.92±0.15	15.32±0.09	Available nitrogen (g/kg)**	0.22±0.02	0.41±0.03
Soil pH**	4.81±0.05	5.91±0.28	Available phosphorus (g/kg) *	0.004±0.00	0.015±0.00

Data present the mean value and standard error.
*, $P<0.05$.
**, $P<0.01$.
doi:10.1371/journal.pone.0093773.t001

Soil organic carbon, oxidizable organic carbon, and microbial biomass carbon were significantly higher ($P<0.01$) in the SEC samples than in the MAT samples. Soil geochemical properties, such as total nitrogen, available nitrogen, available phosphorus, pH, and soil moisture, were also significantly different ($P<0.01$) between the two forest types (Table 1).

Soil microbial community composition and structure

To determine the overall composition of the soil microbial community in the two forest types, soil microbial communities were analyzed by 16S rRNA high throughput sequencing. After preprocessing of all reads, 65,666 and 65,584 high-quality sequences were obtained from the MAT and SEC sites, respectively (Table S1). The number of sequences ranged from 15,521 to 16,919 in MAT samples and from 16,002 to 16,970 in SEC samples. 10,160 OTUs were detected using 97% identity as a cutoff. Phylogenetic analysis showed that one known archaeal phylum and 15 known bacterial phyla as well as unclassified phylotypes were presented in these forest types (Table 2), and that all of the phylotypes were detected in both MAT and SEC, except for BRC1, which was only found in SEC. At the phylum level, the majority of OTUs were derived from *Proteobacteria* (4,048, 39.8%), followed by *Acidobacteria*(1,787, 17.6%), *Actinobacteria* (1,019, 10.0%), and *Verrucomicrobia* (896, 8.82%) (Table 2).

Table 2: Numbers of soil microbial OTUs and composition in two forest types

Domain and phylum	Total[a] (%)	Average[b]		P (unpaired t test)
		MAT	SEC	
Archaea				
Crenarchaeota	19 (0.19)	4.50±0.96	10.00±0.82	0.005
Bacteria				
Acidobacteria	1787 (17.59)	777.75±22.21	665.50±14.06	0.017
Actinobacteria	1019 (10.03)	362.25±19.69	443.75±17.09	0.021
Armatimonadetes	22 (0.22)	5.25±0.63	8.50±1.19	0.065
Bacteroidetes	621 (6.11)	91.50±5.87	379.50±10.81	0.000
BRC1	3 (0.03)	0.00	2.25±0.25	0.000
Chlamydiae	17 (0.17)	8.00±1.47	1.50±0.28	0.019
Chloroflexi	109 (1.07)	26.25±2.32	42.25±3.71	0.014
Cyanobacteria	3 (0.03)	1.00±0.00	2.50±0.29	0.014
Firmicutes	129 (1.27)	45.50±4.17	60.75±5.07	0.051
Gemmatimonadetes	97 (0.95)	23.00±1.08	53.50±5.74	0.011
Nitrospirae	7 (0.07)	1.00±0.41	4.50±0.29	0.001
Planctomycetes	657 (6.47)	288.50±9.91	173.75±9.29	0.000
Proteobacteria	4048 (39.84)	1409.00±31.39	1826.25±52.01	0.001
Verrucomicrobia	896 (8.82)	459.00±41.12	349.75±3.07	0.076
WS3	12 (0.12)	0.50±0.28	6.50±0.87	0.004
Unclassified	714 (7.03)	212.25±5.96	326.75±11.06	0.000
Total	**10160**	**3623.00±77.66**	**4469.75±53.03**	**0.000**

[a]Data represent total numbers of OTUs detected by sequencing across all 8 samples.
[b]Data represent the mean value and standard error of OTUs detected using 4 samples in different forest types.
doi:10.1371/journal.pone.0093773.t002

The Shannon index of 16S rRNA sequences was 8.05 and 8.28 for the MAT and SEC sites, respectively (Table S1). The number of OTUs and the Simpson index were significantly ($P<0.05$) different between MAT and SEC samples (Table S1). DCA was performed using the relative abundance values of sequencing data, resulting in two distinct clusters (Figure 1A). The results of nonparametric multivariate statistical tests, Adonis and MRPP, showed significant differences (R^2=0.688, P=0.001, and δ=0.387, P=0.026, respectively) based on the abundance of all OTUs detected in MAT and SEC samples (Table S2). Significant (P<0.05) differences were also observed at the phylum level for *Acidobacteria, Actinobacteria, Armatimonadetes, Bacteroidetes, Proteobacteria, Planctomycetes*, and BRC1, except for WS3 (Table 2). The differences in microbial diversity at the phylum level in the two forest types varied. Ten known microbial phyla and unclassified taxa were significantly (P<0.05) higher in SEC samples, including the dominant phyla *Proteobacteria* and *Actinobacteria*. However,*Acidobacteria, Chlamydiae*, and *Planctomycetes* were significantly (P<0.05) lower in SEC soils. These results indicated that the overall soil microbial taxonomic composition and structure were significantly different between the SEC and MAT samples.

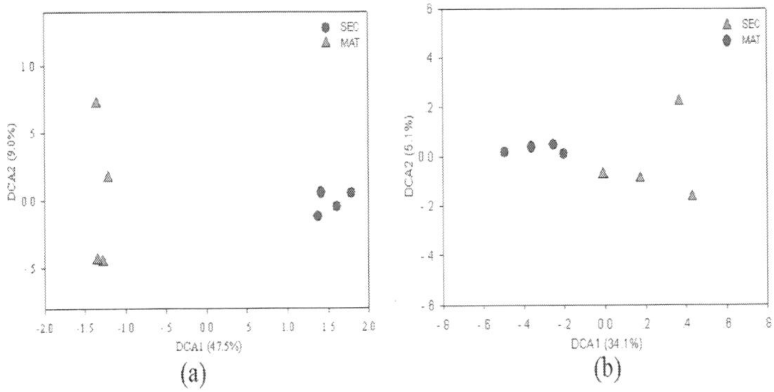

Figure 1: Detrended correspondence analysis (DCA) of soil microbial community structure.

The DCA was analyzed based on the relative abundances of OTUs (A) and functional genes based on GeoChip 4.0 (B).

Soil microbial functional gene diversity

A total of 38,925 and 50,068 gene probes were detected using GeoChip 4.0 in the MAT and SEC samples, respectively (Table S3). The detected gene groups were related to different biogeochemical processes, such as carbon degradation, carbon fixation, methane metabolism, nitrogen cycling, phosphorus utilization, stress responses, sulfur cycling, contaminant remediation, metal remediation, and energy processing (Table S4). The average number of gene probes detected was 43,427.3 for SEC samples and 33,982.8 for MAT samples. The Shannon index for detected functional microbial gene probes were significantly higher (P<0.05) at SEC sites (10.67) than at MAT sites (10.43) (Table S3). DCA showed that MAT samples were well separated from SEC samples (Figure 1B). The regression analysis showed that a strong positive (P<0.05) correlation was existed between the soil microbial functional gene diversity and phylogenetic diversity (Fig. 2).

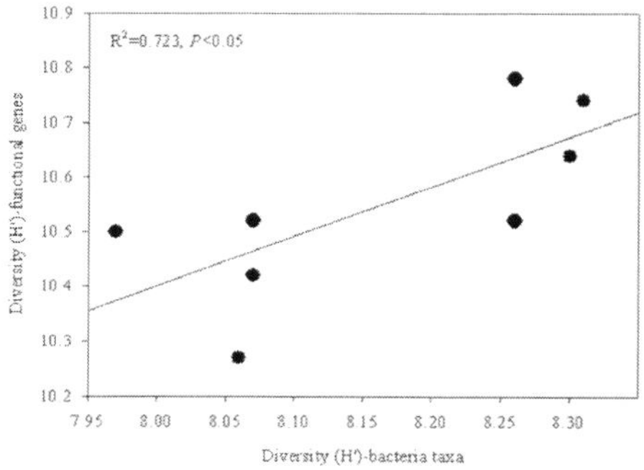

Figure 2: Relationship between soil microbial phylogenetic diversity index and functional gene diversity index.

Microbial functional genes and metabolic potential related to carbon cycling

To gain insight into the soil microbial metabolic potential related to carbon cycle, GeoChip 4.0 data for genes related to carbon cycling was analyzed further. Among the 5,792 gene probes related to carbon degradation, 4,276 (74%) were common between the two forest types, while 59 (1%) were presented only in MAT samples and 1,457 (25%) were presented only in SEC samples. The relative signal intensities of the different gene categories related to degradation of relatively active carbon (e.g. starch, hemicelluloses, pectin, and cellulose) and recalcitrant carbon (e.g. chitin and lignin) were significantly ($P<0.05$) higher in SEC than in MAT samples (Figure 1). This included genes encoding amylase and pullulanase (related to starch degradation), arabinofuranosidase and xylanase (related to hemicellulose degradation), cellobiase and exoglucanase (related to cellulose degradation), pectinase (related to pectin degradation), endochitinase and exochitinase (related to chitin degradation), and glyoxal oxidase and phenol oxidase (related to lignin degradation). These results indicated that the soil microbial metabolic capacity involved in carbon degradation were significantly different between the two different forest types.

Among the 1,085 genes involved in carbon fixation, the relative abundance of three important functional genes, ribulose-1,5-bisphosphate carboxylase/oxygenase (Rubisco), carbon monoxide dehydrogenase (CODH), and propionyl-coA (Pcc), was substantially different. A total of 306, 650 and 230 gene probes were detected for the Rubisco, Pcc, and CODH pathways, respectively. These genes had significantly higher signal intensity in SEC soils than in MAT soils (Figure S1). Geochip 4.0 also detected three enzymes involved in methane metabolism, namely the alpha-subunit of methyl coenzyme M reductase (mcrA), for methane production, and particulate methane monooxygenase (pmoA) and methane monooxygenase (mmoX) for methane consumption. Over 80% of the gene probes were associated with uncultured bacteria and archaea. The signal intensity of all the genes involved in methane metabolism was higher in SEC samples than in MAT samples (Figure 3). These results indicated that the microbial functional capacity related to carbon fixation and methane metabolism may be significantly affected by forest types. The relationships between soil microbial metabolic capacity related to carbon cycling and soil carbon components or enzyme activities were analyzed using the Mantel test (Table 3). The results showed that soil organic carbon and soil microbial biomass carbon were significantly linked ($P<0.05$) to the relative abundance of genes related to carbon cycling. Oxidizable organic carbon was significantly linked ($P<0.05$) to the relative abundance of genes involved in active organic carbon degradation (cellulase, hemicellulase, and starch). As expected, cellulase, glucanase, and amylase activities were significantly linked ($P<0.05$) to the relative abundance of genes encoding cellulase, endoglucanase and exoglucanase, and amylase, respectively.

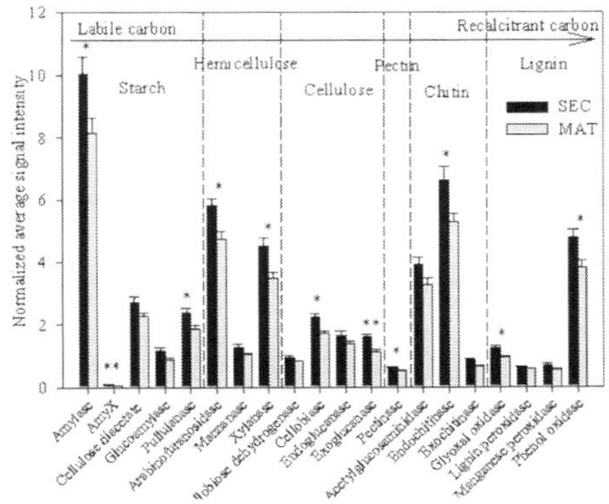

Figure 3: The normalized gene relative signal intensity of the detected key gene categories involved in carbon degradation.

The signal intensities were the sum of detected individual gene for each functional gene, averaged among 4 soil samples. The complexity of carbon is presented in order from active to recalcitrant. All data are presented as mean ± standard error. Significant differences between two forest types are indicated above the bars. *, $P<0.05$, **,$P<0.01$.

Table 3: The relationships between relative abundances of microbial functional genes involved in carbon and nitrogen cycling and soil carbon or nitrogen components and enzyme activity by Mantel-test

Soil carbon parameters	Microbial relative abundance	R value	P
Soil organic carbon	Carbon cycling genes	0.389	**0.039**
Oxidizable organic carbon	Oxidizable organic carbon degradation genes(Cellulose, Hemicellulose, Starch)	0.403	**0.046**
Water dissolved organic carbon	Carbon cycling genes	0.114	0.216
Soil microbial biomass carbon	Carbon cycling genes	0.500	**0.020**
Total soil nitrogen	Nitrogen cycling genes	0.382	**0.050**
NH_4^+-N	Ammonification genes	0.296	**0.089**
No_3^-N	Nitrification genes	0.302	**0.079**
Cellulase activity	Cellulase genes	0.342	**0.046**
Glucanase activity	Endo- and exoglucanase genes	0.351	**0.022**
Polyphenol oxidase activity	Polyphenol oxidase genes	0.046	0.626
Amylase activity	Amylase genes	0.358	**0.027**

doi:10.1371/journal.pone.0093773.t003

Microbial functional genes and metabolic potential related to nitrogen cycling

Among the gene probes detected in the forest soil samples, 4,238 were involved in nitrogen cycling, including 767 related to nitrogen fixation, 513 related to ammonification (including 250 gene probes related to assimilatory N reduction), 1,734 related to denitrification, 351 related to dissimilatory N reduction, and 621 related to nitrification. These key functional gene categories were associated with different phylogenetic groups, for example, almost all gene probes related to the ammonification were derived from cultured bacteria, including *Rhodococcus erythropolis*,*Pseudomonas entomophila*, and *Acinetobacter baumannii*. A total of 1,533 (88%) gene probes related to the denitrification were linked to unclassified bacteria, and 137 (22%) gene probes related to the nitrification were derived from archaea, which suggests that archaea maybe play an important role in nitrification process.

The abundance of several genes involved in nitrogen cycling were significantly (P<0.05) higher in SEC samples than in MAT samples (Figure 4), including *nifH* (related to nitrogen fixation),*amoA* (related to nitrification), *narG*, *nirS*, *nirK*, and *nosZ*, (related to denitrification), *napA* and*nrfA* (related to dissimilatory N reduction to ammonium), *gdh* and *ureC* (related to ammonification), and *niR* and *nirB* (related to assimilatory N reduction). The relationships between microbial functional genes and soil nitrogen components were analyzed using the Mantel test (Table 4). The results showed that soil total nitrogen was significantly linked (P<0.05) to the relative signal intensity of nitrogen cycling genes. The NH_4^+-N and NO_3-N concentrations were linked (P<0.1) to the signal intensity of genes involved in the ammonification and nitrification processes, respectively.

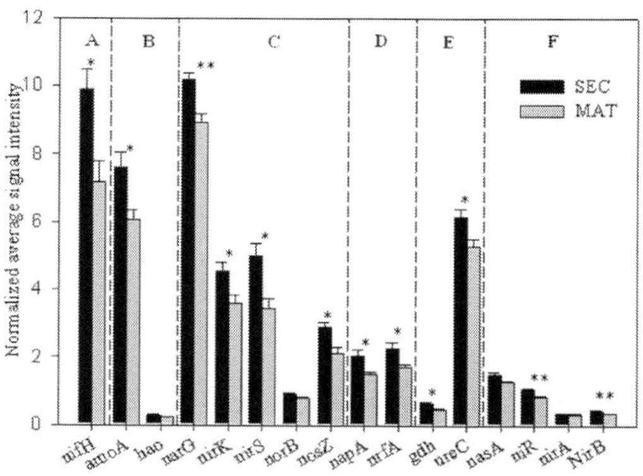

Figure 4: The normalized average signal intensity of detected key gene categories involved in the N cycling between SEC and MAT.

The signal intensities were the sum of all detected individual gene for each gene category, and then averaged among 4 samples. (A). N_2 fixation, including nifH encoding nitrogenase; (B). Nitrification, including amoA encoding ammonia monooxygenase, hao for hydroxylamine oxidoreductase; (C). Denitrification, including narG for nitrate reductase, nirS for nitrite reductase, norB for nitric oxide reductase, and nosZ for nitrate reductase; (D). Dissimilatory N reduction to ammonium, including napA for nitrate reductase, and nrfA for c-type cytochrome nitrite reductase; (E). Ammonification, including gdh for glutamate dehydrogenase and ureC encoding urease; (F). Assimilatory N reduction, including nasA encoding nitrate reductase, niR, nirA and nirB encoding dissimilatory nitrite reductase. All data are presented as the mean±SE (error bars). **,$P<0.01$, *, $P<0.05$.

Table 4: Mantel test and partial Mantel test between 16S rRNA OTUs and Geochip data with environmental properties

Environmental variable	16S OTUs		GeoChip 4.0	
	r value	P value	r value	P value
Whole variable	0.868	0.005	0.412	0.008
Soil organic carbon	0.770	0.003	0.392	0.013
Total nitrogen	0.832	0.003	0.402	0.019
pH	0.622	0.006	0.322	0.043
Temperature at the 10 cm depth	0.463	0.002	0.085	0.272
Shannon Index for trees and shrubs	0.779	0.006	0.337	0.028
Diameter at breast height of trees	0.532	0.008	0.125	0.158

The Mantel test was only performed for whole variable, which included all variables presented in this table.
doi:10.1371/journal.pone.0093773.t004

Linking microbial community structure and environmental factors

Mantel tests and CCA were performed to analyze the major environmental factors responsible for shaping the microbial community structure and functional diversity. The Mantel test showed that microbial community structure and functional gene diversity were significantly ($P<0.05$) linked to soil and plant factors (Table 4). CCA was used to identify the major environmental variables controlling the soil microbial community structure, resulting in a model significant at a confidence level of $P=0.048$. The plant Shannon index and tree DBH appeared to be important environmental factors controlling the microbial community structure, as they had a long projection on Axis 1 ($P<0.01$), which represented the major variation among microbial communities (Figure 5A). Similarly, soil temperature and soil organic carbon were critical for shaping microbial communities.

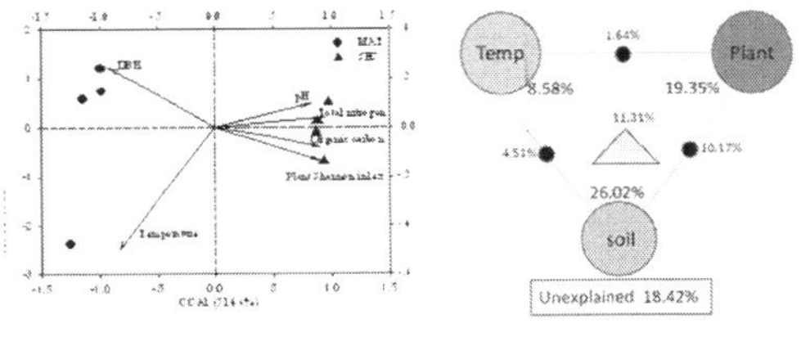

Figure 5: The Canonical correspondence analysis (CCA) and Variation partitioning analysis (VPA) between OTUs data and soil environmental variables.

The left was CCA between microbial OTUs data and soil environmental factors, and the right was the VPA. DBH, diameter at breast height of tree. Soil, including soil organic carbon, total nitrogen and pH, soil temperature (Temp) and plant diversity (Plant, including the Shannon index of trees and shrubs, diameter at breast height of trees), and their relationships. Each diagram represents the biological variation partitioned into the relative effects of each factor or a combination of factors, in which geometric areas were proportional to the percentages of explained variation.

Variance partitioning analysis was used to quantify the contributions of soil chemical factors (Soil), soil temperature (Temp), and plant diversity (Plant) to the variation in the microbial community. A total of 81.58% of the variation was significantly explained by these three environmental variables (Figure 5B). Soil, Temp, and Plant explained 26.02%, 8.58%, and 19.35% of variation, respectively, verifying that they were major factors in shaping the microbial community structure.

DISCUSSION

The development and increasing availably of high-throughput molecular techniques has helped improve our understanding of microbial composition, especially through the use of direct sequencing of metagenomic DNA. Fierer et al. indicated that all biomes were dominated by*Acidobacteria*, *Actinobacteria*, *Proteobacteria* and *Bacteroidetes*, and the bacterial community composition does not vary significantly across different biomes [1]. Our results showed that microbial diversity at the phylum level differed between the two forest types, and*Acidobacteria*, *Protecobacteria*, and *Actinobacteria* were three of most abundant soil bacterial phyla. *Acidobacteria* is known to be dominant in other soil types, although they are often difficult to cultivate [8], [28]. The relative abundance of*Acidobacteria* was also shown to be higher in forest, desert, and prairie soils compared with agricultural soils [28]. Recent studies showed that *Acidobacteria* is in general oligotrophic, consistent with results showing that the proportion of *Acidobacteria* was significantly lower in nutrient-rich rhizosphere and agricultural soils than in bulk soil [4], [29]. Some studies also found that the relative abundance of*Acidobacteria* was negatively linked to the level of nitrogen input [30]. In this study, the relative abundance of *Acidobacteria* was high in the natural forest, which had low soil organic carbon and nitrogen contents.

Due to microbial highly diversity and natural ecosystem complexity, it is a great challenging to set up the link of soil microbial structure and composition to functional activity related to ecosystem functioning in microbial ecology. Many studies have described microbial community structure in a range of different environments, but have not identified critical groups related to detailed functional processes [4]–[8], [31]. Therefore, it remains unclear the relationship between microbial community structure and microbial functional gene diversity[32]–[34]. Functional redundancy is considerable debate in ecology as to how the relationship between the taxonomic diversity and observed functional diversity [35]. This study combines high-throughput sequencing to study the microbial community structure at the phylogenetic level and GeoChip to microbial potential metabolic activities at functional gene level. 16S rRNA gene-based high-throughput sequencing data provide information on microbial phylogenetic structure and composition, and GeoChip-based data provide quantitative information on various microbial functional groups [23]. These integrated technologies provide a reliable method for detection soil microbial structure, composition and potential functional activity [23]. In this study, a strong positive correlation were found between the soil microbial functional and phylogenetic diversity among sampling sites, which suggesting microbial phylogenetic and functional diversity aligned well and a low degree of functional redundancy were existed. Our result was consistent with a recent work on reconstructing the microbial diversity and function of pre-agricultural tallgrass prairie soils in the United States [36]. Furthermore, the higher phylogenetic diversity detected in SEC was consistent with the higher number of functional genes involved in the carbon and nitrogen cycles, unveiling the larger functional potential and capacity of the secondary forest soil.

Soil microbial metabolic potential and ecosystem function have received little attention owing to difficulties in methodology. As we all known, most of the biogeochemical processes were controlled and related to diverse and various functional genes, therefore, the conjunction between microbial community identified by using normal molecular methods (i.e. DGGE, T-RFLP) with soil enzyme activities or components is difficult. GeoChip-based data provides large-scale quantitative information on various biogeochemical functional groups, thus making it possible to link the functional structure of microbial communities with ecosystem processes[23]. Previous studies showed that change observed in the microbial community structure and functional groups has significant effects on soil enzyme activities or components related to carbon and nitrogen cycles. Yergeau et al. [37] and Reeve et al. [34] showed a significant correlation between cellulase enzyme activity and cellulose gene variants, detected using GeoChip, as well as significant correlations between dehydrogenase genes and enzyme activity, urease genes and soil ammonium, and sulfite reduction genes and soil sulfur level[34]. In this study, we also found the linked of soil microbial functional gene diversity to soil enzyme activities and components, such as the soil organic carbon and soil microbial biomass carbon were significantly linked ($P<0.05$) to the relative abundance of genes related to carbon cycling, oxidizable organic carbon was significantly linked ($P<0.05$) to the relative abundance of genes involved in active organic carbon degradation (cellulase, hemicellulase, and starch), respectively. Although the GeoChip data can't directly reflect the actual microbial populations and functional activities, it implied that the microbial populations carrying those genes could exist and have the functional capacity. Our GeoChip data showed that metabolic genes involved in the degradation of starch, cellulose, hemicellulose, chitin, lignin, and pectin, as well as those involved in nitrogen cycling, were significantly higher ($P<0.05$) in SEC than in MAT soils. Therefore, SEC soil had high microbial gene diversity and microbial metabolic potential. The use of GeoChip 4.0 may allow us to rapidly obtain community-wide information and observe differences functional capacity in different environments.

Only about 1% of the microorganisms in soil can be cultured, and researchers have increased our understanding of microbes in the natural environments by use of PCR amplification methods which reduce our dependence on the presence of culturable microbes. In our study, the majority of the genes related to carbon and nitrogen cycles were derived from the uncultured bacteria, suggesting that they are unique and may represent novel genes in the study area. For example, 88% gene probes related to the denitrification process were linked to unclassified bacteria. In previous studies, researchers received the same results, for example, Xie et al found 64.28% sequences were from uncultivated bacteria in the acid mine drainage[38], Zhang et al found the 320/372 genes were derived from the unidentified or uncultured organisms in the alpine meadows soils in the Qinghai-Tibetan plateau [39]. Therefore, a lot of the microbial functional gene resources were existed in natural environments.

Much recent work in soil microbial ecology has showed that soil microbial community structures are affected by specific environmental changes or disturbances [11], [25]. Soil properties and plant diversity are key factors in shaping the microbial community composition and structure [40], [41]. Organic carbon availability limited microbial communities in most soils[37], and addition of labile organic material rapidly alters microbial communities by selecting for populations that are most competitive in terms of growth rates and ability to absorb nutrients[41]. Thus, it was not considered to be the limiting factor for the microbial communities. Our data showed that plant diversity also significantly influenced the microbial community structure at the phylum level. Plant communities can influence the associated soil microbial communities through the types and amounts of organic carbon and nutrient inputs, and by altering the temperature and water content of the soil [42], [43]. Therefore, changes in plant diversity, composition, and production affect the composition and diversity of soil microbial communities [9]. However, our understanding of differences in soil microbial diversity depending on various vegetation types is still poor.

In summary, this work studied the microbial community structure and metabolic potential by combining high-throughput sequencing and GeoChip technologies, and showed the soil microbial community structure and functional metabolic potential had a positive correlation. Although the microbial populations and metabolic activities could not be characterized comprehensively, the detected functional gene groups may represent the *in situ* microbial metabolic potential at the functional cluster level to some degree. Further analysis of the functional activity with different approaches such as mRNA-based analysis is needed to understand the biogeochemical processes related to microbial community.

ACKNOWLEDGMENTS

We gratefully acknowledge Dr. Ye Deng, Dr. Junjun Ding, Mr. Jingyuan Yang, Mr. Xiujiang Su, Dr. Guangliang Li and Dr. Yadong Xue for helping data analysis and field survey the plant plots.

AUTHOR CONTRIBUTIONS

Conceived and designed the experiments: YZ JZ DL. Performed the experiments: YZ JC HL CY. Analyzed the data: YZ JC YY. Contributed reagents/materials/analysis tools: YZ DL. Wrote the paper: YZ.

REFERENCES

1. Fierer N, Jackson RB (2006) The diversity and biogeography of soil bacterial communities. Proc Natl Acad Sci USA 103: 626–631. doi: 10.1073/pnas.0507535103
2. Green JL, Bohannan BJM, Whitaker RJ (2008) Microbial biogeography: from taxonomy to traits,. Science 320: 1039–1043. doi: 10.1126/science.1153475
3. He Z, Nostrnd JDV, Deng Y, Zhou J (2011) Development and applications of functional gene microarrays in the analysis of the functional diversity, composition, and structure of microbial communities. Front. Environ. Sci Engin China 5: 1–20. doi: 10.1007/s11783-011-0301-y
4. López-Lozano NE, Heidelberg KB, Nelson WC, Garcia-Oliva F, Eguiarte LE, et al. (2013) Microbial secondary succession in soil microcosms of a desert oasis in the Cuatro Cienegas Basin, Mexico. PeerJ 1: e47 Available:http://dx.doi.org/10.7717/peerj.47. Accessed 2013 March 5.
5. Fierer N, Nemergut D, Knight R, Craine JM (2010) Changes through time: integrating microorganisms into the study of succession. Res Microbiol 161: 635–642. doi: 10.1016/j.resmic.2010.06.002
6. Kuramae EE, Gamper HA, Yergeau E, Piceno YM, Brodie EL, et al. (2010) Microbial secondary succession in a chronosequence of chalk grasslands. ISME J 4: 711–715. doi: 10.1038/ismej.2010.11
7. Banning NC, Gleeson DB, Grigg AH, Grant CD, Andersen GL, et al. (2011) Soil microbial community successional patterns during forest ecosystem restoration. Appl Environ Microbiol 77: 6158–6164. doi: 10.1128/aem.00764-11
8. Ding G, Piceno YM, Heuer H, Weinert N (2013) Changes of soil bacterial diversity as a consequence of agricultural land use in a semi-arid ecosystem. PLOS one 8: e59497. doi: 10.1371/journal.pone.0059497
9. Zhu W, Cai X, Liu X, Wang J, Cheng S, et al. (2010) Soil microbial population dynamics along a chronosequence of moist evergreen broad-leaved forest succession in southwestern China. J M Sci 7: 327–338. doi: 10.1007/s11629-010-1098-z
10. He Z, Gentry TJ, Schadt CW, Wu L, Liebich J, et al. (2007) Geochip: a comprehensive microarray for investigating biogeochemical, ecological and environmental processes. ISME J 1: 67–77. doi: 10.1038/ismej.2007.2
11. Zhou J, Kang S, Schadt CD, Garten CTJ (2008) Spatial scaling of functional gene diversity across various microbial taxa. Proc Natl Acad Sci USA 105: 7768–7773. doi: 10.1073/pnas.0709016105
12. Aerts R, Honnay O (2011) Forest restoration, biodiversity and ecosystem functioning. BMC Ecol 11: 29. doi: 10.1186/1472-6785-11-29
13. Wang CK, Yang JY (2007) Rhizospheric and heterotrophic components of soil respiration in six Chinese temperature forests. Global Change Biol 13: 123–131. doi: 10.1111/j.1365-2486.2006.01291.x

14. Ma C, Zhu C, Zheng C, Wu C, Guan Y, et al. (2008) High-resolution geochemistry records of climate changes since late-glacial from Dajiuhu peat in Shennongjia Mountains, Central China. Chinese Sci Bull 53(Supp.1): 28–41. doi: 10.1007/s11434-008-5007-6

15. Zhao C, Chen W, Tian Z, Xie Z (2005) Altitudinal pattern of plant species diversity in Shennongjia Mountains, Central China. J Integr Plant Biol 47(12): 1431–1449. doi: 10.1111/j.1744-7909.2005.00164.x

16. Bao SD (1999) Soil and agricultural chemistry analysis. Beijing: China Agriculture Press, 25–150 p.

17. Chen CR, Xu ZH, Mathers NJ (2004) Soil carbon pools in adjacent natural and plantation forests of subtropical Australia. Soil Sci Soc Am J 68: 282–291. doi: 10.2136/sssaj2004.2820

18. Zhou JZ, Bruns MA, Tiedje JM (1996) DNA recovery from soils of diverse composition. Appl Environ Microbiol 62: 316–322.

19. Ahn S, Costa J, Emanuel J (1996) PicoGreen quantitation of DNA: effective evaluation of samples pre- or post-PCR. Nucleic Acids Res 24: 2623–2625. doi: 10.1093/nar/24.13.2623

20. Caporaso JG, Lauber CL, Walters WA, Berg-Lyons D, Lozupone CA, et al. (2011) Global patterns of 16S rRNA diversity at a depth of millions of sequences per sample. Proc Natl Acad Sci USA 108: 4516–4522. doi: 10.1073/pnas.1000080107

21. Caporaso JG, Lauber CL, Walters WA, Berg-Lyons D, Huntley J, et al. (2012) Ultra-high-throughput microbial community analysis on the Illumina HiSeq and MiSeq platforms. ISME J 6: 1621–1624. doi: 10.1038/ismej.2012.8

22. Cole JR, Wang Q, Cardenas E, Fish J, Chai B, et al. (2009) The ribosomal database project: improved alignments and new tools for rRNA analysis. Nucleic Acids Res 37: 141–145. doi: 10.1093/nar/gkn879

23. He Z, Xu M, Deng Y, Kang S, Kellogg L, et al. (2010) Metagenomic analysis reveals a marked divergence in the structure of belowground microbial communities at elevated CO_2. Ecol Lett 13: 564–575. doi: 10.1111/j.1461-0248.2010.01453.x

24. Lu Z, Deng Y, Van Nostrand JD, He Z, Voordeckers J, et al. (2011) Microbial gene functions enriched in the deepwater horizon deep-sea oil plume. ISME J 5: 1–10. doi: 10.1038/ismej.2011.91

25. Liang Y, Nostrand JD, Deng Y, He Z, Wu L, et al. (2011) Functional gene diversity of soil microbial communities from five oil-contaminated fields in China. ISME J 5: 403–413. doi: 10.1038/ismej.2010.142

26. McCune B, Grace JB (2002) Analysis of ecological communities. MJM Software Design, Gleneden Beach, OR.

27. Anderson MJ (2001) A new method for non-parametric multivariate analysis of variance. Aust J Ecol 26: 32–46. doi: 10.1046/j.1442-9993.2001.01070.x

28. Janssen PH (2006) Identifying the dominant soil bacterial taxa in libraries of 16S rRNA and 16S rRNA genes. Appl Environ Microbiol 72: 1719–1728. doi: 10.1128/aem.72.3.1719-1728.2006

29. Kielak A, Pijl AS, van Veen JA, Kowalchuk GA (2009) Phylogenetic diversity of Acidobacteria in a former agricultural soil. ISME J 3: 378–382. doi: 10.1038/ismej.2008.113
30. Fierer N, Lauber CL, Ramirez KS, Zaneveld J, Bradford MA, et al. (2012) Comparative metagenomic, phylogenetic and physiological analyses of soil microbial communities across nitrogen gradients. ISME J 6: 1007–1017. doi: 10.1038/ismej.2011.159
31. Kirk JL, Beaudette LA, Hart M, Moutoglis P, Klironomos JN, et al. (2004) Methods of studying soil microbial diversity. J Microbiol Meth 58: 169–188. doi: 10.1016/j.mimet.2004.04.006
32. Kelly JJ (2003) Molecular techniques for the analysis of soil microbial processes: functional gene analysis and the utility of DNA microarrays. Soil Sci 168: 597–605. doi: 10.1097/01.ss.0000090800.06903.2e
33. Little AEF, Robinson CJ, Peterson SB, Raffa KE, Handelsman J (2008) Rules of engagement: inter-species interaction that regulate microbial communities. Annu Rev Microbiol 62: 375–401. doi: 10.1146/annurev.micro.030608.101423
34. Reeve JR, Schadt CW, Carpenter-Boggs L, Kang S, Zhou J, et al. (2010) Effects of soil type and farm management on soil ecological functional genes and microbial activities. ISME J 4: 1–9. doi: 10.1038/ismej.2010.42
35. Hillebrand H, Matthiessen B (2009) Biodiversity in a complex world: consolidation and progress in functional biodiversity research. Ecol Lett 12: 1405–1419. doi: 10.1111/j.1461-0248.2009.01388.x
36. Fierer N, Ladau J, Clemente JC, Leff JW, Owens SM, et al. (2013) Reconstructing the microbial diversity and function of pre-agricultural tallgrass prairie soils in the United States. Science 342: 621–624. doi: 10.1126/science.1243768
37. Yergeau E, Kang S, He Z, Zhou J, Kowalchuk GA (2007) Functional microarray analysis of nitrogen and carbon cycling genes across an Antarctic latitudinal transect. ISME J 1: 1–17. doi: 10.1038/ismej.2007.24
38. Xie JP, He ZL, Liu XX, Liu XD, Van Nostrand JD, et al. (2011) GeoChip-Based Analysis of the Functional Gene Diversity and Metabolic Potential of Microbial Communities in Acid Mine Drainage. Appl Environ Microbiol 77: 991–999. doi: 10.1128/aem.01798-10
39. Zhang Y, Lu Z, Liu S, Yang Y, He Z, et al. (2013) Geochip-based analysis of microbial communities in alpine meadow soils in the Qinghai-Tibetan plateau. BMC Microbiol 13(1): 72. doi: 10.1186/1471-2180-13-72
40. Carney K, Maston P (2005) Plant communities, soil microorganism, and soil carbon cycling: does altering the world belowground matter to ecosystem functioning? Ecosystem 8: 928–940. doi: 10.1007/s10021-005-0047-0
41. Drenovsky RE, Vo D, Graham KJ, Scow KM (2004) Soil water content and organic carbon availability are major determinants of soil microbial community composition. Microbial Ecol 48: 424–430. doi: 10.1007/s00248-003-1063-2

42. Myers RT, Zak DR, White DC, Peacock A (2001) Landscape-level patterns of microbial community composition and structure in upland forest ecosystems. Soil Sci Soc Am J 65: 359–367. doi: 10.2136/sssaj2001.652359x

43. Waldrop MP, Firestone MK (2004) Microbial community utilization of recalcitrant and simple carbon compounds: impact of oak-woodland plant communities. Oecologia 138: 275–284. doi: 10.1007/s00442-003-1419-9

4

CHAPTER

MICROBIOLOGICAL ANALYSIS OF NECROSOLS COLLECTED FROM URBAN CEMETERIES IN POLAND

Ireneusz Całkosinski[1] Katarzyna Płoneczka-Janeczko[2] Magda Ostapska[1] Krzysztof Dudek[3] Andrzej Gamian[4] and Krzysztof Rypuła[2]

[1]Laboratory of Neurotoxicology and Environmental Diagnosis, Faculty of Health Science, Wroclaw Medical University, 51-618 Wroclaw, Poland

[2]Division of Infectious Diseases and Veterinary Administration, Department of Epizootiology and Clinic of Birds and Exotic Animals, Faculty of Veterinary Medicine, Wroclaw University of Environmental and Life Sciences, 50-366 Wroclaw, Poland

[3]Department of Logistics and Transport Systems, Faculty of Mechanical Engineering, Wroclaw University of Technology, 50-371 Wroclaw, Poland

[4]Department of Immunology of Infectious Diseases, Institute of Immunology and Experimental Therapy, 53-114 Wroclaw, Poland

ABSTRACT

Decomposition of organic matter is the primary function in the soil ecosystem, which involves bacteria and fungi. Soil microbial content depends on many factors, and secondary biological and chemical contaminations change and affect environmental feedback. Little work has been done to estimate the microbiological risk for cemetery employees and visitors. The potential risk of infection for people in the cemetery is primarily associated with injury and wound contamination during performing the work. The aim of this study was to analyze the microbiota of cemetery soil obtained from cemeteries and bacterial composition in selected soil layers encountered by gravediggers and cemetery caretakers. The most common bacterial pathogens wereEnterococcus spp. (80.6%), Bacillus spp. (77.4%), and E. coli (45.1%). The fungi Penicillium spp. and Aspergillusspp. were isolated from 51% and 6.4% of samples, respectively. Other bacterial species were in the ground

Citation: Ireneusz Całkosiński, Katarzyna Płoneczka-Janeczko, Magda Ostapska, Krzysztof Dudek, Andrzej Gamian, and Krzysztof Rypuła, "Microbiological Analysis of Necrosols Collected from Urban Cemeteries in Poland," BioMed Research International, vol. 2015, Article ID 169573, 7 pages, 2015. doi:10.1155/2015/169573.

cemetery relatively sparse. Sampling depth was not correlated with bacterial growth ($P<0.05$), but it was correlated with several differences in microbiota composition (superficial versus deep layer).

INTRODUCTION

Soil microbial content depends on many factors, and changes therein result from secondary biological and chemical contamination. The soil microbiota is affected by water content; amounts of mineral and organic substances; soil structure, composition, and degree of acidity; and gas-phase reactions occurring in soil [1]. Exogenous organic matter penetrates into soil in the form of secretions, excretions, and bacteria from dead animals and humans. Secondary biological contaminants also include manure and human sewage, household and farms beheaded, and precipitation washed from areas inhabited by humans and industrial environments [2,3].

The decomposition of organic matter, which involves primarily bacteria and fungi, is fundamental for the functioning of the soil ecosystem. Soil decomposition contributes to a high degree of heterogeneity in physical, chemical, and/or biological composition [4]. Bacteria in soil are classified in two groups: autochthonous, referring to microorganisms adapted to the presence of minimal nutrients (Arthrobacter spp., Azotobacter spp.,Clostridium spp., Nitrobacter spp., Nitrosomonas spp., Pseudomonas spp., Serratia spp., Bradyrhizobium spp.,Mesorhizobium spp., Rhizobium spp., Sinorhizobium spp., Acidithiobacillus spp., Desulfovibrio spp., andThiobacillus spp.) and zymogenic, encompassing microorganisms showing rapid growth only after the introduction of highly concentrated nutrients (Bacillus spp., Corynebacterium spp., Escherichia coli, Proteusspp., and thermophilic microorganisms) [5, 6].

A 1998 World Health Organization report described the potential impacts of cemeteries on the environment and human health, focusing on soil decomposition and soil and groundwater contamination. The authors pointed out that little research had examined cemetery-related sources of environmental contamination. In the following years some studies dealing with this matter were published [7]. An analysis of cemetery impacts on groundwater contamination conducted in Portugal in 2000-2001 involved hydrological and geographic surveys of cemeteries that took differences in lithological conditions into account [8]. E. coli, including the O157:H7 serotype, Campylobacter spp., Salmonella spp., Listeria spp., Listeria monocytogenes, and Mycobacterium tuberculosis have been isolated from cemetery environments [9–11].

The aim of this study was to analyze the microbiota of cemetery soil obtained from cemeteries in the region of Lower Silesia, Poland. Microbiota composition in selected soil layers encountered by gravediggers and cemetery caretakers was evaluated.

MATERIALS AND METHODS

The Ethical Committee for Animal Experiments in Wrocław, Poland, approved this study and all samples were collected in accordance with the research protocol that is accepted.

Research Area

Samples were obtained from five urban necropolises: four in Wroclaw: Grabiszynski municipal cemetery: A, Osobowice municipal cemetery: B, municipal cemetery at the Bujwida Street: C, and municipal cemetery at the Kiełczowska Street: D, and one municipal cemetery in Oleśnica at the Polish Army Street: E (Figure 1). Cemeteries A–C were established in the 19th century (1881, 1867, and 1866, resp.), cemetery E has been operating since about 1926, and cemetery D, a municipal cemetery built in the postwar period, has held burials since 1996. All of these cemeteries continue to accept burials; in accordance with current Polish law [12], graves can be reused after 20 years if no person objects and the burial fee has not been paid. All cemeteries in the study sample are located in the same climate transition zone of the clearly temperate climate (dominated by oceanic influences), typical of Lower Silesia. The average yearly temperature hovers in the area of 8.5°C. The average amount of rainfall is 500–620 millimeters, with its maximum in July and minimum in February. The snow layer disappears after 45 days. Similar to west part/side of Poland, appearing winds are westerly and south-westerly.

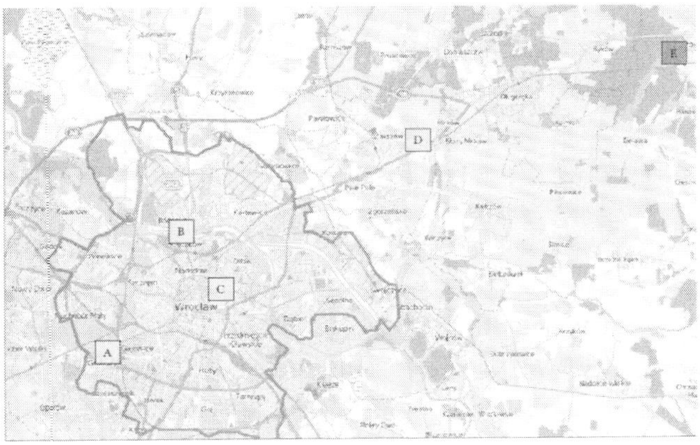

Figure 1: Location of the cemetery in the city of Wrocław; the yellow square indicates cemeteries in the metropolitan city of Wroclaw (A–D); the cemetery in Olesnica is highlighted in blue (E).

Selection of Samples

Samples for microbiological examination were collected from a total of 155 burial sites (A, n=5; B, n=20; C,;n=30 D, n=35; E, n=25) between summer 2013 and spring 2014. Each site was sampled only once. Soil samples were collected during simple grave preparation (reused places after 20 years) using 150 mL containers (Medlab, Poland). At two depths (0.15–0.20 m and 2 m), 200 g soil was taken from each of five points (four grave corners P1–P4 and center P5, identified by the intersection of diagonal lines from the corners; Figure 2). Samples were packed for shipping and transported by automobile to the EPI-Vet Diagnostic Laboratory, Faculty of Veterinary Medicine, Wroclaw, for microbial analysis. The laboratory implements a quality management system (ISO/IEC 17025:2005 + API:2007 + AC:2007).

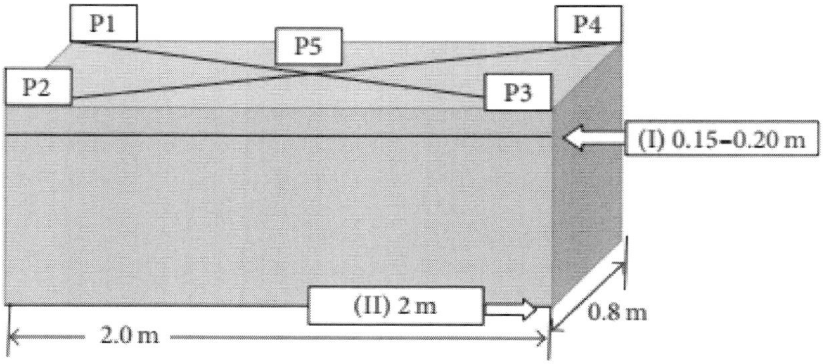

Figure 2: The diagram of sampling for analysis from one place of burial: on each of two depths (I, II), in each case the soil collected at the points P1–P4 (the corners of the grave) and P5 (a point defined at the intersection of the diagonals).

Microbiological Analysis

Samples from P1–P5 were pooled at the laboratory for each depth, and the occurrence of microorganisms at the two depths was compared (comparison in the total sample per grave and per cemetery). All pooled samples from each depth were initially flooded with sterile PBS (IITD, Poland) in the volume ratio 1 : 1 and then with the sterile loop (10 μL of the diameter, Sterbios, Poland) the material was seeded in agar with 5% horse blood and other media. The presence of aerobic and anaerobic gram-positive and gram-negative cocci and rods was evaluated. McConkey agar with crystal violet was used to detect gram-negative oxygen bacteria of the family Enterobacteriaceae, Mannitol Salt agar was used to identify growth of Staphylococcus spp., Enterococcosel agar was used for Enterococcus spp., and ORIE chromogenic substrate was used to detect bacteria in the family Enterobacteriaceae and gram-positive cocci. Sabouraud agar supplemented with chloramphenicol and gentamicin was used to detect growth of yeasts and molds. All culture media were supplied by GRASO Biotech (Poland). Bacterial preparations were incubated for 2 days at 37°C and samples used to detect fungi were incubated for 7 days at 25°C. Microorganism identification was based on morphological characteristics of the colony, gram staining, and biochemical characteristics, as defined using ENTEROtest 16 (Erba La Chema, Czech Republic). Fungal identification was based on the examination of direct preparations in saline under an optical microscope (BIOLAR C; CB, Poland) at 20 Å ~ magnification. Bacteria colonies were classified as very large (>10^4 colony forming units (CFU)/mL), large (10^3–10^4 CFU/mL), or few (<10^3 CFU/mL).

Statistical Analysis

The Pearson chi-squared test (two-tailed, significance level = 0.05) was used to compare proportions and cross table analysis was performed using Statistica 10.0.0 software (StatSoft).

RESULTS

Seven genera of bacteria (Bacillus spp. (Bacillus megaterium and B. cereus), Enterococcus spp. (includingEnterococcus faecalis), Escherichia spp., including E. coli, the Klebsiella-Enterobacter-Serratia (KES group), andStaphylococcus spp. including Staphylococcus epidermidis and other coagulase-negative staphylococci (CNS)) and two genera of fungus (Penicillium spp. and Aspergillus spp.) were isolated from cemetery soil samples. The most common pathogens were Enterococcus spp. (80.6%) and Bacillus spp. (77.4%). All bacterial species exceptE. coli (45.1%) were in the ground cemetery relatively sparse at 6.4% (B. cereus, B. megaterium, E. faecalis, S. epidermidis, and CNS). Penicillium spp. and Aspergillus spp. were isolated from 51% and 6.4% of samples, respectively. The results of microorganism analyses are shown in Table 1.

Table 1: The number and frequency (%) of isolations of individual bacteria and fungi in the ground of the cemetery, collected at the depths of 0.15–0.20 m (I) and 2 m (II) in the five necropolises (A–E)

Pathogens	Cemetery									
	A $n = 45$		B $n = 20$		C $n = 30$		D $n = 35$		E $n = 25$	
	I	II	I	II	I	II	I	II	I	II
Bacillus spp.	30 67%	30 67%	20 100%	20 100%	30 100%	30 100%	25 71%	25 71%	25 100%	25 100%
Bacillus cereus	10 22%	10 22%	nd 0%	nd 0%	nd 0%	nd 0%	nd 0%	nd 0%	nd 0%	nd 0%
Bacillus megaterium	10 22%	10 22%	nd 0%	nd 0%	nd 0%	nd 0%	nd 0%	nd 0%	nd 0%	nd 0%
Enterococcus spp.	30 67%	30 67%	20 100%	20 100%	25 83%	25 83%	35 100%	35 100%	25 100%	25 100%
Enterococcus faecalis	10 22%	10 22%	nd 0%	nd 0%	nd 0%	nd 0%	nd 0%	nd 0%	nd 0%	nd 0%
Enterobacteriaceae spp. *Escherichia coli*	25 56%	25 56%	5 25%	5 25%	15 50%	15 50%	5 14%	5 14%	25 100%	25 100%
KES group *Serratia megaterium*	nd 0%	nd 0%	nd 0%	nd 0%	nd 0%	nd 0%	nd 0%	nd 0%	20 80%	20 80%
Staphylococcus spp.	10 22%	10 22%	nd 0%	nd 0%	nd 0%	nd 0%	10 29%	10 29%	nd 0%	nd 0%
Staphylococcus epidermidis	10 22%	10 22%	nd 0%	nd 0%	nd 0%	nd 0%	nd 0%	nd 0%	nd 0%	nd 0%
CNS	nd 0%	nd 0%	nd 0%	nd 0%	nd 0%	nd 0%	10 29%	10 29%	nd 0%	nd 0%
Penicillium spp.	30 67%	30 67%	20 100%	20 100%	nd 0%	nd 0%	10 29%	10 29%	20 80%	20 80%
Penicillium spp.	nd 0%	nd 0%	nd 0%	nd 0%	nd 0%	nd 0%	10 29%	10 29%	nd 0%	nd 0%

*Bacterial and fungal systematic given at the most detailed taxonomic level.

Figure 3 shows the bacterial growth levels of the two soil layers according to microorganisms identified. Bacillusspp. showed abundant growth (>10^4 CFU/mL) in samples from all cemeteries, independent of sampling depth. Growth of Enterococcus spp. ranged from <10^3 to >10^4 CFU/mL in both soil layers. The growth of Escherichiaspp., predominantly E. coli, was greater in samples from the deep soil layer (from 10^3 to >10^4 CFU/mL) than in those from the superficial layer (<10^3 to 10^3). The KES group was identified in samples from only one grave (at cemetery E), with >10^4 CFU/mL observed in samples from both soil depths.

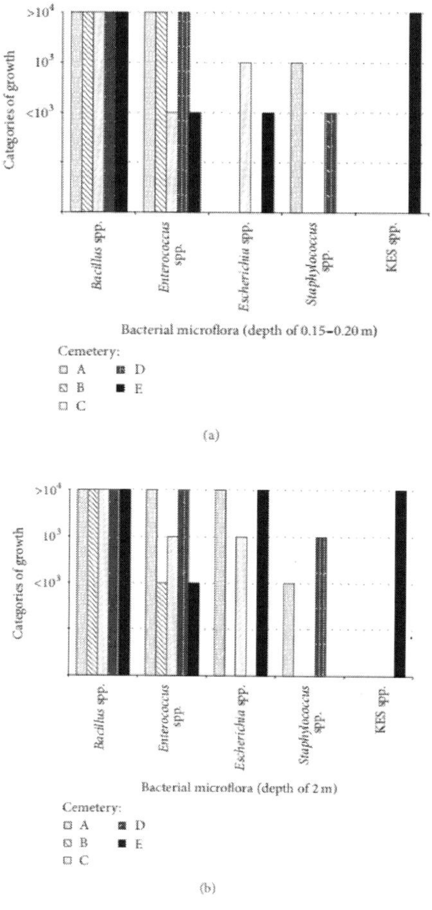

Figure 3: (a) The presence of bacterial microflora in the soil of burial at a depth of 0.15–0.20 m collected from the five necropolises (A–E). (b) The presence of bacterial microflora in the soil of burial at a depth of 2.0 m collected from the five necropolises (A–E). BS:Bacillus spp., EnS: Enterococcus spp., EsS: Escherichia spp., KES: Klebsiella spp.,Enterobacter spp., and Serratia spp., StS: Staphylococcus spp.

Staphylococcus spp. was identified at only two sites (cemeteries A and D); the growth of bacteria from cemetery A with $<10^3$ CFU/mL was observed in samples taken from both soil depths, whereas in cemetery D the superficial layer generated less growth ($<10^3$ CFU/mL) than the deep layer (10^3–10^4 CFU/mL). Three genera were predominant at all cemeteries: Bacillus spp., Enterococcus spp., and Escherichia spp. Penicillium spp. was identified in samples from four of the five (80%) cemeteries. Sampling depth was not correlated with bacterial growth ($P>0.05$; Figure 4), but it was correlated with several differences in microbiota composition (superficial versus deep layer: Bacillus spp. versus Enterococcus spp. ($p=0.021$), Bacillus spp. versus KES group ($p=0.036$),

Bacillus spp. versus Staphylococcus spp. (p=0.002), and Enterococcus spp. versus KES group (p=0.019)).

Cemetery A	I	II
Bacillus spp.	4	4
Enterococcus spp.	4	4
Enterobacteriaceae spp.	0	4
Staphylococcus spp.	3	2
KES spp.	0	0

Depths I versus II: $p = 0.272$

Cemetery B	I	II
Bacillus spp.	4	4
Enterococcus spp.	4	2
Enterobacteriaceae spp.	0	0
Staphylococcus spp.	0	0
KES spp.	0	0

Depths I versus II: $p = 0.938$

Cemetery C	I	II
Bacillus spp.	4	4
Enterococcus spp.	2	3
Enterobacteriaceae spp.	3	3
Staphylococcus spp.	0	0
KES spp.	0	0

Depths I versus II: $p = 0.929$

Cemetery D	I	II
Bacillus spp.	4	4
Enterococcus spp.	4	4
Enterobacteriaceae spp.	0	0
Staphylococcus spp.	2	3
KES spp.	0	0

Depths I versus II: $p = 0.927$

Cemetery E	I	II
Bacillus spp.	4	4
Enterococcus spp.	2	2
Enterobacteriaceae spp.	2	4
Staphylococcus spp.	0	0
KES spp.	4	4

Depths I versus II: $p = 0.915$

Growth (CFU/mL)	
	0
$<10^3$	2
10^3	3
$>10^4$	4

Figure 4: Microbial growth on five cemeteries including the depth and space sampling.

DISCUSSION

Research on cemetery soil emerged in the second half of the 20th century, accompanied by the introduction of the term "necrosol." Subject to transformation of the soil profile, this layer does not exceed a depth of 0.2 m [13]. International research on necrosols has focused on their physicochemical properties, primarily phosphorus, nitrogen, and organic carbon contents [14, 15]. The groundwater in the vicinity of cemeteries showed increased concentrations of intestinal flora, ions, and amino acids, such as putrescine and cadaverine. The composition of the air in these environments is characterized by increased concentrations of unstable gases, such as phosphine and ethylene. Wax formation has also been observed in necrosols [16].

In Poland, according to the 7 March 2008 Regulation of the Minister of Infrastructure regarding cemeteries, graves and other burial sites of human remains and debris must be below ground and hold one individual casket of the dimension of 2 m × 1 m × 1.7 m (excluding children and family graves). This provision does not apply to cemeteries, graves, and other burial sites existing at the time of the regulation's establishment. The location of cemeteries established in the last two centuries did not account for intense urban development, which has led to a shortage of burial plots and permitted reuse of graves for which fees have not been paid after 20 years [17].

Little work has been done to assess the microbiological risk for cemetery employees and visitors. Knight and Dent [18] demonstrated that groundwater near burial sites contained high concentrations of Pseudomonas aeruginosa. Using the decomposition of fats by enzymes, Forbes et al. [19] indirectly showed the presence ofClostridium perfringens in necrosols.

However, microbiological analysis provided only supporting evidence in that study. The context of potential cemetery-related risks, that is, with respect to soil depth, has not been examined. The impacts of soil microbe profiles should be assessed with consideration of human contact, as in the present study. Moreover, analyses of groundwater contamination in/and around cemeteries does not allow confrontation with results achieved in Polish scientific study, as they have been performed in a different geographic location, also to each other [13–15].

The potential risk of infection for cemetery workers is associated primarily with injury and wound contamination during performance of the work. Lesser risks are associated with accidental microorganism ingestion, inhalation, or transfer to the mucous membranes of the eyes [20].

Live isolates from human tissue belong to the most common taxonomic groups, such as Streptococcus spp.,Staphylococcus spp., Clostridium spp., Bacillus spp., and Lactobacillus spp. [7]. The microbiota profiles of cemetery soil samples tested in the present study were similar to this ante mortem human microbial composition, with the addition of the KES group and Aspergillus spp. and Penicillium spp. fungi. This composition largely reflects the decomposition of human remains, as well as the flora and fauna of the burial sites. A limitation of the study is that the risk of contact with cemetery soil was not assessed, as not all genera of isolated microorganisms could be identified based on their biochemical properties. Pathogenic and commensal bacteria were identified.

The results of this study highlight the need for microbial analysis of necrosols using more sensitive laboratory techniques, such as polymerase chain reaction assays; however it would increase the cost of research. The level of some risks posed by identified species of bacteria in the analyzed soil layers could not be avoided. Bacillus cereus and B. megaterium may be primary causes of wound infection [6, 21]. B. cereus is also considered to be the primary pathogen of food poisoning and eye infection, and it participates in progressive pneumonia and sepsis in infections of the nervous system [6, 22]. In recent years, an increased interest of Enterococcus spp. has been noted and the presence of bacteria of this genera has been confirmed within many infectious endogenous pathogenic bacteria isolates; E. faecalis is the dominant species (57%) in clinical cases and was also detected in the analyzed necrosol samples. The presence of this isolate is important in the context of possible injury-related contamination due to contact with cemetery soil. This pathogen produces hyaluronidase, an enzyme that depolymerizes mucopolysaccharides of connective tissue, which may facilitate the spread of bacteria and toxins [23]. Detailed characterization, which is not encompassed by the present research, also requires determination of the potential virulence of E. coli isolates from each grave and depth. This bacterium is the third most frequently isolated (after S. aureus and P. aeruginosa) from wounds and skin and soft tissue infections (SSTIs). These infections are often self-limiting, but some require antibiotic treatment and even hospitalization [24].Staphylococci were detected and identified in only two cases in the present study, but S. intermedius and CNS are known to cause many conditions, such as wound infections, skin abscesses, and inflammatory changes in joints [25]. The KES group comprises opportunistic organisms that commonly cause hospital infections related to manipulation in the respiratory tract (tracheostomy, inhalation) and catheterization. Klebsiella spp. rarely cause SSTIs, but cases of infection in the extremities, accompanied by purulent processes, gas formation, and metastatic lesions, have been reported [26]. Nail infection involving Klebsiella has also been reported [27]. Amin et al. [3] describe Enterobacter as accompanying SSTI. Skin lesions such as granulomas, fasciitis, lumps, abscesses, and ulcers caused by Serratia spp. are prevalent in elderly and immunocompromised individuals; however infections cannot be ruled out also in young and immunocompetent people [28].

Two genera of fungi were detected in this study. Aspergillus spp. and Penicyllium spp. Aspergillus spp. are common in the environment (in soil, food, air, water, decomposing plant, and animal materials) andPenicillium spp. cause infection, most commonly through the inhalation of spores. Like most fungi, these species are sensitizing, potentially causing asthma, allergic rhinitis, and atopic dermatitis [20]. The most serious disease caused by fungi of the genus Aspergillus spp. is pulmonary aspergillosis [29]. The exposure of different groups of people to cemetery soil microbiota should also be assessed. Microorganisms abound in different layers of the soil profile, and the frequency of contact is almost certainly greater among cemetery workers than among regular visitors. Further research in cooperation with human medicine and hospitals will be focused on the epidemiological aspects of human infections, because there are no literature data which currently record the place of infection (including cemetery), site of infection, and its correlation with developed disease stages.

REFERENCES

1. P. Eldor, Soil Microbiology, Ecology and Biochemistry, Elsevier, Oxford, UK, 3rd edition, 2007.
2. A. Usulu, "An ecological approach for the evaluation of an abandoned cemetery as a greenarea: the case of Ankara /Karakusunlar cemetery," African Journal of Agricultural Research, vol. 5, no. 10, pp. 1043–1054, 2010.
3. A. N. Amin, E. A. Cerceo, S. B. Deitelzweig, J. C. Pile, D. J. Rosenberg, and B. M. Sherman, "Hospitalist perspective on the treatment of skin and soft tissue infections," Mayo Clinic Proceedings, vol. 89, no. 10, pp. 1436–1451, 2014. · ·
4. J. A. van Veen and P. J. Kuikman, "Soil structural aspects of decomposition of organic matter by micro-organisms," Biogeochemistry, vol. 11, no. 3, pp. 213–233, 1990. · ·

5. U. Langer, L. Böhme, and F. Böhme, "Classification of soil microorganisms based on growth properties: a critical view of some commonly used terms," Journal of Plant Nutrition and Soil Science, vol. 167, no. 3, pp. 267–269, 2004. · ·

6. E. J. Bottone, "Bacillus cereus, a volatile human pathogen," Clinical Microbiology Reviews, vol. 23, no. 2, pp. 382–398, 2010. · ·

7. A. Üçisik and P. Rushbrook, The Impact of Cemeteries on the Environment and Public Health an Introductory Briefing, EUR/HFA target 23, EUR/ICP/EHNA 010401(A), WHO Regional Office for Europe, Nancy Project Office, Copenhagen, Denmark, 1998.

8. L. Rodrigues and A. Pacheco, "Groundwater contamination from cemeteries cases of study in," inProceedings of the Environmental 2010: Situation and Perspectives for the European Union, pp. 1–6, Porto, Portugal, May 2003.

9. J. B. Russell and G. N. Jarvis, "Practical mechanism for interrupting the oral-faecal lifecycle ofEscharichia coli," Journal of Molecular Microbiology and Biotechnology, vol. 279, no. 52, pp. 53924–53931, 2001.

10. S. Mukherjee, P. Das, and R. Sen, "Towards commercial production of microbial surfactants," Trends in Biotechnology, vol. 24, no. 11, pp. 509–515, 2006. · ·

11. G. F. Wells, C. H. Wu, Y. M. Piceno et al., "Microbial biogeography across a full-scale wastewater treatment plant transect: evidence for immigration between coupled processes," Applied Microbiology and Biotechnology, vol. 98, no. 10, pp. 4723–4736, 2014. · ·

12. Cemeteries and Burials from January 31, 1959, Laws Press, Warszawa, Poland, No. 11, Pos. 62, with amended, 1959.

13. L. Majgier, O. Rahmonov, and R. Bednarek, "Features of abandoned cemetery soils on sandy substrates in Northern Poland," Eurasian Soil Science, vol. 47, no. 6, pp. 621–629, 2014. · ·

14. P. Charzyński, R. Bednarek, M. Świtoniak, and B. Żołnowska, "Ekranic technosols and urbic technosols of Toruń necropolis," Geologija, vol. 53, no. 4, pp. 179–185, 2011. ·

15. L. Majgier and O. Rahmonov, "Selected chemical properties of necrosols from abandoned cemeteries Słabowo and Szymonka (Great Masurian Lakes District)," Bulletin of Geography. Physical Geography Series, vol. 5, no. 1, pp. 43–55, 2012.

16. J. Żychowski, "Impact of cementeries on groundwater chemistry. A review," Catena, vol. 93, pp. 100–114, 2012.

17. "On the requirements to be met by cemeteries, graves and other burial of corpses and debris from March 7, 2008," in Regulation of the Minister of Infrastructure, No. 48, Pos. 284, Laws Press, Warszawa, Poland, 2008.

18. M. J. Knight and B. B. Dent, "Sustainability of waste and groundwater management systems," inGroundwater: Sustainable Solutions Proceedings of the International Association of Hydrogeologists. Melbourne, pp. 359–374, 1998.

19. S. L. Forbes, B. H. Stuart, and B. B. Dent, "The effect of burial environment on

adipocere formation,"Forensic Science International, vol. 154, no. 1, pp. 44–52, 2004.

20. Y. Mou, L. Ye, M. Ye, D. Yang, and M. Jin, "A retrospective study of patients with a delayed diagnosis of allergic bronchopulmonary aspergillosis/allergic bronchopulmonary mycosis," Allergy and Asthma Proceedings, vol. 35, no. 2, pp. e21–e26, 2014. · ·

21. K. O. Duncan and T. L. Smith, "Primary cutaneous infection with Bacillus megaterium mimicking cutaneous anthrax," Journal of the American Academy of Dermatology, vol. 65, no. 2, pp. e60–e61, 2011. · ·

22. F. A. Drobniewski, "Bacillus cereus and related species," Clinical Microbiology Reviews, vol. 6, no. 4, pp. 324–338, 1993. ·

23. K. Fisher and C. Phillips, "The ecology, epidemiology and virulence of Enterococcus," Microbiology, vol. 155, no. 6, pp. 1749–1757, 2009. · ·

24. Ž. Petkovšek, K. Elersič, M. Gubina, D. Žgur-Bertok, and M. S. Erjavec, "Virulence potential ofEscherichia coli isolates from skin and soft tissue infections," Journal of Clinical Microbiology, vol. 47, no. 6, pp. 1811–1817, 2009. · ·

25. N. Wang, A. M. Neilan, and M. Klompas, "Staphylococcus intermedius infections: case report and literature review," Infectious Disease Reports, vol. 5, no. 3, pp. 6–11, 2013. · ·

26. C.-M. Chang, H.-C. Lee, N.-Y. Lee et al., "Community-acquired Klebsiella pneumoniae complicated skin and soft-tissue infections of extremities: emphasis on cirrhotic patients and gas formation," Infection, vol. 36, no. 4, pp. 328–334, 2008. · ·

27. P. Nenoff, U. Paasch, and W. Handrick, "Infections of finger and toe nails due to fungi and bacteria,"Hautarzt, vol. 65, no. 4, pp. 337–348, 2014. · ·

28. M. Carlesimo, A. Pennica, M. Muscianese et al., "Multiple skin ulcers due to Serratia marcescens in an immunocompetent patient," Journal on Dermatology and Sexually Transmitted Diseases, vol. 149, no. 3, pp. 367–370, 2014. ·

29. M. L. Chabi, A. Goracci, N. Roche, A. Paugam, A. Lupo, and M. Revel, "Pulmonary aspergillosis,"Diagnostic and Interventional Imaging, vol. 96, no. 5, pp. 435–442, 2015. ·

5

CHAPTER

WATER MICROBIOLOGY. BACTERIAL PATHOGENS AND WATER

João P. S. Cabral

Center for Interdisciplinary Marine and Environmental Research (C. I. I. M. A. R.), Faculty of Sciences, Oporto University, Rua do Campo Alegre, 4169-007 Oporto, Portugal

ABSTRACT

Water is essential to life, but many people do not have access to clean and safe drinking water and many die of waterborne bacterial infections. In this review a general characterization of the most important bacterial diseases transmitted through water—cholera, typhoid fever and bacillary dysentery—is presented, focusing on the biology and ecology of the causal agents and on the diseases' characteristics and their life cycles in the environment. The importance of pathogenic Escherichia coli strains and emerging pathogens in drinking water-transmitted diseases is also briefly discussed. Microbiological water analysis is mainly based on the concept of fecal indicator bacteria. The main bacteria present in human and animal feces (focusing on their behavior in their hosts and in the environment) and the most important fecal indicator bacteria are presented and discussed (focusing on the advantages and limitations of their use as markers). Important sources of bacterial fecal pollution of environmental waters are also briefly indicated. In the last topic it is discussed which indicators of fecal pollution should be used in current drinking water microbiological analysis. It was concluded that safe drinking water for all is one of the major challenges of the 21st century and that microbiological control of drinking water should be the norm everywhere. Routine basic microbiological analysis of drinking water should be carried out by assaying the presence of Escherichia coli by culture methods. Whenever financial resources are available, fecal coliform determinations should be complemented with the quantification of enterococci. More studies are needed in order to check if ammonia is reliable for a preliminary screening for emergency fecal pollution

Citation: João P. S. Cabral, Water Microbiology. Bacterial Pathogens and Water, doi:10.3390/ijerph7103657.

outbreaks. Financial resources should be devoted to a better understanding of the ecology and behavior of human and animal fecal bacteria in environmental waters.

DRINKING WATER AS A VEHICLE OF DISEASES

Water is essential to life. An adequate, safe and accessible supply must be available to all. Improving access to safe drinking-water can result in significant benefits to health. Every effort should be made to achieve a drinking water quality as safe as possible [1].

Many people struggle to obtain access to safe water. A clean and treated water supply to each house may be the norm in Europe and North America, but in developing countries, access to both clean water and sanitation are not the rule, and waterborne infections are common. Two and a half billion people have no access to improved sanitation, and more than 1.5 million children die each year from diarrheal diseases [2]. According to the WHO, the mortality of water associated diseases exceeds 5 million people per year. From these, more that 50% are microbial intestinal infections, with cholera standing out in the first place.

In general terms, the greatest microbial risks are associated with ingestion of water that is contaminated with human or animal feces. Wastewater discharges in fresh waters and costal seawaters are the major source of fecal microorganisms, including pathogens [1–4].

Acute microbial diarrheal diseases are a major public health problem in developing countries. People affected by diarrheal diseases are those with the lowest financial resources and poorest hygienic facilities. Children under five, primarily in Asian and African countries, are the most affected by microbial diseases transmitted through water [5].

Microbial waterborne diseases also affect developed countries. In the USA, it has been estimated that each year 560,000 people suffer from severe waterborne diseases, and 7.1 million suffer from a mild to moderate infections, resulting in estimated 12,000 deaths a year [6]. The most important bacterial diseases transmitted through water are listed in Table 1.

Table 1: The main bacterial diseases transmitted through drinking water

Disease	Causal bacterial agent
Cholera	*Vibrio cholerae*, serovarieties O1 and O139
Gastroenteritis caused by vibrios	Mainly *Vibrio parahaemolyticus*
Typhoid fever and other serious salmonellosis	*Salmonella enterica* subsp. *enterica* serovar Paratyphi
	Salmonella enterica subsp. *enterica* serovar Typhi
	Salmonella enterica subsp. *enterica* serovar Typhimurium
Bacillary dysentery or shigellosis	*Shigella dysenteriae*
	Shigella flexneri
	Shigella boydii
	Shigella sonnei
Acute diarrheas and gastroenteritis	*Escherichia coli*, particularly serotypes such as O148, O157 and O124

CHOLERA

The Genus Vibrio

Vibrio are small, curved-shaped Gram-negative rods, with a single polar flagellum. Vibrios are facultative anaerobes capable of both fermentative and respiratory metabolism. Sodium stimulates growth of all species and is an absolute requirement for most. Most species are oxidase-positive and reduce nitrate to nitrite. Cells of certain species (*V. cholerae*, *V. parahaemolyticus* and *V. vulnificus*) have pili (fimbriae), structures composed of protein TcpA. TcpA formation is co-regulated with cholera toxin expression and is a key determinant of *in vivo* colonization (see below) [7,8].

Several *Vibrio* species can infect humans (Table 2). *V. cholerae* is, by far, the most important of these species. *V. alginolyticus* has been isolated from several types of soft tissue infections.

Table 2: Main species of *Vibrio* and their occurrence in human clinical specimens[a]

Main species	Occurrence in human clinical specimens	
	Intestinal	Extra-intestinal
Vibrio alginolyticus	+	++
Vibrio cholerae O1 and O139	+++++	+
Vibrio cholerae non O1 or O139	++	++
Aliivibrio fischeri (*Vibrio fischeri*)	-	-
Vibrio fluvialis	++	-
Vibrio furnissii	++	-
Vibrio harveyi	-	+
Grimontia hollisae (*Vibrio hollisae*)	++	-
Vibrio mimicus	++	+
Vibrio natriegens	-	-
Vibrio parahaemolyticus	++++	+
Vibrio vulnificus	+	+++

[a] Adapted from [7,8]. Nomenclature according to [9]. The symbols give the relative frequency of each organism in human clinical specimens, and apply to the whole World, rather than to a particular country.

V. fluvialis, *Grimontia hollisae* (*V. hollisae*), and *V. mimicus* can cause diarrhea or infections of the gastrointestinal tract. *V. furnissii* has been isolated from a few individuals with diarrhea, but there is no evidence that it can actually cause this pathology. *V. parahaemolyticus* is a well-documented causal agent of acute food-borne gastroenteritis, particularly in Japan and South East Asia. Cases are associated with the consumption of raw or undercooked shellfish such as oysters, shrimp, crabs, and lobster. *V. vulnificus* is an important cause of (often fatal) septicemia and wound infections. Other vibrios, namely *Allivibrio fischeri*(*Vibrio fischeri*) and *V. natriegens*, have no relation with humans [7,8].

Vibrios are primarily aquatic bacteria. Species distribution depends on sodium concentration and water temperature. Vibrios are very common in marine and estuarine environments, living free or on the surfaces and in the intestinal contents of marine animals. Species with a low sodium requirement are also found in freshwater habitats [7,8].

The Species Vibrio Cholerae

Vibrio cholerae cells can grow at 40 °C with pH 9–10. The growth is stimulated by the presence of sodium chloride. *Vibrio cholerae* is a very diverse bacterial species (Table 3). It is divided in ca. 200 serovarieties, characterized by the structure of the lipopolysaccharide (LPS) (O antigens). Only serovarieties O1 and O139 are involved in "true" cholera. Some other serovarieties can cause gastroenteritis, but not cholera. The distinction between Classical and El Tor biotypes is based on biochemical and virological characteristics [1,7,8,10,11].

Table 3: Subdivision of *Vibrio cholerae* below the species level[a]

Serovariety	Serotype	Biotype
O1	Inaba	Classical
		El Tor
	Ogawa	Classical
		El Tor
	Hikojima	
O139		
others		

[a] Adapted from [8].

Cholera

Characterization of the disease

The incubation period for cholera is ca. 1–3 days. The disease is characterized by an acute and very intense diarrhea that can exceed one liter per hour. Cholera patients feel thirsty, have muscular pains and general weakness, and show signs of oliguria, hypovolemia, hemoconcentration, followed by anuria. Potassium in blood drops to very low levels. Patients feel lethargic. Finally, circulatory collapse and dehydration with cyanosis occurs [7].

The severity of the disease depends on several factors: (1) personal immunity: this may be conferred by both previous infections and by vaccines; (2) inoculum: the disease only occurs after ingestion of a minimum amount of cells, ca. 10^8 [1,7,8,10,11]; (3) The gastric barrier: *V. cholera* cells likes basic media and therefore the stomach, normally very acidic, is an adverse medium for bacterial survival. Patients consuming anti-acidic medications are more susceptible to infection than healthy people; (4) blood group: for still unknown reasons, people with O-group blood are more susceptible than others [1,7,8,10,11].

In the absence of treatment, the mortality of cholera-patients is ca. 50%. It is mandatory to replace not only lost water but also lost salts, mainly potassium. In light dehydrations, water and salts can be orally-administered, but in severe conditions, rapid and intravenous-administration is obligatory. The most efficient antibiotic is currently doxicyclin. If no antibiotic is available for treatment, the administration of water with salts and sugar can, in many cases, save the patient and help in the recovery [1,7,8,10,11].

There are two main determinants of infection: (1) the adhesion of the bacterial cells to the intestinal mucous membrane. This depends on the presence of pili and adesins at the cell's surface; (2) the production of cholera toxin [1,7,8,10,11].

Cholera toxin

Cholera toxin is an exotoxin with a very precise action on target cells. The toxin attaches to a specific receptor (ganglioside Gl) on the cell membrane of intestinal cells and activates the enzyme adenylate cyclase. This results in a non-stop degradation of internal ATP, with release of cAMP and inorganic phosphate. The rise in the internal concentration of cAMP causes an efflux of water, sodium, potassium, chloride and carbonate ions from the cells of the mucous membrane, and this is the main cause of diarrhea [7].

Cholera pandemics and the emergence of El Tor biotype and O139 serovariety. New facts about cholera epidemiology

Cholera has been a well known disease since the 19th century. In the 19th and 20th centuries, seven major pandemics are recognized. The first six pandemics occurred during the following periods: 1st: 1816–1826, 2nd: 1829–1851, 3rd: 1852–1860, 4th: 1863–1875, 5th: 1881–1896, 6th: 1899–1923. These pandemics all started in Asia, passed through Europe and then reached South America. The Classical biotype was involved. The seventh pandemic, still in course, started in 1961 in the Celebes Isles, in Asia. In the 1960s, the disease spread through Asia, in the 1970s reached the Middle East and Africa, and in 1991 streaked violently across South America. Now El Tor has replaced the Classical biotype. El Tor biotype had been detected before, in 1905, but only in the development of the seventh pandemic did this biotype replace the Classical one and become dominant [1,7,8,10,11].

In 1992, a new serovariety (O139), which was coined the Bengal serovariety, was detected for the first time in Bangladesh. This new serovariety quickly spread to India and to southeastern Asia, displacing O1. Although serovariety O1 El Tor has reappeared in 1994 and 1995, the Bengal serovariety still remains the dominant one. The illness caused by serovarieties O139 and O1 are indistinguishable [8,12,13].

In 1991, the seventh pandemic entered South America through the coastal area of Peru. On 23 January, in Chancay, north Peru, *Vibrio cholerae* O1 El Tor was isolated from patients with cholera symptoms, confirming the disease. In this region, between 24 January and 9 February, 1,859 people were hospitalized and 66 died. From Peru, the disease spread rapidly to other countries in South America. Two routes have been proposed for the entrance of the bacterium in Peru: (1) ballast water from a boat coming from Asia; (2) the El Niño current may have transported zooplankton harboring *V. cholerae* cells. Shellfish and fish nourishing on this zooplankton became contaminated and the bacterium was transmitted to humans who ate these marine foods [14–17].

The misfortune of people who died in the first months of this disastrous South American cholera epidemic appeared to have unleashed scientists to study the disease harder and, indeed, important epidemiological studies were carried out during this outbreak. These studies confirmed that contaminated uncooked food and beverages can also be a vehicle for transmission of cholera [18].

Genes for toxin and pili protein production

The genes responsible for toxin production are harbored in the CTXΦ segment (7–9.7 kb) of the chromosome (only in toxigenic strains). The CTXΦ segment carries at least six genes. In addition to the gene encoding cholera toxin production, this segment (virulence cassette) include an accessory cholera toxin (*ace*), a zonula occludens toxin (*zot*), core encoded pilin (*cep*), and an open reading frame of unknown function. During the replication of the chromosome, the CTXΦ fragment can form an autonomous copy and this can constitute an independent plasmid. The plasmid can give rise to virus-like particles—CTXΦ bacteriophages, which can infect non-toxigenic strains. The CTXΦ segment incorporates into the chromosome of the infected cells which became toxigenic. This process was demonstrated *in vitro* in cell suspensions and *in vivo* in the gut of the rat [8,13,19,20].

Epidemic and pandemic strains of *V. cholerae* contain another chromosomal segment designated as VPI. VPI is 39.5 kb in size and contains two ToxR-regulated genes: a regulator of virulence genes (ToxT) and a gene cluster containing colonization factors, including the toxin co-regulated pili (TCP). The tcp gene encodes for the 20.5-kDa TcpA pili protein. This VPI segment appears to be transferable from *V. cholerae* O1 to non-O1 strains. *V. cholerae* O139 strains, like O1, carry the structural genes encoded by the CTX operon and TCP. *V. cholerae* strains non-O1 or O139 normally lack cholera toxin genes and have never been found to carry TCP [8].

Ecology of the bacterium and the cycle of the disease

V. cholerae non-O1 or O139 strains are common in the environment, especially in estuaries. They have been isolated from many estuarine animals such as birds, frogs, fishes and shellfish, and survive and multiply on the surface of phytoplankton and zooplankton cells [8,21].

V. cholerae O1 and O139 strains are isolable from the environment only in epidemic areas. They survive in the cultivable state in water and aquatic and marine organisms for a considerable period of time [8,12,22–24]. When *V. cholerae* cells face adverse environmental conditions, they reduce cell size, became coccoid and enter a dormant stage inside exopolysaccharide biofilms. Cells display a certain metabolism, but are not able to growth and multiply on the surface of agarized media and give rise to colonies. Cells in this viable but non-culturable state retain viability as well as the potential for pathogenicity for significant periods of time [25–27].

Viable but non-culturable cells can leave their dormant stage and multiply again, resulting in an explosion of their concentration in the environment. Since the presence of non-toxigenic strains is common in aquatic milieu, especially in estuaries, if a horizontal transfer of cholera exotoxin producing genes occurs between toxigenic and non-toxigenic strains, the number of toxigenic cells in the environment can rise rapidly and pronouncedly. The episodic nature and the sudden appearance of violent cholera outbreaks, followed by a rapid slowing down, are probably related with these phenomena.

SALMONELLOSIS

The Genus Salmonella. Pathogenicity of Main Serovars

The genus *Salmonella* was designated by Lignières in 1900 [28,29]. Antigenic analysis began when Castellani described, in 1902, a method for absorbing antisera. The first antigenic scheme for *Salmonella* was published by White in 1926, and subsequently developed extensively by Kauffmann, in two classical works published in 1966 and 1978 [28,29]. The Kauffmann-White antigenic scheme contained, by 1988, about 2,250 different serovars [28,29].

The genus *Salmonella*, a member of the family *Enterobacteriaceae*, include Gram-negative motile straight rods. Cells are oxidase-negative and catalase-positive, produce gas from D-glucose and utilize citrate as a sole carbon source. Salmonellae have several endotoxins: antigens O, H and Vi [28,29].

The concept "one serovar-one species", in use for many years, is no longer acceptable. The taxonomy and nomenclature of the genus *Salmonella* has been subject of debate since Le Minor and Popoff proposed changes in a paper published in 1987. The issue was settled by a decision of the International Committee on the Systematics of Prokaryotes and published in 2005. The current taxonomy of the genus is presented in Table 4. According to the rules of bacterial nomenclature, the names of the serovars are not italicized and the first letter must be a capital [28–30].

Table 4: Current taxonomy and nomenclature of the genus *Salmonella*. Habitat and pathogenicity of main serovars[a]

Species	Sub-species	Main serovars (from a total of ca. 1,443)	Habitat and pathogenicity
Salmonella enterica	*Salmonella enterica* subsp. *enterica*	Abortusovis	Pathogenic to sheeps.
		Choleraesuis	Pathogenic to humans and animals.
		Enteritidis	Ubiquitous and frequently the cause of infections in humans and animals. Very frequent agent of gastroenteritis in humans.
		Gallinarum	Isolated chiefly from chickens and other birds. Causal agent of fowl thyphoid
		Paratyphi A	Pathogenic only to humans. Causes paratyphoid fever
		Paratyphi B	Causes paratyphoid fever in humans and very rarely infects animals
		Paratyphi C	Causes paratyphoid fever in humans
		Typhi	Pathogenic only to humans, causing typhoid fever. Transmitted by water and food contaminated with feces.
		Typhimurium	Ubiquitous and frequently the cause of infections in humans and animals. Very frequently, the causal agent of gastroenteritis in humans.
		Typhisuis	Pathogenic to swines

	Salmonella enterica subsp. *arizonae*	At least 94 serovars.	
	Salmonella enterica subsp. *diarizonae*	At least 323 serovars.	
	Salmonella enterica subsp. *houtenae*	At least 70 serovars.	Isolated mainly from cold-blooded animals and from the environment. Not pathogenic to humans.
	Salmonella enterica subsp. *indica*	At least 11 serovars.	
	Salmonella enterica subsp. *salamae*	At least 488 serovars.	
Salmonella bongori		At least 20 serovars.	

[a] Adapted from [29]. Nomenclature according to [9].

S. enterica subsp. *enterica* serovar Enteritidis is the most frequently isolated serovar from humans all over the world. However, locally, other serovars can be predominant. In the period 1994–2004, Tunisia was exposed to salmonellosis outbreaks in 1997, 1999, 2002 and 2004. In 1997, salmonellosis outbreak was caused by serovar Mbandaka. In 1999, three salmonellosis outbreaks were reported from hospitals located in three different regions. Each outbreak was associated with a different serotype: Mbandaka, Livingstone and Typhi Vi+. In 2002, a *S. enterica* subsp. *enterica* serovar Livingstone infection occurred in the same hospital that reported an outbreak caused by serovar Typhi Vi+ in 1999, but in a different unit. In that year, the Livingstone serovar jumped to the first position in human infection in Tunisia. In 2004, a second outbreak by serovar Typhi Vi+ was reported. The source of isolation was a fermented juice traditionally extracted from palm-tree [31].

Characterization of the Diseases

Salmonellae pathogenic to humans can cause two types of salmonellosis: (1) typhoid and paratyphoid fever (do not confuse with typhus, a disease caused by a rickettsia); (2) gastroenteritis [28]. Low infective doses (less than 1,000 cells) are sufficient to cause clinical symptoms. Salmonellosis of newborns and infants presents diverse clinical symptoms, from a grave typhoid-like illness with septicemia to a mild or asymptomatic infection. In pediatric wards, the infection is usually transmitted by the hands of staff [29].

Food-borne *Salmonella* gastroenteritis are frequently caused by ubiquitous *Salmonella* serovars such as Typhimurium. About 12 h following ingestion of contaminated food, symptoms (diarrhea, vomiting and fever) appear and last 2–5 days. Spontaneous cure usually occurs. *Salmonella* may be associated with all kinds of food. Prevention of *Salmonella* food-borne infection relies on avoiding contamination (improvement of hygiene), preventing multiplication of *Salmonella* in food (constant storage of food at 4 °C), and use of pasteurization (milk) or sterilization when possible (other foods). Vegetables and fruits may carry *Salmonella*when contaminated with fertilizers of fecal origin, or when washed with polluted water [28].

The incidence of typhoid fever decreases when the level of development of a country increases (*i.e.*, controlled water sewage systems, pasteurization of milk and dairy products). Where these hygienic conditions are missing, the probability of fecal contamination of water and food remains high and so is the incidence of typhoid fever [29].

Ecology of Salmonellae and the Cycle of Salmonellosis

The principal habitat of *Salmonella* is the intestinal tract of humans and animals [28]. Salmonellae are constantly found in environmental samples, because they are excreted by humans, pets, farm animals, and wild life. Municipal sewage, agriculture pollution, and storm water runoff are the main sources of these pathogens in natural waters [1,32]. Salmonellae do not seem to multiply significantly in the natural environment, but they can survive several weeks in water and in soil if conditions of temperature, humidity, and pH are favorable [28].

Salmonellae isolated from environmental sources are predominantly non-Typhi or Paratyphi serovars. In a study carried out in Tunisia during 1994–2004, *S. enterica* subsp. *enterica* serovars Anatum, Enteritidis and Corvallis were the most common serotypes isolated from food. The great majority of the strains were isolated from poultry, red meat, milk and dairy products, vegetables and fruits. From environmental sources, 73% of the isolates were from tap water. Serovars Corvallis, Enteritidis, and Anatum were the commonest [31]. Arvanitidou *et al.* [32] reported a comparative study carried out in Rivers Aliakmon and Axios, in northern Greece, during a 1-year period, from May 2002 to April 2003. A total of 29 *Salmonella* species were recovered from the water samples. Many of the isolated *Salmonella* serovars were of non-human animal origin such as Mbantaka, Virchow, Hadar, Infantis and Senftenberg, commonly isolated from poultry farm.

Unlike cholera, humans infected with salmonellae can carry the bacteria in the gut without signs of disease. Infected humans can harbor the bacteria for considerable periods of time. About 5% of patients clinically cured from typhoid fever remain carriers for months or even years. These people can be chronic holders of the bacterium in the gut, and constitute the main reservoir of the bacteria in the environment [29].

The salmonellosis cycle in the environment can involve shellfish. Salmonellae survive sewage treatments if suitable germicides are not used in sewage processing. If effluent from the sewage plant passes into a coastal area, edible shellfish (mussels, oysters) can become contaminated. Shellfish concentrate bacteria as they filter several liters of water per hour. Ingestion by humans of these seafoods (uncooked or superficially cooked) may cause typhoid fever or other salmonellosis. Evidence of such a cycle has been obtained by the use of strain markers, including phage typing [29].

SHIGELLOSIS OR BACILLARY DYSENTERY

The Genus Shigella

Shigella are Gram-negative, non-sporeforming, non-motile, straight rod-like members of the family*Enterobacteriaceae*. Cells ferment sugars without gas production. Salicin, adonitol and myo-inositol are not fermented. Cells do not utilize citrate, malonate and acetate as sole carbon source and do not produce H_2S. Lysine is not decarboxylated. Cells are oxidase-negative and catalase-positive. Members of the genus have a complex antigenic pattern, and taxonomy is based on their somatic O antigens [1,33,34].

Table 5: Current taxonomy and nomenclature of the genus Shigella. Habitat and pathogenicity of species[a]

Species	Main serotypes	Habitat and pathogenicity
Shigella dysenteriae	15 serotypes.	Intestinal pathogens of humans and primates, causing bacillary dysentery. Humans are the primary reservoir. A long-term carrier state occurs in few cases.
Shigella flexneri	8 serotypes 9 subserotypes	*Shigella dysenteriae* serotype 1 causes more severe disease then other serotypes and produces a potent
Shigella boydii	19 serotypes	exotoxin (Shiga toxin). Large epidemics in developing countries are commonly caused by serotype 1. Diseases caused by other serotypes may be mild or severe.
Shigella sonnei	1 serotype	*Shigella sonnei* illness is usually milder than that caused by other *Shigella* species.

[a] Adapted from [34]. Nomenclature according to [9].

Characterization of the Disease

The incubation period is 1–4 days. The disease usually begins with fever, anorexia, fatigue and malaise. Patients display frequent bloody stools of small volume (sometimes grossly purulent) and abdominal cramps. Twelve to 36 hours later, diarrhea progresses to dysentery, blood, mucus and pus appearing in feces that decreases in volume (no more than 30 mL of fluid per kg per day) [34–36].

Although the molecular basis of shigellosis is complex, the initial step in pathogenesis is penetration of the colonic mucosa. The resulting focus of *Shigella* infection is characterized by degeneration of the epithelium and by an acute inflammatory colitis in the lamina propria. Ultimately, desquamation and ulceration of the mucosa cause leakage of blood, inflammatory elements, and mucus into the intestinal lumen. Under these conditions the absorption of water by the colon is inhibited and the volume of stool is dependent upon the ileocecal flow. As a result, the patient will pass frequent, scanty, dysenteric stools [37,38].

In order for *Shigella* to enter an epithelial cell, the bacterium must first adhere to its target cell. Generally, the bacterium is internalized via an endosome, which it subsequently lyses to gain access to the cytoplasm where multiplication occurs [37,38].

Virulence Factors

S. dysenteriae serotype 1 produces high levels of a cytotoxic Shiga toxin. *S. sonnei* and *S. flexneri* produce much lower amounts of this toxin. Shiga toxin binds to Galotl-4Galp (galabiose) glycolipid receptors and inhibits mammalian protein synthesis by cleaving the N-glycosidic bond at adenine 4324 in 28S rRNA. The toxic mechanism is identical to that of the plant toxin ricin, produced by *Ricinus communis*. Shigella also release a LPS endotoxin (O antigens), that cause an inflammatory response [37,38].

Shigella 180- to 230-kb plasmids encode genes essential for virulence, namely for: production of adhesins involved in the adherence of bacteria onto the surface of target epithelial cells; production of invasion plasmid antigens (Ipa) that have a direct role in the *Shigella* invasion process; transport or processing functions that ensure the correct surface expression of the Ipa proteins; induction of endocytic uptake of bacteria and disruption of endocytic vacuoles; regulation of plasmid-encoded virulence genes [37,38].

Shigella emerged from *E. coli* during evolution. The acquisition and evolution of the pathogenicity island which encodes all of the genes required for cell invasion and phagolysosomal lysis, permitted a major alteration in pathogenesis [37,38].

Risk Factors

Many studies have identified risk factors and protective effects for shigellosis incidence and fatality. Despite gradual improvements in water supply, shigellosis continues to be endemic among the disadvantaged populations living in the tropics, often among displaced populations following natural disasters and political crises. In Guatemala, young children, the elderly, and 15–44-year-old males were found to be most susceptible to *S. dysenteriae* serotype 1. In Sierra Leone, the attack rate was higher among children younger than 5 years of age than in the rest of the population. In rural Bangladesh, shigellosis was most common in children aged 1–2 years and in people 60 years or older. In Dhaka, Bangladesh, it was found that shigellosis mortality was most common in severely malnourished people of all ages, in children under 2 who were not being breastfed, and in all children under 1. In a 3-year study carried out in Matlab, Bangladesh, during 1992 to 1994, it was found that the incidence of *S. dysenteriae* serotype 1 and *S. flexneri* was highest in children under 2 followed by children from 2 to 5. The location of *S. dysenteriae* serotype 1 risk varies in time but *S. flexneri* risk areas were persistent in time. Neighborhoods near bazaars with many non-septic latrines were at highest risk for *S. dysenteriae* serotype 1. *S. flexneri* was most common in flood-controlled areas. It was concluded that *S. dysenteriae* serotype 1 risk was more related to hygiene and sanitation whereas *S. flexneri*was more related to the environment [35].

Shigellosis through the World

The total number of *Shigella* episodes that occur each year throughout the World is estimated to be 164.7 million, including 163.2 million cases in developing countries, 1.1 million of which result in death. Children under 5 account for 61% of all deaths attributable to shigellosis [35,36].

Shigella species are not uniformly distributed in the world. *S. dysenteriae* is usually found in densely populated areas of South America, Africa and Asia. Infections usually result in significant epidemic outbreaks. Serotype 1 has been distinguished by both its virulence and its ability to produce ravaging epidemics. It predominates in

India, Malaysia and Guatemala. Serotype 2 predominates in Yemen and Nigeria. *S. flexneri* is usually found in areas where endemic shigellosis occurs. *S. boydii* occurs sporadically, except in the Indian subcontinent where it was first identified. *S. sonnei* usually occurs in Western developed countries, such as France and USA [35,36].

Important epidemics were reported in the last decades: (1) in 1970 in Central America where 112,000 people were affected and 13,000 died; (2) in 1985, in Texas (USA), 5,000 people became infected after ingestion of contaminated lettuce; (3) in May–June 1994, domestic cases of *S. sonnei* infection were detected in several European countries, including Norway, Sweden, and the United Kingdom. Epidemiological evidence incriminated imported iceberg lettuce as the vehicle of transmission; (4) in 1996, in Paris, with 153 reported patients [33].

Ecology of Shigellae and the Cycle of Shigellosis

Shigella is typically an inhabitant of the intestinal tract of humans and other primates [1,33,34,36,39]. It is typically spread by fecal-contaminated drinking water or food, or by direct contact with an infected person. In water, shigellae can survive for at least six months at room temperature, and this high survival favors transmission through water. Flies have been implicated on the transmission of *Shigella* cells from human feces to foods. The hand is an important vehicle for transmission of shigellosis, since *S. dysenteriae* serotype 1 cells survives for up to one hour on a human's skin and a very small inoculum is required to unchain infection and disease. Indeed, studies on American volunteers experimentally infected with *Shigella* have shown that as few as one hundred *Shigella* cells given orally cause the disease in 25–50% of the cases. Resistance of *Shigella* to gastric juice certainly accounts, although not exclusively, for this high infectivity [36,40]. Asymptomatic and inappropriately-treated patients with shigellosis can harbor the bacteria in the gut and these appear to be the main reservoirs of the bacteria in the environment [41].

Recent reports on the ecology of shigellae have brought new elements for the understanding of the cycle of the disease. In environmental waters of regions with high numbers of shigellosis' cases, it has been found that, although numbers of cultivable cells were low, genetic elements such as plasmids and genetic fragments with bacteriophage origin, could be detected. Many of the genes that code for exotoxin production are precisely found in these genetic elements. These results suggest that the sudden rise of the number of virulent strains in the environment can result from the incorporation, by cells with reduced virulence, of this type of genetic elements present in the waters. If this hypothesis is confirmed, there is a certain similarity between the cholera and the shigellosis cycles in the environment. It remains to be elucidated if shigellae can also exist in environmental waters in a viable but non-culturable state, as vibrios [41].

PATHOGENIC *ESCHERICHIA COLI* STRAINS

E. coli strains isolated from intestinal diseases have been grouped into at least six different main groups, based on epidemiological evidence, phenotypic traits, clinical features of the disease and specific virulence factors. From these, enterotoxigenic

(ETEC, namely O148), enterohemorrhagic (EHEC, namely O157) and enteroinvasive serotypes (EIEC, namely O124) are of outstanding importance and can be transmitted through contaminated water [42,43].

Enterotoxigenic E. coli (ETEC) Strains

Enterotoxigenic *E. coli* (ETEC) serotypes can cause infantile gastroenteritis. The number of reports of their occurrence in developed countries is comparatively small, but it is an extremely important cause of diarrhea in the developing world, where there is no adequate clean water and poor sanitation. In developing countries, these strains are the most commonly isolated bacterial enteropathogen in children below 5 years of age, and account for several hundred million cases of diarrhea and several ten of thousand deaths each year [42–44].

Disease caused by ETEC follows ingestion of contaminated food or water and is characterized by profuse watery diarrhea lasting for several days that often leads to dehydration and malnutrition in young children [42–44]. ETEC also are the most common cause of "travelers' diarrhea" that affects individuals from industrialized countries travelling to developing regions of the World [42–44].

Enterohemorrhagic E. coli (EHEC) Strains

Reported outbreaks had been associated mainly with the consumption of contaminated foods, such as raw or undercooked ground meat products and raw milk. The primary reservoir of this bacterium has been found to be healthy cattle [42,45,46].

E. coli serotype O157:H7 causes abdominal pain, bloody diarrhea, and hemolytic uremic syndrome. This bacterium produces Shiga-like toxins. The incubation period is 3–4 days, and the symptoms occur for 7–10 days. It is estimated that 2–7% of *E. coli* O157:H7 infections result in acute renal failure [42,45,46].

Although *E. coli* O157:H7 is not usually a concern in treated drinking water, outbreaks involving consumption of drinking water contaminated with human sewage or cattle feces have been documented. An increasing number of outbreaks are associated with the consumption of fruits and vegetables (sprouts, lettuce, coleslaw, salad) contaminated with feces from domestic or wild animals at some stage during cultivation or handling. EHEC has also been isolated from bodies of water (ponds, streams), wells and water troughs, and has been found to survive for months in manure and water-trough sediments [45,46].

Person-to-person contact is an important mode of transmission through the oral-fecal route. An asymptomatic carrier state has been reported, where individuals show no clinical signs of disease but are capable of infecting others [45,46].

Enteroinvasive E. coli (EIEC) Strains

Enteroinvasive *E. coli* (EIEC) behave in many respects like shigellae. They are capable of invading and multiplying in the intestinal epithelial cells of the distal large bowel in humans. The illness is characterized by abdominal cramps, diarrhea, vomiting, fever, chills, a generalized malaise, and the appearance of blood and mucus in the stools of infected individuals. [42,43,47].

EIEC strains were isolated, for instance, from 28 subjects in the Jesreel district of Israel during a peak period for dysentery. An investigation in Croatia showed that *E. coli* O124 could frequently be isolated from cases of gastroenteritis, enterocolitis, and dysentery. The dysentery was more common among the older age groups, while the two other types of disease occurred equally in all age groups. A 1985 survey was carried out in Bankok, Thailand in which 410 children with diarrhea and an equal number of control children without diarrhea were examined for the presence of strains of *Shigella*, EIEC, and other pathogens. It was found that 17 of the children with diarrhea and six without yielded EIEC [42,43].

Any food contaminated with human feces from an ill individual, either directly or via contaminated water, could cause disease in others. Outbreaks have been associated with hamburger meat and unpasteurized milk [47].

EMERGING WATERBORNE BACTERIAL PATHOGENS

The emerging pathogenic bacteria of concern outlined here have the potential to be spread through drinking water, but they do not correlate with the presence of *E. coli* or with other commonly used drinking water quality indicators, such as coliform bacteria. In most cases, there are no satisfactory microbiological indicators of their presence. More studies are needed in order to understand the real significance and dimension of the diseases caused by water contaminated with these bacteria, and the ecology of these pathogens [45].

Mycobacterium Avium Complex (Mac)

The *Mycobacterium avium* complex (Mac) consists of 28 serovars of two distinct species: *Mycobacterium avium* and *Mycobacterium intracellulare.* The importance of Mac organisms was recognized with the discovery of disseminated infection in immunocompromised people, particularly people with HIV and AIDS. Members of MAC are considered opportunistic human pathogens [45,48].

Mac organisms have been identified in a broad range of environmental sources, including marine waters, rivers, lakes, streams, ponds, springs, soil, piped water supplies, plants, and house dust. Mac organisms have been isolated from natural water and drinking water distribution systems in the USA [45,49,50].

The ubiquitous nature of Mac organisms results from their ability to survive and grow under varied conditions. Mac organisms can proliferate in water at temperatures up to 51 °C and can grow in natural waters over a wide pH range [45]. These mycobacteria are highly resistant to chlorine and the other chemical disinfectants used for the treatment of drinking-water. Standard drinking-water treatments will not eliminate Mac organisms but, if operating satisfactorily, will significantly reduce the numbers that may be present in the source water to a level that represents a negligible risk to the general population. The entryway of these mycobacteria in distribution systems is through leaks. Growth of Mac organisms in biofilms is probably important for their continuous presence in distribution systems. Slow growing mycobacteria can be found at densities greater than 4,000 per cm^2 in the surface biofilm, creating a potentially high level of exposure [48].

The symptoms encountered with Mac infections result from colonization of either the respiratory or the gastrointestinal tract, with possible dissemination to other locations in the body. Exposure to Mac organisms may occur through the consumption of contaminated food, the inhalation of air with contaminated soil particles, or contact with or ingestion, aspiration, or aerosolization of potable water containing the organisms [45].

With respect to water supplies, infection with *M. avium* and *M. intracellulare* has been well documented. Unlike gastrointestinal pathogens, where *E. coli* can be used to indicate their potential presence, no suitable indicators have been identified to signal increasing concentrations of Mac organisms in water systems [45].

Helicobacter Pylori

Helicobacter pylori has been cited as a major etiologic agent for gastritis and has been implicated in the pathogenesis of peptic and duodenal ulcer disease and gastric carcinoma. However, most individuals that are infected by this pathogen remain asymptomatic [45].

Using culture-based methods, *H. pylori* has not been isolated from environmental sources, including water [45,51]. On the contrary, molecular methods have been successful in detecting this pathogen. Fluorescence *in situ* hybridization has been successfully used to detect this pathogen in drinking water distribution systems and other water bodies. Polymerase chain reaction has also been used to detect the presence of *H. pylori*DNA in drinking water, especially associated with biofilms [45,51,52]. In drinking-water biofilms, *H. pylori*cells rapidly lose culturability, entering a viable but non-culturable state. In these biofilms, cells can persist for more than one month, with densities exceeding 10^6 cells per cm^2 [51].

How the organism is transmitted is still not fully understood. However, the fact that it has been recovered from saliva, dental plaques, the stomach, and fecal samples strongly indicates oral-oral or fecal-oral transmission. Water and food appear to be of lesser direct importance, but they can still play a significant role in situations with improper sanitation and hygiene [45].

Aeromonas Hydrophyla

In recent years, *A. hydrophila* has gained public health recognition as an opportunistic pathogen. It has been implicated as a potential agent of gastroenteritis, septicemia, meningitis, and wound infections. It can play a significant role in intestinal disorders in children under five years old, the elderly, and immunosuppressed people. [45,53,54].

Aeromonas hydrophila are Gram-negative, non-sporeforming, rod-shaped, facultative anaerobic bacilli belonging to the family *Aeromonadaceae*. Although *A. hydrophila* is usually the dominant species, other aeromonads, such as *A. caviae* and

A. sobria, have also been isolated from human feces and from water sources [45,54].

Aeromonas species, including *A. hydrophila*, are ubiquitous in the environment. It is frequently isolated from food, drinking water, and aquatic environments [45,53,54]. In clean rivers and lakes, concentrations of*Aeromonas* spp. are usually around 10^2 colony-forming units (CFU)/mL. Groundwaters generally contain less than 1 CFU/mL. Drinking water immediately leaving the treatment plant has been found to contain between 0 and 10^2 CFU/mL. Drinking water in distribution systems can display higher *Aeromonas*concentrations, due to the growth in biofilms [45,55]. *Aeromonas* spp. have been found to grow between 5 °C and 45 °C [44,54]. *A. hydrophila* is resistant to standard chlorine treatments, probably surviving inside biofilms [56].

The common routes of infection suggested for *Aeromonas* are the ingestion of contaminated water or food or contact of the organism with a break in the skin. Drinking or natural mineral water can be a possible source of contamination for humans. No person-to-person transmission has been reported [45,54].

MICROBIOLOGICAL WATER ANALYSIS

The Rationale of the Use of Fecal Indicator Bacteria

The most important bacterial gastrointestinal diseases transmitted through water are cholera, salmonellosis and shigellosis. These diseases are mainly transmitted through water (and food) contaminated with feces of patients. Drinking water can be contaminated with these pathogenic bacteria, and this is an issue of great concern. However, the presence of pathogenic bacteria in water is sporadic and erratic, levels are low, and the isolation and culture of these bacteria is not straightforward. For these reasons, routine water microbiological analysis does not include the detection of pathogenic bacteria. However, safe water demands that water is free from pathogenic bacteria [57].

The conciliation of the two needs was met by the discovery and testing of indicator bacteria. Water contaminated with pathogenic species also has the normal inhabitants of the human intestine. A good bacterial indicator of fecal pollution should fulfill the following criteria: (1) exist in high numbers in the human intestine and feces; (2) not be pathogenic to humans; (3) easily, reliably and cheaply detectable in environmental waters. Additionally, the following requisites should be met if possible: (4) does not multiply outside the enteric environment; (5) in environmental waters, the indicator should exist in greater numbers than eventual pathogenic bacteria; (6) the indicators should have a similar die-off behavior as the pathogens; (7) if human fecal pollution is to be separated from animal pollution, the indicator should not be very common in the intestine of farm and domestic animals [1,4,6,57,58]. The usefulness of indicator bacteria in predicting the presence of pathogens was well illustrated in many studies, namely by Wilkes *et al.* [59].

The Composition of Human and Animal Feces

Microbiological analysis of the human feces was important in order to structure and validate the use of fecal indicator bacteria in environmental waters. Bacteria present in feces are naturally derived from the microbiota of the human gastrointestinal tract.

Although bacteria are distributed throughout the human gastrointestinal tract, the major concentration of microbes and metabolic activity can be found in the large intestine. The upper bowel (stomach, duodenum, and jejunum) has a sparse microbiota with up to 10^5 CFU/ml of contents. From the ileum on, bacterial concentrations gradually increase reaching in the colon 10^{10} to 10^{11} CFU/g [60].

It has been estimated that at least 500–1,000 different microbial species exist in the human gastrointestinal microbiota, although on a quantitative basis 10–20 genera usually predominate (Table 6). The total number of microbial genes in the human gastrointestinal tract has been estimated as 2–4 million. This represents an enormous metabolic potential which is far greater than that possessed by the human host [60,64].

Table 6: Total viable count in feces of healthy humans (children, adults and elderly)[a]

Microbial group	Log_{10} CFU/g feces
Bacteroides	11.3*
Eubacterium	10.7*
Bifidobacterium	10.2*
Ruminococcus	10.2*
Peptostreptococcus	10.1*
Peptococcus	10.0*
Clostridium	9.8*
Lactobacillus	9.6*
Propionobacterium	9.4*
Actinomyces	9.2*
Methanobrevibacter	8.8*
Desulphovibrio	8.4*
Fusobacterium	8.4*
Enterococci	3.5–7.2**
Enterobacteriaceae	5.9–8.0**
Escherichia coli	7.5–7.7**
Citrobacter	3.3**
Klebsiella	2.4**
Yeasts	1.0–2.5**

[a] Adapted from [61-63]** and [64]*. * Values expressed as dry weight.
** Values expressed as wet weight.

The composition of feces from an individual is stable at genus level, but the species composition can vary markedly from day to day. The relative proportion of intestinal bacterial groups can vary between individuals [60,64].

The microflora of the human gastrointestinal tract is dominated by obligate anaerobes, which are ca. 10^3more abundant than facultative anaerobes. The main anaerobic genera are *Bacteroides*, *Eubacterium* and Bifidobacteria. These organisms account for ca. 90% of the cultivable human fecal bacteria. *Bacteroides*(mainly *B. thetaiotaomicron* and *B. vulgatus*) are the most abundant organism in the human feces and account for 20–30% of cultivable bacteria. The most abundant facultative anaerobes are Enterococci and*Enterobacteriaceae*. The main *Enterobacteriaceae* genera are *Escherichia*, *Citrobacter*, *Klebsiella*, *Proteus*and *Enterobacter*. *Citrobacter* and *Klebsiella* are present in most individuals although in low numbers.*Proteus* and

Enterobacter are only present in a minority of humans [64].

A variety of molecular techniques have been used to study the microbial composition of the human gastrointestinal tract. Results yielded by these studies have shown that many microbes detected by molecular techniques are not isolable by conventional culture-based methods. The presence of high proportions of bifidobacteria detected by culture-based methods is not supported by the results of molecular-based studies. However, the results of molecular-based approaches support many of the findings derived from culture-based methods: the dominance of the obligate anaerobes over facultative anaerobes; the presence of high counts of*Bacteroides*, *Clostridium* and *Eubacterium* [64].

Anaerobic bacteria such as *Bacteroides* and *Eubacterium* are not easily cultured by conventional techniques since require incubation chambers with nitrogen atmosphere. *Bifidobacterium* and *Lactobacillus* tolerate some oxygen but are fastidious bacteria growing very slowly in culture media. Therefore, these four genera are not adequate to be used as indicators of fecal pollution (the introduction of molecular techniques may improve the situation). *Citrobacter*, *Klebsiella* and *Enterobacter* are present in low numbers in the human intestine and are widespread in environmental waters, and therefore are also not suitable as indicators of fecal pollution. *Clostridium*, *Streptococcus* and *Escherichia* do not suffer from these drawbacks. Therefore, their suitability as fecal indicators has been tested since several decades.

Fecal Bacteria in Their Hosts and in the Environment

Bacteroides

The traditional genus *Bacteroides* included Gram-negative, non-sporeforming, anaerobic pleiomorphic rods. Many species have been transferred to other genera—*Mitsuokella, Porphyromonas, Prevotella, Ruminobacter*. Bacteroides are the most abundant bacteria in human feces. In animal feces, on the contrary,*Bacteroides* are present at low numbers. Although anaerobic, *Bacteroides* are among the most tolerant to oxygen of all anaerobic human gastrointestinal species. *B. thetaiotaomicron* is one of the most abundant species in the lower regions of the human gastrointestinal tract. *Bacteroides* have a high pathogenic potential and account for approximately two-thirds of all anaerobes isolated from clinical specimens. The most frequently isolated species has been *B. fragilis*. The survival of *Bacteroides* in environmental waters is usually much lower than the survival of coliforms [64,65].

Eubacterium

The traditional genus *Eubacterium* included anaerobic non-sporeforming Gram-positive rods. Some species have been transferred to other genera—*Actinobaculum*, *Atopobium*, *Collinsella*, *Dorea*, *Eggerthella*,*Mogibacterium*, *Pseudoramibacter* and *Slackia*. Cells are not very aerotolerant. Species isolated from the human gastrointestinal tract include: *E. barkeri*, E. *biforme*, *E. contortum*, *E. cylindrioides*, *E. hadrum*, *E. limosum*, *E. moniliforme, E. rectal* and *E. ventricosum* [64].

Bifidobacterium

Bifidobacteria are Gram-positive, non-sporeforming, pleiomorphic rods. Bifidobacteria are anaerobic (some species tolerate oxygen in the presence of carbon dioxide) or facultative anaerobic. The optimum growth temperature is 35–39 °C. The genus *Bifidobacterium* contains ca. 25 species, most of which have been detected in the human gastrointestinal tract [64–66].

Bifidobacteria are present in high numbers in the feces of humans and some animals. Several*Bifidobacterium* species are specific either for humans or for animals. *B. cuniculi* and *B. magnum* have only been found in rabbit fecal samples, *B. gallinarum* and *B. pullorum* only in the intestine of chickens and *B. suis* only in piglet feces. In human feces, the species composition changes with the age of the individual. In the intestine of infants *B. breve* and *B. longum* generally predominate. In the adult, *B. adolescentis*, *B. catenulatum*, *B. pseudocatenulatum* and *B. longum* are the dominant species. In both human and animal feces, bifibobacteria are always much more abundant than coliforms [64–66].

Bifidobacteria have been found in sewage and polluted environmental waters, but appears to be absent from unpolluted or pristine environments such as springs and unpolluted soil. This results from the fact that upon introduction into the environment, bifidobacteria decrease appreciably in numbers, probably due to their stringent growth requirements. Bifidobacteria grow poorly below 30 °C and have rigorous nutrient requirements. Reports on the survival of bifidobacteria in environmental waters indicate that their survival is lower than that of coliforms [64–66].

The presence of bifidobacteria in the environment is therefore considered an indicator of fecal contamination. Since some species are specific for humans and animals, the identification of *Bifidobacterium* species present in the polluted water could, in principle, provide information on the origin of fecal pollution [64–66].

A study carried out in a highly contaminated stream near Bologna, Italy, revealed that *B. adolescentis*, *B. catenulatum*, *B. longum*, *B. pseudocatenulatum* and *B. thermophilum* were the most representative species, whereas *B. angulatum*, *B. animalis* subsp. *animalis (B. animalis)*, *B. breve*, *B. choerinum*, *B. minimum*, *B. pseudolongum* subsp. *globosum* (*B. globosum*) and *B. subtile* occurred only in low numbers [66].

Bifidobacteria are the less studied of all fecal bacteria, due to the technical difficulties in their isolation and cultivation. Other Gram-positive bacteria, such as *Streptococcus* and *Lactobacillus*, which may occur in higher numbers than bifidobacteria, can inhibit their growth. Although selective media has been designed for the isolation of bifidobacteria from environmental waters, the outcome is still unsatisfactory, with appreciable numbers of false positives and low recovery percentages [64–66].

Clostridia

The genus *Clostridium* is one of the largest genera of the prokaryotes containing 168 validly published species. From these, 77 (including *C. perfringens*) are considered to belong to a united group—*Clostridium sensu stricto* [64,67,68].

Clostridia are Gram-positive rods, forming endospores. Most of the clostridial species are motile with peritrichous flagellation. Cells are catalase-negative and do not carry out a dissimilatory sulphate reduction. Clostridia usually produce mixtures of organic acids and alcohols from carbohydrates and proteins. Many species are saccharolytic and proteolytic. Some species fix atmospheric dinitrogen [64,67,68].

The genus *Clostridium* includes psychrophilic, mesophilic, and thermophilic species. The major role of these organisms in nature is in the degradation of organic material to acids, alcohols, CO_2, H_2, and minerals. Frequently, a butyric acid smell is associated with the proliferation of clostridia. The ability to form spores that resist dryness, heat, and aerobic conditions makes the clostridia ubiquitous [64,67,68].

Most species are obligate anaerobic, although tolerance to oxygen occurs. Oxygen sensitivity restricts the habitat of the clostridia to anaerobic areas or areas with low oxygen tensions. Growing and dividing clostridia will, therefore, not be found in air saturated surface layers of lakes and rivers or on the surface of organic material and soil. Clostridial spores, however, are present with high probability in these environments, and will germinate when oxygen is exhausted and when appropriate nutrients are present [64,67,68].

C. perfringens ferment lactose, sucrose and inositol with the production of gas, produce a stormy clot fermentation with milk, reduce nitrate, hydrolyze gelatin and produce lecithinase and acid phosphatase. The species is divided into five types, A to E, on the basis of production of major lethal toxins [68,69].

C. perfringens appears to be a universal component of the human and animal intestine, since has been isolated from the intestinal contents of every animal that has been studied. Humans carry *C. perfringens* as part of the normal endogenous flora. The main site of carriage is the distal gastrointestinal tract. The principal habitats of type A are the soil and the intestines of humans, animals, and birds. Types B, C, D, and E appears to be obligate parasites of animals and occasionally are found in humans [68,69].

Clostridium perfringens is the most frequently isolated *Clostridium* in clinical microbiology laboratories, although it seldom causes serious infections. *C. perfringens* is isolated from infections in humans and the organism most commonly found in gas gangrene in humans. *C. perfringens* is most commonly isolated from infections derived from the colonic flora, namely peritonitis or abdominal abscess [68,69].

This organism is a common cause of food poisoning due to the formation of the enterotoxin in the intestine.*C. perfringens* food poisoning is seldom fatal, being marked by diarrhea and nausea, with no vomiting and no fever [68,69].

Sources yielding *C. perfringens* include soil and marine sediment samples worldwide, clothing, raw milk, cheese, semi-preserved meat products, and venison. Like *E. coli*, *C. perfringens* does not multiply in most water environments and is a highly specific indicator of fecal pollution. Berzirtzoglou *et al.* [70] reported a comparative study on the occurrence of vegetative cells and spores of *Clostridium perfringens* in a polluted station of the lake Pamvotis, in rural North-West Greece. The numbers of *C. perfringens* varied according to the water depth. Sporulated forms were found in all sampling sites with the exception of the surface sampling.

Lactobacillus

Lactobacilli are non-sporeforming Gram-positive long rods. There are more than thirty species in the genus. Most are microaerophillic, although some are obligate anaerobes. Cells are catalase-negative and obtain their energy by the fermentation of sugars, producing a variety of acids, alcohol and carbon dioxide. Lactobacilli have complex nutritional requirements and in agarized media may need the supplementation with aminoacids, peptides, fatty-acid esters, salts, nucleic acid derivatives and vitamins. Lactobacilli very rarely cause infections in humans [64].

Enterococci

Enterococci are Gram-positive, non-sporeforming, catalase-negative ovoid cells. Cells occur singly, in pairs or short chains. Optimal growth for most species is 35–37 °C. Some will grow at 42–45 °C and at 10 °C. Growth requires complex nutrients but is usually abundant on commonly used bacteriological media. Cells are resistant to 40% bile, 0.4% azide, 6.5% sodium chloride, have β-glucosidase and hydrolyze esculin. The enterococci are facultative anaerobic but prefer anaerobic conditions [64,71].

The genus was separated from *Streptococcus* in the 1980s. Enterococci form relatively distinct groups. Members of such groups exhibit similar phenotypic characteristics and species delimitation can be difficult. The *E. faecalis* group contains, among others, *E. faecalis*. The *E. avium* group contains, among others, *E. avium*. The *E. faecium* group contains, among others, *E. faecium*, *E. durans* and *E. hirae*. The *E. gallinarum*group contains, among others, *E. gallinarum* [64,71].

Most species are part of the intestinal flora of mammals, reptiles, birds, and other animals. In the human digestive tract, *E. faecalis* is the prevailing species, although in particular situations, *E. faecium* may predominate. In poultry, *E. cecorum*, *E. durans*, *E. faecalis*, *E. faecium* and *E. hirae* and dominate the intestinal flora [64,71].

Enterococci have been increasingly isolated from a variety of nosocomial and other infections, mainly from the urinary tract and wound infections, bacteremias, and endocarditis [64,71].

Although enterococci are considered only a temporary part of the microflora of plants, in optimal conditions, cells can proliferate on their surfaces. *E. casseliflavus*, *E. faecalis*, *E. faecium*, *E. hirae*, *E. mundtii* and *E. sulfureus* have been isolated from plants. They are generally isolated more often from flowers than from buds or leaves [64,71].

Enterococci are naturally present in many kinds of foods, especially those of animal origin such as milk and milk products, meat and fermented sausages. Enterococci are usually considered secondary contaminants of food, although they often play a positive role in ripening and aroma development of some types of cheeses [64,71]. Although soil is not a natural habitat for enterococci, cells can be found in this habitat due to the transport by rain [64,71].

Environmental waters are not a natural habitat for enterococci and their presence in this milieu is considered the result of fecal pollution. The most common species found in environmental waters are *E. durans, E. faecalis, E. faecium* and *E. hirae*, and less commonly, *E. avium*, *E. cecorum*, *E. columbae* and *E. gallinarum*. However, pristine waters in Finland have been reported to contain *E. casseliflavus* [64,71].

In environmental samples (compost, sewage effluent, harbor sediments, brackish water and swimming pool water), Pinto *et al.* [72] reported the isolation of *E. casseliflavus*, *E. durans*, *E. faecalis*, *E. faecium*, *E. gallinarum* and *E. hirae. E. durans*, *E. faecium* and *E. hirae* were isolated from all sources except from harbor sediments. *E. raffinosus* was only isolated from compost and swimming pool water. *E. faecalis* and *E. faecium* accounted for the vast majority of enterococcal strains.

Escherichia

Escherichia, a member of *Enterobacteriaceae*, are oxidase-negative catalase-positive straight rods that ferment lactose. Cells are positive in the Methyl-Red test, but negative in the Voges-Proskauer assay. Cells do not use citrate, do not produce H_2S or lipase, and do not hydrolyze urea [73]. *E. coli* is a natural and essential part of the bacterial flora in the gut of humans and animals. Most *E. coli* strains are nonpathogenic and reside harmlessly in the colon. However, certain serotypes do play a role in intestinal and extra-intestinal diseases, such as urinary tract infections [43]. In a study of the enteric bacteria present in the feces of Australian mammals, Gordon and FitzGibbon [74] reported that *E. coli* was the commonest species, being isolated from nearly half of the species studied.

Citrobacter

Citrobacter, a member of *Enterobacteriaceae*, are motile straight rods. Cells are oxidase-negative, catalase-positive and positive in the Methyl-Red test. Cells use citrate, are negative in the Voges-Proskauer test and do not decarboxylate lysine [73].

In a study of the enteric bacteria present in the feces of Australian mammals, Gordon and FitzGibbon [74] reported the isolation of *C. amalonaticus, C. freundii* and *C. koseri* (*C. diversus*). *Citrobacter* species can be isolated from different clinical sites. In particular, *C. freundii* is an intestinal inhabitant of humans that may sometimes have—or acquire—the ability to produce an enterotoxin and thus become an intestinal pathogen.*Citrobacter* is reported to occur in environments such as water, sewage, soil and food [75,76].

Klebsiella and Raoultella

Klebsiella and *Raoultella* are *Enterobacteriaceae*, oxidase-negative catalase-positive non-motile straight rods, surrounded by a capsule. Cells decarboxylate lysine, but are ornithine and arginine dihydrolase negative. Cells grow on KCN, do not produce H_2S and ferment most carbohydrates [73].

In humans, *K. pneumoniae* is present as commensal in the nasopharynx and in the intestinal tract. *Klebsiella*spp. can cause human diseases, ranging from asymptomatic colonization of the intestinal, urinary, or respiratory tract to fatal septicemia. *Klebsiella* are mostly considered nosocomial pathogens. *K. pneumoniae*and *Enterobacter aerogenes* (*K. mobilis*) are most frequently involved, although *K. oxytoca* and *R. planticola*, and rarely *R. terrigena*, can be found. In the hospital, the principal reservoir of *K. pneumoniae* is the gastrointestinal tract of patients. The principal vectors are the hands of personnel [77,78]. In a study of the enteric bacteria present in the feces of Australian mammals, Gordon and FitzGibbon [74] reported the isolation of *K. pneumoniae* and *K. oxytoca*.

Klebsiellae are ubitiquous in the environment. They have been found in a variety of environmental situations, such as soil, vegetation, or water, and they influence many biochemical and geochemical processes. They have been recovered from aquatic environments receiving industrial wastewaters, plant products, fresh vegetables, food with a high content of sugars and acids, frozen orange juice concentrate, sugarcane wastes, living trees, and plants and plant byproducts. They are commonly associated with wood, sawdust, and waters receiving industrial effluents from pulp and paper mills and textile finishing plants (see below). *Klebsiella*have been isolated from the root surfaces of various plants. *K. pneumoniae*, *K. oxytoca*, and *R. planticola* are all capable of fixing dinitrogen [77,78].

Enterobacter

Enterobacter a member of *Enterobacteriaceae*, are motile straight rods. Cells are positive in the Voges-Proskauer test VP and in Simmons citrate agar. Cells do not decarboxylate lysine, but are ornithine positive. Malonate is usually utilized and gelatin is slowly liquefied. Cells do not produce H_2S, deoxyribonuclease and lipase [73].

In a study of the enteric bacteria present in the feces of Australian mammals, Gordon and FitzGibbon [74] reported the isolation of *Enterobacter cloacae* subsp. *cloacae* (*E. cloacae*), *E. cancerogenus* (*E. taylorae*) and *E. aerogenes* (*Klebsiella mobilis*).

Before the widespread use of antibiotics, *Enterobacter* species were rarely found as pathogens, but these organisms are now increasingly encountered, causing nosocomial infections such as urinary tract infections and bacteremia. In addition, they occasionally cause community-acquired infections [79,80].

In the USA, the Surveillance and Control of Pathogens of Epidemiological Importance project analyzed 24,179 nosocomial bloodstream infections, from 1995–2002. *Enterobacter* species were the second most common gram-negative organism, behind *Pseudomonas aeruginosa*. Both bacteria were reported to each represent 4.7% of bloodstream infections in intensive care units. *Enterobacter* species represented 3.1% of bloodstream infections in non-intensive care units. Of nearly 75,000 gram-negative organisms collected from intensive care units' patients in the USA, between 1993 and 2004, *Enterobacter* species comprised 13.5% of the isolates. Multidrug resistance increased over time, especially in infections caused by *E. cloacae* [81].

In the USA, the National Healthcare Safety Network reported a study on healthcare-associated infections between 2006 and 2007. They found *Enterobacter* species to be the eighth most common cause of healthcare-associated infections (5% of all infections) and the fourth most common gram-negative cause of these infections [82].

Enterobacter cloacae subsp. *cloacae* (*E. cloacae*) occurs in the intestinal tracts of humans and animals, in hospital environments, the skin, in water, sewage, soil, meat. Nitrogen-fixing strains have been isolated from the roots of rice plants. *E. amnigenus* has been mostly isolated from water, but some strains were isolated from clinical specimens from the respiratory tract, wounds and feces. *E. asburiae* strains were isolated from clinical specimens, mostly urine, respiratory tract, feces, wounds, and blood [79,80].

Origin of the Use of Fecal Indicator Bacteria

Historically, the design and use of indicators of fecal pollution comes from the end of the 19th to beginning of the 20th century. In 1880, von Fritsch described *Klebsiella pneumoniae* and *K. pneumoniae* subsp.*rhinoscleromatis* (*Klebsiella rhinoscleromatis*) as micro-organisms characteristically found in human feces [83]. In 1885, Escherich described several microorganisms in the feces of newborn and suckling babies. This included a motile, rod-shaped microorganism that caused milk to clot, which was named "*Bacterium coli commune*". He observed that within a few weeks after birth, this bacterium became the dominant organism in the infant colon [6]. Also in 1885, Percy and Grace Frankland started the first routine bacteriological examination of water in London, using Robert Koch's solid gelatin media to count bacteria [83]. In 1891, Percy and Grace Frankland came up with the concept that organisms characteristic of sewage must be identified to provide evidence of potentially dangerous pollution [83]. In 1892, Schardinger proposed that since "*Bacterium coli*" was a characteristic component of the fecal flora, its presence in water could be taken as an indication of the presence of fecal pollution and therefore of the potential presence of enteric pathogens [6]. Soon after the description of "*Bacterium coli*", other bacteria were isolated from stools and water—*Klebsiella* in 1882 and *Enterobacter* in 1890 [6]. By 1893, the "Wurtz method" of enumerating "*Bacterium coli*", by direct plating water samples on litmus lactose agar, was being used by sanitary bacteriologists. This was based on the concept of acid and gas production (detected by the Durham tube) from lactose as a diagnostic feature [6]. In 1905, MacConkey described his now famous MacConkey's broth, which was diagnostic for lactose-fermenting bacteria tolerant of bile salts. Coliforms were already considered to be a heterogeneous group of organisms, many of which were not of fecal origin. The origins of the critical observation that "*Bacterium coli*" was largely fecal in origin while other coliforms were not, could be claimed by Winslow and Walker in 1907 [83].

Various classification schemes for coliforms have emerged. The earliest were those of MacConkey in 1909, which recognized 128 different coliform types, while Bergey and Deehan in 1908, identified 256. By the early 1920s, differentiation of coliforms had come to a series of correlations that suggested that indole production, gelatin liquefaction, sucrose fermentation and the Voges–Proskauer reaction were among the more important tests for determining fecal contamination. These developments culminated in the IMViC (Indole, Methyl Red, Voges–Proskauer and Citrate) tests for the differentiation of so-called fecal coliforms, soil coliforms and intermediates [83].

Fecal Indicator Bacteria

Coliforms

Total coliforms are Gram-negative, oxidase-negative, non-sporeforming rods, that ferment lactose with gas production at 35–37 °C, after 48h, in a medium with bile salts and detergents [1,4,6,57,84]. When the test of coliforms is carried out with environmental waters, several species of the four *Enterobacteriaceae* genera*Escherichia*, *Klebsiella*, *Enterobacter* and *Citrobacter* give positive results and therefore are coliforms according to this definition. However, the environmental significance of these four genera is very disparate as discussed in the present text. Therefore, total coliform counts are not necessarily a measure of fecal pollution and indeed can have no relation with this cause [1,4,6,84].

Fecal coliforms (or thermotolerant coliforms) are traditionally defined as coliforms that ferment lactose at 44.5 °C in a medium with bile salts [1,4,57,84]. The range of species detected by the experimental procedure is much lower than that of total coliforms. With environmental polluted waters, only *E. coli*, and *K. oxytoca* and*K. pneumoniae* gave positive results in the test [85].

Traditional tests for total and fecal coliforms are carried out either by the multiple-tube fermentation technique or by filtration through membrane. The multiple-tube fermentation technique is used for medium or highly contaminated waters, and the filtration through membrane for low or very low contaminated waters. Filtration through membrane is a very sensitive technique since can detect one (culturable) cell in 500 or even 1,000 mL of water. However, both methods take several days to complete and do not detect viable but non-culturable bacteria [3,57,86]. These limitations stimulate the discovery of alternative methods, faster and, if possible, less prone to false negative results such as those caused by the viable but non-culturable bacteria.

The detection of β-D-galactosidase activity (at 37 °C) is usually a good marker for total coliforms in environmental waters, since most of these bacteria display this enzymatic activity [1,3,57,87–91]. Most*Escherichia*, *Citrobacter*, *Enterobacter*, *Klebsiella* and *Raoultella* strains have galactosidase. *Hafnia*,*Serratia* and *Yersinia* also possess this enzymatic activity. Most *Proteus*, *Salmonella* and *Edwardsiella* strains do not display β-galactosidase [92–95]. Ca. 10% of the coliform strains isolated from the environment do not have an active formic hydrogenolyase (cleaves formate with the formation of CO_2) and therefore do not produce gas being undetected by the traditional techniques but are detected by the assay of β-galactosidase activity [57,96,97].

β-galactosidase cleaves lactose in glucose and galactose, and can be detected by using colored or fluorescent markers that change color after enzyme action, such as XGAL (5-bromo-4-chloro-3-indol-β-galactopyranoside) and ONPG (O-nitrophenyl-β-D-galactopyranoside) or MUGAL (4-methylumbelliferyl-β-D-galactopyranoside), respectively [57,96,97].

In environmental waters, the presence of *Aeromonas* or *Vibrio cholerae* can be a source of false positives in the β-D-galactosidase assay, since these bacteria have galactosidase, but are not coliforms [93,95,96,98]. Additionally, in particular environments, such as estuaries, β-galactosidase activity can overestimate total coliform count due to UV-stimulated enzymatic activity in certain bacteria such as *E. coli* [86].

The detection of β-D-glucuronidase activity (at 44.5 °C) is, generally, a good marker for fecal coliforms in environmental polluted waters and very specific for *E. coli* [13,85,87–91,97,99–101]. In Gram-negative bacteria, this enzymatic activity if found in most *E. coli* strains and in some *Salmonella* and *Shigella* strains [92–95,97,102]. *Aeromonas*, *Citrobacter*, *Enterobacter*, non-*coli Escherichia*, *Hafnia*, *Klebsiella*, *Proteus*,*Serratia*, *Vibrio*, *Yersinia*, and most *Salmonella* strains do not display β-glucuronidase activity [93–95,101,102].

β-D-glucuronidase activity can be detected by using colored or fluorescent markers that change color after enzyme action, such as XGLUC (5-bromo-4-chloro-3-indoxyl-β-d-glucuronide), IBDG (indoxil-β-glucuronide), and MUGLU (4-methylumbelliferyl-β-d-glucuronide), respectively [57,97,100].

The presence of this enzyme in some strains of *Bacteroides*, *Flavobacterium*, *Staphylococcus*, *Streptococcus*, in anaerobic corynebacteria and *Clostridium*, has also been reported [93,95–97,102]. β-D-glucuronidase activity in fecal bacteria other then *E. coli* (*Bacteroides*, bifidobacteria, clostridia, enterococci and*Lactobacillus*) is very limited [61]. Although all these glucuronidase positive bacteria could lead to false positive detections in the fecal coliform test, experimental results for environmental polluted waters indicate a significant correlation between fecal coliform detection using conventional techniques and the glucuronidase assay, suggesting that false positives are not significant [85,96].

The detection of total coliforms and fecal coliforms by enzymatic methods are much less time consuming than traditional techniques. With fluorescent markers and the use of a spectrofluorimeter the detection of coliforms can be performed in minutes [57,101]. However, in very low contaminated waters, enzymatic methods might not be able to detect coliform cells. Moreover, on-line monitoring of glucuronidase activity is currently too insensitive to replace culture based detection of *E. coli*. Nevertheless, on-line enzymatic methods can be a valuable complementary tool for high temporal resolution monitoring. More research is needed in order to enhance sensitive and lower detection limits of available on-line glucuronidase techniques.

The seminal work of Leclerc *et al.* [103] clarified the diversified roles that coliforms have in the environment and the real meanings of the tests on total coliforms and fecal coliforms. It was shown that*Enterobacteriaceae* encompass three groups of bacteria with very different roles in the environment. Group I harbored only *E. coli*. Since this species usually do not survive for long periods outside this environment (but see topic 10), it was considered a good and reliable indicator of fecal pollution (both animal and human). Group II, the "ubiquitary" group, encompassed several species of *Klebsiella* (*K. pneumoniae* and *K. oxytoca*), *Enterobacter* (*Enterobacter cloacae* subsp. *cloacae*, *E. aerogenes*) and *Citrobacter* (*C. amalonaticus*, *C. koseri* and *C. freundii*). These bacteria live in the animal and human gut, but also in the environment, and are easily isolated from the soil, polluted water and plants. Their presence in polluted waters does not necessarily indicate fecal pollution. Finally Group III was composed of *Raoultella planticola*,*R. terrigena, Enterobacter amnigenus* and *Kluyvera intermedia* (*Enterobacter intermedius*), *Serratia fonticola*, and the genera *Budvicia*, *Buttiauxella*, *Leclercia*, *Rahnella*, *Yersinia*, and most species of *Erwinia*and *Pantoea*. These bacteria live in fresh waters, plants and small animals. They grow at 4 °C, but not at 41 °C. They are not indicators of fecal pollution, although can be detected in the total coliform test. Leclerc *et al.*concluded that: (1) in the enterobacteria, *E. coli* is the only true and reliable indicator of fecal pollution in environmental waters; (2) the traditional total coliform test should be abandoned, since it can detect bacteria that have no connection with fecal pollution; (3) the detection of fecal coliforms must be carried out at 44.5 °C, and positive results confirmed by identification to species levels in order to exclude false positives such as*K. pneumoniae*.

Streptococci and Enterococci

Fecal streptococci also belong to the traditional indicators of fecal pollution. Fecal streptococci are Gram-positive, catalase-negative, non-sporeforming cocci that grow at 35 °C in a medium containing bile salts and sodium azide. Cells hydrolyze esculin [1,4,57]. Azide is a strong inhibitor of the respiratory chain. Since streptococci are one of the very few bacteria that have no respiratory chain, the test is very specific for this group, and false positives are rarely found [104,105].

Fecal enterococci (*E. faecalis*, *E. faecium*, *E. avium* and *E. gallinarum*) are fecal streptococci that grow in the presence of 6.5% NaCl at 45 °C. Selective media use these particular characteristics in order to separate enterococci from the other streptococci [104,105].

Several studies [104,106] have reported on the microbiological composition of human and animal (cattle, chicken, deer, dog, fowl, goose, and swine) feces. *E. faecalis* and *E. faecium* were present in human and animal feces. However, whereas human feces almost have only these two enterococci, in the animals others species co-occur, like *E. avium*, *E. cecorum*, *E. durans*, *E. gallinarum* and *E. hirae*. It was concluded that in urban areas where contamination with dog and chicken feces is not likely, the best marker for human fecal pollution was *E. faecalis*.

The intestinal enterococci group has been used as an index of fecal pollution. In human feces, the numbers of intestinal enterococci are generally about an order of magnitude lower than those of *E. coli* (Table 6). Most species do not grow in environmental waters. In this milieu, fecal enterococci are able to survive longer, are more resistant to drying and chlorination, than *E. coli* [1,84].

However, caution should be taken with interpreting the results obtained by the enterococci procedure in water analysis. Enterococci and other group D-streptococci are present in many foods, especially those of animal origin. The isolation of *E. faecalis* and *E. faecium* was used to indicate fecal contamination of food. However, enterococci are now also considered as normal parts of the food microflora and not only as indicators for poor hygiene [104]. In addition, agricultural soils and crops with added with manure also harbor enterococci [105].

The use of ratios between indicator counts

The ratio of counts of fecal coliforms to fecal streptococci has been proposed as a means to differentiating between contamination from human and animal sources. Ratios greater than 4 have been suggested to indicate a human source whereas ratios less than 0.7 suggest an animal source. This results from the fact that streptococcal concentrations in human feces are generally less than coliforms (Table 6). In contrast, in animal feces fecal streptococci generally outnumber fecal coliforms (Table 7). In urban sewage, fecal streptococci tend to be present in concentrations 10–100 times less than fecal coliforms [65].

Table 7: Bacteria in the feces of farm and domestic warm-blooded animals[a]

Animal	Log_{10} cells/g wet weight feces		
	Fecal coliforms	Fecal streptococci	*Clostridium perfringens*
Chicken	5.4	6.1	2.3
Duck	7.5	7.7	-
Horse	4.1	6.8	< 0
Pig	6.5	7.9	3.6
Sheep	7.2	7.6	5.3
Turkey	5.5	6.4	-
Cat	6.9	7.4	7.4
Dog	7.1	9.0	8.4

[a] Adapted from [83].

Geldreich [107] summarized the information available on the fecal coliforms to fecal streptococci ratios in the feces of warmblooded animals, and reported the following values: human feces, 4.3; cattle, sheep, and poultry, from 0.104 to 0.421; and wild animals (including rabbits, field mice, chipmunks, and birds), 0.0008 to 0.043. Fecal coliforms to fecal streptococci ratios for the feces of wild animals appear to be at least 10-fold lower than those of domestic livestock.

Doran and Linn [108] reported a study of the runoff from a cow-calf pasture in eastern Nebraska (USA), monitored during a three-year period. It was concluded that the fecal coliforms to fecal streptococci ratio in pasture runoff was useful in identifying the relative contributions of cattle and wildlife and in evaluating the effects of cattle management and distribution on runoff water quality. Ratios below 0.05 were indicative of wildlife sources and ratios above 0.1 were characteristic of grazing cattle. Fecal coliforms to fecal streptococci ratios of diluted cattle waste in excess of 1 were interpreted as the result of differential aftergrowth and die-off between fecal coliforms and fecal streptococci. Ratios between 0.7 and 4.0 may indicate situations where cattle are localized close to sampling or outflow points.

However, the interpretation of this ratio should be cautious. It has been observed a shift in the ratio with time and distance from the fecal pollution source. This resulted from the fact that both in surface and groundwaters, fecal streptococci are more persistent than fecal coliforms. Therefore increasing the distance from the pollution point and with passing time, the ratio tends to decrease without a change in the nature of the pollution source [65]. For these reasons, this ratio has been considered by some authors as too unreliable to be useful in characterizing pollution sources [65,84].

The ratio of fecal enterococci to fecal streptococci differs among vertebrate species. Humans have a predominance of enterococci, whereas animals contain appreciable amounts of streptococci. However, since enterococci are also present in animals and are more persistent in the environment than other fecal streptococci, the identification of the enterococci and streptococci species present in polluted waters, and the concomitant calculation of this ratio is generally considered unreliable as an indicator of the source of fecal pollution [65].

Limitations of coliform and enterococcus counts as indicator of fecal pollution

An extreme case of uselessness of the determination of total and fecal coliforms and enterococci in the assessment of fecal pollution has been demonstrated by several authors studying the microbiology of pulp and paper mill effluents.

Caplenas and Kanarek [109] reported a study of pulp and paper mills located in Wisconsin (USA). Fresh water supplies, re-cycled water within mills, treated effluent wastewater and waters receiving effluent wastes downstream, were assessed for the presence of fecal coliforms and *Klebsiella*. Wastewaters prior to treatment contained fecal coliforms and *Klebsiella*. Up to 84% of the fecal coliforms (detected by the standard test procedure) were indeed *Klebsiella*. In treated effluent wastewaters this value reached 90%. Treatment of the wastewater lowered the concentration of "true" fecal bacterial contamination, but since *Klebsiella* grew rapidly in the wastewaters, fecal coliform counts were high, although no true fecal contamination was involved. The source of *Klebsiella* was traced to the early pulping stages in the mills. *Klebsiella* maintains a wood, bark or soil reservoir. It was concluded that: (1) *Klebsiella* are ubiquitous in the pulp and paper mill industry processing stages; (2) the standard procedure for fecal coliform estimation is useless to assess the microbiological quality of the effluents of these industries; (3) The assay of *E. coli* should replace the fecal coliform detection procedure.

Gauthier *et al.* [110] and Gauthier and Archibald [58] reported two studies of seven pulp and paper mills in Ontario and Quebec, Canada. Total and fecal coliforms and enterococci were detected in nearly all the biotreatment, biosolids (sludges), and in-mill water system samples. In the mill samples, the majority of the fecal coliforms (detected by the standard test procedure) were *K. pneumoniae*, *Raoultella terrigena* and*Raoultella planticola*, with *E. coli* in minority. *E. faecalis* and *E. faecium* were detected in relatively large numbers in most samples from all of the seven mills examined. Other coliforms such as *Enterobacter* spp. and *Citrobacter freundii* were occasionally recovered from total and fecal coliform tubes. Biofilms established in the piping, tanks, and machinery where thermal and pH conditions permit were the most likely source of these bacteria. Analyses using two independent *Salmonella* detection/ enumeration methods showed no detectable *Salmonella* cells in the sludges and final effluents of the five mills tested. It was concluded that for these particular systems, the determination of total and fecal coliforms and enterococci is useless and have no relationship with real fecal pollution. These studies also demonstrated the importance of checking the identities of bacteria causing the positive results in the tests. Both *Escherichia* and *Klebsiella* can give positive results in the fecal coliform test, but their ecological meaning is opposite.

Another important case of failure of the use of coliforms to detect fecal pollution was the 1993*Cryptosporidium* outbreak in Milwaukee (USA).

Cryptosporidium parvum, a protozoan parasite that causes gastrointestinal illness, is transmitted by ingestion of oocysts excreted in human or animal feces. Typical modes of transmission include person to person, animal to person, by exposure to contaminated food or water [111,112].

From 1990 to 2000, at least 10 cryptosporidiosis outbreaks associated with contaminated drinking water were reported in the USA. In 1993, an estimated 403,000 residents of the greater Milwaukee area (Wisconsin, population, ca. 1.61 million) became ill when an ineffective filtration process led to the inadequate removal of*Cryptosporidium* oocysts in one of two municipal water-treatment plants [111,112].

It was the largest waterborne disease outbreak in documented USA history. Over the span of approximately two weeks, people became ill with stomach cramps, fever, diarrhea and dehydration caused by the pathogen. More than half the people who received residential drinking water from the southern water-treatment plant became ill, which was twice the rate of illness among people whose residential drinking water came mainly from the northern water-treatment plant. Over 54 deaths were attributed to this outbreak, mostly among the elderly and immunocompromised people, such as AIDS patients [111,112].

Standard microbiological water analysis was ineffective in detecting this parasite. Indeed, throughout the period from February to April, samples of treated water from both plants were negative for coliforms. The origin of the contamination was determined as water from Lake Michigan. No specific source of the*Cryptosporidium* was ever identified but runoff from abnormally heavy spring rains most likely carried the parasite to the lake [111,112].

Clostridium perfringens

Sulphite-reducing clostridia, namely *Clostridium perfringens*, are spore-forming Gram-positive, non-motile, anaerobic, sulfite-reducing rods. *C. perfringens* is present in higher numbers in the feces of some animals, such as dogs, than in the feces of humans and less often in the feces of many other warm-blooded animals. The numbers excreted in feces are normally substantially lower than those of *E. coli.*

Clostridium spores are exceptionally resistant to unfavorable conditions in water environments, including UV irradiation, temperature and pH extremes, and disinfection processes, such as chlorination. Although clostridia probably do not growth in surface waters, the high resistance of their spores makes their presence ubiquitous in environmental waters [1,4,84,113].

The presence of chlorine in water rapidly inactivates indicator bacteria such as *E. coli* and coliforms, but it leaves the most resistant pathogens almost unaffected for several hours. This creates a false sense of security by providing negative coliform and negative *E. coli* results to authorities responsible for water testing.*Giardia* cysts, *Crystosporidium* oocysts, and human enteric viruses all have higher resistance to disinfectants and constitute a major public health risk if distribution system integrity is breached. *C. perfringens* spores are less affected by the residual concentrations of chlorine. Testing for the spores of this bacterium can probably provide an added margin of safety in the evaluation of treatment [114].

Correlations between parameters used to assess fecal pollution

In environmental waters, several studies have reported significant correlations between indicators of fecal pollution and between indicators and pathogenic gastrointestinal bacteria.

Charriere *et al.* [115] reported a study of deep aquifer waters (raw waters and piped chlorinated waters) in Normandy, France. In heavily contaminated raw waters and in slightly contaminated treated waters, fecal coliforms and enterococci were correlated.

Martins *et al.* [116] reported a study of 60 public outdoor swimming pools in Sao Paulo city, Brazil. Total coliforms, fecal coliforms and fecal streptococci levels increased with number of bathers and water temperature, and decreased with chlorine levels. All these indicators were significantly correlated with each other.

Ferguson *et al.* [117] reported a study carried out in Georges River, in the Sydney region, Australia. In the water column, concentrations of fecal coliforms, fecal streptococci, *C. perfringens* spores were all positively correlated with each other. Isolation of *Salmonella* spp. were most frequent during rainfall and sewage overflow events. In the water column, 55% of samples contained *Salmonella* when fecal coliform densities exceeded 2,000 CFU/100 mL.

Medema *et al.* [118] reported a study of seven different fresh water sites normally used for triathlon competitions. Sites were small rivers, channels, lakes and harbors, and were influenced by sewage effluents and agricultural run-off. When data from all triathlons were pooled, geometric mean densities of fecal coliforms and *E. coli*, of *E. coli* and fecal enterococci and of fecal coliforms and fecal enterococci, were significantly correlated.

Polo *et al.* [119] reported a study of water samples obtained from 213 beaches, eight rivers and 14 freshwaters in north-eastern Spain. In freshwaters and heavily contaminated seawaters, *Salmonella* and fecal coliforms were correlated, while in less contaminated seawaters, the highest correlation was with between*Salmonella* and fecal streptococci.

Byamukama *et al.* [99] reported a study of the microbiology of Nakivubo channel, Uganda. This channel receives raw sewage from slums, industrial effluents, and discharges from a sewage treatment plant and from a complex of slaughterhouses. Water from eight sampling sites was assessed for the presence of total and fecal coliforms, *E. coli* and sulphite-reducing clostridia. All microbiological parameters were significantly correlated.

Noble *et al.* [120] reported a comparative determination of total coliforms, fecal coliforms and enterococci in 108 sites along the southern California coastline, USA. Results by traditional and enzymatic methods and from all three parameters were correlated.

Harwood *et al.* [121] reported a study on indicator and pathogenic microorganisms carried out in six wastewater reclamation facilities in the USA. Data from disinfected effluent (reclaimed water) samples were analyzed separately (by facility) and as a pooled data set (all facilities). Significant correlations between indicator organism concentrations were observed in the pooled data sets, namely for total and fecal coliforms.

Cabral and Marques [85] reported a microbiological study of a polluted river (Febros) in the Great Oporto area, northwest Portugal. Total and fecal coliforms, fecal streptococci and enterococci were all significantly correlated with each other.

Touron *et al.* [122] reported a study carried out in the Seine estuary, France. Water was sampled at nine stations (along an upstream/downstream transect of 156 km), during nine years, for fecal coliforms, *E. coli*, enterococci and *Clostridium perfringens* spores. At the upstream part of the estuary (at Poses), *Salmonella*and fecal coliforms, and *E. coli* and enterococci counts were correlated. At the mouth of the estuary (at Honfleur), significant correlation was found for *Salmonella* and enterococci counts. No significant correlation between concentrations of any combination of indicator organism and pathogen was observed.

Wilkes *et al.* [59] reported a comparative study on the presence and concentration of several pathogenic and indicator bacteria in the surface water of a Canadian river. Surface water was collected within the South Nation River basin in eastern Ontario, from the river proper, and from several lower stream order tributaries. Using data aggregated during the entire multi-year study, significant correlations were found among all indicator bacteria - total and fecal coliforms, *E. coli*, *Enterococcus*, and *C. perfringens*.

However, in others studies no correlation was found between the different fecal indicator bacteria. Garrido-Pérez *et al.* [123] reported a study of the bathing seawater quality in 18 Spanish beaches near the Strait of Gibraltar. Sample locations were selected as a single point located in the area of highest bather density of each beach. No significant correlation was found between fecal coliforms and *Clostridium perfringens*counts in the bathing seawater.

FECAL INDICATOR CHEMICAL COMPOUNDS

Several chemical substances have been used as markers of fecal pollution in environmental waters. Caffeine is present in several beverages and in many pharmaceutical products. It is excreted in the urine of individuals who have ingested the substance. The main source of caffeine in domestic wastewaters is excretion following consumption of coffee, tea, soft drinks, or medication. Levels of caffeine in domestic wastewater have been measured to be between 20 and 300 μg/L. Levels in receiving waters are much lower due to significant dilution. Due to its high solubility, low octanol-water partition coefficient, insignificant volatility and clear anthropogenic origin, the presence of caffeine in environmental waters can be a good marker for human fecal pollution [124–129].

However, relationships between fecal indicator bacteria and caffeine are variable. Wu *et al.* [129] reported a study carried out in the Rochor Canal and Marina Bay, Singapore. In Rochor Canal, the highest concentration of caffeine (1.35 ng/mL) was found at downstream, and the lowest (0.68 and 0.37 ng/mL) were determined at middle and upstream points. At Marina Bay, the concentration of caffeine was in the range of 0.41–0.96 ng/mL. Fecal coliform concentrations were very high, exceeding 5,000 CFU/100 mL. Caffeine and fecal coliform concentrations were significantly correlated in Rochor Canal samples, but no significant correlation was observed in Martina Bay water samples.

The use of caffeine as marker of fecal pollution has additional important limitations. Caffeine is often present in the urban environment from numerous plant species debris as well as from human "dumping" of coffee wastes. In addition, the current analytical methods used are relatively complex and expensive [124].

Coprostanol (5β-cholestan-3β-ol) is a fecal stanol that is formed by indigenous bacteria present in the gut of humans and higher animals, during catabolism of cholesterol. It is the main stanol present in human feces (24 to 89% of total steroids) and in domestic wastewater. Based on these facts, it has been proposed as a chemical indicator of human fecal pollution. Feces from pigs and cats also contain coprostanol, but at much lower levels. Additional fecal stanols, such as 24-ethylcoprostanol, were found to be predominant in herbivores, such as cows, horses, and sheep, suggesting potential use of this chemical as an indicator of fecal pollution from these sources. Reported half-lives of coprostanol in aerobic conditions are generally lower than 10 days at 20 °C. Thus, the presence of coprostanol in an aerobic environment can be considered an indication of recent fecal input to the waters. [124,128,130].

Isobe *et al.* [130] reported a study carried out in the Mekong Delta (Vietnam) and in the Tokyo metropolitan area. During the wet season in the Mekong Delta, higher bacterial densities were observed in rivers, probably due to the higher bacterial inputs from soil particles with runoff. In Tokyo, higher bacterial densities were usually observed during summer, followed by those in the typhoon aftermath and winter. Significant correlations between the concentrations of *E. coli* and coprostanol (log scale) were found in all surveys. It was concluded that the determination of coprostanol can improve standard microbiological assays of fecal pollution.

Reports from several regions throughout the world indicate variable quantitative relationships between fecal coliform densities and coprostanol concentrations. In the Derwent Estuary and Sydney region (Australia), coprostanol concentration of 400 ng/L corresponded to 1,000 CFU of fecal coliforms/100 mL. In the Mekong Delta, during the wet season, this coliform density corresponded to 30 ng coprostanol/L, and in the dry seasons, 100 ng/L. In Tokyo metropolitan area, these values were 30 ng coprostanol/L, in summer, and 100 ng coprostanol/L, in a typhoon aftermath. These differences were interpreted as a result of differences in water temperature and soil particle concentration [130].

Fecal sterol analysis, although expensive and complex, has resolved problems of source attribution in urban and rural environments not possible with use of traditional fecal indicator bacteria [124]. These chemical indicators are especially useful in environments which allow survival and growth of fecal bacteria. For instance, in tropical regions, characterized by high temperatures and frequent rainstorms that facilitate erosion of soils, fecal bacteria can proliferate in environmental waters reaching densities that are not representative of real sewage inputs in the environment [130].

SOURCES OF FECAL BACTERIAL POLLUTION OF ENVIRONMENTAL WATERS

Sources of Surface and Groundwater Contamination

Determinations carried out in the sewage systems of urbanized areas have confirmed the presence of high numbers of intestinal bacteria. Treatment of sewage reduces the concentration of these bacteria by 1–2 logs, but effluent still contains high levels of intestinal bacteria (Table 8). Effluents from sewage treatment plants can be a source of contamination of surface waters with fecal bacteria.

Table 8: Typical concentrations of selected bacteria in raw and treated domestic wastewater[a]

Bacterial group	Raw sewage (cells/ml)	Treated effluent (cells/ml)
Salmonella	$10^{-1} - 10^{1}$	$10^{-1} - 10^{1}$
Total coliforms	$10^{4} - 10^{6}$	$10^{3} - 10^{5}$
Fecal coliforms	$10^{3} - 10^{5}$	$10^{2} - 10^{4}$
Enterococci	$10^{3} - 10^{4}$	$10^{1} - 10^{3}$
Clostridium perfringens	$10^{2} - 10^{3}$	$10^{1} - 10^{2}$

[a] Adapted from Medema *et al.* [131].

Septic tanks, cesspools, latrines and other on-site systems are widely used for wastewater storage and treatment. The water percolating from these facilities contains bacteria that may contaminate groundwater supplies.

Many farmers use cellars, tanks or landfills to store manure. Water leaching from these storage sites may also contaminate groundwater, especially during periods of rainfall. The application of animal manure to agricultural lands as fertilizer is common practice throughout the world. Bacteria present in the manure may leach into the groundwater.

The potential for bacteria present in human and animal wastes to contaminate water in nearby wells needs special attention [131]. An important source of contamination of surface and ground waters is runoff water from agricultural and pasture lands, and urban areas.

Fecal bacteria enter surface water by direct deposit of feces and by overland runoff. The movement of animal wastes into surface waters can be a major factor contributing to the pollution of available water in many regions. Over one-third of the land area of USA is used for grazing livestock and receives 50% of all livestock wastes.

In a study reported by Doran and Linn [108], runoff from a cow-calf pasture in eastern Nebraska was monitored during a three-year period. Rainfall runoff from the grazed area contained 5 to 10 times more fecal coliforms than runoff from the fenced, ungrazed area. However, fecal streptococci counts were higher in runoff from the ungrazed area and reflected the contributions from wildlife.

Urban and suburban areas are dominated by impervious cover. During storms, rainwater flows across these impervious surfaces, mobilizing contaminants. The pollutants carried in runoff originate from a variety of urban and suburban nonpoint sources. Contaminants commonly found in stormwater runoff include fecal and pathogenic bacteria. Stormwater transports pollutants to water bodies such as lakes and streams [132].

Enterococci and *E. coli* can be found in high numbers in most storm drains and creeks. In Southern California (USA), Ferguson *et al.* [133] found high levels of enterococci (*Enterococcus faecalis,Enterococcus faecium*, *Enterococcus hirae*, *Enterococcus casseliflavus* and *Enterococcus mundtii*) in intertidal sediments in a seasonal river, and near a storm drain outlet.

Survival in Surface Water

Most intestinal bacteria that contaminate environmental waters are not able to survive and multiply in this environment. Survival rates vary widely among fecal bacteria introduced in environmental waters. Pathogenic enteric bacteria and *E. coli* display low survival rates (Table 9).

Table 9: Reduction times for fecal bacteria in surface waters[a]

Bacterial group	Time for 50% reduction in concentration (days)
Total coliforms	0.9
E. coli	1.5–3
Enterococci	0.9–4
Clostridium perfringens	60 – > 300
Salmonella	0.1–0.67
Shigella	1

[a] Adapted from Medema *et al.* [131].

The ability of fecal bacteria to survive in environmental waters generally increases as the temperature decreases. Others factors that influence survival include dissolved organic carbon concentration, sunlight intensity and the ability to enter the viable but non-culturable state [131].

In a comparative study on the survival of 10 different coliform species (*E. coli*, *Citrobacter freundii,Citrobacter youngae*, *Klebsiella pneumoniae*, *K. oxytoca*, *Enterobacter amnigenus*, *Enterobacter cloacae*subsp. *cloacae*, and *Pantoea* *agglomerans* (*Enterobacter agglomerans*)) inoculated in sterilized river water with different concentrations of dissolved organic carbon, Boualam *et al.* [134] found that only *C. freundii,K. pneumoniae* and *E. cloacae* subsp. *cloacae* remained cultivable after 96 hours of incubation. In a posterior study, using the same bacteria and medium, Boualam *et al.* [135] found that after 28 days, only *C. freundii*and *E. cloacae* subsp. *cloacae* survived.

Baudišová [136] reported a comparative study on the survival of total coliforms, fecal coliforms and *E. coli*, in sterile and non-sterile river water. In sterile water, all bacteria survived for many months. However, in non-sterile conditions (closer to true environmental conditions), the elimination rate of all bacteria was considerably faster. Total coliforms survived the longest and *E. coli* the shortest.

Survival in Groundwater

Survival of bacteria in groundwater is influenced by several factors, namely the survival in soil, since in order to reach the groundwater bacteria have to percolate through the soil. Generally, survival in soil (and concomitantly in groundwater) is enhanced by low temperatures, high soil humidity, neutral or alkaline soil pH and the presence of organic carbon [131].

WHICH INDICATORS OF FECAL POLLUTION SHOULD BE USED?

Several fecal indicator bacteria in environmental waters are in current use. From these stand out fecal coliforms, *E. coli* and enterococci [1,6,137]. In environmental waters, most fecal coliform strains are *E. coli.*

In particular situations, the presence of *E. coli* is definitively not associated with fecal pollution. These situations were firstly detected in some African countries, namely Nigeria, Ivory Coast and New Guinea (although not in others, such as Uganda) [42,99].

Recent studies carried out in temperate zones indicated that *E. coli* can persist in secondary, nonhost habitats, outside the hot tropical areas, and become naturalized in these habitats. Byappanahalli *et al.* [138] reported that *E. coli* could be isolated from coastal temperate forest soils in Indiana (USA). The aquatic alga*Cladophora glomerata* (L.) from several Lake Michigan beaches was shown to harbor high densities of *E. coli* [139].

Ishii *et al.* [140] reported a study on the survival of *E. coli* in temperate riverine soils of northern Minnesota (USA). Viable *E. coli* populations were repeatedly isolated from northern temperate soils in three Lake Superior watersheds. Seasonal variation in the population density of soilborne *E. coli* was observed; the greatest cell densities were found in the summer to fall, and the lowest numbers, occurred during the winter to spring months. Horizontal, fluorophore-enhanced repetitive extragenic palindromic PCR (HFERP) DNA fingerprint analyses indicated that identical soilborne *E. coli* genotypes, overwintered in frozen soil and were present over time, and that these strains were different from *E. coli* strains obtained from wildlife commonly found in the studied habitats or river water. Soilborne *E. coli* strains had HFERP DNA fingerprints that were unique to specific soils and locations. In laboratory studies, naturalized *E. coli* strains had the ability to grow and replicate to high cell densities, in nonsterile soils when incubated at 30 or 37 °C and survived longer than 1 month when soil temperatures were lower than 25 °C. It was concluded that these *E. coli* strains became naturalized, autochthonous members of the soil microbial community.

In a latter paper, Ksoll *et al.* [141] studied epilithic periphyton communities at three sites on the Minnesota shoreline of Lake Superior (USA). Fecal coliform densities increased up to 4 orders of magnitude in early summer, and decreased during autumn. HFERP DNA fingerprint analyses indicated that waterfowl (geese, terns, and gulls) were the major primary source of periphyton *E. coli* strains that could be identified. Periphyton and sewage effluent were also major potential sources. Several periphyton *E. coli* isolates were genotypically identical, repeatedly isolated over time. Inoculated *E. coli* rapidly colonized natural periphyton in laboratory microcosms and persisted for several weeks, and some cells were released to the overlying water. It was concluded that *E. coli* had became a naturalized member of the bacterial periphyton communities.

The presence, persistence, and possible naturalization of *E. coli* in these habitats can confound the use of fecal coliforms as a reliable indicator of recent fecal contamination of environmental waters. Future studies should consider other nonhost habitats as potential sources of fecal coliform bacteria in aquatic environments.

Considering these limitations, it appears to be advisable, in order to check the microbiological quality of drinking water, to complement the determination of *Escherichia coli* with the assay of enterococci. This rationale that has been followed, for many years, in the making of the drinking-water legislation in the European Union.

However, for many developing countries, where limited financial resources are the norm and reality, the routine determination of these two parameters can be difficult to implement. In these circumstances, it appears common sense that is better to determine a (good) parameter, such as *Escherichia coli*, than have no analysis done.

In this context, USA legislation emerges as a pragmatic approach to the problem. According to the American legislation, total coliforms are the routine parameter to be determined. Only when these determinations are repeatedly positive, it is mandatory to assess fecal coliforms [142,143]. Although total coliforms are not necessarily fecal bacteria, the rationale behind this system is correct, since: (1) a positive test in fecal coliforms (which is our target) is necessarily positive in the total coliform procedure; (2) the inverse is not necessarily true; (3) total coliforms are easily and cheaply assayed in waters.

As an alternative to the determination of both *E. coli* and enterococci, the assay of ammonia in environmental waters can be useful and complement the determination of fecal coliforms.

Ammonia is one of the key molecules in the nitrogen cycle. The presence of ammonia in surface waters can be due to direct contamination by agricultural fertilizers, and/or to microbial degradation of proteins, nucleic acids and urea, implying therefore the presence of a considerable concentration of organic matter in the water. Ammonia is rapidly oxidized in the environment and is typically found in natural waters at concentrations less than 0.1 mg/L. Concentrations significantly above this indicate gross contamination by fresh sanitary waste, where ammonia levels are typically very high (tens or hundreds of mg/L) [84].

Espigares *et al.* [144] reported a comparative study of chemical and microbiological indicators (total and fecal coliforms, fecal streptococci and sulphite-reducing clostridia) in a stretch of the Guadalquivir River (Spain) and its affluents. Total coliforms were correlated with fecal coliforms, but were not correlated with fecal streptococci and clostridia. Fecal coliforms were correlated with the other indicators. Fecal streptococci and sulphite-reducing clostridia were correlated with the other indicators except for total coliforms. All these microbiological indicators were correlated with dissolved oxygen (negatively), dissolved organic carbon and ammonia (positively). Cabral and Marques [85] in a study of a polluted river (Febros) in the Great Oporto area (northwest Portugal) found that ammonia was significantly correlated with all the microbiological assayed parameters—total and fecal coliforms, fecal streptococci and enterococci. These correlations are most probably due to the carry-over of organic matter in wastewaters, and to a high microbial ammonification activity [85].

Table 10: Disappearance rates of fecal bacteria in groundwaters[a]

Bacterial group	Disappearance rate (per day)
E. coli	0.063–0.36
Fecal streptococci	0.03–0.24
Clostridium bifermentans spores	0.00
Salmonella enterica subsp. *enterica* serovar Typhimurium	0.13–0.22

[a] Adapted from Medema *et al.* [131].

Simple and rapid in-field tests and automated and continuous systems are available for the assay of ammonia in environmental waters. More studies are needed in order to confirm the use of ammonia as a reliable parameter in a preliminary screening for emergency fecal pollution outbreaks.

CONCLUSIONS

1. Safe drinking water for all is one of the major challenges of the 21st century.
2. Microbiological control of drinking water should be the norm everywhere.
3. Routine basic microbiological analysis of drinking water should be carried out by assaying the presence of *Escherichia coli* by the culture methods. On-line monitoring of glucuronidase activity is currently too insensitive to replace culture based detection of *E. coli* but is a valuable complementary tool for high temporal resolution monitoring. Whenever financial resources are available, coliform determinations should be complemented with the quantification of enterococci.
4. More studies are needed in order to check if ammonia is reliable for a preliminary screening for emergency fecal pollution outbreaks.

5. Financial resources should be devoted to a better understanding of the ecology and behavior of human and animal fecal bacteria in environmental waters.

REFERENCES

1. WHO (World Health Organization) Guidelines for Drinking-water Quality, Incorporating 1st and 2nd Addenda, Volume 1, Recommendations. 3rd ed. WHO; Geneva, Switzerland: 2008.
2. Fenwick A. Waterborne Diseases—Could they be Consigned to History? Science. 2006;313:1077–1081.
3. George I, Crop P, Servais P. Use of β-D-Galactosidase and β-D-Glucuronidase Activities for Quantitative Detection of Total and Faecal Coliforms in Wastewater. Can. J. Microbiol. 2001;47:670–675.
4. Grabow WOK. Waterborne Diseases: Update on Water Quality Assessment and Control. Water SA.1996;22:193–202.
5. Seas C, Alarcon M, Aragon JC, Beneit S, Quiñonez M, Guerra H, Gotuzzo E. Surveillance of Bacterial Pathogens Associated with Acute Diarrhea in Lima, Peru. Int. J. Infect. Dis. 2000;4:96–99.
6. Medema GJ, Payment P, Dufour A, Robertson W, Waite M, Hunter P, Kirby R, Anderson Y. Assessing Microbial Safety of Drinking Water Improving Approaches and Method. WHO & OECD, IWA Publishing; London, UK: 2003. Safe drinking water: an ongoing challenge; pp. 11–45.
7. Farmer JJ, Hickam-Brenner FW. The Genus *Vibrio* and *Phtotobacterium*. In: Dworkin M, Falkow S, Rosenberg E, editors. The Prokaryotes: An Evolving Electronic Resource for the Microbiological Community. 3th ed. Springer-Verlag; New-York, NY, USA: 2003. electronic release 314,
8. Farmer JJ, Janda JM, Brenner FW, Cameron DN, Birkhead KM. Genus *Vibrio*. In: Brenner DJ, Krieg NR, Staley JT, editors. Bergey's Manual of Systematic Bacteriology. 2nd ed. Part B. Vol. 2. Springer; New York, NY, USA: 2005. pp. 494–546.
9. DSMZ (Deutsche Sammlung von Mikroorganismen und Zellkulturen GmbH) Bacterial Nomenclature Up-to-Date (Approved Lists; Validation Lists) DSMZ; Braunschweig, Germany: 2010. List 05/2010.
10. Sack DA, Sack RB, Nair GB, Siddique AK. Cholera. Lancet. 2004;363:223–233.
11. Todar K. *Vibrio cholerae* and asiatic cholera. Todar's Online Textbook of Bacteriology. 2009. Available online: http://www.textbookofbacteriology.net/cholera.html (assessed on 4 May 2010)
12. Ali M, Emch M, Yunus M, Sack RB. Are the Environmental Niches of *Vibrio cholerae* 0139 Different from those of *Vibrio cholerae* 01 El Tor? Int. J. Infect. Dis. 2001;5:214–219.
13. Ramamurthy T, Yamasaki S, Takeda Y, Nair GB. *Vibrio cholerae* O139 Bengal: Odyssey of a Fortuitous Variant. Microbes Infect. 2003;5:329–344.

14. Arbora S, Crum S. Medical Geography and Cholera in Perú. The Geographer's Craft Project, Department of Geography, The University of Colorado at Boulder; Boulder, CO, USA: 1996. Available online: http://www.colorado.edu/geography/gcraft/warmup/cholera/cholera_f.html (assessed on 4 May 2010).

15. Glass RI, Claeson M, Blake PA, Waldmann RJ, Pierce NF. Cholera in Africa: Lessons on Transmission and Control for Latin America. Lancet. 1991;338:791–795.

16. Glass RI, Libel M, Brandling-Bennet AD. Epidemic Cholera in the Americas. Science. 1992;256:1524–1525.

17. Tamplin ML, Parodi CC. Environmental Spread of *Vibrio cholerae* in Peru. Lancet. 1991;338:1216–1217.

18. Steinberg EB, Greene KD, Bopp CA, Cameron DN, Wells JG, Mintz ED. Cholera in the United States, 1995–2000: Trends at the End of the Twentieth Century. J. Infect. Dis. 2001;184:799–802.

19. Meibom KL, Blokesch M, Dolganov NA, Wu C-Y, Schoolnik GK. Chitin Induces Natural Competence in *Vibrio cholerae*. Science. 2005;310:1824–1827.

20. Waldor MK, Mekalanos JJ. Lysogenic Conversion by a Filamentous Phage Encoding Cholera Toxin.Science. 1996;272:1910–1914.

21. Hervio-Heath D, Colwell RR, Derrien A, Robert-Pillot A, Fournier JM, Pommepuy M. Occurrence of Pathogenic Vibrios in Coastal Areas of France. J. Appl. Microbiol. 2002;92:1123–1135.

22. Alam M, Hasan NA, Sadique A, Bhuiyan NA, Ahmed KU, Nusrin S, Nair GB, Siddique AK, Sack RB, Sack DA, Huq A, Colwell RR. Seasonal Cholera Caused by *Vibrio cholerae* Serogroups O1 and O139 in the Coastal Aquatic Environment of Bangladesh. Appl. Environ. Microbiol. 2006;72:4096–4104.

23. Bhanumathi R, Sabeena F, Isac SR, Shukla BN, Singh DV. Molecular Characterization of *Vibrio cholerae* O139 Bengal Isolated from Water and the Aquatic Plant *Eichhornia crassipes* in the River Ganga, Varanasi, India. Appl. Environ. Microbiol. 2003;69:2389–2394.

24. Munro PM, Colwell RR. Fate of *Vibrio cholerae* O1 in Seawater Microcosms. Water Res. 1996;30:47–50.

25. Alam M, Sultana M, Nair GB, Sack RB, Sack DA, Siddique AK, Ali A, Huq A, Colwell RR. Toxigenic *Vibrio cholerae* in the Aquatic Environment of Mathbaria, Bangladesh. Appl. Environ. Microbiol.2006;72:2849–2855.

26. Chaiyanan S, Chaiyanan S, Huq A, Maugel T, Colwell RR. Viability of the Nonculturable *Vibrio cholerae* O1 and O139 System. Appl. Microbiol. 2001;24:331–341.

27. Molloy S. Biofilms in the Time of Cholera. Nature Rev. Microbiol. 2007;5:906–907.

28. Le Minor . The genus *Salmonella*. In: Dworkin M, Falkow S, Rosenberg E, editors. The Prokaryotes: An Evolving Electronic Resource for the Microbiological

Community. 3th ed. Springer-Verlag; New York, NY, USA: 2003. electronic release 314,

29. Popoff MY, Le Minor LE. Genus *Salmonella*. In: Brenner DJ, Krieg NR, Staley JT, editors. Bergey's Manual of Systematic Bacteriology. 2nd ed. Part B. Vol. 2. Springer; New York, NY, USA: 2005. pp. 764–799.

30. Tindall BJ, Grimont PAD, Garrity GM, Euzéby JP. Nomenclature and Taxonomy of the Genus*Salmonella*. Int. J. Syst. Evolut. Microbiol. 2005;55:521–524.

31. Ben Aissa R, Al-Gallas N, Troudi H, Belhadj N, Belhadj A. Trends in *Salmonella enterica* Serotypes Isolated from Human, Food, Animal, and Environment in Tunisia, 1994–2004. J. Infect. 2007;55:324–339.

32. Arvanitidou M, Kanellou K, Vagiona DG. Diversity of *Salmonella* spp. and Fungi in Northern Greek Rivers and their Correlation to Faecal Pollution Indicators. Environ. Res. 2005;99:278–284.

33. Kapperud G, Rørvik LM, Hasseltvedt V, Høiby EA, Iversen BG, Staveland K, Johnsen G, Leitão J, Herikstad H, Andersson Y, Langeland G, Gondrosen B, Lassen J. Outbreak of *Shigella sonnei* Infection Traced to Imported Iceberg Lettuce. J. Clin. Microbiol. 1995;33:609–614.

34. Strockbine NA, Maurelli AT. Genus *Shigella*. In: Brenner DJ, Krieg NR, Staley JT, editors. Bergey's Manual of Systematic Bacteriology. 2nd ed. Part B. Vol. 2. Springer; New York, NY, USA: 2005. pp. 811–823.

35. Emch M, Ali M, Yunus M. Risk Areas and Neighborhood-Level Risk Factors for *Shigella dysenteriae* 1 and *Shigella flexneri*. Health Place. 2008;14:96–105.

36. Germani Y, Sansonetti PJ. The Genus *Shigella*. In: Dworkin M, Falkow S, Rosenberg E, editors. The Prokaryotes: An Evolving Electronic Resource for the Microbiological Community. 3th ed. Springer-Verlag; New York, NY, US: 2003. electronic release 314,

37. Hale TL. Genetic Basis of Virulence in *Shigella* Species. Microbiol. Rev. 1991;55:206–224.

38. Todar K. *Shigella* and shigellosis. Todar's Online Textbook of Bacteriology. 2009. Available online:http://www.textbookofbacteriology.net/Shigella.html (assessed on 1 September 2010).

39. Tetteh GL, Beuchat LR. Survival, Growth, and Inactivation of Acid-Stressed *Shigella flexneri* as Affected by pH and Temperature. Int. J. Food Microbiol. 2003;87:131–138.

40. Chompook P, Todd J, Wheeler JG, von Seidlein L, Clemens J, Chaicumpa W. Risk Factors for Shigellosis in Thailand. Int. J. Infect. Dis. 2006;10:425–433.

41. Faruque SM, Khan R, Kamruzzman M, Yamasaki S, Ahmad QS, Azim T, Nair GB, Takeda Y, Sack DA. Isolation of *Shigella dysenteriae* type 1 and *S. flexneri* Strains from Surface Waters in Bangladesh: Comparative Molecular Analysis of Environmental *Shigella* Isolates *versus* Clinical Strains. Appl. Environ. Microbiol. 2002;68:3908–3913.

42. Bettelheim KA. The genus *Escherichia*. In: Dworkin M, Falkow S, Rosenberg E, editors. The Prokaryotes: An Evolving Electronic Resource for the Microbiological Community. 3th ed. Springer-Verlag; New York, NY, USA: 2003. electronic release 314,

43. Scheutz F, Strockbine NA. Genus *Escherichia*. In: Brenner DJ, Krieg NR, Staley JT, editors. Bergey's Manual of Systematic Bacteriology. 2nd ed. Part B. Vol. 2. Springer; New York, NY, USA: 2005. pp. 607–623.

44. World Health Organization Enterotoxigenic *Escherichia coli* (ETEC) Diarrhoeal Diseases. Available online: http://www.who.int/vaccine_research/diseases/diarrhoeal/en/index4.html (assessed on 4 September 2010).

45. Health Canada . Guidelines for Canadian Drinking Water Quality: Guideline Technical Document. Bacterial Waterborne Pathogens. Current and Emerging Organisms of Concern. Health Canada; Ottawa, ON, Canada: 2006.

46. World Health Organization Enterohaemorrhagic *Escherichia coli* (EHEC) Fact sheet N°125; Available online: http://www.who.int/mediacentre/factsheets/fs125/en/ (assessed on 4 September 2010).

47. Enteroinvasive *Escherichia coli* (EIEC); US Department of Health & Human Services, U.S. Food and Drug Administration. Available online:http://www.fda.gov/Food/FoodSafety/FoodborneIllness/FoodborneIllnessFoodbornePathogensNaturalToxins/BadBugBook/ucm071298.htm (assessed on 4 September 2010).

48. WHO (World Health Organization) In: Pathogenic Mycobacteria in Water: A Guide to Public Health Consequences, Monitoring and Management. Pedley S, Bartram J, Rees G, Dufour A, Cotruvo J, editors. IWA Publishing; London, UK: 2004.

49. Holtzman A, Aronson T, Froman S, Berlin OGW, Dominguez P, Kunkel KA, Overturf G, Stelma G, Smith C, Yakrus M. The Isolation and Identification of *Mycobacterium avium* Complex (MAC) Recovered from Los Angeles Potable Water, a Possible Source of Infection in AIDS Patients. Int. J. Environ. Health Res. 1994;4:63–72.

50. Von Reyn CF, Maslow JN, Barber TW, Falkinham JO, Arbeit RD. Persistent Colonisation of Potable Water, a Source of *Mycobacterium avium* Infection in AIDS. Lancet. 1994;343:1137–1141.

51. Gião MS, Azevedo NF, Wilks SA, Vieira MJ, Keevil CW. Persistence of *Helicobacter pylori* in Heterotrophic Drinking Water Biofilms. Appl. Environ. Microbiol. 2008;74:5898–5904.

52. Hulten K, Han SW, Enroth H, Klein PD, Opekun AR, Gilman RH, Evans DG, Engstrand L, Graham DY, El-Zaatari FA. *Helicobacter pylori* in the Drinking Water in Peru. Gastroenterology. 1996;110:1031–1035.

53. Handfield M, Simard P, Couillard M, Letarte R. *Aeromonas hydrophila* Isolated from Food and Drinking Water: Hemagglutination, Hemolysis, and Cytotoxicity for a Human Intestinal Cell Line (HT-29) Appl. Environ. Microbiol. 1996;62:3459–3461.

54. Daskalov H. The Importance of *Aeromonas hydrophila* in Food Safety. Food Control. 2006;17:474–483.

55. Chauret C, Volk C, Creason R, Jarosh J, Robinson J, Warnes C. Detection of *Aeromonas hydrophila* in a Drinking-Water Distribution System: A Field and Pilot Study. Can. J. Microbiol. 2001;47:782–786.

56. Fernández MC, Beatriz N, Giampaolo SB, Ibañez M, Guagliardo V, Esnaola MM, Conca L, Valdivia P, Stagnaro SM, Chiale C, Frade H. *Aeromonas Hydrophila* and its Relation with Drinking Water Indicators of Microbiological Quality in Argentine 2000. Genetica. 2000;108:35–40.

57. George I, Servais P. Sources et Dynamique des Coliformes dans le Basin de la Sein. C. N. R. S.; Paris, France: 2002. Rapport de Synthèse; Programme PIREN-Seine 1998–2001, Sources et dynamique des coliformes dand le bassin de la Seine;

58. Gauthier F, Archibald F. The Ecology of "Faecal Indicator" Bacteria Commonly Found in Pulp and Paper Mill Water Systems. Water Res. 2001;35:2207–2218.

59. Wilkes G, Edge T, Gannon V, Jokinen C, Lyautey E, Medeiros D, Neumann N, Ruecker N, Topp E, Lapena DR. Seasonal Relationships Among Indicator Bacteria, Pathogenic Bacteria, *Cryptosporidium*Oocysts, *Giardia* Cysts, and Hydrological Indices for Surface Waters Within an Agricultural Landscape.Water Res. 2009;43:2209–2223.

60. Saarela M, Lähteenmäki L, Crittenden R, Salminen S, Mattila-Sandholm T. Gut Bacteria and Health Foods—The European Perspective. Int. J. Food Microbiol. 2002;78:99–117.

61. Hawksworth G, Drasar BS, Hill MJ. Intestinal Bacteria and the Hydrolysis of Glycoside Bonds. J. Med. Microbiol. 1971;4:451–459.

62. Hopkins MJ, Sharp R, Macfarlane GT. Variation in Human Intestinal Microbiota with Age. Digest Liver Dis. 2002;34(Suppl.2):S12–S18.

63. Leclerc H, Moriamez JC. Étude Quantitative de la Flore Fécale de l'Adulte et du Nourrison Alimenté Artificiellement. Path. Bio. 1980;28:217–226.

64. Wilson M. Microbial Inhabitants of Humans Their Ecology and Role in Health and Disease. Cambridge University Press; Cambridge, UK: 2005.

65. Sinton LW, Finlay RK, Hannah DJ. Distinguishing Human from Faecal Contamination in Water: A Review. New Zealand J. Marine Freshwater Res. 1998;32:323–348.

66. Biavati B, Mattarelli P. The family *Bifidobacteriaceae*. In: Dworkin M, Falkow S, Rosenberg E, editors.The Prokaryotes: An Evolving Electronic Resource for the Microbiological Community. 3th ed. Springer-Verlag; New York, NY, USA: 2003. electronic release 314,

67. Hippe H, Andreesen JR, Gottschalk G. The genus *Clostridium*—Nonmedical. In: Dworkin M, Falkow S, Rosenberg E, editors. The Prokaryotes: An Evolving Electronic Resource for the Microbiological Community. 3th ed. Springer-Verlag; New York, NY, USA: 2003. electronic release 314,

68. Rainey FA, Hollen BJ, Small A. Genus *Clostridium*. In: De Vos P, Garrity GM, Jones D, Krieg NR, Ludwig W, Rainey FA, Schleifer K-H, Whitman WB, editors. Bergey's Manual of Systematic Bacteriology.2nd ed. Vol. 3. Springer; New York, NY, USA: 2009. pp. 738–828.

69. Smith LDS. The genus *Clostridium*—Medical. In: Dworkin M, Falkow S, Rosenberg E, editors. The Prokaryotes: An Evolving Electronic Resource for the Microbiological Community. 3th ed. Springer-Verlag; New York, NY, USA: 2003. electronic release 314,

70. Bezirtzoglou E, Panagiou A, Savvaidis I, Maipa V. Distribution of *Clostridium perfringens* in Polluted Lake Environments. Anaerobe. 1997;3:169–172.

71. Švec P, Devriese LA. Genus *Enterococcus*. In: DE Vos P, Garrity GM, Jones D, Krieg NR, Ludwig W, Rainey FA, Schleifer K-H, Whitman WB, editors. Bergey's Manual of Systematic Bacteriology. 2nd ed. Vol. 3. Springer; New York, NY, USA: 2009. pp. 594–607.

72. Pinto B, Pierotti R, Canale G, Reali D. Characterization of 'Faecal Streptococci' as Indicators of Faecal Pollution and Distribution in the Environment. Lett. Appl. Microbiol. 1999;29:258–263.

73. Holt JG, et al., editors. Bergey's Manual of Determinative Bacteriology. 9th ed. Williams & Wilkins; Baltimore, MD, USA: 1994. pp. 175–190.

74. Gordon DM, FitzGibbon F. The Distribution of Enteric Bacteria from Australian Mammals: Host and Geographical Effects. Microbiology. 1999;145:2663–2671.

75. Frederiksen W, Søgaard P. The genus *Citrobacter*. In: Dworkin M, Falkow S, Rosenberg E, editors. The Prokaryotes: An Evolving Electronic Resource for the Microbiological Community. 3th ed. Springer-Verlag; New York, NY, USA: 2003. electronic release 314,

76. Frederiksen W. Genus *Citrobacter*. In: Brenner DJ, Krieg NR, Staley JT, editors. Bergey's Manual of Systematic Bacteriology. 2nd ed. Part B. Vol. 2. Springer; New York, NY, USA: 2005. pp. 651–656.

77. Grimont F, Grimont PAD, Richard C. The Genus *Klebsiella*. In: Dworkin M, Falkow S, Rosenberg E, editors. The Prokaryotes: An Evolving Electronic Resource for the Microbiological Community. 3th ed. Springer-Verlag; New York, NY, USA: 2003. electronic release 314,

78. Grimont PAD, Grimont F. Genus *Klebsiella*. In: Brenner DJ, Krieg NR, Staley JT, editors. Bergey's Manual of Systematic Bacteriology. 2nd ed. Part B. Vol. 2. Springer; New York, NY, USA: 2005. pp. 685–693.

79. Grimont F, Grimont PAD. The Genus *Enterobacter*. In: Dworkin M, Falkow S, Rosenberg E, editors.The Prokaryotes: An Evolving Electronic Resource for the Microbiological Community. 3th ed. Springer-Verlag; New York, NY, USA: 2003. electronic release 314,

80. Grimont PAD, Grimont F. Genus *Enterobacter*. In: Brenner DJ, Krieg NR, Staley JT, editors. Bergey's Manual of Systematic Bacteriology. 2nd ed. Part B. Vol. 2. Springer; New York, NY, USA: 2005. pp. 661–669.

81. Lockhart SR, Abramson MA, Beekmann SE, Gallagher G, Riedel S, Diekema DJ, Quinn JP, Doern GV. Antimicrobial Resistance Among Gram-negative Bacilli Causing Infections in Intensive Care Unit Patients in the United States between 1993 and 2004. J. Clin. Microbiol. 2007;45:3352–3359.

82. Hidron AI, Edwards JR, Patel J, Horan TC, Sievert DM, Pollock DA. NHSN Annual Update: Antimicrobial Resistant Pathogens Associated with Healthcare-Associated Infections: Annual Summary of Data Reported to the National Healthcare Safety Network at the Centers for Disease Control and Prevention, 2006–2007. Infect. Control Hosp. Epidemiol. 2008;29:996–1011.

83. Ashbolt NJ, Grabow OK, Snozzi M. Indicators of microbial water quality. In: Fewtrell L, Bartram J, editors. Water Quality: Guidelines, Standards and Health. World Health Organization (WHO), IWA Publishing; London, UK: 2001. pp. 289–316.

84. Payment P, Waite M, Dufour A. Assessing Microbial Safety of Drinking Water Improving Approaches and Method. WHO & OECD, IWA Publishing; London, UK: 2003. Introducing parameters for the assessment of drinking water quality; pp. 47–77.

85. Cabral JP, Marques C. Faecal Coliform Bacteria in Febros river (Northwest Portugal): Temporal Variation, Correlation with Water Parameters, and Species Identification. Environ. Monit. Assess.2006;118:21–36.

86. Villarino A, Toribio AL, Brena BM, Grimont PAD, Bouvet OMM. On the Relationship Between the Physiological State of Bacteria and Rapid Enzymatic Assays of Faecal Coliforms in the Environment.Biotechnol. Lett. 2003;25:1329–1334.

87. George I, Petit M, Servais P. Use of Enzymatic Methods for Rapid Enumeration of Coliforms in Freshwaters. Lett. Appl. Microbiol. 2000;88:404–413.

88. George I, Petit M, Theate C, Servais P. Distribution of Coliforms in the Seine River and Estuary (France) Studied by Rapid Enzymatic Methods and Plate Counts. Estuaries. 2001;24:994–1002.

89. Nelis H, van Poucke S. Enzymatic Detection of Coliforms and *Escherichia coli* Within 4 Hours. Water, Air, and Soil Pollution. 2000;123:43–52.

90. Rompré A, Servais P, Baudart J, de-Roubin M-R, Laurent P. Detection and Enumeration of Coliforms in Drinking Water: Current Methods and Emerging Approaches. J. Microbiol. Method. 2002;49:31–54.

91. Van Poucke SO, Nelis HJ. Rapid Detection of Fluorescent and Chemiluminescent Total Coliforms and*Escherichia coli* on Membrane Filters. J. Microbiol. Method. 2000;42:233–244.

92. Kämpfer P, Rauhoff O, Dott W. Glycosidase Profiles of Members of the Family *Enterobacteriaceae*. J. Clin. Microbiol. 1991;29:2877–2879.

93. Kilian M, Bülow P. Rapid Diagnoses of *Enterobacteriaceae*. I. Detection of Bacterial Glycosidases. Acta Pathol. Microbiol. Scandinavica. 1976;84:245–251.

94. Muytjens HL, van der Ros-van de Repe J, nan Druten HAM. Enzymatic Profiles of *Enterobacter sakazakii* and Related Species with Special Reference to the α-Glucosidase Reaction and Reproducibility of the Test System. J. Clin. Microbiol. 1984;20:684–686.

95. Tryland I, Fiksdal L. Enzyme Characteristics of β-D-Galactosidase- and β-D-Glucuronidase-Positive Bacteria and their Interference in Rapid Methods for Detection of Waterborne Coliforms and *Escherichia coli*. Appl. Environ. Microbiol. 1998;64:1018–1023.

96. Köster W, Egli T, Ashbolt N, Botzenhart K, Burlion N, Endo T, Grimont P, Guillot E, Mabilat C, Newport L, Niemi M, Payment P, Prescott A, Renaud P, Rust A. Assessing Microbial Safety of Drinking Water Improving Approaches and Method. WHO & OECD, IWA Publishing; London, UK: 2003. Analytical methods for microbiological water quality testing; pp. 237–292.

97. Manafi M, Kneifel W, Bascomb S. Fluorogenic and Chromogenic Substrates Used in Bacterial Diagnostics. Microbiol. Rev. 1991;55:335–348.

98. Geissler K, Manafi M, Amorós I, Alonso JL. Quantitative Determination of Total Coliforms and*Escherichia coli* in Marine Waters with Chromogenic and Fluorogenic Media. J. Appl. Microbiol.2000;88:280–285.

99. Byamukama D, Kansiime F, Mach RL, Farnleitner AH. Determination of *Escherichia coli*Contamination with Chromocult Coliform Agar Showed a High Level ofDiscrimination Efficiency for Differing Faecal Pollution Levels in Tropical Waters of Kampala, Uganda. Appl. Environ. Microbiol.2000;66:864–868.

100. Eccles JP, Searle R, Holt D, Dennis PJ. A Comparison of Methods Used to Enumerate *Escherichia coli*in Conventionally Treated Sewage Sludge. J. App. Microbiol. 2004;96:375–383.

101. Farnleitner AH, Hocke L, Beiwl C, Kavka GG, Zechmeister T, Kirschner AKT, Mach RL. Rapid Enzymatic Detection of *Escherichia coli* Contamination in Polluted River Water. Lett. Appl. Microbiol.2001;33:246–250.

102. Edberg SC, Kontnick CM. Comparison of β-Glucuronidase-Based Substrate Systems for Identification of *Escherichia coli*. J. Clin. Microbiol. 1986;24:368–371.

103. Leclerc H, Mossel DAA, Edberg SC, Struijk CB. Advances in the Bacteriology of the Coliform Group: their Suitability as Markers of Microbial Water Safety. Ann. Rev. Microbiol. 2001;55:201–234.

104. Klein G. Taxonomy, Ecology and Antibiotic Resistance of Enterococci from Food and the Gastro-Intestinal Tract. Int. J. Food Microbiol. 2003;88:123–131.

105. Kühn I, Iversen A, Burman LG, Olsson-Liljequist B, Franklin A, Finn M, Aarestrup F, Seyfarth AM, Blanch AR, Vilanova X, Taylor H, Caplin J, Moreno MA, Dominguez L, Herrero IA, Möllby R. Comparison of Enterococcal Populations in Animals, Humans, and the Environment—An European Study.Int. J. Food Microbiol. 2003;88:133–145.

106. Wheeler AL, Hartel PG, Godfrey DG, Hill JL, Segars WI. Potentital of *Enterococcus faecalis* as a Human Fecal Indicator for Microbial Source Tracking. J. Environ. Qual. 2002;31:1286–1293.

107. Geldreich EE. Fecal Coliform and Fecal Streptococcus Density Relationships in Waste Discharges and Receiving Waters. Crit. Rev. Environ. Control. 1976;6:349–369.

108. Doran JW, Linn DM. Bacteriological Quality of Runoff Water from Pasteureland. Appl. Environ. Microbiol. 1979;37:985–991.

109. Caplenas NJ, Kanarek MS. Thermotolerant Non-Fecal Source of *Klebsiella pneumoniae*: Validity of the Fecal Coliform Test in Recreational Waters. Am. J. Public Health. 1984;74:1273–1275.

110. Gauthier F, Neufeld JD, Driscoll BT, Archibald FS. Coliform Bacteria and Nitrogen Fixation in Pulp and Paper Mill Effluent Treatment Systems. Appl. Environ. Microbiol. 2000;66:5155–5160.

111. Corso PS, Kramer MH, Blair KA, Addiss DG, Davis JP, Haddix AC. Cost of Illness in the 1993 Waterborne Cryptosporidium Outbreak, Milwaukee, Wisconsin. Emerg. Inf. Dis. 2003;9:426–431.

112. MacKenzie WR, Hoxie NJ, Proctor ME, Gradus MS, Blair KA, Peterson DE, Kazmierczak JJ, Addiss DG, Fox KR, Rose JB, Davis JP. A Massive Outbreak in Milwaukee of *Cryptosporidium* Infection Transmitted Through the Public Water Supply. N. Engl. J. Med. 1994;331:161–167.

113. Vilanova X, Manero A, Cerdà-Cuéllar M, Blanch AR. The Composition and Persistence of Faecal Coliforms and Enterococcal Populations in Sewage Treatment Plants. J. Appl. Microbiol. 2004;96:279–288.

114. Araujo M, Sueiro RA, Gómez MJ, Garrido MJ. Enumeration of *Clostridium perfringens* Spores in Groundwater Samples: Comparison of Six Culture Media. J. Microbiol. Methods. 2004;57:175–180.

115. Charriere G, Mossel DAA, Beaudeau P, Leclerc H. Assessment of the Marker Value of Various Components of the *Coli-Aerogenes* Group of *Enterobacteriaceae* and of a Selection of *Enterococcus* spp. for the Official Monitoring of Drinking Water Supplies. J. Appl. Bacteriol. 1994;76:336–344.

116. Martins MT, Sato MIZ, Alves MN, Stoppe NC, Prado VM, Sanchez PS. Assessment of Microbiological Quality for Swimming Pools in South America. Water Res. 1995;29:2417–2420.

117. Ferguson CM, Coote BG, Ashbolt NJ, Stevenson IM. Relationships Between Indicators, Pathogens and Water Quality in an Estuary System. Water Res. 1996;30:2045–2054.

118. Medema GJ, van Asperen IA, Havelaar AH. Assessment of the Exposure of Swimmers to Microbiological Contaminants in Fresh Waters. Water Sci. Technol. 1997;35:157–163.

119. Polo F, Figueras MJ, Inza I, Sala J, Fleisher JM, Guarro J. Relationship Between Presence of*Salmonella* and Indicators of Faecal Pollution in Aquatic Habitats. FEMS Microbiol. Lett. 1998;160:253–256.

120. Noble RT, Leecaster MK, McGee CD, Weisberg SB, Ritter K. Comparison of Bacterial Indicator Analysis Methods in Stormwater-Affected Coastal Waters. Water Res. 2004;38:1183–1188.

121. Harwood VJ, Levine AD, Scott TM, Chivukula V, Lukasik J, Farrah SR, Rose JB. Validity of the Indicator Organism Paradigm for Pathogen Reduction in Reclaimed Water and Public Health Protection.Appl. Environ. Microbiol. 2005;71:3163–3170.

122. Touron A, Berthe T, Gargala G, Fournier M, Ratajczak M, Servais P, Petit F. Assessment of Faecal Contamination and the Relationship Between Pathogens and Faecal Bacterial Indicators in an Estuarine Environment (Seine, France) Mar. Pollut. Bull. 2007;54:1441–1450.

123. Garrido-Pérez MC, Anfuso E, Acevedo A, Perales-Vargas-Machuca JA. Microbial Indicators of Faecal Contamination in Waters and Sediments of Beach Bathing Zones. Int. J. Hyg. Environ. Health.2008;211:510–517.

124. Ashbolt N, Fujioka R, Glymph T, McGee C, Schaub S, Sobsey M, Toranzos G. Report of the Experts Scientific Workshop on Critical Research Needs for the Development of New or Revised Recreational Water Quality. U.S. Environmental Protection Agency, Office of Water, Office of Research and Development; Warrenton, VA: 2007. Pathogen indicators, and indicators of fecal contamination; pp. 35–56. EPA 823-R-07–006; Chapter 2,

125. Ferreira AP. Caffeine as an Environmental Indicator for Assessing Urban Aquatic Ecosystems. Cad. Saúde Pública. 2005;21:1884–1892.

126. Gardinali PR, Zhao X. Trace Determination of Caffeine in Surface Water Samples by Liquid Chromatography-Atmospheric Pressure Chemical Ionization–Sass spectrometry (LC-APCI-MS) Environ. Int. 2002;28:521–528.

127. Peeler KA, Opsahl SP, Chanton JP. Tracking Anthropogenic Inputs Using Caffeine, Indicator Bacteria, and Nutrients in Rural Freshwater and Urban Marine Systems. Environ. Sci. Technol. 2006;40:7616–7622.

128. Scott TM, JB, Rose TM, Jenkins S, Farrah R, Lukasik J. Microbial Source Tracking: Current Methodology and Future Directions. Appl. Environ. Microbiol. 2002;68:5796–5803.

129. Wu J, Yue J, Hu R, Yang Z, Zhang L. Use of Caffeine and Human Pharmaceutical Compounds to Identify Sewage Contamination. World Acad. Sci. Eng. Technol. 2008;44:438–442.

130. Isobe KO, Tarao M, Chiem NH, Minh LY, Takada H. Effect of Environmental Factors on the Relationship between Concentrations of Coprostanol and Fecal Indicator Bacteria in Tropical (Mekong Delta) and Temperate (Tokyo) Freshwaters. Appl. Environ. Microbiol. 2004;70:814–821.

131. Medema GJ, Shaw S, Waite M, Snozzi M, Morreau A, Grabow W. Assessing Microbial Safety of Drinking Water Improving Approaches and Method. WHO & OECD, IWA Publishing; London, UK: 2003. Catchment characteristics and source water quality; pp. 111–158.

132. EPA . Source Water Protection Practices Bulletin. Managing Stormwater Runoff to Prevent Contamination of Drinking Water. United States Environmental Protection Agency (EPA); Washington, DC, USA: 2009. Office of Water (4606);

133. Ferguson DM, Moore DF, Getrich MA, Zhowandai MH. Enumeration and Speciation of Enterococci Found in Marine and Intertidal Sediments and Coastal Water in Southern California. J. Appl. Microbiol.2005;99:598–608.

134. Boualam M, Mathieu L, Fass S, Cavard J, Gatel D. Relationship between Coliform Culturability and Organic Matter in Low Nutritive Waters. Water Res. 2002;36:2618–2626.

135. Boualam M, Fass S, Saby S, Lahoussine V, Cavard J, Gatel D, Mathieu L. Organic Matter Quality and Survival of Coliforms in Low-Nutrient Waters. J. AWWA. 2003;95:119–126.

136. Baudišová D. Evaluation of *Escherichia coli* as the Main Indicator of Faecal Pollution. Water Sci. Technol. 1997;35:333–336.

137. Edberg SC, Rice EW, Karlin RJ, Allen MJ. *Escherichia coli*: The Best Biological Drinking Water Indicator for Public Health Protection. J. Appl. Microbiol. 2000;88:106S–116S.

138. Byappanahalli MN, Whitman RL, Shively DA, Sadowsky MJ, Ishii S. Population Structure, Persistence, and Seasonality of Autochthonous *Escherichia coli* in Temperate, Coastal Forest Soil from a Great Lakes Watershed. Environ. Microbiol. 2006;8:504–513.

139. Ishii S, Yan T, Shively DA, Byappanahalli MN, Whitman RL, Sadowsky MJ. *Cladophora*(Chlorophyta) spp. Harbor Human Bacterial Pathogens in Nearshore Water of Lake Michigan. Appl. Environ. Microbiol. 2006;72:4545–4553.

140. Ishii S, Ksoll WB, Hicks RE, Sadowsky MJ. Presence and Growth of Naturalized *Escherichia coli* in Temperate Soils from Lake Superior Watersheds. Appl. Environ. Microbiol. 2006;72:612–621.

141. Ksoll WB, Ishii S, Sadowsky MJ, Hicks RE. Presence and Sources of Fecal Coliform Bacteria in Epilithic Periphyton Communities of Lake Superior. Appl. Environ. Microbiol. 2007;73:3771–3778.

142. Hecq P, Hulsmann A, Hauchman FS, McLain JL, Schmitz F. Drinking water regulations. In: Quevauviller P, Thompson KC, editors. Analytical Methods for Drinking Water. John Wiley; London, UK: 2006. Chapter 1.

143. Total Coliform Rule: A Handbook for Small Noncommunity Water Systems Serving less than 3,300 Persons. Office of Water (4606M), EPA; Washington, DC, USA: 2006. EPA 816-B-06–001;

144. Espigares M, Coca C, Fernández-Crehuet M, Moreno O, Gálvez R. Chemical and Microbiologic Indicators of Faecal Contamination in the Guadalquivir (Spain)

6
CHAPTER

ABUNDANCE OF SULFUR-DEGRADING BACTERIA IN A BENTHIC BACTERIAL COMMUNITY OF SHALLOW SEA SEDIMENT IN THE OFF-TERENGGANU COAST OF THE SOUTH CHINA SEA

Zahar Marziah[1], Akbariah Mahdzir[1,2], Md. Nor Musa[2], Abu Bakar Jaafar[3], Azran Azhim[4] and Hirofumi Hara[1]

[1]Malaysia-Japan International Institute of Technology, Universiti Teknologi Malaysia, Jalan Sultan Yahya Petra, Kuala Lumpur 54100, Malaysia

[2]Ocean Thermal Energy Centre (OTEC-KL), Universiti Teknologi Malaysia, Jalan Sultan Yahya Petra, Kuala Lumpur 54100, Malaysia

[3]Perdana School of Science, Technology and Innovation Policy, Universiti Teknologi Malaysia, Jalan Sultan Yahya Petra, Kuala Lumpur 54100, Malaysia

[4]Kuliyyah of Science, International Islamic University Malaysia, Kuantan 25200, Malaysia

ABSTRACT

This study for the first time provides insight into the bacterial community in the benthic region of the Off-Terengganu Coastline, which is considered to be anthropogenically polluted due to heavy fishing vessel commotion. Subsurface bacteria were randomly collected from two locations at different depths and were examined using the 16S rDNA V3-V4 marker gene on the Illumina™ Miseq platform. In addition, the physiochemical parameters of the sediment were also measured. Surprisingly, the results show a high diversity of sulfur-oxidizing bacteria in the surveyed area, where Sulfurovum sp. was identified to predominate the overall bacterial community. The physiochemical parameters reveal insufficient evidence of hydrothermal vents in the surveyed area. However, there are traces of hydrocarbon pollutants such as gasoline, diesel, and mineral oil in this

Citation: Zahar Marziah, Akbariah Mahdzir, Md. Nor Musa, Abu Bakar Jaafar, Azran Azhim andHirofumi Hara, Abundance of sulfur-degrading bacteria in a benthic bacterial community of shallow sea sediment in the off-Terengganu coast of the South China Sea, DOI: 10.1002/mbo3.380.

area. It is assumed that sediment accumulation in the lee of breakwater plays an important role in trapping the runoff from the nearby harbor, which includes oil spills. Based on the common knowledge, Sulvurofum sp. is a native bacterium that exists in deep hydrothermal vents and volcanic territories. Although the reason for the abundance of Sulfurovum sp. in the surveyed area is still unclear, there is a possibility that metabolic adaptation plays an important role in regulating hydrocarbon pollutants for survival. The work presented in this paper therefore has profound implications for future studies on Sulfurovum sp. versatility. However, future research is needed to strengthen the findings of this study and to provide a better evidence regarding the metabolic response of this bacterium toward hydrocarbon pollutants.

INTRODUCTION

Marine life has existed in oceans since almost 3.5 billion years when microbes were the only form of life in two-thirds of the planet's existence. Although their diversity and metabolic contribution in marine microbiology remains indecisive (Munn 2011), it is widely assumed that marine microbes play a vital role in regulating marine ecological processes, which include the biogeochemical cycles (Danovaro et al. 2015). In recent years, marine microbial research is inclined to investigate the microbial community profiles in order to expand the species identification and variance in different environments (Zhu et al. 2013; Bokulich et al., 2014; Wang et al., 2015b). In recent years, marine microbial deviation and its mechanism are mainly investigated in local communities and may usually include studies on physiochemical parameters to justify the microbial metabolic activities (Van der Gucht et al. 2007 and Wang et al. 2015).

In this study, the region of interest was located in the Southern South China Sea (SSCS) (refer to Fig. 1). Geographically, the SSCS has a shallow, ±50 m neritic epipelagic seabed and, presumably has effective photosynthesis, which contributes to a high coral distribution globally (Taylor and Hayes 1983; Morton and Blackmore 2001; Wang et al. 2007a,b). In addition, the SSCS is signified as the heart that connects Eurasia with the Americas, as the largest shipping port in the world is located here (Fan et al., 2016). Furthermore, the SSCS clenches huge reserves of Tapis-grade crude oil beneath its seabed (Ismail et al. 2015).

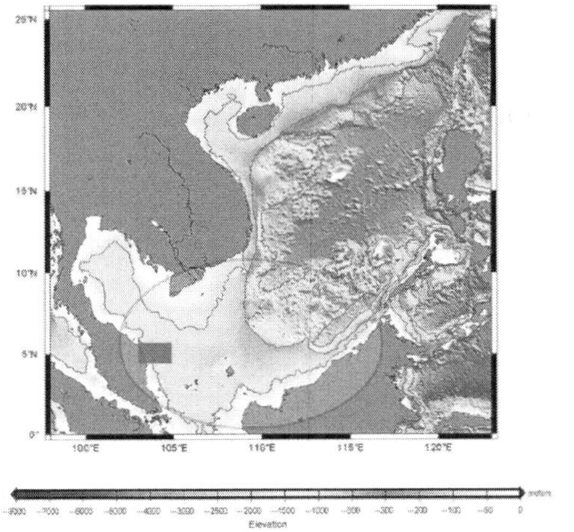

Figure 1: Illustration of South China Sea seafloor bathymetry. The Southern part of the South China Sea is marked in an ellipse. The area bounded by the rectangle indicates the region of interest; Terengganu State in Peninsular Malaysia (5°N 103°E) (Image courtesy of Liu and Dittert 2010).

This study was conducted off the Terengganu coastline area (5°20›N, 103°09›E) as a part of bacterial diversity research to contribute toward the understanding of Malaysian waters, which is also a part of the SSCS region. In addition to the neritic epipelagic bed, this area has an immense collection of tropical marine life forms (Arai 2015), besides being surrounded by a high-quality crude oil cluster (Ismail et al. 2015). One prominent feature that represents this particular Off-Terengganu coastline is the presence of a stern-curve breakwater (refer to Fig. 2) that was constructed to protect the Kuala Terengganu jetty and the nearby estuaries.

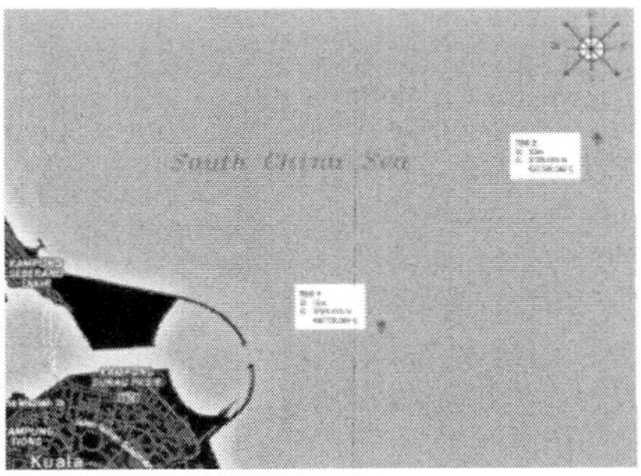

Figure 2: Illustration of sampling points and breakwater structure located in the Off-Terengganu coastline.

A few conditions might influence the outcomes of the bacterial phylogenetic profile in the Off-Terengganu coastline. In a recent study, high values of biochemical oxygen demand (BOD), chemical oxygen demand (COD), total suspended solids (TSS), and ammonical nitrogen (AN) were recorded in the Kuala Terengganu river, which is directly connected to the Off-Terengganu coastline (Suratman et al. 2015). Another study specifies that anthropogenic sources such as municipal waste, surface runoff, agricultural runoff, organic pollution, and urban storm runoff have polluted this location (Kamaruddin et al. 2015). Given the context of possible sediment amiability toward the anthropogenic pollutant potency in the surveyed areas, the bacterial phylogenetic profile might not represent an abundance of the native marine bacterial community. Instead, it may actually illustrate a unique bacterial community with the capability to utilize inorganic compounds such as sulfur as its food sources, or, perhaps, a water-borne bacterium that poses a threat in causing disease to the marine community and to humans.

The aim of this study is mainly to create a steadfast foundation of the marine bacterial community in the SSCS region – specifically in Malaysian seawater – since no published phylogenetic profile has been conducted in the surveyed areas. This study also comprises of several physico-geochemical analyses to investigate the sediment constituents.

MATERIALS AND METHODS

Based on Table 1, sampling activity was conducted on 30th November, 2014 using the Smith McIntyre grab sampler (0.1 m^2 of grab size) at two points (Labeled as "TDS1" and "TDS2") in a north-eastern direction. This sampling method was based on the methodology recommended by Holme and McIntyre (1984) where all samples must be handled carefully, kept in a double-layered polyethylene bag, and stored in a freezer at −25°C until further analysis. No specific permits were required for the described sampling because it did not involve any endangered species nor did it occur within a designated marine-protected area or private reserved park.

Table 1: Location information for sampling sites

Sampling Sites	Longitude (E)	Latitude (N)	Time of sampling (hours)	Depth (m)	Approximate distance initial points[a] (km)
[a]The initial points located in the Pulau Duyong Harbour, Kuala Terengganu. Approximately 4.01 km from the first sampling point.					
TSD1	103°09.954'E	5°20.413'N	09:45	±15	4.01
TSD2	103°09.342'E	5°20.603'N	13:50	±55	8.27

Isolation and bacterial characterization

Genomic DNA was extracted from 1 g of sediment using the PowerSoil®DNA Isolation Kit (MO BIO, Carlsbad, CA) according to the manufacturer's protocols. The extracted DNA product was assessed for its integrity and concentration by a standard agarose gel analysis. Qubit® 2.0 DNA Kit (Invitrogen by Thermo-Scientific Inc. Waltham, MA) was used to quantify the precise DNA product to ensure sufficient DNA amounts for the Polymerase Chain Reaction (PCR). The primer used for PCR was a universal primer for 16SrDNA that targeted the V3 region: -341F (5′ CCTACGGGN GGCWGCAG 3′) and 805R (5′ GACTACHVGGGTATCTAATCC 3′). The amplified product was then analyzed by agarose gel electrophoresis and recovered with the Sangon agarose recovery kit (Sangon Biotech Co., Ltd., Shanghai, China). Subsequently, the recovered DNA products were quantified and mixed to 1:1 ratio based on DNA concentration determined by Qubit® 2.0 Fluorometer (Invitrogen by Thermo-Scientific Inc.). Pyrosequencing analysis was conducted on the Illumina® Miseq platform (Illumina Inc., San Diego, CA) at Sangon Biotech Co., Ltd..

DNA sequence analysis

A total of 37,363 sequences that span the 16S rDNA V3-V4 hypervariable region were identified and filtered using the Illumina Miseq™platform (Illumina Inc.,). Random sequences, ambiguous residues, and sequence lengths of than 150 bp were eliminated. Quality control (QC) for the raw sequences was performed with PRINSEQ-lite 0.19.5 to truncate the low-quality data and improve the merge ratio for subsequent sequences. Using Flash v1.2.7 (University of Maryland, MD, USA), the raw sequence fragment was merged in a dual terminal to form a single primer. Subsequently, short, low-complexity, and low-quality primer fragments were eliminated by PRINSEQ-lite 0.19.5 software (Soundforge Media, La Jolla, CA). Correction of sequencing errors was performed with precluster software and was integrated by the Mothur software (Michigan, MI, USA). Subsequently, chimeras and extraterritorial sequences of the target area were removed with the Uchime software using SILVA data as the template. By the time the QC ended, primer length was successfully aligned between 400–500 bp, with an average of 450 bp. All V3 and V4 optimized sequence reads were determined by RDP classifier 16S (Wang et al. 2007a,b) and Silva 16S (Quast et al.2013).

Diversity and statistical analysis

The sequence parameter for similarity and operational taxonomic unit (OTU) was set to 97%, which is close to genus probability. For the first step, OTU clustering was performed using UCLUST v.1.1.579 to select the longest reads from the clean sequence as seed sequences (Edgar2010). In the second step – a sequence with similarity to the seed sequence within the threshold range – was then selected. Finally, all the sequences obtained from the first and second steps were classified into one OTU category. All three steps of the above process were repeated until all the sequences were successfully classified. The taxonomic unit was classified with the RDP classifier based on Bergey's taxonomy using Bayesian assignment calculation to calculate the probability of each sequence being assigned to the rank on the genus level. One representative sequence with the highest OTU abundance was automatically distinguished by the RDP classifier to categorize the species, with the default value of taxonomy threshold being 0.8/0.5. A cluster of multiple sequences based on the distance between sequences, OTU classifications, and the similarity of the sequence threshold value was determined by the Mothur™ software. Subsequently, all sequence clusters were calculated based on the α-diversity index analysis (based on Richness index, Shannon index, ACE index, Chao1 index). The rarefaction curve value and graph were generated based on

97% of the sequence similarity threshold on every species, genus, and family level analyzed. β-diversity index analysis was excluded from the diversity study due to data deficiency (which requires at least three samples to generate a satisfactory -diversity index).

Physico-chemical analysis

In situ water quality analysis

The Hydrolab Multiparameter Sonde DS5X was used to evaluate the in situ water quality, with seven parameters analyzed: temperature, pH, specific conductivity, salinity, total dissolve solid (TDS), turbidity, and luminescent dissolved oxygen (LDO). The multiparameter probe was cleaned and calibrated prior to each sampling session. Eleven to 12 readings for each parameter were obtained in a single point where every output was directly linked (by GPS) and recorded into the Aqualab Hydras 3 LT Software for Microsoft® Windows 7. Statistical analysis was performed with SPSS 16.0 (SPSS Inc. Chicago, U.S) for Microsoft® Windows 7. The results were interpreted based on Pearson correlation with $P \leq 0.05$ and $P \leq 0.01$ being considered as significant.

CHNS elemental analysis

Rapid identification of carbon (C), hydrogen (H), nitrogen (N), and sulfur (S) in the sedimentary sample was performed using the Vario MACRO™ cube CHNS is acronym that is combined from Carbon (C), Hydrogen (H), Nitrogen (N) and, Sulfur (S). (Elementary, Deutschland). The sediment samples were air-dried in a 50°C oven and then ground, sieved (<2 mm), and homogenized according to the ISO 2004 protocol. Sulfur determination was conducted according to the ISO 2005 protocol.

Oil and grease analysis

Oil and grease (O&G) determination was conducted using a partition-gravimetric method. Specifically, the Hexane Extractable Method – USEPA 1664 (EPA 1999) was used. The oil and grease in the sediment were extracted from water and then attached to *n*-Hexane solvent. The solvent was allowed to evaporate slightly before transferring it to a preweighed culture tube. The solvent was further evaporated completely until dry. The culture tubes were then weighed again (EPA 1999 and Bucci et al. 2015).

TPH analysis

Total petroleum hydrocarbon (TPH) was measured based on the USEPA 8015B test method (EPA 2000). A quantity of 10 g of chilled fresh sediment was transferred into vials with a solid cap and a Teflon septum. A quantity of 20 mL of *n*-Pentane solution was added to the same vial and mixed homogenously by centrifugation for 15 min. The mixture was allowed to settle for 1 hour at room temperature and then considered ready for gas chromatography with flame ionization detector (GC-FID) analysis. Each sample mixture was passed through the Agilent J&W Capillary (DB-5

30 m × 0.25 mm × 0.25 μm) into the Agilent 7890A GC-FID with a carrier gas (Helium) flow rate of 40 cm/sec. Internal quality control was performed by considering ±5% as the acceptance criteria.

RESULTS

A total of 15,268 cleaned effective sequences based on 16S rDNA V3-V4 were grouped into 2,156 unique OTUs (Operational Taxonomic Unit), where one OTU denotes a sequence with an identity value equal to or higher than 97% (Zhu et al. 2013; Wang et al. 2015). Based on the α-diversity index report in Table 2, TSD1 demonstrated higher species richness compared to TSD2 by 70%. This was most likely contributed by the lush sea grass vegetation in the TSD1 area that enhanced organic matter accretion in the sediment (Baden et al. 2010; Jankowska et al.2015). The previously abundant marine bacteria gradually decline with increasing depth of the seawater. This is assumed to be due to the reduced seagrass vegetation at these depths (García-Martínez et al. 2008; Jankowska et al. 2015). In general, the accumulation of organic and inorganic matters in seagrass allows marine bacteria to regulate nitrogen and phosphorous cycling, and thereby, maintain seagrass productivity in seawater (Donnelly and Herbert 1998).

Table 2: The list of alpha diversity index cumulative results for TSD1 and TSD2

Sample ID	Seq. num	OTU num	Shannon index	ACE index	Chao1 index	Coverage index
OTU, operational taxonomic unit.						
TSD1	8210	1496	3.8097	8407.942456	4711.624	0.865652
TSD2	7058	660	2.0567	5949.750563	2538.676	0.92675

Phylogenetic profile report

The bacterial phylogenetic profiles are as depicted in Figure 3. The results indicate that the phylum Proteobacteria tops the overall phylum abundance in the surveyed area with 85% (TSD 1) and 95% (TSD 2). In a general perspective, Epsilonproteobacteria predominates in the bacterial community with 60% (TSD 1) and 88% (TSD2), respectively. This is followed by Gammaproteobacteria with 15% and 5%, Deltaproteobacteria with 5% and 3%, and Alphaproteobacteria with 4% and 1%, respectively. These results also show that an unclassified phylum is the second most abundant phylum detected with 5% and 3%, respectively, followed by the phylum Chloroflexi with 3% and 1%, respectively.

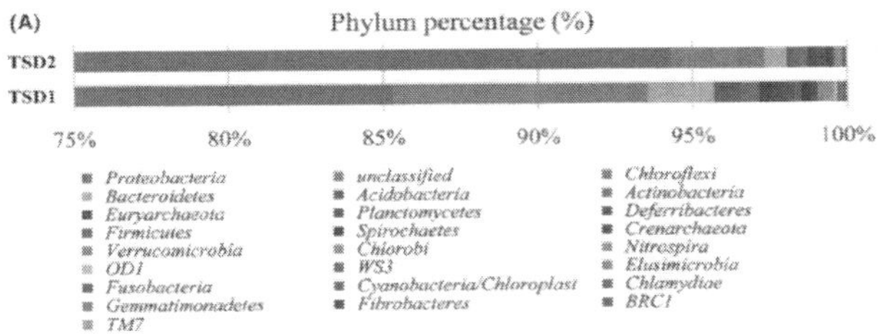

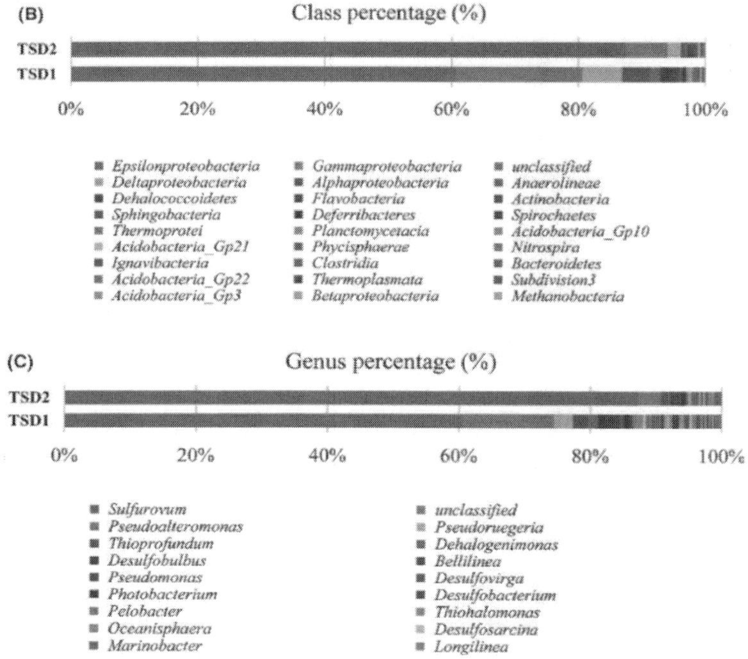

Figure 3: An Illustration of bacterial diversity and phylogenetic distribution based on RDP classifier 16S (Wang et al. 2007a,b) and Silva 16S (Quast et al.2013). The RDP bar was generated based on operational taxonomic unit with the range of 21–27 phylum levels, 35–51 class levels, 46–62 order levels, 76–108 family levels, and 134–217 of genus levels identified in two sampling locations, respectively.

This study reveals that *Sulfurovum* sp. was the only genus that predominantly covered the Epsilonproteobacteria genera in both surveyed areas with 60% (TSD1) and 88% (TSD2). Other than *Sulfurovum* sp., TSD1 also contained several sulfur-oxidizing genera from Gammaproteobacteria, which include *Thioprofundum, Desulfobulbus, Desulfovirga, Desulfobacterium, Desulfosalsimonas, Sulfurimonas, Sulfuricurvum,* and *Thermodesulfovibrio* (Fig. 3 section C). No inclusion data were obtained to demonstrate the species composition of*Sulfurovum* sp. since the 16S rDNA phylogenetic profile is limited to identifying bacterial genus within the V3–V4 hypervariable region. Therefore, further investigation is required to determine the species composition of *Sulfurovum* sp.

Physicochemical reports

In situ water quality analysis

The bottom of the seawater level in both surveyed areas was measured by in situ water quality analysis (refer to Table 3). The results indicate that the temperature, pH, Sp.

conductivity, salinity, TSD, and LDO values in both surveyed areas are significant. However, seawater turbidity in TSD1 indicates an insignificant value (P = 1.03573), whereas the value of TSD2 is significant (P = 0.0000). Despite a constantly irregular, high turbidity value in TSD1, the probability sediment disruption was caused by wave dissipation in the lee of the breakwater, perhaps resulting in particulate matter retention from water runoff in TSD1 (refer to Fig. 2). A stable sediment layer probably contributed to the zero turbidity value in TSD2 because it is remotely located from the breakwater (Qi and Gao 2015).

Table 3: Result of environmental conditions and physio-geochemical analyses

No	Parameters	TSD1	TSD2
	Bottom seawater in situ water quality analysis		
1	Temp (°C)	30.5227 ± 0.00647	30.0683 ± 0.00389
2	pH	7.8200 ± 0.02	7.9325 ± 0.00452
4	Specific conductivity (mS/cm)	60.8091 ± 0.5394	60.700 ± 0.4264
5	Salinity (ppt)	40.8655 ± 0.2734	40.7958 ± 0.3288
6	TDS (g/L)	38.9091 ± 0.3015	38.8917 ± 0.2887
7	LDO (mg/L)	6.6482 ± 00874	6.7717 ± 0.06726
	Sediment TOC analysis		
8	TOC ppm	4600	5200
	Sediment elemental analysis (CHNS)		
9	Carbon (%)	1.86	1.25
10	Hydrogen (%)	1.017	0.035
11	Nitrogen (%)	0.99	0.58
12	Sulfur (%)	0.916	0.212
	Sediment oil and grease analysis		
13	HEM (%)	0.47	0.08
	Sediment TPH analysis		
14	C_6–C_9 (ppm)	0.05	[a]ND
15	C_{10}–C_{19} (ppm)	0.10	0.11
16	C_{20}–C_{36} (ppm)	0.22	0.29
17	C_{37}–C_{44} (ppm)	[a]ND	[a]ND
[a] ND, Not detected. TOC, Total Organic Carbon, HEM, Hexane extractable method; TPH, Total Petroleum Hydrocarbon; LDO, Luminescent dissolved oxygen, TDS, Total dissolve solid.			

Total organic carbon

The results indicate that the total organic carbon (TOC) sedimentary values would increase with depth. In correspondence to the zero water turbidity value in TSD2, it is assumed that sunlight effectively penetrates into the clear water column. This supports photosynthesis, thus creating a better marine food chain environment and generating high organic matter from cell remains (Bell et al. 2015 and Saraswathy et al.2015) where it demonstrates high TOC value (Bendtsen et al. 2015). However, there is no concrete evidence that links *Sulfurovum* sp. abundance with the high TOC in both seawater and the sedimentary layer.

Elemental (CHNS) analysis

The CHNS analysis was mainly performed to observe the overall elemental composition in the surveyed area. Based on Table 3, all four of the main elemental ratios, including sulfur, were scarcely identified. Further analysis is therefore necessary to investigate sulfur concentration in order to demonstrate a convincing association of the *Sulfurovum* sp. with the sulfur content in the surveyed area.

Hexane extracted method and total petroleum analysis

Based on the Hexane Extracted Method (HEM) analysis, the oil and grease fragments were fairly identified to be 0.47% (TSD1) and 0.08% (TSD2). Since HEM assessment showed a promising value, it was necessary to thoroughly quantify the hydrocarbon compounds using TPH analysis (Bucci et al. 2015). The outcome of TPH analysis confirmed the existence of gasoline (C_4–C_9), diesel (C_{10}–C_{19}), and organic oil (C_{20}–C_{36}) fractions in TSD1 at 0.05 ppm, 0.10 ppm, and 0.22 ppm, respectively. Conversely, TSD2 sediment only traced diesel fraction (C10–C19), and organic oil fraction (C20–C36) at 0.11 ppm and 0.29 ppm, respectively. No asphalt / bitumen fraction (C37–C44) was detected in both samples.

DISCUSSION

One of the interesting features that signify the off-Terengganu coastline is the presence of a breakwater structure. This breakwater was built mainly to reduce the wave intensity in the inshore water as part of the coastal defense and as an anchorage protection from the weather and longshore drift effects. However, the dissipation of energy and the formation of relative calm water often result in sediment accretion and salient build-up in the lee of a breakwater (Van Rijn 2010). Furthermore, excessive rainfall inside the breakwater area would cause a runoff that is eventually trapped within the breakwater (Butt 2013). Our finding suggests that turbidity plays an important role in visual prediction of the water condition. It represents a mixture of sediment and xenobiotic compounds that have accumulated for decades. Most of the time, this occurrence would result in unfavorable consequences toward the marine sediment (Jonsson et al. 2006). For example, a 13.4 km breakwater structure was built along the Long Beach, CA coastline to protect the U.S naval ships during World War II. Although the harbor is no longer in use since 1996, the harmful xenobiotic compounds that remained trapped in the lee of breakwater, made the place inaccessible for recreational purposes (Butt 2013). There is a strong probability that TSD1 is positioned in an active wave dissipation area because of water turbidity. Therefore, the sedimentary layer of TSD1 may have richer particulate matter retention from water runoff as compared to TSD2, which ultimately intensifies the concentration of anthropogenic pollutants in the breakwater opening.

Given the context that the majority of bacterial communities in both sampling points are a sulfur-degrading group, it is expected that the marine sediment possibly contain a sulfur component. The lack of sulfur ratio in the elemental analysis indicates that there are no hydrothermal vents or volcanic rock structures in both surveyed areas. However, aromatic hydrocarbon, which has high affinity for sediment (McElroy et al. 1989) and mainly contributes to marine coastline pollutants (Mitsch 2010; Suárez-Suárez et al. 2011), was discovered at both sampling points. Both areas were polluted with gasoline, diesel, and mineral oil (excluding TSD2) – perhaps due to oil spills from fishing vessels and high-speed boats. Based on the available information, 2216 units of fishing vessels were registered in the surveyed area (Kuala Terengganu district) in the year 2001, with 316 units being outboard-powered vessels while the rest were inboard-powered vessels (Information of Fisheries Management in Malaysia 2001). The inboard-powered motors were mainly fueled by gasoline or diesel while the outboard motors were fueled by gasoline with 0–10% of ethanol blended fuel.

Marine bacteria have played an essential role in regulating the oceanic ecosystem for millions of years by controlling the geochemical processes (He et al. 2009). In the pelagic realm, bacteria are indispensable for two major reasons: they are consumed by other organisms and they degrade the organic matter to sustain ecological systems. Bacteria are thus positioned at both the start and the end of the food chain, where they contribute to the production of particulate foodstuff by converting dissolved organic substrates. They are also responsible for the ultimate breakdown of organic matter that leads to the return of nutrients to the sea (Li and Dickie 2003). Bacteria may be the crucial link or sink between detritus, dissolved organic matter, and higher trophic levels. For these reasons, bacteria occupy a central role in two interconnected environmental issues of global concern, namely the sustenance of harvestable living resources and the mitigation of climate change by sequestration of carbon in the deep ocean (Li and Dickie 2003).

To date, Proteobacteria remains the most abundant environmental microbe phylum that is broadly discovered and cultured. It has been widely detected in various environmental conditions such as in cold seawater regions (Sapp et al. 2010 and Stibal et al. 2015), marine sediment (Zhu et al. 2013; Wang et al. 2015a,b), marine sponges (Schmitt et al. 2012), and organic compounds (Kleinsteuber et al. 2008 and Lin et al. 2014). It is also the predominant phylum identified in both surveyed areas at the Off-Terengganu coastline. However, certain regions demonstrate the dominance of a different clade of Proteobacteria. For example, the Alphaproteobacteria genera dominate the benthic bacterial community in the Atlantic Ocean at 55.7% Zinger et al. 2011 and in the East China Sea at 20.1% (Wang et al. 2015). Meanwhile, Gammaproteobacteria flourish in the Northern South China Sea (NSCS) at 53.4% (Zhu et al. 2013). The results from this study provide unique findings, where they exhibit the first insight into the abundance of the Epsilonproteobacteria group in a shallow, nonhydrothermal feature coastline. Generally, the Epsilonproteobacteria group is widely known for its ecologically significant bacterial community in deep hydrothermal vents (López-García et al. 2003 and Zhu et al. 2013). Furthermore, it is also abundant in shallow hydrothermal vents (Sunamura et al. 2004; Takai et al. 2004; Giovannelli et al. 2013; Wang et al. 2015b), sulfite compounds (Nakagawa et al. 2005), and deep sea vent metazoans (López-García et al.2003; Goffredi et al. 2004). The abundance of Epsilonproteobacteria in anthropogenic pollutants, however, is still not well-understood (Inagaki et al. 2004; Takai et al. 2004; Nakagawa et al. 2005) even though few recent studies have discovered their potential to survive in certain organic pollutants (Bolhuis et al. 2014; Lin et al. 2014).

Biological activity in marine sediments is vital because it is partially responsible for the primary marine production and the overall geochemical process (Danovaro et al. 2015). It is assumed that most of the bacterial communities that dwell in coastal sediments have a specific purpose. For instance, 50% of the deposited minerals in the coastline setting are mineralized via sulfate reduction (Jørgensen 1982), which is controlled by sulfur-degrading bacteria (Suárez-Suárez et al. 2011).

The outcome of this study describes that *Sulfurovum* sp. is the only genus that covers the entire Epsilonproteobacteria phylogeny profile in both surveyed areas. This genus is characterized by its egg-like coccoidal shape and is capable of oxidizing sulfur for food and survival. (Inagaki et al. 2004 and Takai et al. 2004). *Sulfurovum* sp. is a gram negative, nonmotile genus that is categorized under sulfur-oxidizing chemoautotrophic genera and was first isolated from a deep sea hydrothermal vent in Okinawa, Japan (Inagaki et al. 2004). It prefers a moderate temperature of between 20°C to 45°C and a medium salinity (Willey et al. 2008). Although the metabolic properties for the majority of the *Sulfurovum* genus remain indecisive, one of the strains, *Sulfurovum* 42BKTT grew chemolithoautotrophically with elemental sulfur or thiosulfate as the sole electron donor and oxygen (optimum 5 % in gas phase) or nitrate as the electron acceptor. The G + C content of its genomic DNA was 48.0 mol% (Inagaki et al. 2004). In a recent finding, the *Sulfurovum* genus inhibits its growth in fresh water rivers (Hubert et al. 2012), high turbidity waters, and acidic conditions (Bolhuis et al. 2014). Therefore, it is probable that the high turbidity value at TSD1 may likely be the main cause of truncated *Sulfurovum* sp. abundance as compared to TSD2.

Overall, *Sulfurovum* is a predominant species identified in deep hydrothermal vents (Inagaki et al. 2004; Wright et al. 2013; Dahle et al. 2015), shallow hydrothermal vents (Giovannelli et al. 2013), volcanic regions (Wang et al. 2015b), caves, sinkholes, and sulfide compounds (Nakagawa et al. 2005; Handley et al. 2012; Jones et al. 2012). Its metabolic versatility was recently recognized where several studies indicate its role in degrading aromatic hydrocarbons (Paissé et al. 2008; Païssé et al. 2010; Håvelsrud et al. 2011; Lin et al. 2014), such as benzene, phenols, and toluene (Kleinsteuber et al. 2008). Furthermore, *Sulfurovum* sp. together with other sulfur-oxidizing bacteria has the capability to produce active surfactants (Grabowski et al. 2005; Xiu et al. 2010). It is suggested that the majority of the sulfur-oxidizing bacteria abundant in oil reservoirs are mainly affected by temperature, mineralization, permeability, and water displacement (Lin et al. 2014) and are stimulated by certain heavy metals effluents such as barium, iron, and manganese, which are discharged from hydrocarbon energy plants (Yeung et al. 2011). In a natural environment, the *Sulfurovum* sp. was discovered in hydrocarbon-polluted coastal seawater such as at a coal oil point in California, USA (Håvelsrud et al. 2011), Berre lagoon in France (Paissé et al. 2008), and Busan Northport in South Korea (Subha et al. 2014). However, none of the above studies exhibit high *Sulfurovum* sp. abundance in a hydrocarbon pollutant compared to its abundance in this particular report at the Off-Terengganu coastline.

Novel findings of this study

This research exhibits an alarming possibility that Off-Terengganu is vulnerable to the impact of anthropogenic pollution. Moreover, no exact calculations have been made to evaluate the anthropogenic-prone areas along the Off-Terengganu coastline and in its surrounding areas. Hypothetically, anthropogenic pollution in a marine ecosystem is dependent on the history of environmental pollution itself. It is believed that the bacterial community adapted to a previous oil spill would then recur faster than in a pristine environment due to its metabolic readiness to utilize the hydrocarbon compounds (Paissé et al. 2008).

The novelty of *Sulforuvum* identification at the Off-Terengganu coastline is evident because this genus was found out of its native preference such as hydrothermal vents and volcanic regions. Furthermore, this study depicted one of the highest *Sulfurovum* sp. distributions ever reported in a natural environment, showing the broadening versatility of its genus in adapting to a different environmental condition. Based on the biodiversity index, *Sulfurovum* sp. abundance shows increase with depth and requires additional effort to generate better species coverage. In the future, this study will require an extensive *Sulfurovum* composition identification and abiotic analysis such as Polycyclic Aromatic Hydrocarbon (PAH) parameter to investigate the interaction of *Sulfurovum* with carcinogenic and toxic compounds in the environment. This will generate better geochemical information, dispersion scale, community variation, and environmental anthropogenic influence.

Data repository

The sequence data from this research have been deposited in the NCBI's Sequence Read Archive database (http://www.ncbi.nlm.nih.gov/sra) with the temporary submission ID of (SUB1112034).

ACKNOWLEDGMENT

We are hugely appreciative of the financial support by the Malaysia-Japan International Institute of Technology (MJIIT) University Teknologi Malaysia, Ministry of Education (MOE), Ocean Thermal Energy Centre and Integrated Envirotech Sdn. Bhd. (IESB). This research was funded partly by the Assessment of Macrobenthos & Heavy

Metal Distribution in Deep Seawater Grant (R.K130000.7643.4C052), An Attempt to Replicate the Composition of Zamzam Water by Processing Deep Sea Water (Q.J130000.2424.00G62), An Integrated Approach to Heavy Metal & Biological Assessment in Deep Seawater Grant (Q.K130000.3043.00M43) and Matching Grant (Q.K130000.3043.00M46). We are grateful to Sangon Biotech Co., Ltd. for their wonderful sequence analyses and various anonymous reviewers for their various constructive advice.

REFERENCES

1. Arai, T. 2015. Diversity and conservation of coral reef fishes in the Malaysian South China Sea. *Rev. Fish Biol. Fisheries* **25**:85–101.
2. Baden, S., C. Böstrom, S. Tobiasson, H. Arponen, and P. O. Moksnes. 2010. Relative importance of thropic interaction and nutrient enrichment in seagrass ecosystems: a broad-scale field experiment in the Baltic-Skagerrak area. *Limnol. Oceanogr.* **55**:1435–1448.
3. Bell, J. J., E. McGrath, A. Biggerstaff, T. Bates, H. Bennett, J. Marlow, et al. 2015. Sediment impacts on marine sponges. *Mar. Pollut. Bull.* **94**:5–13.
4. Bendtsen, J., K. M. Hilligsøe, J. L. Hansen, and K. Richardson. 2015. Analysis of remineralisation, lability, temperature sensitivity and structural composition of organic matter from the upper ocean. *Prog. Oceanogr.* **130**:125–145.
5. Bolhuis, H., H. Schluepmann, J. Kristalijn, Z. Sulaiman, and D. J. Marshall. 2014. Molecular analysis of bacterial diversity in mudflats along the salinity gradient of an acidified tropical Bornean estuary (South East Asia). *Aquat. Biosyst.* **10**:1–13.
6. Bokulich, N. A., J. H. Thorngate, P. M. Richardson, and D. A. Mills. 2014. Microbial biogeography of wine grapes is conditioned by cultivar, vintage, and climate. *Proc Natl Acad Sci* USA **111**:E139–E148.
7. Bucci, R., A. M. Girelli, S. Tafani, and A. M. Tarola. 2015. Oils and grease determination by FT-IR and n-hexane as extraction solvent. *J. Anal. Chem.* **70**:316–319.
8. Bui, V. D., A. Schimanski, K. Stattegger, P. Van Phung, T. H. Nguyen, T. T. Nguyen, et al. 2009. Sandwaves on the Southeast Vietnam shelf recorded by high resolution seismic profiles: formation and mechanism. *Front Earth Sci. China* **3**:9–20.
9. Butt, J.. 2013. LONGBEACH-2013: Coastal and Ocean Engineering ENGI.8751 Undergraduate Student Forum Faculty of Engineering and Applied Science, Memorial University, St. John's, NL, Canada. March 2013.
10. Dahle, H., I. Økland, I. H. Thorseth, R. B. Pederesen, and I. H. Steen. 2015. Energy landscapes shape microbial communities in hydrothermal systems on the Arctic Mid-ocean ridge. *ISME J.* **9**:1593–1606. doi:10.1038/ismej.2014.247.
11. Danovaro, R., C. Corinaldesi, E. Rastelli, and A. Dell'Anno. 2015. Towards a better quantitative assessment of the relevance of deep-sea viruses, bacteria and

archaea in the functioning of the ocean seafloor. *Aquat. Microb. Ecol.* **75**:81–90.

12. Donnelly, A. P., and R. A. Herbert. 1998. Bacterial interactions in the rhizosphere of seagrass communities in shallow coastal lagoons. *J. Appl. Microbiol.* **85**(S1):151S–160S.
13. Edgar, R. C. 2010. Search and clustering orders of magnitude faster than BLAST. *Bioinformatics* **26**:2460–2461.
14. EPA. 1999. Method 1664, Revision A: N-Hexane Extractable Material (HEM; Oil and Grease) and Silica Gel Treated N-Hexane Extractable Material (SGT-HEM; Non-polar Material) by Extraction and Gravimetry. EPA-821-R-98-002; February 1999. Avaliable at: Url: http://water.epa.gov/scitech/methods/cwa/oil/1664.cfm (accessed: 4 February 2016)
15. EPA. 2000. Method 8015C (SW-846): Nonhalogenated Organics Using GC/FID. Revision 3.
16. Fan, Q., Y. Zhang, W. Ma, H. Ma, J. Feng, Q. Yu, et al. 2016. Spatial and seasonal dynamics of ship emissions over the Yangtze River Delta and East China Sea and their potential environmental influence. *Environ. Sci. Technol.* **50**:1322–1329.
17. García-Martínez, M., A. López-López, M. L. Calleja, N. Marbà, and C. M. Duarte. 2008. Bacterial community dynamics in a seagrass (*Posidonia oceania*) meadow sediment. *Estuar. Coast. Shelf Sci.* **32**:276–286.
18. Goffredi, S. K., A. Warén, V. J. Orphan, C. L. Van Dover, and R. C. Vrijenhoek. 2004. Novel forms of structural integration between microbes and a hydrothermal vent gastropod from the Indian Ocean. *Appl. Environ. Microbiol.* **70**:3082–3090.
19. Grabowski, A., O. Nercessian, and F. Fayolle. 2005. Microbial diversity in production waters of a low-temperature biodegraded oil reservoir. *FEMS Microbiol. Ecol.* **54**:427–443.
20. Handley, K. M., K. C. Wrighton, Y. M. Piceno, G. L. Andersen, T. Z. DeSantis, K. H. Williams, et al. 2012. High-density PhyloChip profiling of stimulated aquifer microbial communities reveals a complex response to acetate amendment. *FEMS Microbiol. Ecol.***81**:188–204.
21. Håvelsrud, O. E., T. H. Haverkamp, T. Kristensen, K. S. Jakobsen, and A. G. Rike. 2011. A metagenomic study of methanotrophic microorganisms in Coal Oil Point seep sediments. *BMC Microbiol.* **11**:221.
22. He, L., K. Yin, X. Yuan, D. Li, D. Zhang, and P. J. Harrison. 2009. Spatial distribution of viruses, bacteria and chlorophyll in the northern South China Sea. *Aquat. Microb. Ecol.* **54**:153–162.
23. Holme, N. A., and A.D. McIntyre. 1984. *Methods for the study of marine benthos*. 2nd ed. Blackwell Scientific Publication, Oxford. 387 hlm.
24. Hubert, C. R., T. B. Oldenburg, M. Fustic, N. D. Gray, S. R. Larter, K. Penn, et al. 2012. Massive dominance of Epsilonproteobacteria in formation waters from a Canadian oil sands reservoir containing severely biodegraded oil. *Environ. Microbiol.* **14**:387–404.

25. Inagaki, F., T. Ken, K. H. Nealson, and K. Horikoshi. 2004. Sulfurovum lithotrophicum gen. nov., sp. nov., a novel sulfur-oxidizing chemolithoautotroph within the ε-Proteobacteria isolated from Okinawa Trough hydrothermal sediments. *Int. J. Syst. Evol. Microbiol.* **54**:1477–1482. doi:10.1099/ijs.0.03042-0. ISSN 1466-5026.

26. Information of Fisheries Management in Malaysia. 2001. Available at: URL http://www.fao.org/fi/oldsite/fcp/en/mys/body.htm (accessed 1 April 2001).

27. Ismail, A. S. I., I. Ismail, M. Zoveidavianpoor, R. Mohsin, A. Piroozian, M. S. Misnan, et al. 2015. Experimental investigation of oil–water two-phase flow in horizontal pipes: pressure losses, liquid holdup and flow patterns. *J. Petrol. Sci. Eng.* **127**:409–420.

28. Jankowska, E., K. Jankowska, and M. Włodarska-Kowalczuk. 2015. Seagrass vegetation and meiofauna enhance the bacterial abundance in the Baltic Sea sediments (Puck Bay). *Environ. Sci. Pollut. Res.* **22**:14372–14378.

29. Johnson, M., I. Zaretskaya, Y. Raytselis, Y. Merezhuk, S. McGinnis, and T. L. Madden. 2008. NCBI BLAST: a better web interface. *Nucleic Acids Res.* **36**:W5–W9.

30. Jones, D. S., H. L. Albrecht, K. S. Dawson, I. Schaperdoth, K. H. Freeman, Y. Pi, et al. 2012. Community genomic analysis of an extremely acidophilic sulfur-oxidizing biofilm. *ISME J.* **6**:158–170.

31. Jørgensen, B. B. 1982. Mineralization of organic in the seabed – role of sulphate reduction. *Nature* **296**:643–645.

32. Kamaruddin, A. F., M. E. Toriman, H. Juahir, S. M. Zain, M. N. A. Rahman, M. K. A. Kamarudin, et al. 2015. Spatial characterization and identification sources of pollution using multivariate analysis at Terengganu river basin, Malaysia. *Jurnal Teknologi* **77**:269–273

33. Kleinsteuber, S., K. M. Schleinitz, J. Breitfeld, H. Harms, H. H. Richnow, and C. Vogt. 2008. Molecular characterization of bacterial communities mineralizing benzene under sulphate reducing conditions. *FEMS Microbiol. Ecol.* **66**:143–157.

34. Li, W. K. W., and P. M. Dickie. 2003. Distribution and abundance of bacteria in the Ocean. Fisheries and oceans Canada. Government of Canada. Available at: http://www2.mar.dfompo.gc.ca/science/review/1996/Li/Li_e.html (accessed: 30 January 2003).

35. Lin, J., B. Hao, G. Cao, J. Wang, Y. Feng, X. Tan, et al. 2014. A study on the microbial community structure in oil reservoirs developed by water flooding. *J. Petrol. Sci. Eng.* **122**:354–359.

36. Liu, K.-K., and N. Dittert. 2010. Web-based electronic supplements, appendix C. *in* K.-K. Liu, L. Atkinson, R. A. Quiñones and L. Talaue-McManus, eds. *Carbon and nutrient fluxes in continental margins*. Springer, Berlin.

37. López-García, P., S. Duperron, P. Philippot, J. Foriel, J. Susini, and D. Moreira. 2003. Bacterial diversity in hydrothermal sediment and epsilonproteobacterial dominance in experimental microcolonizers at Mid-Atlantic Ridge. *Environ. Microbiol.* **5**:961–976.

38. McElroy, A. E., J. W. Farrington, and J. M. Teal. 1989. Bioavailability of polycyclic aromatic hydrocarbon. Pp. 2–39 *in* U. Varanasi, ed.*Metabolism of polycyclic aromatic hydrocarbon in the aquatic environment*. CRC Press, Inc., Boca Rton, FL, USA.

39. Mitsch, W. J. 2010. The 2010 oil spill in the Gulf of Mexico: what would mother nature do? *Ecol. Eng.* **36**:1607–1610.

40. Morton, B., and G. Blackmore. 2001. South China Sea. *Mar. Pollut. Bull.* **42**:1236–1263.

41. Munn, C. B.. 2011. Marine microbiology: ecology and applications. 2nd ed. *Garland science*. Taylor and Francis Group, New York, NY. ISBN: 9780815365174.

42. Nakagawa, S., K. Takai, F. Inagaki, H. Hirayama, T. Nunoura, K. Horikoshi, et al. 2005. Distribution, phylogenetic diversity and physiological characteristics of epsilon-Proteobacteria in a deep-sea hydrothermal field. *Environ. Microbiol.* **7**:1619–1632.

43. Paissé, S., F. Coulon, M. Goñi-Urriza, L. Peperzak, T. J. McGenity, and R. Duran. 2008. Structure of bacterial communities along a hydrocarbon contamination gradient in a coastal sediment. *FEMS Microbiol. Ecol.* **66**:295–305. doi:10.1111/j.1574-6941.2008.00589.x.

44. Païssé, S., M. Goñi-Urriza, F. Coulon, and R. Duran. 2010. How a bacterial community originating from a contaminated coastal sediment responds to an oil input. *Microb. Ecol.* **60**:394–405.

45. Qi, W. G., and F. P. Gao. 2015. A modified criterion for wave-induced momentary liquefaction of sandy seabed. *Theor. Appl. Mech. Lett.* **5**:20–23.

46. Quast, C., E. Pruesse, P. Yilmaz, J. Gerken, T. Schweer, P. Yarza, et al. 2013. The SILVA ribosomal RNA gene database project: improved data processing and web-based tools. *Nucl. Acids Res.* **41**(D1):D590–D596.

47. Sapp, M., E. R. Parker, L. R. Teal, and M. Scratzberger. 2010. Advancing the understanding of biogeography-diversity relationship of benthic microorganism in the North Sea. *FEMS Microbiol. Ecol.* **74**:410–429. doi:10.1111/j.1574-6941.2010.00957.x.

48. Saraswathy, R., M. Muralidhar, J. K. Sundaray, N. Lalitha, and P. Kumararaja. 2015. Water quality management in fish hatchery and grow-out systems. Pp. 217–225 *in Advances in marine and brackishwater aquaculture*. Springer, India.

49. Schmitt, S., P. Tsai, J. Bell, J. Fromont, M. Ilan, et al. 2012. Assessing the complex sponge microbiota: core, variable and species-specific bacterial communities in marine sponges. *ISME J.* **6**:564–576. doi:10.1038/ismej.2011.116.

50. Stibal, M., M. Schostag, K. A. Cameron, L. H. Hansen, D. M. Chandler, J. L. Wadham, et al. 2015. Different bulk and active bacterial communities in cryoconite from the margin and interior of the Greenland ice sheet. *Environ. Microbiol. Rep.* **7**:293–300.

51. Suárez-Suárez, A., A. López-López, A. Tovar-Sánchez, P. Yarza, A. Orfila, J. Terrados, et al. 2011. Response of sulfate-reducing bacteria to an artificial oil-spill in a coastal marine sediment. *Environ. Microbiol.* **13**:1488–1499.

52. Subha, B., Y. C. Song, and J. H. Woo. 2014. The Influence of Slow Release Biostimulant Ball on Sulphate Reducing Bacteria in Coastal Sediment. International Conference on Emerging Trends in Computer and Image Processing (ICETCIP'2014) Dec. 15-16, 2014 Pattaya (Thailand).

53. Sunamura, M., Y. Higashi, C. Miyako, J. Ishibashi, and A. Maruyama. 2004. Two bacterial phylotypes are predominant in the Suiyo seamount hydrothermal plume. *Appl. Environ. Microbiol.* **70**:1190–1198.

54. Suratman, S., M. M. Sailan, Y. Y. Hee, E. A. Bedurus, and M. T. Latif. 2015. A preliminary study of water quality index in Terengganu River basin, Malaysia. *Sains Malays.* **44**:67–73.

55. Takai, K., T. Gamo, U. Tsunogai, N. Nakayama, H. Hirayama, K. H. Nealson, et al. 2004. Geochemical and microbiological evidence for a hydrogen-based, hyperthermophilic subsurface lithoautotrophic microbial ecosystem (HyperSLiME) beneath an active deep-sea hydrothermal field. *Extremophiles* **8**:269–282.

56. Taylor, B., and D. E. Hayes. 1983. Origin and history of the South China sea basin. *Tectonic Geologic Evol. SE Asian Seas Islands: Part***2**:23–56.

57. Van der Gucht, K., K. Cottenie, K. Muylaert, N. Vloemans, S. Cousin, S. Declerck, et al. 2007. The power of species sorting: local factors drive bacterial community composition over a wide range of spatial scales. *Proc Natl Acad Sci* USA **104**:20404–20409.

58. Van Rijn, L. C.. 2010. Coastal erosion control based on the concept of sediment cells. Conscience project Final Report. Avaliable at: URL: http://www.leovanrijn-sediment.com/papers/Coastalerosion2012.pdf(accessed 1 May 2016)

59. Wang, Y. H., C. F. Dai, and Y. Y. Chen. 2007a. Physical and ecological processes of internal waves on an isolated reef ecosystem in the South China Sea. *Geophys. Res. Lett.* **34**:(18).

60. Wang, Q., G. M. Garrity, J. M. Tiedje, and J. R. Cole. 2007b. Naïve bayesian classifier for rapid assignment of rRNA sequences into the new bacterial taxonomy. *Appl. Environ. Microbiol.* **73**:5261–5267.

61. Wang, K., X. Ye, H. Chen, Q. Zhao, C. Hu, J. He, et al. 2015a. Bacterial biogeography in the coastal waters of northern Zhejiang, East China Sea is highly controlled by spatially structured environmental gradients. *Environ. Microbiol.* **17**:3898–3913. doi:10.1111/1462-2920.12884.

62. Wang, L., M. K. Cheung, H. S. Kwan, J. S. Hwang, and C. K. Wong. 2015b. Microbial diversity in shallow-water hydrothermal sediments of Kueishan Island, Taiwan as revealed by pyrosequencing. *J. Basic Microbiol.* **55**:1308–1318.

63. Willey, J. M., L. Sherwood, Christopher. J. Woolverton, and Lansing. M. Prescott. 2008. *Prescott, Harley, and Klein's microbiology*. McGraw-Hill Higher Education, Print, New York.

64. Wright, K. E., C. Williamson, S. E. Grasby, J. R. Spear, and A. S. Templeton. 2013. Metagenomic evidence for sulfur lithotrophy by Epsilonproteobacteria as the major energy source for primary productivity in a sub-aerial arctic glacial deposit, Borup Fiord Pass.*Front. Microbiol.* **4**:63.

65. Xiu, J. L., L. Yu, and C. G. Zheng. 2010. Application of microbial community structure analysis in microbial enhanced oil recovery. *Oil-Gas field Surf. Eng.* **29**:48–50.

66. Yeung, C. W., B. A. Law, T. G. Milligan, K. Lee, L. G. Whyte, and C. W. Greer. 2011. Analysis of bacterial diversity and metals in produced water, seawater and sediments from an offshore oil and gas production platform. *Mar. Pollut. Bull.* **62**:2095–2105.

67. Zhu, D., S.-H. Tanabe, C. Yang, W. Zhang, and J. Sun. 2013. Bacterial community composition of South China sea sediments through pyrosequencing-based analysis of 16S rRNA Genes. *PLoS ONE* **8**:e78501.

68. Zinger, L., L. A. Amaral-Zetter, J. A. Furhman, M. C. Horner-Devine, S. M. Huse, et al. 2011. Global pattern of bacteria beta-diversity in seafloor and seawater ecosystems. *PLoS ONE* **6**:e24570. doi:10.1371/journal.pone.0024570.

7

CHAPTER

IDENTIFICATION OF BACTERIA IN DRINKING AND PURIFIED WATER DURING THE MONITORING OF A TYPICAL WATER PURIFICATION SYSTEM

Vessoni Thereza Christina Penna, Silva Alzira Maria Martins and Priscila Gava Mazzola

Department of Biochemical and Pharmaceutical Technology, School of Pharmaceutical Sciences, University of São Paulo, São Paulo, Brazil

ABSTRACT

Background

A typical purification system that provides purified water which meets ionic and organic chemical standards, must be protected from microbial proliferation to minimize cross-contamination for use in cleaning and preparations in pharmaceutical industries and in health environments.

Methodology

Samples of water were taken directly from the public distribution water tank at twelve different stages of a typical purification system were analyzed for the identification of isolated bacteria. Two miniature kits were used: (i) identification system (api 20 NE, Bio-Mérieux) for non-enteric and non-fermenting gram-negative rods; and (ii) identification system (BBL crystal, Becton and Dickson) for enteric and non-fermenting gram-negative rods. The efficiency of the chemical sanitizers used in the stages of the system, over the isolated and identified bacteria in the sampling water, was evaluated by the minimum inhibitory concentration (MIC) method.

Citation: Vessoni Thereza Christina Penna, Silva Alzira Maria Martins and Priscila Gava Mazzola, Identification of bacteria in drinking and purified water during the monitoring of a typical water purification system, DOI: 10.1186/1471-2458-2-13

Results

The78isolatedcolonieswereidentifiedasthefollowingbacteriagenera:Pseudomonas, Flavobacterium andAcinetobacter. According to the miniature kits used in the identification, there was a prevalence of isolation ofP. aeruginosa 32.05%, P. picketti (Ralstonia picketti) 23.08%, P. vesiculares 12.82%,P. diminuta 11.54%, F. aureum 6.42%, P. fluorescens 5.13%, A. lwoffi 2.56%, P. putida 2.56%, P. alcaligenes 1.28%, P. paucimobilis1.28%, and F. multivorum 1.28%.

Conclusions

We found that research was required for the identification of gram-negative non-fermenting bacteria, which were isolated from drinking water and water purification systems, since Pseudomonas genera represents opportunistic pathogens which disperse and adhere easily to surfaces, forming a biofilm which interferes with the cleaning and disinfection procedures in hospital and industrial environments.

BACKGROUND

Water is one of the most important elements for all forms of life and is indispensable in the maintenance of life on Earth and essential for the composition and renewal of cells. Water represents 70% of our body, participates in the composition of our tissues, and transports the most diverse substances throughout our organism. Notwithstanding, human beings increasingly continue to pollute the reserves which still remain, provoking illnesses that can jeopardize the population [1].

Bioaccumulation, sewerage, agricultural, industrial, radioactive, and thermal residues are the principal polluters of water on our planet.

Purified Water

Water is essential for industrial, pharmaceutical and hospital purposes, in the preparation and processing of medicines and other health products. In the majority of cases, water is an input, which should be incorporated into the product during processing. At other times, even if it is not present in the preparation, it is especially used for cleaning and hygiene purposes. It is recognized that the greatest demand on water is destined for human consumption, its quality being relatively guarantied up to the point where the pipe transportation network terminates.

Although potable water is suitable for human consumption, it does not guarantee that the so-called potable quality could also be used in industrial installations, equipment, preparation of medicines, foodstuffs, cosmetics, chemopharmaceutical materials, or health center units for the cleaning and washing of semi-critical areas or devices prior to the application of disinfection or sterilization procedures.

For this reason, every industrial or pharmaceutical plant related to health products must rely on appropriate water purification systems, allowing it to meet its particular requirements, especially as to the problems related to storage and internal distribution.

This procedure must guarantee supply according to the volume required and pursuant to the demanded quality consumption points.

Purified water is used as an expedient in the production of official preparations, pharmaceutical (industrial, and in health center units) applications such as the cleaning of semi-critical devices, areas and equipment, as well as in the preparation of pharmaceutical chemicals and bacteriological media. It is also used as the principal component in peritoneal dialyses solutions in hospitals, in nutrient solutions (including nursing bottles) and liquid enteral nutrient solutions, prepared in the hospital lactarium and administered to children and weak patients who can eat regular solid food.

Purified water is obtained from drinking water [1] through a typical water purification system of unit operations, meeting the standards set forth by the 1978/1990 directives issued by the Brazilian Ministry of Health [2]. Purified water systems must be validated in order for water to meet the requirements [3]: for the purity of ionic (conductivity ≤ 1.3 μS/cm, at 25°C) and for total organic compounds (TOC < 0.5 mg/L), and must even be protected from microbial proliferation (total enumeration < 100 CFU/mL), preventing pyrogen formation. The bacteriological standard ≤ 1.0 Endotoxin Unit (EU/mL) is required for sterile purified water used in the washing of critical devices before autoclaving.

The purified water system that produces, stores, and circulates water under background conditions is "susceptible to the establishment of adhesive biofilms or microorganisms", which can be the source of undesirable levels of viable microorganisms or endotoxins in the effluent water. Recent studies have shown that nearly all large water purification systems can cause the formation of biofilm in the piping. This biofilm can spread microorganisms within the system and contribute to an increase in particles, bacteria, and the level of total organic carbon (TOC). Contamination can affect the whole process in the pharmaceutical industry or hospital environment. These systems require frequent sanitation and microbiological monitoring to ensure water of the appropriate microbiological quality (microbial limit at the points of use) [3].

Monitoring data should be analyzed on an ongoing basis so as to ensure that the process continues to be performed within acceptable limits. It should be recognized that the microbial alert and action levels established for any pharmaceutical water system are obligatorily linked to the monitoring method chosen. Using the recommended methodologies generally considered, appropriate action levels are 500 colony-forming units(CFU) per mL of drinking water, 100 CFU/mL of purified water, and 10 CFU/mL of water for injection (WFI); the limits for pyrogen are < 1.0 Endotoxin Unit/mL of purified water, < 0.5 EU/mL of WFI and < 0.25 EU/mL of sterile WFI [2, 3]. For gram-negative fermenting bacteria in drinking water, the standards show that total coliforms must be fewer than 10^{-2} CFU/mL to that of drinking water. Neither the Brazilian Federal standards nor the USP 24 [3] includes levels for gram-negative non-fermenting bacteria, such as the*Pseudomonas* species, which are the principal cause of biofilms and enterotoxins [4] in purified water.

During the performance of a typical water purification system, the purposes of the present work were: (i) to examine the efficiency of each treatment stage; (ii) to characterize the isolated bacteria at each stage, from the storage tank of publicly supplied water up to the final point of purified water consumption. The minimum inhibitory concentration (MIC) of the main chemical agents used in the disinfection of the water purification system over the isolated and identified bacteria was also studied.

METHODS

Water Samples

Samples from thirteen points of a typical water purification system (10,000 L/day) were analyzed in triplicate. Three analysts collected samples of water from every point indicated in the Flow Sheet of a typical USFilter water purification system [5], placed in the Edwards Lifesciences Macchi, SP, Br.

Before sampling the water, every sampling tap (point) was made sanitary with ethyl 70% alcohol, the valve was opened and the water was allowed to flow (approximately 3 L/min) freely for about 60 seconds, followed by the taking of a sample of 100 mL of the water in a sterile polyethylene bag (Millipore, Bedford, MA, USA).

Flow sheet of a typical water purification system

Point 1: Storage Tank (feed water)

Point 2: Two multimedia Filters (primary filtration)

Point 3: Two water softeners (hardness reduction)

Point 4: One filter of activated carbon (chlorine removal)

Point 5: One 5.0 μm filter (removal of particle materials)

Point 6: One reverse osmosis membrane system (removal of organic and inorganic substances)

Point 7: One continuous deionization column (removal of dissolved minerals and salts)

Point 8: One storage tank (treated water)

Point 9: Light UV: 254 nm (reduce TOC)

Point 10: Three 0.05 μm filters in parallel (removal of particles and bacteria)

Points 11, 12, 13: Loop of distribution of purified water for consumption

The water purification system presented in the Flow Sheet [5] included the following stages and subsequent apparatus, from where the water points were sampled:

Point 1: The storage tank was fed with drinking water distributed by SABESP [1] (Basic Sanitation Company for the State of São Paulo). The polyethylene storage tank has a closed cover and a capacity of 10,000 L H_2O designed with a siphon valve fitted at the bottom of the tank outlet. It is also fitted with a centrifugal pressure pump (stainless steel, 40 L/min/ 414 kPa), which pressurizes the water to point 02 across the multimedia filters.

Point 2: Two multimedia filters parallel to each other (US Filter, rlzms12 fxxfa, Warrendale, PA, USA) offers a highly efficient removal of suspended fragmented matter from the water. The three layers of media (sand, anthracite and quartz) are selected in accordance with their particular size, specific gravity, and ability to trap particles of specific size ranges (≥ 10 micra). As the water flows downwards through the bed, it finds a layer of media with decreasing porousness/permeability so that successively smaller particles are trapped in each layer, providing depth filtration.

Point 3: Two water softeners of an alternated sodium resin (US Filter, rlzsd12fxxya) which remove hard minerals from the water. Ion exchange water softening exchanges the calcium and magnesium cations from water with an equivalent number of sodium cations.

Point 4: One filter of activated carbon (US Filter relzcs12fxxfa) is used to remove chlorine, chloramines, and dissolved organic substances from the water. Carbon filters are frequently used as the pretreatment to osmosis membranes and ion exchange resins, avoiding damage by oxidant substances, such as chlorine.

Point 5: One 5.0 micra polyethylene microporous depth screen filter (US Filter, fcrof2005) is often used ahead of other water purification operations, such as deionization, and reverse osmosis, as a polishing filter for removing resin, carbon fine colloids, and microorganisms.

Point 6: One reverse osmosis membrane system (polyamide polymers, US Filter, rsolv204021133)- The natural process of osmosis occurs when a solution with different concentrations of salts is separated by a semi-permeable membrane. As osmotic pressure drives the water through the membrane, the water dilutes more concentrated solutions, until an equilibrium is achieved. The permeate pure water is collected on the downstream side of the membrane. Reverse osmosis removes 90% – 99% of particles, colloids, bacteria, pyrogens, dissolved organic and inorganic substances greater than 200–300 molecular weight (MW) range or larger than the membrane's pore size of 150 to 200 angstroms. The conductivity of the water at the inlet is 150 μS and at the outlet is 5 μS. Four membranes (10.16 cm × 101.6 cm) with an outlet capacity of 16 L/min are used. The membrane's shelf life is between 2 and 3 years and sanitation is carried out through the association of peracetic acid with a hydrogen peroxide solution.

Point 7: One continuous deionization (US Filter, cdi = ccdis20 × 1) column that removes dissolved minerals and salts, as well as some dissolved organic matter, from the water stream crossing ion exchange resins. The ion exchange operation removes positively charged cations, such as calcium, magnesium, and sodium from the water by cation exchange resins, which are replaced by hydrogen ions. Negatively charged anions such as chloride, nitrate, and silica are removed from the water by strong based anion exchange resins, and hydroxide ions then form water molecules. The water stream passes through a mixed bed of cation and anion exchange resins, which produces a very high quality of water with a resistance of up to 1.3 μs/cm (18.3 megohm-cm) at 25°C.

Point 8: A storage polyethylene tank (1500 L) with the treated water, of which the reservoir has a closed cover design with a spray ball at the top to spread the water, keeping it in continuous movement so as to prevent any hazard of contamination. The tank is also provided with a siphon valve fitted at the bottom of the tank outlet, a pressure relief, and a 0.2 μm ventilation filter. The loop of distribution for the points of use is accomplished through two pumps.

Point 9: Ultraviolet Light (λ = 254 nm, US Filter, SL-1) is used as a final step in the treatment for the purpose of preventing the growth of microorganisms, and reducing total organic carbon (TOC).

Point 10: Three 0.05 μm filters are set parallel to each other (US Filter, zha 153107). Microporous filters are used to remove particles, and bacteria, ranging from 0.05 to 0.5 μm contaminants, which would not ordinarily be removed by depth filtration.

Points of use 11,12, and 13: Every point of use is provided with 3 filters of 0.05 μm set parallel to each other. From those points, the purified water for consumption is provided by a loop of distribution and is used for the cleaning of critical devices (membrane oxygenators and PVC tubes), the washing of semi critical areas, the preparation of chemical solutions, and the culture bacteriological media.

Culture media, enzymatic and biochemical reactions efficiency test

The culture media efficiency test was carried out with standard strains: *Aspergillus niger* ATCC 16404, *Bacillus subtilis* ATCC 6633, *Candida albicans* ATCC 10231, and *Pseudomonas aeruginosa* ATCC 27853, *Escherichia coli*ATCC 25922, pursuant to USP 24 [3]. The evaluation of the enzymatic and biochemical reactions, performed in the identification of the isolated colonies, were carried out with standard strains of *Pseudomonas aeruginosa*ATCC 27853, *Aeromonas hydrophila* ATCC 35654, *Alcaligenes faecalis* ATCC 35655, and *Flavobacterium multivorum* ATCC 35656.

Filtration of sample

The samples were submitted to the filtration method by 0.45 μm membranes in a class 100 laminated flow chamber. This was followed by an incubation period of 72 hours at 30–35°C (Tryptic Soy Agar, TSA, Difco, Detroit, Michigan, USA) for total bacteria enumeration (heterotrophic), and for 22 hours at 34.5–35.5°C (m Endo Agar Less, Difco) for total coliform enumeration, in accordance with the United States Pharmacopoeia National Formulary, in the chapter: "Water For Pharmaceutical Purposes USP 24/ NF 2000" [3]. Every sampling tap (point) was analyzed in triplicate for heterotrophic bacterial counts and total coliforms. In order to stimulate the growth of stressed and chlorine-tolerant bacteria, we just started using R2A Agar to improve the enumeration of heterotrophic organisms in treated potable water.

Isolation of colonies

From each TSA culture, the colonies were transferred to the surface of the Cetrimide (Difco) in plates and incubated at 30–35°C for 18–24 h. The developed colonies were then submitted to oxidase and indol tests, followed by biochemical genera and species identification tests. In every plate, all the developed colonies were isolated for identification. The criteria used were colony morphology and colony color. The total number of colonies found at each sampling point (CFU/100 mL) and picked for identification, are shown in Tables 01 &2.

Table 1: Total number (and percentage) of identified genera and species of gram-negative bacteria through the percentage of the positive reactions from the isolated colonies

Microorganisms	strains	% of identification
Pseudomonas aeruginosa	25	32.05
Pseudomonas picketti	18	23.08
Pseudomonas vesiculares	10	12.82
Pseudomonas diminuta	09	11.54
Flavobacterium aureum	05	6.42
Pseudomonas fluorescens	04	5.13
Acinetobacter lwoffi	02	2.56
Pseudomonas putida	02	2.56
Pseudomonas alcaligenes	01	1.28
Pseudomonas paucimobilis	01	1.28
Flavobacterium multivorum	01	1.28
Enumeration	78	100.0

Table 2: Total enumeration (CFU/100 mL) and identification of Gram-negative non-fermenting bacteria from the genera:*Pseudomonas, Flavobacterium, Acinetobacter* identified by the BBL Crystal kit (Becton Dickinson) and API 20 NE kit (bioMérieux), at every sampling point

Sampling points/Flow sheet	***CFU/100 mL Average ± SD***	***Colonies***		***identification***	
		Morphology & color	***Number of isolates***	***BBL Crystal***	***API 20 NE***
Point 1 Feed Water (storage)	507 ± 50	Shining beige	02	*F. multivorum*	*P. vesiculares*
		Circular Brown	02	*P. aeruginosa*	*P. alcaligenes*
		Circular pumpkin	02	*P. diminuta*	*P. diminuta*
Point 2 Two multimedia Filters	33 ± 12	Circular pumpkin	02	*P. diminuta*	*P. diminuta*
		Circular pumpkin	02	*P. diminuta*	*P. diminuta*
		Circular pumpkin	02	*P. diminuta*	*P. diminuta*
Point 3 Two water softeners	173 ± 50	Circular pink	02	*F. aureum*	*P. vesiculares*

		Circular pink	02	*F. aureum*	*P. vesiculares*
		Circular pink	02	*F. aureum*	*P. vesiculares*
Point 4 Activated carbon filter	897 ± 823	*Circular brown*	02	*P. aeruginosa*	*P. pcketti*
		Circular white	02	*F. aureum*	*P. vesiculares*
		Circular pink	02	*P. vesiculares*	*P. vesiculares*
Point 5 One 5 micron filter	100 ± 20	Circular white	02	*F. aureum*	*P. vesiculares*
		Circular brown	02	*P. aeruginosa*	*P. picketti*
		Circular brown	02	*P. aeruginosa*	*P. aeruginosa*
Point 6 Reverse Osmosis	8 ± 2	Circular beige	02	*P. vesiculares*	*P. vesiculares*
		Circular brown	02	*P. aeruginosa*	*P. picketti*
		Circular brown	02	*P. aeruginosa*	*P. picketti*
Point 7 Continuous deionization	653 ± 344	Circular brown	02	*P. aeruginosa*	*P. fluorescens*
		Circular brown	02	*P. aeruginosa*	*P. picketti*
		Circular brown	02	*P. aeruginosa*	*P. aeruginosa*
Point 8 Storage tank for the purified water.	40 ± 35	Circular brown	02	*P. aeruginosa*	*P. picketti*
		Pointed yellow	02	*A. lwoffi*	*P. diminuta*
		Pointed cream	02	*P. fluorescens*	*P. putida*
Point 9 Light UV	27 ± 12	Circular brown	02	*P. aeruginosa*	*P. picketti*
		Pointed cream	02	*P. paucimobilis*	*P. putida*
		Circular brown	02	*P. aeruginosa*	*P. picketti*
Point 10 0.05 micron Filter	87 ± 31	Circular brown	02	*P. aeruginosa*	*P. picketti*

		Circular brown	02	*P. aeruginosa*	*P. picketti*
		Circular brown	02	*P. aeruginosa*	*P. picketti*
Point 11: use	27 ± 12	Circular brown	02	*P. aeruginosa*	*P. fluorescens*
		Circular brown	02	*P. aeruginosa*	*P. picketti*
		Circular brown	02	*P. aeruginosa*	*P. picketti*
Point 12: use	27 ± 12	Circular brown	02	*P. aeruginosa*	*P. picketti*
		Circular brown	02	*P. aeruginosa*	*P. picketti*
		Circular brown	02	*P. aeruginosa*	*P. picketti*
Point 13: use	27 ± 12	Circular brown	02	*P. aeruginosa*	*P. picketti*
		Circular brown	02	*P. aeruginosa*	*P. picketti*
		Circular brown	02	*P. aeruginosa*	*P. picketti*

SD = standard deviation ($n = 3$; $p < 0.05$)

Identification tests

The identification tests for genera and species were performed by using two standardized micro-method systems: (i) API 20 NE kit from bioMérrieux that was used for the identification of non-fastidious gram-negative rods [6] not belonging to the *Enterobacteriacea* family; (i) the BBL crystal enteric/non-fermenting ID kit from Becton & Dickinson that was used for the identification of aerobic gram-negative bacteria that belong to the family of *Enterobacteriacea* as well as some of the more frequently isolated glucose fermenting and non-fermenting gram-negative bacilli.

The micro-methods employed are modifications of the classical methods and the basic principles of the reactions follow the biochemical and enzymatic standards described for the enteric/non fermenting bacteria by Murray *et al.*[4].

The more specific API 20 NE kit is comprised of 20 microtubes containing dehydrated media and/or substrates, combining 8 conventional tests and 12 assimilation tests. The interpretation of the reactions were made according to the reading tables using the identification software with BBL crystal or API 20 NE data base.

The biochemical tests were employed in the bioMerieux identification system and the respective reactions were: (a) potassium nitrate (NO_3) for the reduction of nitrate to nitrite, (b) tryptophan (trp) for indol production, and (c) glucose (glu) for medium acidification. The substrates used for enzymatic hydrolysis reactions were: (d) arginine (adh) for arginine dihydrolase, (e) urea (ure) for urease, (f) esculin (esc) for β-glucosidase, (g) gelatin (gel) for protease, and (h) p-nitrophenyl-β-D-galactopyranoside (PNPG). For assimilation tests, the substrates employed were the following: glucose (GLU), arabinose (ARA), mannose (MNE), mannitol (MAN), N-acetyl-glucosamine (NAG), maltose (MAL), gluconate (GNT), caprate (CAP), adipate (ADI), malate (MLT), citrate (CIT), and phenyl-acetate (PAC). When the substrate was assimilated by the microorganism, its growth was detected by visible turbidity in the corresponding medium. The reactions were visualized by the addition of reactive agents and a change in the indicator.

The BBL crystal kit is a miniature method of identification that comprises modified conventional and chromogenic substrates, which are distributed in 30 dehydrated enzymatic and biochemical substrata in panels. The biochemical tests used the following carbohydrates: glucose, arabinose, mannose, sucrose, melibiose, rhamnose, sorbital, mannitol, and inositol. The utilization of carbohydrate results in lower pH and a change in the indicator. Chromagenic tests, in turn, used nitrophenyl phosphate, proline nitroanilide, nitrophenyl xyloside, arabinoside, glucoronide and glucosamine, esculin, phenylalanine, arginine, lysin, urea, glycine, citrate, and malonate.

Minimal Inhibitory Concentration (MIC test)

The reference bacteria used were *Pseudomonas aeruginosa, Pseudomonas fluorescens, Pseudomonas alcaligenes, Pseudomonas picketti, Flavobacterium aureum,* and *Acinetobacter lowffi* (Tables 1 &2), which were isolated from the selected points of the typical water purification system, and maintained on a tryptic soy agar slant (TSA, Difco) at 4°C, with monthly transfers. The 24-hour cultures developed on TSA at 30–35°C were harvested in a tryptic soy broth (TSB, Difco), centrifuged (1000 g/ 15 min/ 4°C), and suspended in saline (0.95% NaCl plus 0.1% peptone) to a final population (by pour plate) of 10^6 CFU/mL, and the suspensions were applied to the MIC tests.

The minimal inhibitory concentration (MIC) was determined by using the two-fold broth dilution method [7]. Starting from a chemical agent solution, serial dilutions were prepared in TSB inoculated with the test bacterial populations between 1.2 and 3.0×10^6 CFU/mL. The MIC was identified as having the lowest concentration of the chemical agent, which resulted in the confirmed inhibition of the growth of the tested microorganism, after 24 h of optimum incubation conditions. The chemical agent solutions, started concentrations and the pH values employed are shown in Table 03.

Table 3: The minimum inhibitory concentrations (MIC) for chemical agents for the reduction of bacteria populations over 6-$\log_{10}$

Bacteria	Pseudomonas aeruginosa		Pseudomonas diminuta		Pseudomonas fluorescens		Pseudomonas alcaligenes		Pseudomonas picketti		***Flavobacterium aureum***		***Acinetobacter lowffi***	
MIC test	**mg/L**	**%**	**mg/L**	**%**	**mg/L**	**%**	**mg/L**	**%**	**mg/L**	**%**	**mg/L**	**%**	**mg/L**	**%**
Agent (initial concentration & pH)														
Ethanol (70%; pH = 7.2)	175000	17.5	87500	8.75	87500	8.75	87500	8.75	87500	8.75	87500	8.75	87500	8.75
Sodium Hypochlorite (0.5%; pH = 11.9)	2500	0.25	2500	0.25	2500	0.25	2500	0.25	2500	0.25	2500	0.25	2500	0.25
PAA (0.45%) + H_2O_2 (2.2%) (pH = 2.3)	1125	0.11	562	0.056	140	0.014	280	0.028	1125	0.11	1125	0.11	1125	0.11
	5500	0.55	2750	0.275	680	0.068	1368	0.14	5500	0.55	5500	0.55	5500	0.55
Sodium Hydroxide (0.4%; pH = 12.8)	3000	0.3	1500	0.15	1500	0.15	750	0.075	3000	0.3	3000	0.3	4000	0.4
Citric acid (0.5%; pH= 2.5)	2500	0.25	2500	0.25	2500	0.25	2500	0.25	5000	0.5	2500	0.25	600	0.06
Hydrochloric Acid (0.3%; pH= 0.3)	1560	0.16	1560	0.16	780	0.078	1560	0.16	3125	0.31	1560	0.16	390	0.039
Sodium Bisulfite (1.0%; pH= 4.0)	780	0.078	780	0.078	780	0.078	390	0.039	780	0.078	780	0.078	780	0.078

PAA – peracetic acid H_2O_2 – hydrogen peroxide

Chemical agents, respective solutions, and concentrations used in the MIC test

1. Ethyl alcohol 70 % v/v (Ferreira, SP, Br). 70% Ethanol solution (pH = 7.2) was used as the desinfectant on the external surfaces of sampling taps and valves.
2. Sodium Hypochlorite NaOCl (Nuclear, SP, Br). Since the concentration of commercial sodium hypochlorite is variable, in order to calculate the amount of sodium hypochlorite (H) to be diluted in purified water and to perform approximately 0.5% v/v, the following formula was applied: H = (5.4 * V) / (% NaOCl concentration); where H is the volume of NaOCl at 10% (w/v) to be diluted and V is the final desired volume of the NaOCl solution at 0.5% (approximately 1% of the volume of the tank); and NaOCl is at an initial concentration of 10%. The concentration of the total available chlorine was determined by the iodometric method [7]. The solution of NaOCl at 0.5% (pH= 11.9) for 60 min exposure is used in the clearance of the tanks of feeding water (point 01), storage (point 08) and in the loop of distribution (before points 11, 12 &13).
3. A solution of 4.5 % (v/v) peracetic acid and 22% (v/v) hydrogen peroxide plus 10 mg/L acetic acid (Minncare, pH = 1.3, Minntech Corporation, Minneapolis, MN, USA) was used by dissolving Minncare into purified water, performing 0.45% of PAA and 2.2% of H_2O_2, reaching a final pH of between 2.0 and 2.3 in the tank. This 1% Minncare solution (PAA + H_2O_2) is applied to the cleaning of reverse osmosis membranes (point 6) and the continuous deionization (point 7) apparatus for three hours, so as to obtain purified water, and 18 hours for WFI which is used to prepare parenteral solutions, including peritoneal dialysis solutions.
4. A sodium hydroxide (Nuclear, SP. Br) solution was prepared by dissolving 200 g of NaOH in 50 liters of purified water, performing 0.4 % (w/v) with a final pH of 12.8 in the tank. A solution of 0.4% sodium hydroxide is used to adjust the pH value of acidic water in the reverse osmosis (point 06) and the continuous deionization (point 07).
5. A citric acid (Ciro, SP, Br) solution was prepared by dissolving 225 g of citric acid in 45 liters of purified water, performing 0.5 % (w/v) with a final pH of between of 2.0 and 2.5 in the tank. A solution of 0.5% citric acid is used to acidify the water in the reverse osmosis membranes (point 06), before the application of 1% of the (PAA+H_2O) solution.
6. A solution of hydrochloric acid (Synth, SP, Br) was prepared by adding 5.5 liters of hydrochloric acid at 37% v/v to 40 liters of purified water, performing 5.0% (pH = 0.3). The final solution of 0.3% HCl is applied in the pH adjustment of water in deionization columns (point 7).
7. A solution of sodium bisulfite (Ciro, SP, Br) was prepared by dissolving 1 kg of sodium bisulfite in 100 liters of purified water, attaining 1.0% w/v (pH of 4.0). The solution of sodium bisulfite at 0.5% concentration is used to preserve the multimedia filters (point 02), water softeners (point 03) and in the dechlorinating of carbon bed filters (point 04).

RESULTS AND DISCUSSION

Isolation of colonies

Of the thirteen points, which were analyzed in triplicate, seventy eight (78) colonies were isolated, with different aspects: circular viscosity, sharp pointed and brilliant, opaque, from light brown to dark brown, pumpkin, pink, white and cream colored. All the colonies were confirmed as bacteria of the non-fermentative gram-negative bacilli (NFGNB) group, strictly aerobic, positive for indole and negative for oxidase, with the exception of the *Acinetobacter lwoffi*, which had a negative reaction for oxidase.

The gram-negative non-fermenting rods isolated and identified from the genera *Pseudomonas, Flavobacterium, and Acinectobacter* are considered opportunistic pathogens and very common in nature: soil, water, plants (including fruits and vegetables), animals and organic material in decomposition (sewage). They are also frequently found in water treatment systems, demonstrating an adaptation to environments with a low concentration of nutrients, and to a large range of temperature, from 4°C to 42°C [4, 8].

According to the miniature kits used in the identification of gram-negative non-fermenting bacteria (Tables 1&2), there was a prevalence of isolation by *P. aeruginosa*, 25 strains (32.05%); *P. picketti (Ralstonia picketti)*, 18 strains (23.08%); *P. vesiculares*, 10 strains (12.82%); *P. diminuta*, 09 strains (11.54%); *F. aureum*, 5 strains (6.42%); *P. fluorescens*, 4 strains (5.13%); *A. lowffi*, 02 strains (2.56%); *P. putida*, 02 strains (2.56%); *P. alcaligenes*, 01 strain (1.28%); *P. paucimobilis*, 01 strain (1.28%); and *F. multivorum*, 01 strain (1.28%). The two identification systems complemented each other, since *P. aeruginosa* was identified in all stages by the Becton and Dickinson kit, whereas *P. picketti* was identified by the Bio-Mérieux kit.

The chemical and microbiological characteristics of the water from the sampling points were in compliance with the relevant laws and standards [1, 2, 3]. For the points 1, 4, 11, 12 and 13, respectively, the pH values were 7.15, 7.39, 5.60, 5.60, 5.70; conductivity was 155.2 µs, 148.2 µs, 1.2 µs, 0.8 µs, 0.9 µs. The total organic compounds (TOC) were < 0.5 mg/L for the points 11, 12 and 13.

The enumeration of heterotrophic bacteria was observed to increase from one to two logarithmic cycles for the water crossing through points 3 & 4, respectively, the two resins of softeners and the bed of activated carbon filter (Table 2). This increase in total bacteria enumeration cancelled out the multimedia (point 02) effect by reducing one cycle of the initial population (average of 507 CFU/ 100 mL) from public drinking water (point 01). The reverse osmosis (point 06) showed the best filtration reduction to a maximum residual population of 10 CFU/100 mL, although gram-negative non-fermenting bacteria were still there crossing and reaching the next stage of the continuous deionization. Although in the storage tank (point 08) the total enumeration was maintained at around 40 CFU/ 100 mL, the diversity of the gram-negative non-fermenting bacteria still existed. The ultraviolet (UV) light and the 0.05 µm filter did not alter the heterotrophic bacteria population or the *Pseudomonas* species found in the preceding points of the Flow Sheet. Both the UV light and the last 0.05 μm filter were found to keep the heterotrophic population as low as 100 CFU/ 100 mL of purified water, although those two apparatus showed no efficiency over the gram-negative non-forming bacteria, such as the species of *Pseudomonas, Flavobacterium*, and *Acinetobacter.*

The dynamic flow of water, passing through the purification stages, carried and spread bioburden throughout every spot of the operational units. Although the heterotrophic enumeration was reduced by at least one log cycle from the 7th to the 8th stage in the storage tank, the diversity of gram-negative bacteria was found to be higher than that at the other points. At point 08, the renewing of the water is slower and the room temperature of the tank favored microorganism adjustment. After exposure to the UV light, the diversity of gram-negative bacteria dropped drastically in the species *P. aeruginosa* and *P. paucimobilis*, (BBL Crystal identification kit) and *P. picketti* (API 20 NE identification kit), the populations of which were kept constant in the water flowing through the 0.05 μm filter. The UV light oxidative process and the 0.05 μm filter at the point – of – use were able to guarantee the maintenance of total enumeration lower than the maximum of 40 CFU detected in 100 mL of water and restrict the distribution of gram-negative bacteria. Maybe a biofilm was established and bacteria might continue to be released from the biofilm into bulk water. Appropriate desinfection of the water treatment system would eliminate the biofilm. This is one of the main reasons why we are studying the best performance of every chemical disinfectant applied at each sampling point.

Procedures should be applied to the system at short-term intervals, in order to control the gram-negative entrance in the system from the municipal source of drinking water (Point 01).

The Minimal Inhibitory Concentration (MIC)

The MIC intervals were expressed in percentage and in mg/L of the chemical agent in contact with the bacteria tested, and are shown in Table 3.

1. Ethyl alcohol (70%, pH= 7.2)

Regarding the exposure to 70% ethanol, *P. aeruginosa* showed the greatest MIC at 17.5%, higher than the MIC of 8.75% which was found for *B. subtilis* and *B. stearothermophilus*, obtained by Vessoni Penna *et al.*[7], emphasizing the importance in the validation of the water purification system for the identification of *P. aeruginosa*.

Trautmann *et al.*[9] did not obtain acceptable long-term results by using chlorination and filtration to eliminate the strains of *P. aeruginosa* from every water tap in the ICU. However, the authors noticed that strains of *P. aeruginosa* were not regularly isolated from hospital personnel, whose hands were disinfected with alcohol (ethanol 70%) before and after contact with patients.

2. Sodium hypochlorite (NaOCl, 0.5%; pH11.9)

All the bacteria studied showed the same level of resistance, i.e. a MIC of 0.25% (2500 mg/L). According to Vessoni Penna et al. [7], *E. coli* presented half a MIC of 0.156% (1560 mg/L) in chlorinated compounds, making it clear that the monitoring of chlorinated waters for *P. aeruginosa* is necessary and mainly critical for warm potable waters, including the cleaning of the water circuits, hydrotherapy, baths, and pools. The *P. aeruginosa›s* resistance to chlorine releasing agents (CRAs) is well reported. Wirtanen *et al.*[10] studied the effects of four commercial disinfectants: (100% alcohol-based) containing isopropanol, a peroxide-base containing hydrogen peroxide (0.5–2.0%) and peracetic acid in formulation, and a chlorine-base containing sodium hypochlorite (0.3–0.8%, pH>9.0). The authors found that the peroxide-based disinfectant was the most efficient on biofilm bacteria, and they confirmed that prolonged treatment with the chlorine-base disinfectant was especially efficient on *Pseudomonas* biofilms.

Therefore, there is an urgent need for seeking the range of CRAs and the adequate time contact that would not induce resistance in bacteria, which inhibits biofilm formation and adherence to the porous apparatus of the system. However, the disinfectants should be chosen according to the operation in process, considering the interference of organic substances in the disinfectant›s activity and effectiveness.

3. Association of peracetic acid (PAA, 0.45%) plus hydrogen peroxide (H_2O_2 2.2%). Solution at 1% (PAA+ H_2O_2; pH = 2.3)

For a solution of peracetic acid (0.45%) + hydrogen peroxide (2.2%), *P. aeruginosa, P. picketti, F. aureum,* and*A. lowffi* presented the highest MIC range from 0.11% to 0.55%, an interval of which was shown to be two, four and eight times greater than that for *P. diminuta* (MIC = 0.056% & 0.275%), *P. alcaligenes* (MIC = 0.028% & 0.137%) and *P. fluorescens* (MIC = 0.014% & 0.068%), respectively.

Considering that the set conditions for MIC were kept constant, independent of the bacteria tested, according to Vessoni Penna *et al.*[7], when vegetative cells such as *Acinetobacter calcoaceticus, Enterobacter cloacae, Escherichia coli, Serratia marcescens*, and *Staphylococcus aureus* were submitted to a H_2O_2 solution at 4.0%,*Escherichia coli* showed the greatest resistance among them with a MIC of 0.25%, which is half that obtained for *P. aeruginosa*. When the same bacteria were tested against peracetic acid, *E. coli* showed a MIC of 0.23%, which is twice as much as that for *P. aeruginosa* in a mixture of peracetic acid + hydrogen peroxide, showing that *P. aeruginosa* and *E. coli* should both be required in the monitoring of treated waters. The solution of PAA + H_2O_2 at 1% is applied to the cleaning of the reverse osmosis membranes and the continuous deionization apparatus for three hours, so as to obtain purified water, and 18 hours for WFI that will be used to prepare parenteral solutions, including peritoneal dialysis solutions.

4. Sodium Hydroxide (0.4%; pH = 12.8)

P. fluorescens and *P. alcaligenes* were the bacteria which were the least resistant to sodium hydroxide, exhibiting a MIC of 0.15%, in relation to the resistance shown by the other bacteria (MIC >0.4%). The principal activity of this chemical compound is the pH adjustment of acidic water in the reverse osmosis (point 06) and the continuous deionization (point 07), with some expected sanitation activity.

5. Citric Acid (0.5%; pH = 2.5)

The bacterium least resistant to citric acid solution was *A. lowffi* (0.06%= 600 mg/L). The most resistant was *P. picketti* (0.5%= 5000 mg/L); the remainder showed a MIC of 0.25% = 2500 mg/L. This chemical agent is used in the reverse osmosis (point 06) in adjusting the pH of cleaning water before disinfection with a 1% solution of PAA + H_2O_2.

6. Hydrochloric Acid (0.3%; pH = 0.3)

The bacteria least resistant were *A. lowffi* (MIC 0.039% = 390 mg/L) and *P. fluorescens* (MIC 0.078% = 780 mg/L); the remainder had a MIC of 0.156% (1560 mg/L).

It is noteworthy to mention that *P. aeruginosa* was not affected by the chemical adjuvant compounds such as NaOH (0.4%), citric acid (0.5%) and HCl (0.3%) in the concentration applied to the units of the purification system. However, these chemical solutions are expected to show some disinfecting activity, which is not demanding.

7. Sodium Bisulfite (0.5%; pH= 4.0)

All the bacteria showed the same level of resistance (MIC = 0.078%) to sodium bisulfite. This chemical adjuvant is used for the purpose of dechlorinating and preserving multimedia filters; it is further used in water softeners and activated carbon bed filter.

Profile of the identified microorganisms

The gram-negative water bacteria can be significant contaminants in haemodialysis systems. Ferreira *et al.*[13], performed analyses on samples taken from water supplies, after each treatment step and dialysate in haemodialysis units in Rio de Janeiro from 1999 to 2001. They found that the most commonly isolated bacteria were: *Pseudomonas aeruginosa, Stenotrophomonas maltophilia, Acinetobacter anitratus, Acinetobacter lowffi, Pseudomonas (Brevundimonas) diminuta, Pseudomonas fluorescens, Achromobacter xylosoxidans, Moraxella atlantae, Moraxella osloensis, Enterobacter cloacae,* and *Enterobacter aerogenes*. These isolated bacteria pointed out a real risk for haemodialysis patients with developing gram-negative bacteremia, pyrogenic reactions, and peritonitis. The authors emphasized the need for an urgent program for the disinfection of the water systems and dialysis machines [14], in health care centers.

Pseudomonas have been isolated from a great variety of materials, including soils, fresh or sea water, sewage, many types of clinical specimens, and elements commonly handled in clinical laboratories (including distilled water and antiseptic solutions), assorted foods and food industry wastes, flowers, fruits, vegetables, and diseased plants and animals. The *Pseudomonas* species rather than *P. aeruginosa* do not frequently cause infections [4, 8, 11]. *P. aeruginosa* has been found occasionally on the skin, isolated from clinical material [15] removed from the throat (5%), and in feces (3%) of non-hospitalized people. The percentage of patients with gastrointestinal problems from this bacteria increases by 20% after 72 hours of hospitalization. From the frequency with which this opportunistic pathogen is involved in human illnesses, it has been considered the most important human pathogen with respect to the number and types of infections which they cause and their association with morbidity and mortality. *P. aeruginosa* can be also responsible for an infection outbreak in a non-immune endangered community [4].

P. aeruginosa is frequently associated with contaminated potable or even treated water or solutions, such as gastroenteritis through contaminated food, otitis in swimmers, eye infections through contaminated solutions and tap water during lens care, wound infections caused by tap water, and respiratory tract infections caused by contaminated equipment. The most severe community which acquired infections are endocarditis in intravenous drug users, resulting from the use of contaminated parenteral solutions with drugs added, and peritonitis in individuals undergoing contaminated peritoneal dialysis solutions.

The principal reservoirs of *P. aeruginosa* in a hospital environment are [4, 8, 11, 12]: disinfectants, artificial breathing equipment, hydrotherapy equipment and tanks, food, sewers, drains, taps, soap bars, floor cloths, bed frames, air, chairs, cloth towels and tower hangers, mattresses, hoses, and tubes. Dissemination occurs through the hands of personnel (or staff) resulting from direct contact with taps and water reservoirs. The contamination of such water may be the result of clinical, fecal and urine materials.

CONCLUSION

Kawai *et al.*[16] analyzed, by denaturing gradient gel electrophoresis (DGGE), "the bacterial community in partially purified water, which is prepared by an ion exchange from tap water and is used in pharmaceutical manufacturing processes". The authors verified that "the dominant bacterium in purified water" (lower than that of other aquatic environments) could not be detected by soybean casein digest (SCD) and R2A media, although they presented esterase activity. Therefore, the authors emphasized "the importance of culture-independent methods of quality control for pharmaceutical water."

The water used in health center areas and in pharmaceutical industries should be periodically analyzed as a preventive measure against the spreading of microorganisms, allowing measures of improvement to be taken rapidly, as required.

The analysis of treated water for heterotrophic bacteria including *Pseudomonas* species is valuable in the prevention of the formation of biofilms and in the reduction of the amount of pyrogen.

Several solid surfaces can harbor biolfilms in water purification systems, such as stainless steel, nylon materials (deionization apparatus), and polyamide polymers (osmose reverse membranes). Porous surfaces found on the deionization resins and the osmose reverse membranes provide an excellent opportunity to trap inorganic particles and bacteria which begin the formation of biofilms. In general, all cleaning chemical agents perform better in soft water (after water softner apparatus). Chlorine agents have been shown ineffective at removing biofilms. Chlorine agents are used to keep the storage tanks and distribution loop in low microbial densities. Hydrogen peroxide, peroxide containing sanitizers and peracetic acid chemical agents have been found to be highly effective in the prevention and removal of biofilms, in a short period of contact (1–2 minutes) and they are relative non-corrosive.

This work emphasized the urgent need in the continuous monitoring of the water purification performance, including the evaluation of the effectual removal of non-fermenting gram-negative bacteria, with special attention to *Pseudomonas sp*, which showed the most resistance to the chemical agents, at the concentrations used in the system. Maybe the contact time between the sanitary agents (NaOCl and PAA+H_2O_2) and the system points should be reviewed. The 0.5% NaOCl after 18 h contact was shown sufficient to reduce bacteria population over 6-log_{10}. As all the bacteria tested showed similar MIC, maybe the short-time contact of 60 min of this solution (0.5% NaOCl) in the tank with the feeding water should have been greater to guarantee that the gram-negative bacteria that entered the system through the municipal source of drinking water would not survive. Even the contact-time of this sanitary agent (0.5% NaOCl) in the loop of distribution should have been longer to assure 10^{-2} CFU/mL of purified water with less chance of the*Pseudomonas sp* crossing through the consumption point filters. The loop of distribution should have been cleaned first with PAA+H_2O_2 and then with NaOCl to prevent the survival of any microorganism and possible development. Therefore, the contact time of the sanitizers and the different points of application should be reviewed and evaluated in accordance with the compatibility of the equipment and device material.

The effectual application of PAA+H_2O_2 in the osmose reverse membranes required a larger time-contact in the deionization columns, which showed greater heterotrophic enumeration. The double application of first PAA+H_2O_2 followed by NaOCl solutions (except the reverse osmosis membranes) would both reinforce the efficacy over the non-fermenting gram-negative bacteria. The PAA+H_2O_2 sanitary agent is known for its efficiency in preventing biofilm formation and its low pH may enlarge the NaOCl activity favoring microbial effectiveness.

Sodium hydroxide, hydrochloric acid and citric acid were used as chemical adjuvants on the pH adjustment of acid or alkaline water flow in the reverse osmosis membranes, the continuous deionization and the water softeners. At the concentrations used, no sanitation activity was demanded from their application, of which the ability over the tested microbial population was not confirmed, even though after 18 h of contact. Some efficacy shown by those chemical adjuvants was due to the alteration of environmental pH values. Even for extended exposures, *Pseudomonas sp* was shown to be the most resistant to those chemical agents at the tested concentrations.

Therefore the washing of (storage tanks) reservoirs, deionization columns, reverse osmosis membranes, as well as the sanitation of distribution circuits should be carried out by the determination of a schedule established for quality control (bacteriological and chemical) of water systems in the risky areas.

DECLARATIONS

Acknowledgements

We gratefully acknowledge undergraduate scholarships and funds offered by Brazilian Committees FAPESP, CAPES and CNPq, including the assistance of Biologist Irene Machoshvili and Edwards Lifesciences Macchi.

REFERENCES

1. [http://www.sabesp.com.br]
2. Brazilian Ministry of Health Directives: n° 15/MS/ANVS n°211/MS/ANVS 1999, n ° 196/MS/AVNS 1983, n° 36/MS/AVNS 1991, n° 113/MS/AVNS 1993, n° 122/ DTN 1993, n° 211/MS/ANVS 1999, and Health Sanitation Code n° 12342,. 1988
3. The United States Pharmacopoeia National Formulary: Water For Pharmaceutical Purposes USP 24/NF. 2000, 19: 2154-2165.
4. Schreckenberger PC, Janda JM, Wong JD, Baron EJ: Algorithms for identification of aerobic gram-negative bacteria. In: Manual of Clinical Microbiology. Edited by: Murray PR. 1999, Washington, American Society For Microbiology, 438-452. 7
5. [http://http//:www.us.filter.com]
6. Palmieri MJ, Caritó SL, Meyer RF: Comparison of rapid NFT and API 20 NE with conventional methods for the identification of gram-negative non-fermentative bacilli from pharmaceutics and cosmetics. Appl Environ Microbiol. 1988, 54: 2838-2841.
7. Vessoni Penna TC, Mazzola PG, Martins AM: The efficacy of chemical agents in cleaning and desinfection programs for hospital nurseries. J BMC Infectious Diseases. 2001, 1: 16-10.1186/1471-2334-1-16.
8. Silva CHPM: Bacteriologia um texto ilustrado (Bacteriology an illustrated text). Rio de Janeiro, Eventos Livraria. 2000
9. Traumann M, Michalsky T, Wiedeck H, Radosavljevic V, Ruhnke M: Tap water colonization with *Pseudomonas aeruginosa* in a surgical intensive care unit (ICU) in relation to *Pseudomonas* infections of ICU patients. Infect Control Hosp Epidemiol. 2001, 22: 49-52.
10. Wirtanen G, Salos S, Helander , Mattila Sandholm T: Microbiological methods for testing disinfectant efficiency on*Pseudomonas* biofilm. Colloids and Surfaces B: Biointerfaces. 2001, 20: 37-50. 10.1016/S0927-7765(00)00173-9.
11. Fernandes AT, Fernandes MOV, Filho NR: Infecção Hospitalar e suas Interfaces na Área da Saúde (Hospital Infection and its Consequences). São Paulo, Atheneu. 2000
12. Arnow PM, Flaherty JP: Non-fermentative gram-negative bacilli. In: Hospital Epidemiology and Infection Control. Edited by: Mayhall CG. 1996, Baltimore, Williams & Wilkins, 162-163.

13. Ferreira JAB, Silva Z, Oliveira RC, Branquinho MR: Bacterial contaminants in water and dialysate used for haemodialysis therapy in Rio de Janeiro (1999–2001). XXI Congresso Brasileiro de Microbiologia (Brazilian Congress of Microbiology). 2001, IH 040: 148-

14. Ferreira JAB, Silva Z, Oliveira RC, Branquinho MR: Microbiological quality of the water used for haemodialysis therapy in Rio de Janeiro (1999–2001). XXI Congresso Brasileiro de Microbiologia (Brazilian Congress of Microbiology). 2001, IH-039: 148-

15. Frota CC, Moreira JLB: Frequency of non-fermentative gram-negative bacilli isolated from clinical materials of patients at the Universidade Federal do Ceará Hospital Complex-Brazil. Brazilian J Microbiol. 1998, 29: 179-182.

16. Kawai M, Matsutera E, Kanda H, Yamaguchi N, Tani K, Nasu M: 16S Ribosomal DNA-Based Analysis of Bacterial Diversity in Purified Water Used in Pharmaceutical Manufacturing Processes by PCR and denaturing Gradient Gel Electrophoresis. Applied and Environmental Microbiology. 2002, 68 (2): 699-704. 10.1128/AEM.68.2.699-704.2002.

8
CHAPTER

A METAGENOMIC FRAMEWORK FOR THE STUDY OF AIRBORNE MICROBIAL COMMUNITIES

Shibu Yooseph[1], Cynthia Andrews-Pfannkoch[2], Aaron Tenney[1], Jeff McQuaid[3], Shannon Williamson[3], Mathangi Thiagarajan[4], Daniel Brami[1], Lisa Zeigler-Allen[3], Jeff Hoffman[3], Johannes B. Goll[4], Douglas Fadrosh[3], John Glass[2], Mark D. Adams[3], Robert Friedman[3], J. Craig Venter[3]

[1] Informatics, J. Craig Venter Institute, San Diego, California, United States of America

[2] Synthetic Biology and Bioenergy, J. Craig Venter Institute, Rockville, Maryland, United States of America

[3] Microbial and Environmental Genomics, J. Craig Venter Institute, San Diego, California, United States of America

[4] Informatics, J. Craig Venter Institute, Rockville, Maryland, United States of America

ABSTRACT

Understanding the microbial content of the air has important scientific, health, and economic implications. While studies have primarily characterized the taxonomic content of air samples by sequencing the 16S or 18S ribosomal RNA gene, direct analysis of the genomic content of airborne microorganisms has not been possible due to the extremely low density of biological material in airborne environments. We developed sampling and amplification methods to enable adequate DNA recovery to allow metagenomic profiling of air samples collected from indoor and outdoor environments. Air samples were collected from a large urban building, a medical center, a house, and a pier. Analyses of metagenomic data generated from these samples reveal airborne communities with a high degree of diversity and different genera abundance profiles. The identities of many of the taxonomic groups and protein families also allows for the identification of the likely sources of the sampled airborne bacteria.

Citation: Yooseph S, Andrews-Pfannkoch C, Tenney A, McQuaid J, Williamson S, Thiagarajan M, et al. (2013) A Metagenomic Framework for the Study of Airborne Microbial Communities. PLoS ONE 8(12): e81862. doi:10.1371/journal.pone.0081862.

INTRODUCTION

Air is an important transport medium for microbes, allowing them to overcome geographic barriers and disperse over long distances [1]–[3]. While airborne microbes were initially studied in the context of disease transmission [3], the role of bacteria in atmospheric processes like cloud formation is becoming evident [4], [5]. Furthermore, it is also evident that air can serve as a habitat for certain microbes that are able to metabolize compounds found in the atmosphere for growth and reproduction [6]–[8]. While most of our understanding of aerosol microbiology has been obtained from the study of bacteria and fungi that can be cultivated, it is also known that cultivable microbes make up only a small fraction of the total airborne microbes [1]. Overall, airborne microbial community composition and dynamics remain poorly characterised, particularly in the indoor environment where humans in industrialized countries spend nearly 90% of their time [9]. The development of new approaches for comprehensive characterization of airborne microbial communities thus has the potential to impact various disciplines, including studies of microbial diversity and biogeography, public health, building design, and understanding microbial role in biogeochemical cycling and climate processes.

Airborne environments, however, have extremely low density of biological material. Concentrations of atmospheric bacteria over land are estimated to be 10^4 cells/m^3 (with concentrations over sea lower by a factor of 100–1000) [1], [10]. In contrast, the number of bacteria in surface marine environments has been estimated to be 10^9 cells/m^3 [11], in soil 10^{10}–10^{11} per gram of soil [12], and in the human distal gut 10^{13}–10^{14} per human [13]. Fungal concentrations in outdoor environments have been reported to be in the 10^3–10^4 spores/m^3range [14], [15]. Primarily because thousands of liters have to be collected for a representative sample, the majority of non-culture based studies conducted on airborne communities to date have relied on amplification and sequencing of the 16S (for bacteria) or 18S (for fungi) ribosomal RNA gene to characterize their taxonomic composition [16]–[25].

In contrast to targeted sequencing of taxonomic markers, the metagenomic paradigm based on whole genome shotgun (WGS) sequencing of environmental DNA samples, allows for examination and analysis of both taxonomic composition and metabolic potential of the sampled microbial communities [26]. This paradigm has been used to study microbial composition and diversity in many environments [13], [27]–[29].

Here we present a metagenomic framework for studying airborne microbial communities, including protocols for sample collection, and for isolation and amplication of DNA from these samples. A WGS study of airborne microbes requires millions of liters of air be sampled to obtain enough material for sequencing; this requires long sampling times, an efficient means for capture of microbial constituents, and processes to limit post-capture growth which would skew the representation of sampled microbes. We applied this framework to study the composition of indoor and outdoor air samples collected from locations in New York City (NYC), New York and in San Diego (SD), California. Since DNA yields were inadequate for sequencing, the DNA isolated from the samples was amplified to generate sufficient amounts for library construction. Sequence data generated from the libraries were used to study the taxonomic composition of the sampled microbial communities and the functional composition of their bacterial constituents.

MATERIALS AND METHODS

Samplers and Sample collection

Indoor and outdoor air environments were sampled in NYC and SD. Permissions were obtained to collect air samples at each of the locations. Sampling was carried out using SpinCon®PAS 450-10A Wet Cyclone Portable Air Samplers (Sceptor Industries, Kansas City, MO), modified as described below. The samplers were operated at a flow rate of 450 L/min. The sampling buffer for the SpinCon air samplers was comprised of 1X PBS pH 7.0 supplemented with 0.25 mg/ml tetracycline, 0.25 mg/ml gentamycin and 0.125 mg/ml chloramphenicol made up in one liter of sterile water for injection (Baxter, Deerfield, IL). Evaporative loss of the water inside the sampler was off-set with sterile water for irrigation (Baxter, Deerfield, IL), supplied from a 10 L bag. Approximately 10 mls of sample buffer was maintained in the sampling chamber throughout the sampling cycle. The sample cycle time was two hours. After each two hour collection cycle, the sampler was programmed to turn off and drain the sample into a 250 ml corning bottle containing sodium azide (0.02% final concentration after the 12 hour sampling cycle) placed in a modified 4°C mini-refrigerator. The sampler would then refill the sample chamber with sample buffer and restart the collection cycle. Every 12 hours the aggregate sample, comprised of six discrete samples, was removed from the 4°C refrigerator, molecular biology grade glycerol was added to a final concentration of 10% and the sample was stored at −20°C for transport back to the lab and subsequent storage at −80°C. Every 24 hours the samplers were subjected to a clean in place (CIP) protocol described below.

Modification of the SpinCon®PAS 450-10A Portable Air Samplers

All manufacturer supplied tubing and fittings were replaced with antimicrobial fittings and tubing (Cole Parmer, Vernon Hills, IL) where appropriate or completely removed from the sampler. Anti-microbial tubing and fittings were used to feed sample buffer and the make-up water to the sampling chamber.

CIP Protocol

The SpinCons were disassembled and cleaned as follows. The leads from the makeup water and the sample buffer were closed via a stopcock, capped and removed from the unit. The units were turned off and unplugged. The air inlet and outlet pipes were removed, one end was capped with aluminum foil, and the interior of the pipes were sprayed with 70% isopropanol. The other end was capped and the pipes were set aside until reassembly. The collection chamber and blower unit were removed from the box. All gaskets were removed washed with Liquinox, and rinsed with DI water and air dried. The collector/concentrator was cleaned thoroughly with Liquinox solution, rinsed well in DI water, and sprayed with 70% isopropanol, wrapped in foil and set aside. The peristaltic pump head was removed from the bottom of the sampler and the tubing was replaced with a new piece (30 cm) of autoclaved Pharmed (Cole Parmer, Vernon Hills, IL) tubing (1/8″ OD) with foil wrapped ends. The pump head was then remounted on pump axle. The sampler was reassembled and rinsed as follows: Two manual collections and two rinses were done with 10% Chlorox. The second Chlorox rinse was allowed to remain in the collector/concentrator chamber for 10 minutes before being drained. Two manual collection and two rinses were done with 0.5% sodium thiosulphate. The second sodium thiosulphate rinse was allowed to remain in the sample chamber for 10 minutes before being drained. Two manual collection and two rinses were done with 70% isopropanol. The second isopropanol rinse was allowed to remain in the collector/concentrator chamber for 10 minutes before being drained. The sample buffer was reattached to the unit and one manual collection and one rinse was done. The makeup water was reattached to the unit and the sampling cycle was restarted.

Material Handling and Controls Used During Sample Preparation

Large volumes of buffers and reagents were UV treated by placing them in a biosafety cabinet with the UV lamp on for 2 hours. When appropriate, caps were removed to allow maximum exposure to the UV light. Liquid volumes less than 500 μl were UV treated in a Stratalinker 1800 for 10 minutes. Tubes used during the extraction of the filters were treated with Sigmacote, rinsed with UV treated water and air dried in a biosafety cabinet. The tubes were then UV treated in a Stratalinker. Low binding, UV treated microfuge tubes (Ambion, Grand Island NY) were used during the purification and concentration of the DNA. All gel rigs were treated with DNAaway (Molecular BioProducts, San Diego, CA) and rinsed with UV treated water. Dedicated pipets and low binding pipet tips were used for all of the protocols. All of the enzymes used in the purification and amplification of the DNA were treated with 0.125-0.5 units of nuclease (Roche Applied Science, Indianapolis, IN) for one hour at 30°C in the presence of 5 mM $CaCl_2$. Treated enzymes were stored at −20°C in UV treated tubes. Prior to addition of the enzyme to a reaction, the enzyme was incubated at room temperature for 10 minutes in an appropriate reaction buffer supplemented with EGTA to inactivate the nuclease. DNA and/or random hexamers was then added to the reaction. Positive controls during the MDA and linker amplification steps consisted of purified genomic DNA, while no template negative controls were used to assess spurious amplification and DNA contamination. No template control lanes in verification gels were post-stained with SYBR Gold and visualized using a Typhoon (GE Healthcare Life Sciences, Piscataway, NJ), and the gels were overexposed to show no detectable background amplification. Controls were also employed during the amplification of the 16S V3–V5 variable region.

Fractionation of the Particulates

Individual 12 hour indoor or outdoor samples were thawed overnight at 4°C. The samples were then transferred to a biosafety cabinet and the outside of the sample bottle was sprayed with 70% isopropanol and allowed to air dry. Samples were transferred to 50 mL conical tubes. De-aggregation of the particulates was accomplished by adding Tween-20 to a final concentration of 0.003%, incubation with inversion on a LabQuake Rotisserie at 37°C for 30 minutes, sonication at room temperature for 10 minutes on the highest setting in a bath sonicator (Aquasonic Model 50D, VWR) and a final incubuation with inversion at 37°C for 30 minutes. Samples were then filtered through an autoclaved 25 mm Versapor 3.0 μm filter (Pall Life Sciences, Ann Arbor, MI) followed by filtration through an autoclaved 25 mm Supor 0.1 μm filter (Pall Life Sciences, Ann Arbor, MI). The filters were folded in half sample side in and transferred to vacuum seal bags and stored at −80°C. The 0.1 μm filtrate was stored in 50 ml conical tubes at −80°C and not used for the studies presented here.

Our pilot sampling efforts conducted in the spring and summer months of 2005–2006 used a DNA isolation method based on that used by the Global Ocean Sampling (GOS) project [27],[30]. DNA isolated from this sampling attempt was not refractory to enzymatic treatment. However, initial attempts at DNA isolation from the NYC November 2007 samples using the GOS DNA isolation method yielded DNA that was refractory to enzymatic manipulation including amplification by PCR or MDA. This was likely a result of the presence of magnetic particulates, possibly iron, in the samples. It has been reported [31] that auto-oxidation of iron in the presence of Tris buffers is greatly enhanced, leading to free radical formation which can potentially damage DNA. It has also been noted [32] that heavy metals present in both coarse and fine particulate matter in urban air samples caused DNA damage. In light of these observations, the DNA extraction protocol was changed to the method described below, which removed the Tris and the EDTA from the extraction buffer. Out of an abundance of caution, the extracted DNA was placed on a magnet to remove particulates and drop dialyzed. The DNA was then treated with PreCR (New England BioLabs, Beverly, MA) to repair any DNA damage. In order to maximize DNA recovery, the filters were dissoved and the DNA was pooled with the DNA from the initial filter extraction.

DNA Isolation

Individual filters were removed from the −80°C and allowed to thaw at room temperature inside a biosafety cabinet. Thawed filters were aseptically cut into small pieces and placed in individual silanized tubes. Powersoil bead reagent (MoBio Laboratories, Inc. Carlsbad, CA) was added to just cover the filter pieces (300–400 μl). Sixty μl of Powersoil C1 and Proteinase K (200 μg/ml final concentration) were added and the sample was incubated at room temperature for one hour. The sample was gently mixed every twenty minutes. An aliquot of Proteinase K equal to the first aliquot of Proteinase K was added (a final concentration of 400 μg/ml) and the sample was subjected to three freeze/thaw cycles in dry ice/ethanol and a 55°C water bath. Samples were incubated at 55°C for two hours with rotation at 175 rpm. The sample was recovered from the filter pieces and transferred to microfuge tubes. Recovery of nucleic acids was accomplished by adding sodium chloride to a final concentration of 0.25 M, 15 mg of glycoblue (Ambion Grand Island, NY) and an equal sample volume of isopropanol. The precipitation was carried out overnight at room temperature. The DNA was recovered from the isopropanol by spinning at >20 K rpm at room temperature for one hour. Pellets were washed twice with 70% ethanol and resuspended in TE. The samples were phenol extracted twice followed by extraction with phenol:chloroform:isoamyl alcohol. The nucleic acids were isopropanol precipitated overnight at room temperature in the presence of sodium acetate (0.3 M final concentration). The DNA was recovered from the isopropanol by spinning at >20 K rpm at room temperature for one hour. Pellets were washed with 70% ethanol and resuspended in TE. Pellets were pooled with the DNA recovered from the filter dissolution (see below).

Filter Dissolution

Extracted filters were resuspended in a volume of Powersoil bead reagent (MoBio Laboratories, Inc. Carlsbad, CA) to just cover the filters and 0.5 volume of phenol was added. For the 0.1 μm filters 1.5 volumes of chloroform was added, for the 3.0 μm filter, 2 volumes of chloroform was added. The filters were then incubated with inversion at 65°C for 10 minutes. The phases were separated by centrifugation and the aqueous phase was recovered. The aqueous phase was extracted twice with phenol:chloroform:isoamyl alcohol (25:24:1 v/v). DNA was precipitated as described above and pellets were combined with the initial DNA extraction.

Removal of Magnetic Particulates (NYC samples only)

The combined sample (from the initial extraction and the filter dissolution) was placed on a magnet (Life Technologies, Grand Island, NY) for 20 minutes at room temperature. The solution containing the DNA was then carefully removed and the tube was removed from the magnet. The magnetic particulates were washed with 50 μl of TE and placed on the magnet for an additional 20 minutes. This wash was pooled with the initial solution and drop dialyzed.

Drop Dialysis (NYC samples only)

Approximately 100 mls of MilliQ water (enough to cover the bottom) was placed in the bottom of an oblong Pyrex dish and heated on high power in a microwave for 1 minute. The dish was covered and allowed to cool slightly. A small petrie dish was placed in the pyrex dish and 30 mls of 1/5X TE was added to the petrie dish. A 0.025 μm Millipore (Millipore, Billerica, MA) disc was floated on top of the 1/5X TE and the DNA sample was carefully added to the disc. Dialysis was carried out at RT for 45 minutes. Samples were recovered from the disc and the spot where the sample was on the disc was washed with an additional 50 μl of 1/5 X TE. The sample was then placed in a Speed-Vac and evaporated until approximately 10 μl remained in the tube. Samples were pooled according to filter size and location. Yield was estimated on a 1 mm thick vertical 1% agarose gel post-stained with Sybr gold (Life Technologies, Grand Island, NY) and visualized on a Typhoon.

Repair of Damaged DNA

Gel purified DNA was treated with Pre-CR (New England Biolabs, Beverly, MA) as follows: a mastermix comprised of 5 μl of 5X Thermopol buffer supplemented with 50 mM EGTA and 2.5 μl of treated PreCR enzyme mix was incubated at room temperature for 10 minutes. 0.5 μl of 100X NAD+, 0.5 μl of 10 mM dNTP's, and 12.5 μl of water was added to the mastermix. 5 μl of the mastermix was added to 5 μl of DNA. The reaction was incubated at 37°C for 20 minutes followed by overnight incubation at 4°C. The treated DNA was phenol extracted, followed by a phenol:chloroform:isoamyl alcohol (25:24:1 v/v) extraction, followed by ethanol precipitation. For post MDA and debranching this reaction was scaled by a factor of 5 and the reaction was carried out at 37°C for one hour before extraction and precipitation.

phi29 Amplification of Samples (NYC samples only)

DNA was amplified using the Illustra Genomophi (GE Healthcare Life Sciences, Piscataway, NJ) enzyme cocktail following a modified version of a previously described protocol [33]. The final reaction buffer (G-Buffer) was comprised of 37 mM Tris-Cl pH 7.5, 10 mM $MgCl_2$, 5 mM $(NH_4)_2SO_4$, 50 mM KCl, 1 mM each dNTP's, 10 mM EGTA, 4 mM DTT, 0.04 mM random hexamers (IDT) 0.2% Tween 20 and 100–1000 pg of input template. Ten reactions were set up for each sample to minimize any bias that might result from the amplification. The positive control consisted of 100 pg of DNA isolated from a 0.1 μm marine metagenomic filter [30]. The amplification protocol was performed in a thermocycler and was modified as follows: samples were heated to 70°C for 5 minutes and then quickly cooled to 4°C. Samples were removed to ice and opened in a biosafety cabinet where the enzyme master mix was added. The samples were briefly mixed and returned to the thermocycler. It has been reported that high G + C% regions of DNA are preferentially amplified during amplification with phi29 polymerase [34]. One possible explanation for this is that the 30°C annealing and amplification temperature that is used favors G + C hexamer annealing. To minimize this effect, samples were incubated at 4°C for 10 minutes. A temperature ramp of 0.1°C/sec was applied to the samples until 10°C was reached; incubation at 10°C was for 10 minutes. This ramp and incubation was repeated at 15°C, 20°C, and 25°C with a final 0.1 C/sec ramp to 30°C. One hundred picograms of DNA isolated from a 0.1 μm marine metagenomic filter served as the positive contol; a no template control was also included. Samples were then incubated at 30°C for 1.5 hours, followed by a 10 minute 65°C incubation. The yield from individual reactions were assessed by agarose gel electrophoresis. Approximatedly 1/10th of each reaction containing template or the entire no template control was visualized on a 0.8% e-gel. DNA from lanes containing input template were clearly visible with ethidium bromide staining. To assess the no template control lanes the gel was post-stained with SYBR Gold and visualized with a Typhoon. Over exposure of the gel showed no detectable background amplification. Amplified DNA was pooled, phenol:chloroform:isoamyl alcohol (25:24:1 v/v) extracted and ethanol precipitated overnight at room temperature in the presence of 0.3 M sodium acetate and 15 mg glycoblue. Pooled samples were washed with 70% ethanol and resuspended in TE. The yield from the MDA was determined by reading the A_{260} using a Nanodrop ND-1000 spectrophotometer.

Debranching (NYC samples only)

Debranching has been reported to improve library quality [35]. All samples from the

MDA reaction including the positive control were debranched as follows. Residual hexamers in the sample were removed with approximately 30 units of $RecJ_f$ (New England Biolabs, Beverly MA). Incubation was for one hour at 37°C followed by phenol:chloroform:isoamyl alcohol extraction and ethanol precipitation. The samples were then incubated in a 50 μl reaction as follows: 25 μl 2X G-Buffer 4 μl Illustra Genomiphi enzyme cocktail, 0.5 mM dNTP›s and 4 mM DTT for two hours at to 30°C. Forty units of Bst DNA polymerase large fragment was added to the reaction and incubation proceeded for an additional 2 hours at 65°C. Samples were extracted with phenol:chloroform:isoamyl alcohol and ethanol precipitated. To remove any remaining branched structures, the samples were incubated in a reaction containing 1X NEB buffer 2 and 10 units of T7 endonuclease I (New England Biolabs, Beverly MA) for 30 minutes at 37°C. The samples were phenol:chloroform:isoamyl alcohol (25:24:1 v/v) extracted and ethanol precipitated overnight at room temperature.

Linker Adapted Amplification of Samples (SD samples only)

DNA was sheared using the Covaris S series (Woburn, MA) to 500–800 bp and subsequently blunt-end repaired using BAL-31 nuclease (New England Biolabs, Beverly MA) and T4 DNA Polymerase (New England Biolabs, Beverly MA). I-ceu1 adapters[36] with 5′-phosphate groups were then appended using T4 ligase (New England Biolabs, Beverly MA) followed by amplification using Phusion DNA polymerase (Finnzymes, Espoo Finland). Excess adapters were removed from each sample using gel excision (1% LMP agarose) purification. The amplified DNA fragments were excised from the agarose and recovered with beta-agarase (New England Biolabs, MA), phenol extraction and isopropanol precipitation. The positive control consisted of purified E. coli DNA.

Gel Purification of DNA

Pooled DNA was gel purified on a 1% LMP gel post stained with SYBR gold. For each sample, an area of the gel corresponding to greater than 500 bp–1000 bp was excised and used for library construction. DNA was recovered from the gel slices with beta-agarase (New England Biolabs, Beverly MA), phenol extraction and isopropanol precipitation.

Shotgun Library Construction

For NYC samples, two methods were used to produce WGS libraries from the phi29 amplified DNA isolated from the indoor and outdoor filters. In both, DNA was sheared, size selected, and the ends repaired. In the first method, A/B adaptors (Roche technical bulletin 004-2009) were added in an overnight 4°C ligation reaction. Excess adaptors were removed and the adaptorized fragments were recovered as described above. Biotinylated B adaptor for 454 FLX sequencing was incorporated using Phusion DNA polymerase and 15 cycles of PCR. Three PCR reactions were set up for each sample and the reactions consisted of ~50 ng of adaptorized DNA, 1X Phusion Buffer0.2 mM dNTP's, 50 μm 454 FLX A, 50 μm 454 FLX BioB and 5 units of Phusion DNA polymerase. The biotinylated fragments were recovered from the PCR reaction using streptavidin coated M-280 Dynabeads (Invitrogen, Grand Island, NY) following the manufacturer's supplied protocol. In the second method, libraries were constructed using the 454 rapid library construction kit according to the manufacturer's protocols. For SD samples, the linker adapted DNA libraries were constructed using the 454 rapid library construction kit according to the manufacturer's protocols.

16S PCR Amplification (NYC samples only)

Prior to 16S ribosomal PCR, the MDA products were debranched. This step proved necessary to achieve reliable PCR amplification. Post-MDA 16S PCR products were generated using the V3–V5 variable region of the 16S gene. The V3_357F PCR primer sequence was 5′ CCTACGGGAGGCAGCAG and the V5_926R primer sequence was 5′ CCGTCAATTCMTTTRAGT. Three reactions for each sample, including the positive control described in the phi29 amplification section and a no template control, were performed using Phusion DNA polymerase (New England Biolabs, Beverly MA) as described above. The reactions for each sample were pooled, phenol:chloroform:isoamyl alcohol (25:24:1 v/v) extracted, ethanol precipitated and sequenced using the 454 Titanium platform.

Initial processing of metagenomic data

Raw 454 reads were processed to identify and remove artificial replicate sequences [37]. Then, low complexity regions in the remaining reads were identified using DUST [38] and a read was excluded from further analysis if >50% of the length of the read was identified as low complexity. RepeatMasker [39] was used to identify known eukaryotic repeat families (using the RepBase library) on the remaining reads. A read was excluded from further analysis if >50% of the length of the read contained a match to a repeat family.

Identification of chimeric metagenomic reads

The goal was to identify sequences containing one of the two chimeric rearrangement patterns[40] - inverted sequences and transposed direct sequences. As part of this, we first flagged any sequence that contained an exact ≥25 bp repeat within it (in either the forward or reverse direction) and any sequence with >90% match (over >95% of its length) to this sequence, as chimeras (containing repeats). We chose to be conservative in the flagging process and acknowledge that this constraint may be too strict and can eliminate sequences that contain true repeats (as may be the case in eukaryotes). An all-vs-all BLASTN [41] of the sequences in the sample was then carried out to identify match regions between pairs of sequences. Sequence pairs that had match regions corresponding to the two types of chimeric rearrangements (inverted sequences and transposed direct sequences) were noted as a chimeric pattern. Subsequently, any sequence that had more sequences similar to it with chimeric patterns than those without chimeric patterns, was also flagged as a chimera. This criterion invokes the assumption that chimeric sequences account for only a small fraction of the total reads.

Taxonomic assignment of metagenomic reads

We used a combination of nucleotide and amino acid searches to assign taxonomy to the metagenomic reads. The nucleotide level searches allowed classification of non-protein coding sequences and were particularly important for identifying human, mouse, and other large eukaryotic components of the data set, while the protein level searches allowed greater sensitivity to detect homology. Thus, first, reads were searched against NCBI non-redundant nucleotide database [42] using BLASTN. A maximum of ten blast hits with an e-value < 1e-5 were retained for each read. If all of the matching database sequences were from eukaryotic genomes, the read was classified as eukaryotic in origin. Then, for all other reads, peptides were predicted and these predictions were annotated (see below). Taxonomic classifications for the peptides (using BLASTP [41]) were obtained as part of the annotation process (including assignments possibly to eukaryotes). The peptide taxonomies were transferred onto the reads: for a read that had multiple peptides with the same taxonomy, that taxonomic classification was transferred onto the read; when the taxonomies conflicted, then the read was assigned a "Mixed" status; if a read had multiple peptides with consistent taxonomy along with unclassified peptides, then the read was assigned that taxonomy. Together with the results from the initial BLASTN step, this process assigned the metagenomic reads to following categories: Archaea, Bacteria, Eukaryota, Virus, Other (synthetic sequences), Mixed, and Unclassified.

Functional annotation of metagenomic reads

Peptides were predicted on reads using a method that was designed to deal with pyrosequencing errors that result in frame shifts [43]. These peptides were annotated using the JCVI metagenomic annotation pipeline [44] modified to use the Uniref100 database [45] as the subject database for the BLASTP searches.

The bacterial peptides were also searched against KEGG Orthologs (KO) [46], COG [47], Antibiotic Resistance Genes database (ARDB) [48] and Virulence Factor database (VFDB)[49]. Abundance of each KO in a sample was calculated from raw counts and after accounting for differences in read lengths (between samples) and in gene lengths [50]. Individual KO abundances were summed to generate abundances for KEGG functional categories in each sample. COGs were processed in a similar fashion. For ARDB and VFDB searches, only matches with ≥90% identity over ≥90% of the peptide length were considered.

Mixture modeling

Let U, E, and PV denote the %(G+C) distributions of the reads in the Unclassified, Eukaryotic, and Prokaryotic/Viral groups respectively. Define a new distribution X such that $X=p*E + (1-p)*PV$. We computed an optimal value of p for which the symmetric Kullback-Leibler distance [51] between X and U is minimized.

Assembly of metagenomic reads

Metagenomic assemblies were computed using Newbler GS De Novo Assembler, version 2.3 (Roche, 454).

NYC 16S data analysis

Sequences were processed using in-house scripts that deconvoluted the barcoded samples, trimmed barcodes and primers, and removed low quality sequences. Chimeras were detected using Chimeraslayer [52] and removed. Remaining sequences were assigned taxonomy using Ribosomal Database Project classifier [53]. Sub-genus diversity was analyzed using mothur[54]. For this, sequences were aligned using the silva template and the alignment was subsequently trimmed and short sequences removed to produce a master alignment. Since the Operational Taxonomic Unit (OTU) analysis on the full dataset from this master alignment was computationally prohibitively expensive, we used subsampling to calculate rarefaction curves and diversity measures. We generated 25 random samples where each random sample consisted of 10,000 indoor sequences and 10,000 outdoor sequences. These sequences were preclustered and a distance-matrix was generated and subsequently used to produce OTUs at 97% identity (using average linkage clustering).

RESULTS

Collection and DNA Isolation

The NYC samples were collected from a modern high rise office building in Midtown Manhattan, while the SD samples were collected from a hospital medical center, a single family ranch-style home with no pets, and a marine pier. The NYC collection effort sampled 5.9×10^6 liters of outdoor air using two modified SpinCon wet cyclone samplers on an outside covered walkway at the 22nd floor air intake plenum (101 m above street level) of the building. In addition, 10.8×10^6 liters of indoor air was sampled using three modified SpinCon samplers in the mixing room of the same building where filtered outdoor air was mixed with returning indoor air prior to being circulated through the building. As part of the SD sampling, indoor and outdoor air of the hospital was sampled using five SpinCon devices deployed in an HVAC mixing room: three of the devices sampled the recirculating indoor air, while the other two devices sampled the incoming fresh air from the outdoors; 7.2×10^6 and 3.1×10^6 liters of indoor and outdoor air were sampled at this site. Indoor air of the house was sampled by running three air samplers placed in the main living area; 0.9×10^6 liters of air were sampled at this location. The pier sample was collected using four collection devices deployed at the end of the pier, sampling 5.4×10^6 liters of marine air.

Extensive precautions were taken to reduce the possibility of sample contamination or bacterial growth during sampling. These included a daily decontamination of the samplers between sample batches, inclusion of bacteriostatic antibiotics in the sampling buffer, and UV treatment of all reagents used for DNA isolation and manipulation.

Following collection into liquid, samples were filtered onto 3.0 μm and 0.1 μm filters using serial filtration.

Despite the large volumes of air that were sampled, DNA yields were less than 15 ng for each sample, which was inadequate for standard 454 library preparation methods. Therefore it was necessary to perform initial amplification to produce sufficient quantity of DNA for library construction. Amplification requires DNA that is relatively free from non-biological particulates; contamination with foreign DNA either from the sampling or the DNA extraction and isolation should be avoided. To ensure that foreign DNA was not introduced during DNA isolation or amplification, strict protocols were followed, with extensive assessment of potential contamination at each stage. Two methods of amplification were chosen: the NYC samples were amplified by Multiple Displacement Amplification (MDA) using phi29 DNA polymerase [55]and the SD samples were amplified by linker amplification [36]. The amplification products for these samples were quantified using spectroscopy and varied from 100 ng to 1000 ng (Table 1). For these samples, in general, the DNA yields from the 3.0 μm filter could be expected to be at least as high as the yields from the 0.1 μm filter. This is due to a comparatively higher amount of starting material expected on this filter resulting from the capture of large cells (eukaryotic and prokaryotic) in combination with incomplete disaggregation during fractionation, due to free DNA clumping together or with abiotic particles. Table 1 shows this to be the trend for all samples except the indoor hospital sample (likely attributable as a stochastic event).

Table 1: DNA yields after amplification

Sample	Estimated final DNA yield (ng)
NY_INDOOR (0.1 μm)	100
NY_INDOOR (3.0 μm)	200
NY_OUTDOOR (0.1 μm)	500
NY_OUTDOOR (3.0 μm)	1000
SD_IHOSP (0.1 μm)	468
SD_IHOSP (3.0 μm)	259
SD_OHOSP (0.1 μm)	235
SD_OHOSP (3.0 μm)	327
SD_IHOUS (0.1 μm)	110
SD_IHOUS (3.0 μm)	308
SD_SCRPP (0.1 μm)	294
SD_SCRPP (3.0 μm)	370

The MDA products for the NYC samples were quantified by UV spectroscopy and the linker amplification products for the SD samples were quantified by fluorescence spectroscopy.
doi:10.1371/journal.pone.0081862.t001

Sequence data

Libraries were constructed from the corresponding 3.0 μm and 0.1 μm filters, and these libraries were sequenced using 454 pyrosequencing platform [56]. Reads from the two filters for each location were pooled, generating six metagenomic datasets: NYC indoor (NY_INDOOR), NYC outdoor (NY_OUTDOOR), SD indoor hospital (SD_IHOSP), SD outdoor hospital (SD_OHOSP), SD indoor house (SD_IHOUS), and SD pier (SD_SCRPP). In addition, we were also able to generate 16S rRNA sequence tags from the NYC samples after PCR amplification of the MDA products. Sequence data from all samples have been deposited in the sequence read archive at NCBI [57] and are available under BioProject accession PRJNA218551.

Metagenomic data

The WGS reads were processed to generate data suitable for downstream analysis. As part of this, three categories of reads were identified and removed: artificial replicates generated by pyrosequencing, reads of low-complexity, and reads containing repeats. An examination of the resulting reads from the MDA samples (NY_INDOOR and NY_OUTDOOR) revealed the presence of chimeric reads. Amplification using phi29 DNA polymerase has been shown to generate chimeric sequences when applied to whole genome amplification from low input DNA amounts; a model based on branch migration can be used to explain the mechanism of chimera formation [40]. Thus, a method using sequence similarity was implemented to detect and remove chimeric sequences from these data. This method was applied to each of the six datasets. At the end of these processing steps, a total of 5,342,939 reads remained, and only these were used in subsequent analysis.

Taxonomy based on metagenomic data

A BLAST based method was used to assign taxonomy to the reads. On average, 54% of the reads in a sample could be assigned taxonomy, with SD_IHOSP having the highest (74%) and NY_OUTDOOR having the lowest (35%) assignment proportions. Sequences from eukaryota and bacteria dominate the portion of data that can be assigned taxonomy. Eukaryotic sequences make up 71%, 82%, and 88% respectively of assigned sequences in NY_INDOOR, NY_OUTDOOR, and SD_SCRPP, with the remaining assigned sequences in these samples primarily from bacteria. On the other hand, SD_IHOSP, SD_OHOSP, and SD_IHOUS are dominated by bacteria (81%, 60%, and 79% respectively), with the remaining assigned sequences in these samples almost exclusively from eukaryota. For the eukaryotic portion of these samples, human and fungal (*Ascomycota* and*Basidiomycota*) sequences are predominant in indoor samples, while outdoor samples contain these together with a more diverse mix of rodent (mouse), plant (*Streptophyta*), fish, and insect (*Arthropoda*) sequences.

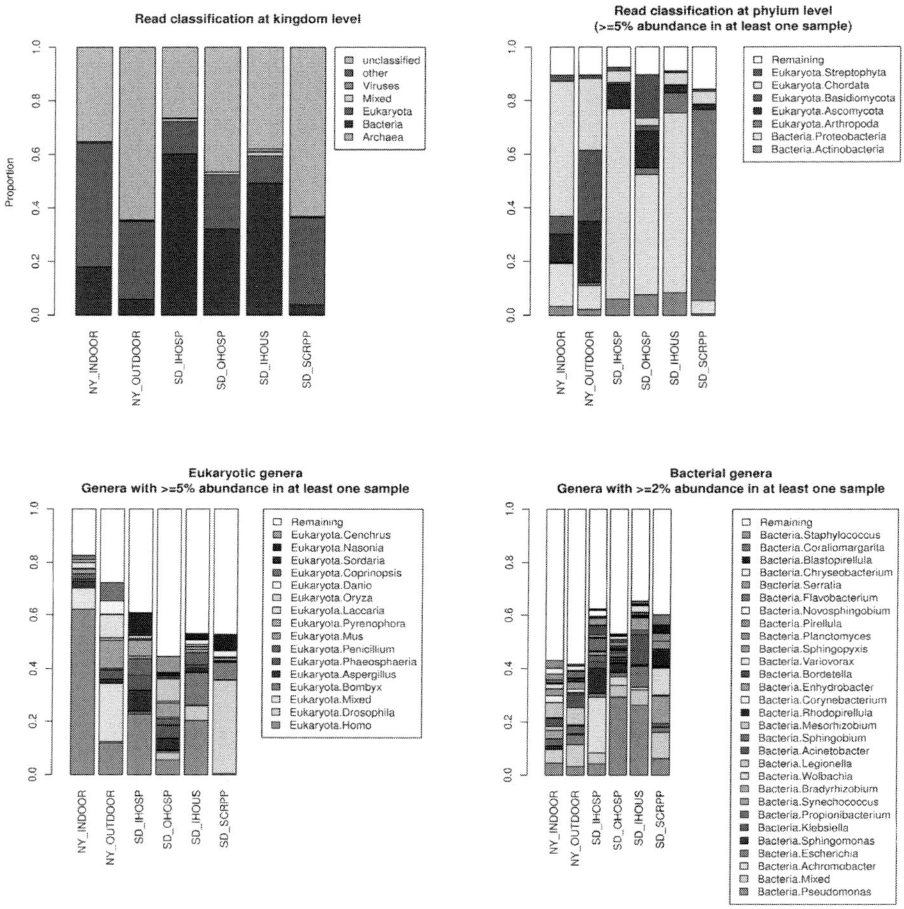

Figure 1: Taxonomic classification of metagenomic reads.

Table 2: Taxonomic classification of metagenomic reads

	NY_INDOOR	NY_OUTDOOR	SD_IHOSP	SD_OHOSP	SD_IHOUS	SD_SCRPP
Archaea	1,048	269	286	606	90	324
Bacteria	256,691	55,681	347,562	263,396	192,326	44,092
Eukaryota	663,225	281,601	68,998	167,883	40,334	375,198
Mixed	6,012	1,915	5,975	7,010	5,936	2,356
Viruses	958	495	1,926	514	4,186	2,026
Other	1,084	1,389	491	167	206	67
Unclassified	504,660	620,628	152,468	384,538	148,683	729,639
Total	1,433,678	961,978	577,706	824,114	391,761	1,153,702

Kingdom taxonomy: *Mixed* refers to a read that had matches to multiple kingdoms and could not be definitively assigned to one kingdom. *Other* refers to sequences that could not be identified (NCBI taxonomy ID 32644), mostly synthetic constructs. *Unclassified* refers to a read that could not be assigned to any kingdom (i.e. had no match in the reference databases).
doi:10.1371/journal.pone.0081862.t002

Proteobacteria is the dominant bacterial phylum in all samples, ranging from 46% of bacterial sequences (in SD_SCRPP) to 87% of bacterial sequences (in SD_IHOSP). *Actinobacteria* and*Bacteroidetes* phyla are also well represented in most of these samples. SD_SCRPP has a high representation (>10% of bacterial sequences) from two marine associated bacterial phyla (*Cyanobacteria* and *Planctomycetes*). At the genus level, however, the bacterial genera representation and abundance vary between samples. The top three represented genera in the samples (as a proportion of all bacterial sequences in the sample) are: NY_INDOOR – *Mesorhizobium* (5.5%), *Pseudomonas* (4.5%), and *Bradyrhizobium* (3.1%); NY_OUTDOOR – *Legionella* (6.4%), *Acinetobacter* (4.6%), and *Escherichia* (3.5%); SD_IHOSP – *Achromobacter* (20.9%), *Sphingomonas* (9.5%), and *Sphingobium* (4.7%); SD_OHOSP – *Pseudomonas* (29.3%), *Sphingomonas* (3.3%), and *Achromobacter* (3.2%); SD_IHOUS – *Pseudomonas* (26.2%), *Klebsiella* (11.2%), and *Escherichia* (7.9%); SD_SCRPP – *Synechococcus* (10.3%), *Wolbachia* (10%), and *Pseudomonas* (6.1%). The bacterial populations represented by these samples have a long tail of low abundant genera. This measure of diversity is supported by the observation that bacterial genera that have <2% abundance (in the corresponding sample) account for 64%, 66%, 43%, 60%, 42%, and 46% respectively of the bacterial sequences in NY_INDOOR, NY_OUTDOOR, SD_IHOSP, SD_OHOSP, SD_IHOUS, and SD_SCRPP.

A substantial number of metagenomic reads in these samples (45% on average) have no taxonomic assignments (Table 2). To explore the likely taxonomic origin of these unclassified reads, we compared GC composition of reads in the unclassified group (U) to that of reads in the eukaryotic group (E) and the prokaryotic/viral group (PV). Assuming that reads from the unclassified group actually belong to one of E or PV groups, the U distribution can be interpreted as a mixture distribution obtained by sampling from the E and PV distributions. We computed the probability p of sampling from the E distribution (and thus probability $1-p$ of sampling from the PV distribution) so as to minimize the (symmetric) Kullback-Leibler distance[51] between the resulting mixture distribution and the U distribution. These calculations suggest that a vast majority of the unclassified reads in NY_OUTDOOR ($p=0.93$) and SD_SCRPP ($p=0.89$), and to a lesser extent unclassified reads in SD_OHOSP ($p=0.78$) and NY_INDOOR ($p=0.75$), may be eukaryotic in origin.

The six datasets were assembled individually to assess recovery of long genomic sequences from these metagenomes. However, the assemblies were fragmented, with average contig sizes less than 1 kbp for each sample. At 96% identity threshold, SD_IHOSP had the largest proportion of assembled reads (65.2%) while NY_INDOOR had the lowest proportion (7.1%). The low degree of these assemblies attest to the diverse nature of the metagenomes being sampled and combined with an insufficient depth of sequencing of the associated microbial communities.

16S rRNA based taxonomic profiling of NYC samples

410,373 and 230,506 sequences respectively were generated from NY_INDOOR and NY_OUTDOOR samples. A small fraction of these sequences were identified as chimeras: 12,503 (3.0%) and 5,287 (2.2%) respectively in NY_INDOOR and NY_OUTDOOR, and these were removed. Chloroplast sequences comprised approximately 6% of the 16S sequences in NY_INDOOR and 8% of the sequences from NY_OUTDOOR, and these were also removed prior to further analysis. At the phylum level, NY_INDOOR is dominated by *Proteobacteria*(60%) and *Firmicutes* (14%), while NY_OUTDOOR is dominated by *Proteobacteria* (52%) and*Bacteroidetes* (22%). *Actinobacteria* are nearly equally abundant in both samples comprising 5% of NY_INDOOR and 7% of NY_OUTDOOR. *Mesorhizobium* (8.2%), *Staphylococcus*(4.0%), and *Pseudomonas* (3.0%) are the most abundant genera in NY_INDOOR, while*Hymenobacter* (17.6%), *Acinetobacter* (8.4%), and *Propionibacterium* (4.8%) are the most abundant genera in NY_OUTDOOR (Figure 2). Overall, 48% of NY_INDOOR sequences and 40% of NY_OUTDOOR sequences were assigned to genera that have <2% abundance in both samples. NY_INDOOR has a higher richness and diversity compared to NY_OUTDOOR (Figure 2).

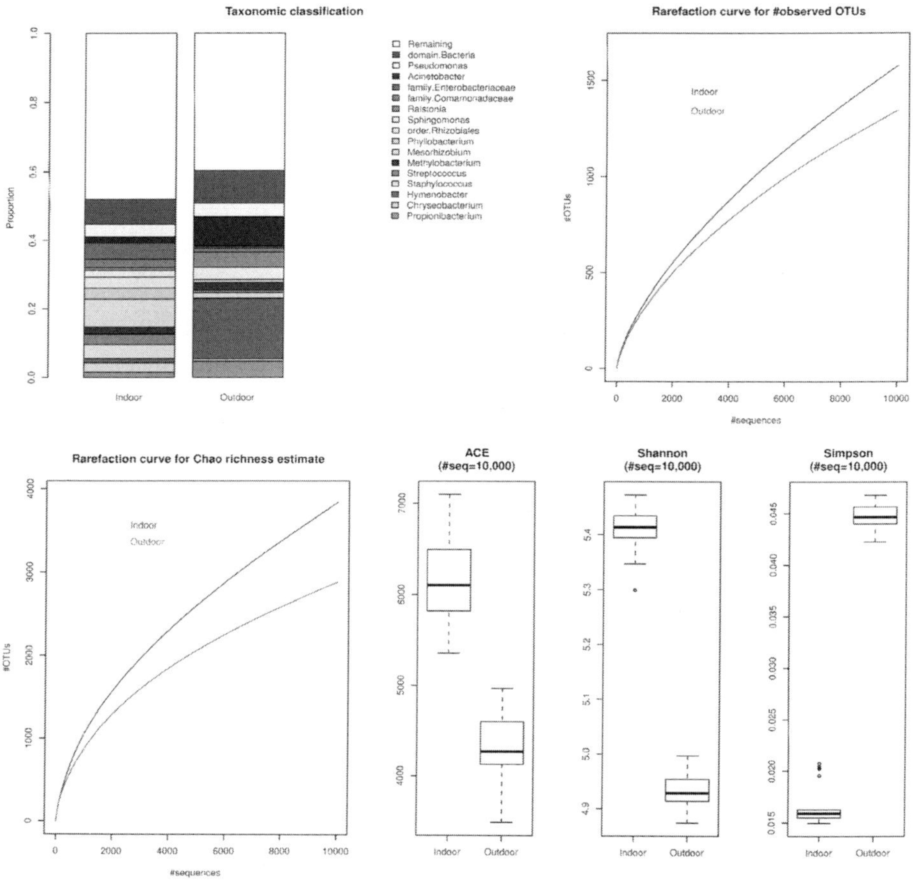

Figure 2: Taxonomic classification and diversity of the NYC 16S data.

In the stacked barcharts, only those taxonomic groups that have ≥2% abundance in at least one of the samples is reported. The rarefaction curves along with the richness and diversity estimates were calculated using mothur and were based on averages of 25 random samples of 10,000 sequences from each dataset.

For these two samples, the 16S based abundances of some taxonomic groups are different from their abundances inferred from metagenomic data. This can be primarily attributed to a lack of sequenced genomes that can serve as adequate near-neighbour references, but for which 16S sequences are available. These groups include *Hymenobacter (Bacteriodetes), Paracoccus (Alphaproteobacteria)*, *Enhydrobacter (Alphaproteobacteria)*, and *Phyllobacterium (Alphaproteobacteria)*.

Functional analysis of metagenomic data

A total of 3,777,158 proteins were predicted from reads in prokaryotic, viral, and unclassified groups, and subsequently annotated. 45% of them could be assigned a function or had a match to a hypothetical protein in the reference database. SD_IHOSP had the highest assignment proportion (71%) while NY_OUTDOOR had the lowest assignment proportion (14%). Those samples that had a larger predicted contribution of eukaryotes to the unclassified group, tended to have lower proportions of functional assignments.

In order to assess the functional composition of bacteria in these samples, KEGG Orthology (KO) [46] abundances were calculated using the bacterial proteins. The six samples show similar abundance profiles of KEGG categories (obtained by summing KO abundances), with an average correlation of 0.95 between pairwise sample profiles (Figure 3). In general, metabolism and transport (amino acid, carbohydrate and energy conversion) are the top functional categories followed by translation and signal transduction. A higher resolution ordination of the samples was also carried out by a principal component analysis using KO abundance profiles. While the first three principal components PC1, PC2, and PC3 explain 83.1% of the total variation, the ordination does not suggest obvious sample groupings or a separation across indoor and outdoor environments (Figure 4). It is however possible to identify the KOs that contribute the most along each principal component axis and the corresponding outlier samples. Notably, SD_IHOUS is an outlier along PC1 and this is primarily driven by a higher proportion of DNA replication proteins (K02314 and K02315) in this sample. Along PC3, SD_SCRPP has a higher abundance of arylsulfatases (K01130), while NY_OUTDOOR has a higher abundance of beta-lactamase (K01467) and tetracycline resistance proteins (K08151). A similar observation is made by an analysis of COGs [47].

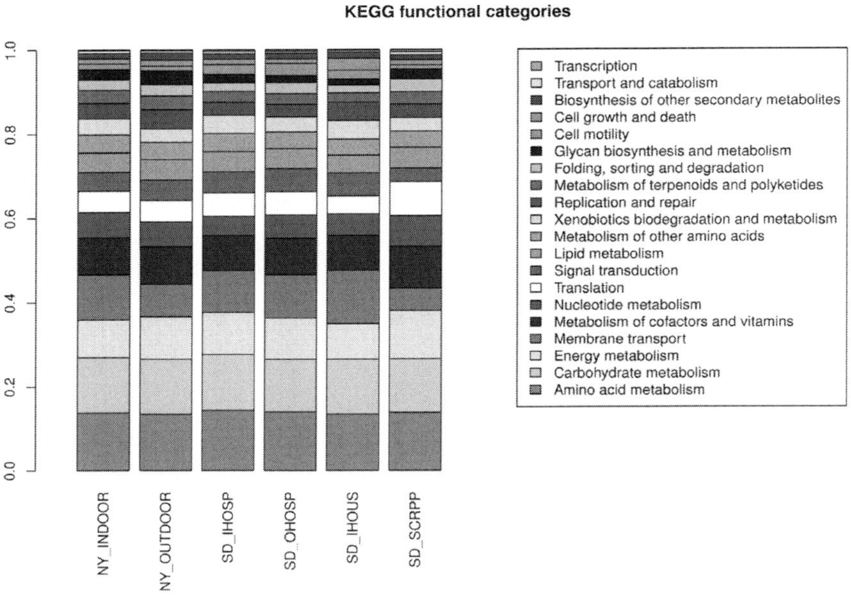

Figure 3: Abundances of KEGG functional categories.

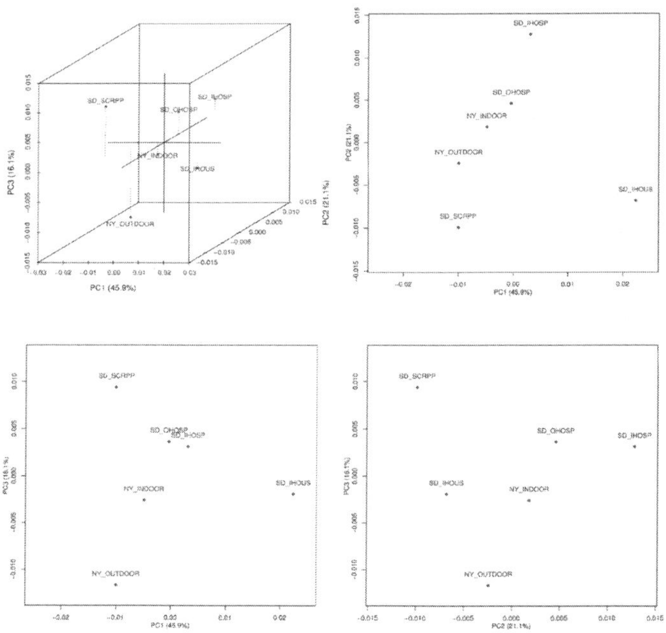

Figure 4: Ordination of the six metagenomic samples by principal component analysis based on KO abundance profiles.

The first three principal components (PC1, PC2, and PC3) account for 83.1% of the total variation. The dashed lines in the 3D plot shows the height (PC3) of the sample points when projected on to the PC1-PC2 plane.

Bacterial proteins were searched for presence of virulence genes and antibiotic resistance genes in these samples. There was no evidence of significant matches to true virulence genes [58] directly implicated in pathogenicity. The antibiotic resistance gene searches, however, return matches to beta-lactamases and tetracycline resistance genes in several samples, with NY_OUTDOOR having comparatively higher amounts (consistent with the KEGG based analysis).

DISCUSSION

Air is a low biomass environment and therefore the application of WGS methods to study airborne microbial communities is a challenge. A previously reported metagenome study of airborne microbes in indoor environments consisted of the extraction and sequencing of DNA collected on air filters that are part of air-handling units in buildings [59]. Here we describe protocols for the direct sampling of air from indoor and outdoor environments. As obtaining enough nucleic acid to conduct sequencing of these samples requires long sampling times, suitable collection protocols are necessary to avoid growth during sample collection, which would otherwise reflect culture biases rather than the source environment. Abiotic particulates present in the samples contribute to the challenge of isolating DNA that is not refractory to amplification. Despite these limitations we were able to develop sampling and amplification strategies to enable the study of this low biomass environment. We presented two different amplication strategies, phi29 MDA and linker amplification, to generate enough material for pyrosequencing to assess the genomic content of airborne metagenomes. Transposon based methods, such as Nextera, for simultaneously fragmenting and tagging DNA recently became available. These methods have been used to sequence phage and viral genome samples containing <1 ng of DNA [60], [61]. The utility of this method for sequencing low biomass metagenomic samples is promising, however, this type of method may miss genomes present in the sample which contain few or no recognition sites for the transposon resulting in DNA that is not fragmented or poorly fragmented, or contain multiple recognition sites for the transposon causing over-fragmentation of the DNA. The methods presented here avoid this issue.

Analysis of sequence data generated from these samples reveals a high degree of diversity of the sampled microbes. This is supported by their taxonomic profiles (both metagenomic and 16S profiles contain long tails of many low-abundant organisms), a high number of unclassified sequences and hypothetical proteins, and the highly fragmented nature of nucleotide assemblies. Furthermore, sequenced bacterial genomes do not serve as adequate near-taxonomic neighbor references for these samples, as evidenced by a low-number of bacterial reads with high identity nucleotide matches to these genomes.

While the six samples have different genera abundance profiles, they share common features as well. These data include human associated bacteria in many of the samples. For instance, *Propionibacterium*, a group associated with human skin microbiome, is found in all indoor samples. In addition, the hospital samples had notable numbers of human associated medically relevant organisms, several of which are known hospital associated organisms, including *Klebsiella* and *Bordetella*. On the other hand, many of the abundant bacterial genera from the SD_SCRPP sample were of marine origin (including *Synechococcus*, *Plantomyces*,*Pirellula*, and *Rhodopirellula*), showing that the environment influences the types of microbes found in the overlying air.

Human, fungal, insect, and rodent sequences constitute the eukaryotic portion of these data with human and fungal sequences present in relatively larger amounts in the indoor samples. The SD_SCRPP sample had a sizeable number of sequences from *Wolbachia*, an endosymbiont of insects including *Drosophila*, which constitutes a large portion of the eukaryotic sequences in SD_SCRPP. There are several potential sources of eukaroytic DNA including cells, sloughed tissue (such as skin), and DNA adhered to particulates that were captured on the filters.

The data presented here support the notion that sources of airborne bacteria in indoor and outdoor environments include water, soil, vegetation, and fauna [1], [16], [25]. However, at the functional level, no obvious grouping by environments (indoor and outdoor) or sampling locations is evident for these samples, and sample separation is driven by taxonomic composition. For instance, the higher abundance of arylsulfatases in the SD_SCRPP sample is due to their occurrence in the corresponding marine organisms.

The framework presented here, in combination with further experiments including repeated sampling and gene expression data generation, can be used to identify airborne bacteria that are metabolically active, and to differentiate between transient members and those that use air as a habitat.

ACKNOWLEDGMENTS

We are grateful to Michael Geimer for help with sample collection and to Monika Bihan for informatics support.

AUTHOR CONTRIBUTIONS

Conceived and designed the experiments: JCV RF. Performed the experiments: CAP JMQ JH JG SW LZA DF. Analyzed the data: SY CAP AT JMQ SW MT DB LZA JBG MDA. Contributed reagents/materials/analysis tools: SY CAP AT SW. Wrote the paper: SY CAP JMQ MDA AT MT LZA.

REFERENCES

1. Burrows SM, Elbert W, Lawrence MG, Poschl U (2009) Bacteria in the global atmosphere - Part 1: Review and synthesis of literature data for different ecosystems. Atmospheric Chemistry and Physics 9: 9263–9280. doi: 10.5194/acp-9-9263-2009
2. Creamean JM, Suski KJ, Rosenfeld D, Cazorla A, Demott PJ, et al.. (2013) Dust and Biological Aerosols from the Sahara and Asia Influence Precipitation in the

Western U.S. Science.

3. Womack AM, Bohannan BJ, Green JL (2010) Biodiversity and biogeography of the atmosphere. Philosophical transactions of the Royal Society of London Series B, Biological sciences 365: 3645–3653. doi: 10.1098/rstb.2010.0283
4. Maki LR, Galyan EL, Chang-Chien MM, Caldwell DR (1974) Ice nucleation induced by pseudomonas syringae. Applied microbiology 28: 456–459.
5. Mohler O, DeMott PJ, Vali G, Levin Z (2007) Microbiology and atmospheric processes: the role of biological particles in cloud physics. Biogeosciences 4: 1059–1071. doi: 10.5194/bg-4-1059-2007
6. Dimmick R, Straat P, Wolochow H, Levin G, Chatigny M, et al. (1975) Evidence for metabolic activity of airborne bacteria. Journal of Aerosol Science 6: 387–393. doi: 10.1016/0021-8502(75)90054-3
7. Amato P, Parazols M, Sancelme M, Laj P, Mailhot G, et al. (2007) Microorganisms isolated from the water phase of tropospheric clouds at the Puy de Dome: major groups and growth abilities at low temperatures. FEMS microbiology ecology 59: 242–254. doi: 10.1111/j.1574-6941.2006.00199.x
8. Sattler B, Puxbaum H, Psenner R (2001) Bacterial growth in supercooled cloud droplets. Geophysical Research Letters 28: 239–242. doi: 10.1029/2000gl011684
9. Klepeis NE, Nelson WC, Ott WR, Robinson JP, Tsang AM, et al. (2001) The National Human Activity Pattern Survey (NHAPS): a resource for assessing exposure to environmental pollutants. Journal of exposure analysis and environmental epidemiology 11: 231–252. doi: 10.1038/sj.jea.7500165
10. Prospero JM, Blades E, Mathison G, Naidu R (2005) Interhemispheric transport of viable fungi and bacteria from Africa to the Caribbean with soil dust. Aerobiologia 21: 1–19. doi: 10.1007/s10453-004-5872-7
11. Williamson SJ, Yooseph S (2012) From bacterial to microbial ecosystems (metagenomics). Methods in molecular biology 804: 35–55. doi: 10.1007/978-1-61779-361-5_3
12. van der Heijden MG, Bardgett RD, van Straalen NM (2008) The unseen majority: soil microbes as drivers of plant diversity and productivity in terrestrial ecosystems. Ecology letters 11: 296–310. doi: 10.1111/j.1461-0248.2007.01139.x
13. Gill SR, Pop M, Deboy RT, Eckburg PB, Turnbaugh PJ, et al. (2006) Metagenomic analysis of the human distal gut microbiome. Science 312: 1355–1359. doi: 10.1126/science.1124234
14. Codina R, Fox RW, Lockey RF, DeMarco P, Bagg A (2008) Typical levels of airborne fungal spores in houses without obvious moisture problems during a rainy season in Florida, USA. Journal of investigational allergology & clinical immunology : official organ of the International Association of Asthmology 18: 156–162.
15. Neas LM, Dockery DW, Burge H, Koutrakis P, Speizer FE (1996) Fungus spores, air pollutants, and other determinants of peak expiratory flow rate in children. American journal of epidemiology 143: 797–807. doi: 10.1093/oxfordjournals.

aje.a008818

16. Bowers RM, Sullivan AP, Costello EK, Collett JL Jr, Knight R, et al. (2011) Sources of bacteria in outdoor air across cities in the midwestern United States. Applied and environmental microbiology 77: 6350–6356. doi: 10.1128/aem.05498-11
17. Brodie EL, DeSantis TZ, Parker JP, Zubietta IX, Piceno YM, et al. (2007) Urban aerosols harbor diverse and dynamic bacterial populations. Proceedings of the National Academy of Sciences of the United States of America 104: 299–304. doi: 10.1073/pnas.0608255104
18. Fierer N, Liu Z, Rodriguez-Hernandez M, Knight R, Henn M, et al. (2008) Short-term temporal variability in airborne bacterial and fungal populations. Applied and environmental microbiology 74: 200–207. doi: 10.1128/aem.01467-07
19. Kembel SW, Jones E, Kline J, Northcutt D, Stenson J, et al.. (2012) Architectural design influences the diversity and structure of the built environment microbiome. The ISME journal.
20. Radosevich JL, Wilson WJ, Shinn JH, DeSantis TZ, Andersen GL (2002) Development of a high-volume aerosol collection system for the identification of air-borne micro-organisms. Letters in applied microbiology 34: 162–167. doi: 10.1046/j.1472-765x.2002.01048.x
21. Taubel M, Rintala H, Pitkaranta M, Paulin L, Laitinen S, et al. (2009) The occupant as a source of house dust bacteria. Journal of Allergy and Clinical Immunology 124: 834–840. doi: 10.1016/j.jaci.2009.07.045
22. Fahlgren C, Hagstrom A, Nilsson D, Zweifel UL (2010) Annual variations in the diversity, viability, and origin of airborne bacteria. Applied and environmental microbiology 76: 3015–3025. doi: 10.1128/aem.02092-09
23. Urbano R, Palenik B, Gaston CJ, Prather KA (2011) Detection and phylogenetic analysis of coastal bioaerosols using culture dependent and independent techniques. Biogeosciences 301: 301–309. doi: 10.5194/bg-8-301-2011
24. Wu Z, Tsumura Y, Blomquist G, Wang XR (2003) 18S rRNA gene variation among common airborne fungi, and development of specific oligonucleotide probes for the detection of fungal isolates. Applied and Environmental Microbiology 69: 5389–5397. doi: 10.1128/aem.69.9.5389-5397.2003
25. Hospodsky D, Qian J, Nazaroff WW, Yamamoto N, Bibby K, et al. (2012) Human occupancy as a source of indoor airborne bacteria. PLoS One 7: e34867. doi: 10.1371/journal.pone.0034867
26. Handelsman J (2004) Metagenomics: application of genomics to uncultured microorganisms. Microbiol Mol Biol Rev 68: 669–685. doi: 10.1128/mmbr.68.4.669-685.2004
27. Venter JC, Remington K, Heidelberg JF, Halpern AL, Rusch D, et al. (2004) Environmental genome shotgun sequencing of the Sargasso Sea. Science 304: 66–74. doi: 10.1126/science.1093857

28. DeLong EF, Preston CM, Mincer T, Rich V, Hallam SJ, et al. (2006) Community genomics among stratified microbial assemblages in the ocean's interior. Science 311: 496–503. doi: 10.1126/science.1120250
29. Huttenhower C, Gevers D, Knight R, Abubucker S, Badger JH, et al. (2012) Structure, function and diversity of the healthy human microbiome. Nature 486: 207–214.
30. Rusch DB, Halpern AL, Sutton G, Heidelberg KB, Williamson S, et al. (2007) The Sorcerer II Global Ocean Sampling expedition: northwest Atlantic through eastern tropical Pacific. PLoS biology 5: e77. doi: 10.1371/journal.pbio.0050077
31. Welch KD, Davis TZ, Aust SD (2002) Iron autoxidation and free radical generation: effects of buffers, ligands, and chelators. Archives of biochemistry and biophysics 397: 360–369. doi: 10.1006/abbi.2001.2694
32. Shi T, Knaapen AM, Begerow J, Birmili W, Borm PJ, et al. (2003) Temporal variation of hydroxyl radical generation and 8-hydroxy-2'-deoxyguanosine formation by coarse and fine particulate matter. Occupational and environmental medicine 60: 315–321. doi: 10.1136/oem.60.5.315
33. Hutchison CA, Smith HO, Pfannkoch C, Venter JC (2005) Cell-free cloning using phi 29 DNA polymerase. Proceedings of the National Academy of Sciences of the United States of America 102: 17332–17336. doi: 10.1073/pnas.0508809102
34. Allen LZ, Ishoey T, Novotny MA, McLean JS, Lasken RS, et al. (2011) Single virus genomics: a new tool for virus discovery. PloS one 6: e17722. doi: 10.1371/journal.pone.0017722
35. Zhang K, Martiny AC, Reppas NB, Barry KW, Malek J, et al. (2006) Sequencing genomes from single cells by polymerase cloning. Nature biotechnology 24: 680–686. doi: 10.1038/nbt1214
36. Andrews-Pfannkoch C, Fadrosh DW, Thorpe J, Williamson SJ (2010) Hydroxyapatite-mediated separation of double-stranded DNA, single-stranded DNA, and RNA genomes from natural viral assemblages. Applied and environmental microbiology 76: 5039–5045. doi: 10.1128/aem.00204-10
37. Teal TK, Schmidt TM (2010) Identifying and removing artificial replicates from 454 pyrosequencing data. Cold Spring Harbor protocols 2010: pdb prot5409.
38. Hancock JM, Armstrong JS (1994) SIMPLE34: an improved and enhanced implementation for VAX and Sun computers of the SIMPLE algorithm for analysis of clustered repetitive motifs in nucleotide sequences. Computer applications in the biosciences : CABIOS 10: 67–70. doi: 10.1093/bioinformatics/10.1.67
39. Smit AFA, Hubley R (1996–2010) RepeatMasker Open-3.0.http://www.repeatmasker.org. Accessed 2010 Jan 25.
40. Lasken RS, Stockwell TB (2007) Mechanism of chimera formation during the Multiple Displacement Amplification reaction. BMC biotechnology 7: 19. doi: 10.1186/1472-6750-7-19

41. Altschul SF, Gish W, Miller W, Myers EW, Lipman DJ (1990) Basic local alignment search tool. Journal of molecular biology 215: 403–410. doi: 10.1016/s0022-2836(05)80360-2

42. Wheeler DL, Barrett T, Benson DA, Bryant SH, Canese K, et al. (2007) Database resources of the National Center for Biotechnology Information. Nucleic Acids Res 35: D5–12. doi: 10.1093/nar/gkl1031

43. Allen LZ, Allen EE, Badger JH, McCrow JP, Paulsen IT, et al.. (2012) Influence of nutrients and currents on the genomic composition of microbes across an upwelling mosaic. The ISME journal.

44. Tanenbaum DM, Goll J, Murphy S, Kumar P, Zafar N, et al. (2010) The JCVI standard operating procedure for annotating prokaryotic metagenomic shotgun sequencing data. Standards in genomic sciences 2: 229–237. doi: 10.4056/sigs.651139

45. Suzek BE, Huang H, McGarvey P, Mazumder R, Wu CH (2007) UniRef: comprehensive and non-redundant UniProt reference clusters. Bioinformatics 23: 1282–1288. doi: 10.1093/bioinformatics/btm098

46. Kanehisa M, Araki M, Goto S, Hattori M, Hirakawa M, et al. (2008) KEGG for linking genomes to life and the environment. Nucleic acids research 36: D480–484. doi: 10.1093/nar/gkm882

47. Tatusov RL, Galperin MY, Natale DA, Koonin EV (2000) The COG database: a tool for genome-scale analysis of protein functions and evolution. Nucleic acids research 28: 33–36. doi: 10.1093/nar/28.1.33

48. Liu B, Pop M (2009) ARDB—Antibiotic Resistance Genes Database. Nucleic acids research 37: D443–447. doi: 10.1093/nar/gkn656

49. Chen L, Xiong Z, Sun L, Yang J, Jin Q (2012) VFDB 2012 update: toward the genetic diversity and molecular evolution of bacterial virulence factors. Nucleic acids research 40: D641–645. doi: 10.1093/nar/gkr989

50. Sharon I, Pati A, Markowitz VM, Pinter RY (2009) A Statistical Framework for the Functional Analysis of Metagenomes. Proceedings of the 13th Annual International Conference on Research in Computational Molecular Biology. Tucson, Arizona: Springer-Verlag. pp. 496–511.

51. Kullback S, Leibler RA (1951) On Information and Sufficiency. Annals of Mathematical Statistics 22: 79–86. doi: 10.1214/aoms/1177729694

52. Haas BJ, Gevers D, Earl AM, Feldgarden M, Ward DV, et al. (2011) Chimeric 16S rRNA sequence formation and detection in Sanger and 454-pyrosequenced PCR amplicons. Genome research 21: 494–504. doi: 10.1101/gr.112730.110

53. Maidak BL, Cole JR, Lilburn TG, Parker CT Jr, Saxman PR, et al. (2001) The RDP-II (Ribosomal Database Project). Nucleic acids research 29: 173–174. doi: 10.1093/nar/29.1.173

54. Schloss PD, Westcott SL, Ryabin T, Hall JR, Hartmann M, et al. (2009) Introducing mothur: open-source, platform-independent, community-supported software for describing and comparing microbial communities. Applied and environmental

microbiology 75: 7537–7541. doi: 10.1128/aem.01541-09

55. Binga EK, Lasken RS, Neufeld JD (2008) Something from (almost) nothing: the impact of multiple displacement amplification on microbial ecology. The ISME journal 2: 233–241. doi: 10.1038/ismej.2008.10
56. Margulies M, Egholm M, Altman WE, Attiya S, Bader JS, et al. (2005) Genome sequencing in microfabricated high-density picolitre reactors. Nature 437: 376–380.
57. Leinonen R, Sugawara H, Shumway M (2011) International Nucleotide Sequence Database C (2011) The sequence read archive. Nucleic Acids Res 39: D19–21. doi: 10.1093/nar/gkq1019
58. **58.** Wassenaar TM, Gaastra W (2001) Bacterial virulence: can we draw the line? FEMS microbiology letters 201: 1–7. doi: 10.1111/j.1574-6968.2001.tb10724.x
59. Tringe SG, Zhang T, Liu X, Yu Y, Lee WH, et al. (2008) The airborne metagenome in an indoor urban environment. PloS one 3: e1862. doi: 10.1371/journal.pone.0001862
60. Marine R, Polson SW, Ravel J, Hatfull G, Russell D, et al. (2011) Evaluation of a transposase protocol for rapid generation of shotgun high-throughput sequencing libraries from nanogram quantities of DNA. Appl Environ Microbiol 77: 8071–8079. doi: 10.1128/aem.05610-11
61. Willner D, Hugenholtz P (2013) From deep sequencing to viral tagging: recent advances in viral metagenomics. Bioessays 35: 436–442. doi: 10.1002/bies.201200174

9

CHAPTER

PHYLOGENETIC IDENTIFICATION OF MARINE BACTERIA ISOLATED FROM DEEP-SEA SEDIMENTS OF THE EASTERN SOUTH ATLANTIC OCEAN

Marcus Adonai Castro da Silva[1], Angélica Cavalett[1], Ananda Spinner[1], Daniele Cristina Rosa[1], Regina Beltrame Jasper[1], Maria Carolina Quecine[3], Maria Letícia Bonatelli[3], Aline Pizzirani-Kleiner[3], Gertrudes Corção[2] and André Oliveira de Souza Lima[1]

[1]Centro de Ciências Tecnológicas da Terra e do Mar, Universidade do Vale do Itajaí (UNIVALI), Rua Uruguai, 458, Itajaí, SC CEP 88302202, Brazil

[2] Departamento de Microbiologia, Imunologia e Parasitologia, Instituto de Ciências Básicas da Saúde, Universidade Federal do Rio Grande do Sul (UFRGS), Rua Sarmento Leite 500, Cidade Baixa CEP 90050-170Porto AlegreRS, Brazil

[3] Departamento de Genética, Universidade de São Paulo, Escola Superior de Agricultura "Luiz de Queiroz", Av. Padua Dias, 11, Piracicaba, SP CEP 13418-260, Brasil

ABSTRACT

The deep-sea environments of the South Atlantic Ocean are less studied in comparison to the North Atlantic and Pacific Oceans. With the aim of identifying the deep-sea bacteria in this less known ocean, 70 strains were isolated from eight sediment samples (depth range between 1905 to 5560 m) collected in the eastern part of the South Atlantic, from the equatorial region to the Cape Abyssal Plain, using three different culture media. The strains were classified into three phylogenetic groups, Gammaproteobacteria, Firmicutes and Actinobacteria, by the analysis of 16s rRNA gene sequences. Gammaproteobacteria and Firmicutes were the most frequently identified groups, withHalomonas the most frequent genus among the strains. Microorganisms belonging to Firmicutes were the only ones observed in

Citation: Marcus Adonai Castro da Silva, Angélica Cavalett, Ananda Spinner, Daniele Cristina Rosa, Regina Beltrame Jasper, Maria Carolina Quecine, Maria Letícia Bonatelli, Aline Pizzirani-Kleiner, Gertrudes Corção and André Oliveira de Souza Lima, Phylogenetic identification of marine bacteria isolated from deep-sea sediments of the eastern South Atlantic Ocean, DOI: 10.1186/2193-1801-2-127.

all samples. Sixteen of the 41 identified operational taxonomic units probably represent new species. The presence of potentially new species reinforces the need for new studies in the deep-sea environments of the South Atlantic.

BACKGROUND

Deep-sea environments are the largest continuous ecosystems on our planet. Except for specific places considered diversity hot spots, such as hydrothermal vents and cold seeps, other deep-sea environments including the abyssal plains are in most cases scarcely studied (Jorgensen and Boetius 2007).

The abyssal plains are located between 3000 and 5000 meters deep, and cover 54% of the planet's surface. A big part of this region is covered with biogenic sediments, although hard substrata such as manganese nodules may be found in some places. Characteristically, the biota of abyssal plains is limited by food, since this environment is devoid of primary production and depends on the input of organic matter from surface waters. Because of its extent and difficulty of access, the abyssal plains are among the least understood ecosystems of our planet, despite the great diversity that these environments may harbor (Smith et al. 2008).

Most of the research on microbial diversity conducted on sediment samples from abyssal plains are based on cultivation-independent methods, which are considered more appropriate, since they allow the detection of the non-cultivable microorganism that are generally constitute the largest fraction of the total microbial community (Polymenakou et al. 2009; Schauer et al. 2010). However, cultivation-based studies may provide information on the physiological characteristics of the organisms living in the sediments, and allow the description of new species and the prospecting for microorganisms of biotechnological interest (Zengler et al. 2002; Pettit 2011). Gärtner et al. (2011) isolated and identified 107 bacteria from deep sediments of the Mediterranean Sea, and most of the strains belonged to two phylogenetic groups, Firmicutes and Actinobacteria. The most frequently identified genus was*Bacillus*, and 13 possibly new species were also detected. Most of the isolated strains were capable of growing under conditions similar to those prevailing at the source of the samples, relating to temperature and nutrient concentrations. Aiming to find microorganisms with potential for lipase production, Zeng et al. (2004) isolated lipolytic microorganisms from samples of deep marine sediments from the Eastern Pacific Ocean. Most of the bacteria isolated belonged to the genera *Halomonas*, *Psychrobacter*, *Stenotrophomonas*, *Pseudomonas* and*Pseudalteromonas* of Gammaproteobacteria. Some of the strains produced several types of hydrolytic enzymes, which demonstrated the potential of microorganisms from deep-sea sediments.

The Atlantic Ocean is younger and less tectonically active than the Pacific Ocean. The mid-oceanic ridge is the most important feature of the Atlantic Ocean, and separates it in two halves. Each half of the South Atlantic is further separated in other basins by smaller ridges and other features. On the eastern side, which is the focus of this work, two features are important, the Romanche Trench, in the equatorial region, and the Walvis Ridge, located further south. Between these features is the Angola Abyssal Plain and, south of the Walvis Ridge, the Cape Abyssal Plain (Levin and Gooday 2003).

The Walvis Ridge is a discontinuous submarine feature that intercepts the African continental margin at a latitude of 20°S, extending to the mid-ocean ridge at 37°S. The Walvis Ridge has a characteristic oceanic crust composition, with twice the normal thickness, about 12 to 15 km (Hekinian 1974; Salters and Sachi-Kocher 2010). The origin of the Walvis Ridge is still in debate, but it seems to be associated with the interaction of the Tristão da Cunha hot spot and the movement of the overlying tectonic plate. It is believed that this aseismic ridge represents the beginning of a hot spot trail with southward movement, formed from a fissure of the oceanic crust during the movement of the American and African tectonic plates (Hekinian 1974; Elliott et al. 2009). The Angola and Cape Abyssal Plains are separated by the Walvis Ridge, which functions as a barrier that limits the water flow between these two oceanic basins. The North Atlantic Deep Water, which originates in the Arctic, dominates the Angola Basin. The sediments in this region are typical of deep-sea environments, comprising mainly siliceous and calcareous microfossils. The Low Circumpolar Deep Water, which originates in the Antarctic continent, dominates the Cape Basin (Jansen et al.1984; Schauer et al. 2010). All these features may influence the distribution of microbial species living in the sea floor and at the deep-sea ecosystems in general.

The deep-sea environments are less studied concerning their microbial diversity, in comparison to other ecosystems, and this applies to the South Atlantic. Recently, Schauer et al. (2010) studied microbial diversity and its biogeography in some basins of the Eastern South Atlantic. These investigators reported the dominance of Proteobacteria, Gammaproteobacteria in particular, in the Guinea, Cape and Angola Basins. Other phylogenetic groups were also identified. The communities of cultivable bacteria were not studied. In short, there is a paucity of studies on cultivable bacteria from the abyssal plains, despite the importance to biotechnology (Arahal and Ventosa 2006; Simon-Colin et al. 2008; Pettit 2011), for the description of new species and to understand physiological aspects that are relevant to the survival of deep-sea microorganisms. Therefore, the aim of this study was the isolation of microorganisms from sediment samples collected on the abyssal plains of the Eastern South Atlantic Ocean and their phylogenetic identification.

RESULTS

Colony counts of the samples, as determined in MA, varied between 23 and 1.41×10^6 CFU/g of sediment. The number of isolated strains per sample varied between 4 (sample 5) and 15 (sample 1) (Table 1).No tendencies were observed in these counts, in relation to the depth and location of the sampling stations. Seventy microorganisms were isolated from the three culture media employed. Thirty-five were isolated from MA plates, 18 from MA+Tween 40 plates and 17 MA+CMC plates.

Table 1: Colony counts (MA plates) and number of strains obtained from the three culture media employed

Sample	Number of isolates	CFU/g
1	15	1410000
2	5	1172500

3	10	1533
4	7	467
5	4	43111
6	10	193333
7	6	23
8	13	27

According to the partial sequencing of the 16S rRNA gene, the 70 isolated microorganisms were identified as belonging to three phylogenetic groups, Gammaproteobacteria, Firmicutes (Gram-positives with low G+C percentage) and Actinobacteria (Gram-positives with high G+C percentage) (Figures 1 and 2). Most of the strains belonged to Gammaproteobacteria (39 strains) and Firmicutes (27 strains). Only four strains belonged to Actinobacteria. The only phylogenetic group detected in all samples was Firmicutes. Microorganisms from the Gammaproteobacteria class were not detected in sample 5 (sample from 70 cm below the surface of the sediment), in which only Firmicutes were detected. Actinobacteria were detected in samples 1, 3 and 7.

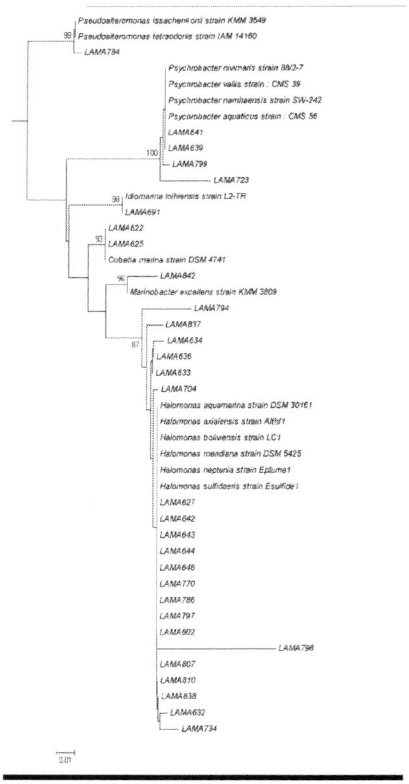

Figure 1: Neighbor-joining tree showing the phylogenetic relationships of 16S rRNA gene se-

quences of Gammaproteobacteria strains. Legend: Bootstrap support values over 70% are shown. The scale bar indicates evolutionary distance.

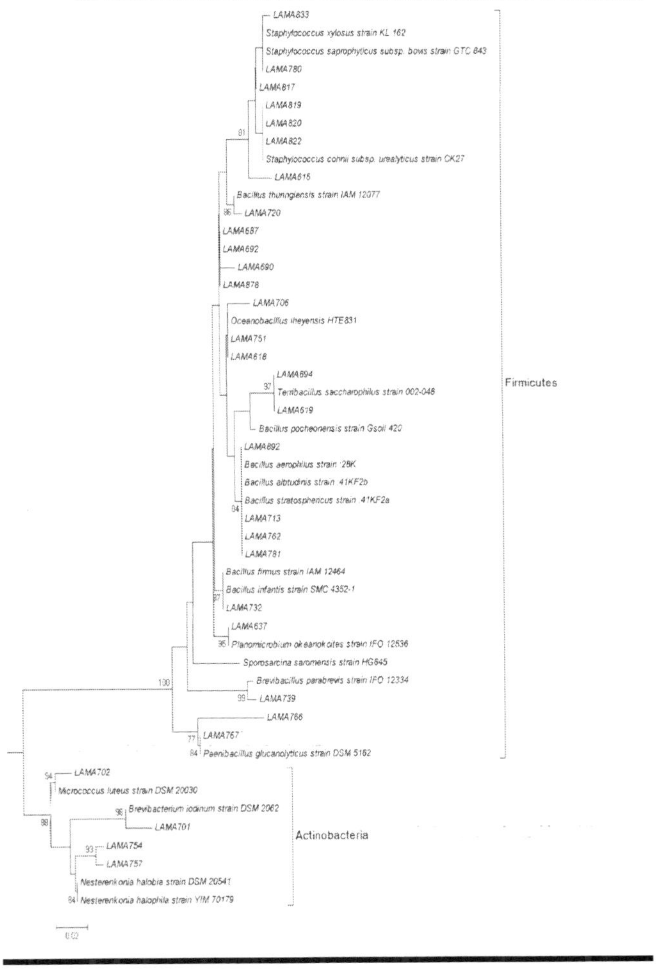

Figure 2: Neighbor-joining tree showing the phylogenetic relationships of 16S rRNA gene sequences of Firmicutes/Actinobacteria strains. Legend: Bootstrap support values over 70% are shown. The scale bar indicates evolutionary distance.

Forty-one different OTUs were defined among the 70 isolated strains Additional file 1. Sixteen OTUs may represent new species, since they showed less than 97% similarity to sequences in the database / published sequences, each of them containing only one strain. Nineteen OTUs were identified in the Gammaproteobacteria class, with 10 belonging to the genus *Halomonas*. Six OTUs of *Halomonas* may represent new species. Three OTUs were attributed to *Psychrobacter*, with two representing new species, and three OTUs were related to *Idiomarina*, with two possibly new species.

Eighteen OTUs were identified in Firmicutes, six belonging to *Bacillus* and four to *Staphylococcus*. Two OTUs were related to the genera *Paenibacillus* and two others to *Oceanobacillus*. Five OTUs of Firmicutes may represent new species, belonging to the genera *Bacillus*, *Oceanobacillus*, *Paenibacillus* and *Staphylococcus*. Each strain of Actinobacteria represented a different OTU, with one possibly being a new species. Two OTUs of Actinobacteria were related to the genus *Nesterenkonia*.

The number of OTUs per sample reflected in an approximate way the number of strains isolated. A small number of OTUs were identified in samples with a small number of strains. For instance, two OTUs were obtained from sample 5, from which only four strains were obtained. This was the only sample collected below the surface of the sediment (70 cm below the surface). A higher number of strains (n = 15) were isolated from sample 1, which also showed the highest number of OTUs. This sample was collected at the Cape Abyssal Plain, at the highest latitude sampled in this study.

Only nine out the 41 identified OTUs were detected in more than one sample (Table 2). The OTUs 7 (related to*Halomonas boliviensis* strain LC1) and 23 (related to *Bacillus aerophilus* strain :28K/*B. altitudinis* strain :41KF2b/*B. stratosphericus* strain :41KF2a) were observed in four samples. The highest numbers of unique OTUs were observed in samples 1 and 6 (n = 8 and 6). No unique OTUs were observed in sample 5.

Table 2: Numbers of each OTU identified in more than one sample analyzed

OTUs (number and description)	Sample							
	1	2	3	4	5	6	7	8
1, *Cobetia marina* strain DSM 4741						1	1	
7, *Halomonas boliviensis* strain LC1			4	1		1		4
9, *Halomonas sulfidaeris* strain Esulfide1			1			1		1
13, *Idiomarina loihiensis* strain L2-TR	5					1		
21, *Bacillus firmus* strain IAM 12464				1				4
23, *Bacillus pocheonensis* strain Gsoil 420	1			1	1			1
27, *Oceanobacillus iheyensis* HTE831	1			1				
31, *Planomicrobium okeanokoites* strain IFO 12536		1					1	
37, *Terribacillus saccharophilus* strain 002-048					3			1

By cluster analysis, the samples could be separated into three groups (Figure 3). The first group comprised samples 2 (Angola Abyssal Basin) and 7 (South Equatorial Region), both originating from further north. The second group included samples 3 (Cape Abyssal Plain) and 6 (Angola Abyssal Basin). Finally, the third group comprised samples 1, 4, 5 (all originating from the Cape Abyssal Plain) and 8 (Walvis Ridge Region). Most of the samples collected south of the Walvis Ridge (except sample 3), were in the third group of the cluster. This group also included samples 4 and 5, collected at the same station, but at different depths of the sediment.

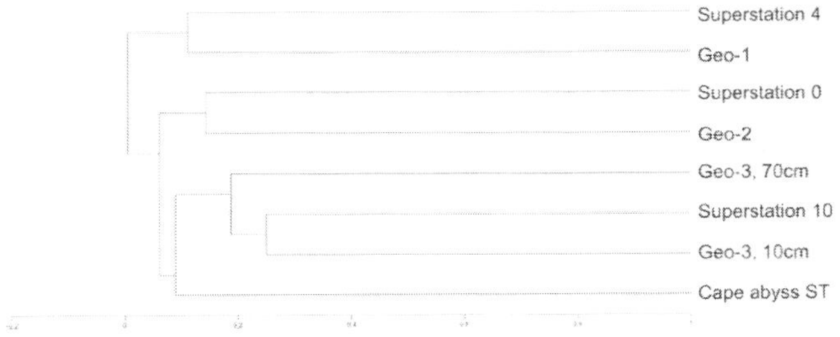

Figure 3: Dendogram showing the similarity of the samples, accordingly to the presence/absence of the established OTUs. Legend: Similarities were calculated using the Jaccard coefficient and UPGMA linkage. Group 1 compromise samples 2 and 7, group 2, samples 3 and 6, and group 3 samples 1, 4, 5 and 8.

DISCUSSION

Microorganisms belonging to Gammaproteobacteria were dominant in the samples studied. This is not unexpected considering that bacteria from this phylogenetic group are among the most known and readily cultivable microorganisms from the marine environment (Fuhrman and Hagström 2008), and this result is in agreement with reports from other studies on diverse marine environments (; Kobayashi et al. 2008; Ettoumia et al. 2011; Finnegan et al. 2011). In the culture–independent study of Schauer et al. (2010) conducted at the Angola, Guinea and Cape Basins, Gammaproteobacteria was also the dominant group, although most of the phylotypes identified belonged to not yet cultivated species.

The only sample in which we could not detect Gammaproteobacteria was sample 5, collected at 70 cm below the surface of the sediment. This observation may be related to the fact that most of the marine bacteria from this phylogenetic group are aerobic or facultative anaerobic (Fuhrman and Hagström 2008), and their occurrence is favored on the surface of the sediment, where oxygen is more abundant. We did not measure the oxygen concentration in the samples studied, but it is reasonable to presume that this element was absent in sample 5, considering the distribution of dissolved oxygen in marine sediments (Hensen et al. 2006).

Firmicutes was the second group most frequently isolated from the samples studied. This group and Actinobacteria comprise the Gram-positive bacteria. Recent studies have reported a greater presence of Gram-positive bacteria in marine sediments (Zhuang et al. 2003; Gontang et al. 2007; Sass et al. 2008), and in several cases these microorganisms are dominant among the cultivable bacteria (Toffin et al. 2004; Gärtner et al. 2011; Velmurugana et al. 2011). Most of the 27 strains identified in Firmicutes were related to endospore-forming bacteria (EFB, 19 strains), mainly *Bacillus*. The ability to produce endospores and the great metabolic and physiological diversity are two characteristics of the EFB that allow their distribution in all environments of our planet (Priest 1993), and this may explain the detection of these bacteria in all samples analyzed in our study. Some of the OTUs (25 identified as*Bacillus thuringiensis*, for instance) were identified as non-marine species of the EFB, and it is probable that these strains were latent in the sediments studied. However, other OTUs (27 and 28) were identified in the genus*Oceanobacillus*, which was originally described from deep-sea sediments (Lu et al. 2001). Therefore, some of the strains of EFB are probably true members of the communities of cultivable bacteria of deep-sea sediments.

The genus most frequently identified among the isolated strains was *Halomonas* (Gammaproteobacteria). This result was expected considering that this genus is widespread in the marine environment, and its cultivation has been reported in several types of samples collected from deep-sea ecosystems, including sediments from hydrothermal vents, from abyssal plains and sub-surface, invertebrates and water, sometimes as dominant microorganisms (Xu et al. 2005; Kaye and Baross 2000; Kaye et al. 2004; Simon-Colin et al. 2008; Kaye et al.2011). The widespread occurrence of *Halomonas* has been associated with its great physiological and metabolic versatility, which was demonstrated by their capability to assimilate a great variety of carbon sources, and by the growth of these bacteria in a wide range of salinities, temperatures and pressures (Okamoto et al. 2004; Kaye and Baross 2004). These characteristics may not only explain the dominance of this genus, but also its occurrence in the samples studied. The metabolic versatility of *Halomonas* has also been associated with a great biotechnological potential, for production of enzymes and polyhydroxyalcanoates, for instance (Arahal and Ventosa 2006; Simon-Colin et al. 2008).

Psychrobacter and *Idiomarina* were the two other genera of Gammaproteobacteria more frequently identified among the strains. The detection of these genera in our samples is consistent with other studies. The genus*Idiomarina*, for instance, was originally described for strains isolated from deep-sea waters (Ivanova et al. 2000 , and lately, another species has been described from a submarine volcano nearby the Hawaii Islands, and was named *Idiomarina loihiensis* (Donachie et al. 2003). Other investigators also reported the occurrence of*Psychrobacter* in deep-sea environments (Maruyama et al. 2000; Xu et al. 2005). Therefore, bacteria from these two genera may be considered native members of deep-sea communities.

Fifteen percent of the strains were isolated from the MA plates, without supplementation with CMC or Tween 40. The incorporation of CMC or Tween 40 in the culture media could have favored the growth of few organisms especially adapted to the utilization of these carbon sources or adapted to a higher concentration of carbon. These organisms may have inhibited the growth of other species, a phenomenon reported especially in Gammaproteobacteria (Gärtner et al. 2011), the dominant group among our strains. This hypothesis may explain the lower number of strains isolated from MA+CMC and MA+Tween40 plates.

The highest numbers of OTUs were detected in samples 1, collected at the Cape Abyssal Plain, and 6, collected at the Angola Abyssal Plain. Both samples also had the highest number of unique OTUs. These observations may be associated with local enrichments of the sediments with organic matter and nutrients in general, which are associated with the import of primary production from the overlying waters (Orcutt et al. 2011) and this region is characterized by a higher annual primary production

(Gregg et al. 2003) in comparison to more tropical or central areas, resulting in a greater exportation of carbon to the sediments. This enrichment may result in a higher diversity of cultivable bacteria since these microorganisms, which are typically adapted to environments richer in organic matter in comparison to non-cultivable species (Fuhrman and Hagström 2008), may become dominant members of the community.

According to Schauer et al. (2010), the Walvis Ridge forms a barrier between the Cape Abyssal Plain and Angola Abyssal Plain, limiting the northward and southward flow of water at depths greater than 3000 m. Although these authors did not detect any influence of the Walvis Ridge on microbial dispersal, they reported differences between the bacterial communities of the Angola and Cape Abyssal Plains. In our cluster analysis, the two samples originating from the Angola Basin grouped separately from most of the samples collected at the Cape Basin, forming two groups (group 1 and 2) with samples 7 from the Equatorial Region and 3 from the Cape Abyssal Plain, respectively. These differences may reflect different water masses or other environmental factors that occur north and south of the Walvis Ridge. To confirm and further elucidate these differences, we need to analyze a greater number of samples from both regions.

As already mentioned, sixteen OTUs could represent new species, considering a limit of similarity of 97% in the 16S rRNA gene sequences. However, this number could be even higher if we consider a limit of 98.5 to 99%, as revised and proposed by Stackebrandt and Ebers (2006). Concerning the methodologies adopted in this study, we did not employ special conditions of cultivation, such elevated hydrostatic pressures and low temperatures, similar to that existing at the origin of the samples. Therefore, we can say that other new microorganisms can be cultured from the source of the sediment samples. We also believe that a high number of strains would allow testing hypothesis related to the distribution of the different species identified.

CONCLUSIONS

The results of the present study demonstrated that a great part of the cultivable bacterial diversity of deep-sea environments is still unknown, and the implementation of new studies with this fraction of the microbial communities is justified. This is reinforced by the enormous biotechnological potential of these microorganisms, already demonstrated in studies by other authors (Arahal and Ventosa 2006; Simon-Colin et al. 2008; Pettit 2011). In future studies, efforts should be directed to emulate the prevailing conditions at the deep-sea floor, and to isolate a higher number of strains.

METHODS

Sampling

The samples studied were collected during a cruise carried out between 10/25/2009 and 11/29/2009 with the vessel Akademik Loffe (Academy of Sciences – Russia). A total of eight samples of deep-sea sediments (50 g) were collected (Table 3 and Figure 4). Samples 4 and 5 were originated from the same sampling station, but from different depths in the sediment (top 10 cm layer and 70 cm below the surface). All the remaining samples were collected at the surface of the sediments. Samples were collected with trawls (samples 1, 5, 7 and 8) or push-core samplers (samples 2 and 4), aseptically transferred to sterile 15-ml Falcon tubes on board and kept at 4°C until processing, a month after collection. The in-situ temperature was not measured at the time of sampling, but Schauer et al. (2010) mentions values between 1.14 and 2.1°C in the deep-water of the Cape and Guinea Abyssal Plains.

Table 3: Localization and depths of sediment samples studied

Sample	Station	Latitude	Longitude	Depth (m)	Region	Sampler
1	Cape abyss ST	36°21'22'S	05°52'09'E	1107	CAP	Trawls
2	Geo-1	23°30'52'S	04°17'19'W	5000	AAP	Push-core
3	Geo-2	35°20'00'S	03°01'00'W	5083	CAP	Push-core
4	Geo-3, 10cm	35°50'18'S	03°26'38'E	5200	CAP	Push-core
5	Geo-3, 70cm	35°50'18'S	03°26'38'E	5200	CAP	Trawls
6	Superstation 0	25°41'70'S	02°20'72'W	4640	RT	Push-core
7	Superstation 4	04°40'22'S	12°16'20'W	1900	SEMS	Trawls
8	Superstation 10	33°40'17'S	02°35'12'E	4400	WRS	Trawls

Legend: *CAP*, Cape Abyssal Plain; *AAP*, Angola Abyssal Plain; *SEMS*, South Equatorial Region; *WRS*, Walvis Ridge Region.

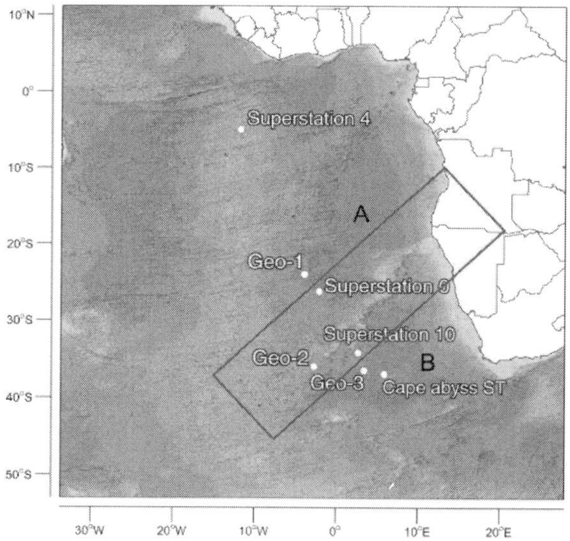

Figure 4: Localization of the sampling stations. Legend: The Walvis Ridge is highlighted in red. A, Angola Abyssal Plain; B, Cape Abyssal Plain.

Isolation of microorganisms

For the isolation of microorganisms, samples were serially and decimally diluted from 10^{-1} to 10^{-5} with filtered sterile seawater as diluent. Plates of Marine Agar (MA), Marine Agar supplemented with Tween 40 (MA+Tween 40) and Marine Agar supplemented with 1% carboxymethylcellulose (MA+CMC) were inoculated with 100 μl of the different dilutions, in triplicate, and incubated for three weeks at 10°C. This temperature was chosen since it supports the growth of a wider range of bacteria (psychrophiles and cold-tolerant mesophiles). After incubation, plates were examined and colonies of different morphology were selected and transferred to new agar plates several times, for their isolation and purification. Bacterial isolates were stored in MA slants at 4°C, where they were transferred to new agar slants every four months (Bowman 2001).

DNA extraction and 16S gene amplification

For DNA extraction, the microorganisms were cultivated in marine broth for 2 to 7 days at 15°C. The extraction was conducted with the Genomic DNA Extraction kit from Real Genomics, following the instructions of the manufacturer. The 16S rRNA gene was amplified using the primers 27F (5'- AGAGTTTGATCMTGGCTCAG-3') and 1492R (5'TACGGYTACCTTGTTACGACTT3'). The amplification was performed in a reaction volume of 35 μl with 5 μl of 10X concentrated buffer, 1.5 mM $MgCl_2$, 200 μM dNTP mix, 0.1 μM each primer, 2.5 U Taq polymerase and 20 to 40 ng template DNA. The PCR conditions were initial denaturation of 2 min at 94°C, followed by 35 cycles of 1 min at 94°C, 1.5 min at 55°C and 1 min at 72°C, and a final extension at 72°C for 3 min. The reaction product was visualized on an agarose gel (1%) under UV light after ethidium bromide staining.

16S gene sequencing and analysis

The amplified PCR products of bacterial gene fragments were purified and sequenced at MACROGEN sequencing company, Seoul, Korea using the automated sequencer ABI 3100 (Applied Biosystems) with BigDye Terminator Kit v. 3.1 (Applied Biosystems). Primers 27F (5'AGAGTTTGATCMTGGCTCAG3') and 1492R (5'TACGGYTACCTTGTTACGACTT3') were used for sequencing. The sequences obtained were edited with the software Vector NTI Suite 9, and compared with the NCBI database through BLAST searches. In this comparison, sequences of type strains most closely related to the sequences of the isolates were searched. For the definition of operational taxonomic units (OTUs), a similarity limit of 97% was adopted. The sequences were aligned with Muscle, and the trees were constructed from the evolutionary distances by the neighbor joining method with the software Mega (Tamura et al., 2011). The 16S rRNA gene sequences obtained in this study have been deposited in the NCBI GenBank under accession numbers JX860193 to JX860262.

Comparison of samples according to operational taxonomic units (OTUs)

The Shannon-Wiener diversity index and evenness were calculated according to the distribution of the OTUs among the sediment samples. The samples were also compared according to the presence or absence of the different OTUs by cluster analysis. Accordingly, a dendrogram representation was generated using the Jaccard coefficient of similarity and the unweighted pair-wise grouping with mathematical averages (UPGMA) linkage method. This analysis was conducted with the Multi-Variate Statistical Package (MVSP) software (Kovach Computing Services).

DECLARATIONS

Acknowledgments

We would like to thank the captain, the crew of the R/V Akademik Loffe and the Academy of Sciences, Russia, for the support during the cruise. We are grateful for Dr. José Angel Alvarez Perez coordinator of the project "Patterns and Processes of the Ecosystems of the Southern Mid-Atlantic" – Census of Marine Life (CoML), for the support to the cruise and to Dr André Silva Barreto for the collection of the samples. We also acknowledge the International Centre for Genetic Engineering and Biotechnology (ICGEB) and the Brazilian National Council for Scientific and Technological Development (CNPq, Process 577915/2008-8) for financial support.

Authors' contributions

MACS carried out the isolation and identification of the strains, the analysis of molecular data and drafted the manuscript; AC carried out the molecular analysis of part of the stains; AS, DCR and RBJ carried out the isolation of bacteria; MCQ, MLB and AP-K carried out the identification of part of the strains; GC participated in the design and coordination of the study, and help to draft the manuscript. AOSL participated in the design and coordination of the study, and help to align sequences and draft the manuscript. All authors read and approved the final manuscript.

REFERENCES

1. Arahal DR, Ventosa A: The Family Halomonadaceae. In *The Prokaryotes. A handbook on the biology of bacteria*. 3rd edition. Edited by: Dworkin M, Schleifer KH, Stackebrandt . New York: Springer-Verlag; 2006.
2. Bowman JP: Methods for psychrophilic bacteria. *Method Microbiol* 2001, 30: 591-614.
3. Donachie SP, Hou S, Gregory TS, Malahoff A, Alam M: *Idiomarina loihiensis* sp. nov., a halophilic gamma-Proteobacterium from the Lō›ihi submarine volcano, Hawai›i. *Int J Syst Evol Microbiol* 2003, 53: 1873-1879. 10.1099/ijs.0.02701-0

4. Elliott GM, Berndt C, Parson LM: The SW African volcanic rifted margin and the initiation of the Walvis Ridge, South Atlantic.*Mar Geophys Res* 2009, 30: 207-214. 10.1007/s11001-009-9077-x

5. Ettoumia B, Bouhajjaa E, Borinb S, Daffonchiob D, Boudabousa A, Cherif A: Gammaproteobacteria occurrence and microdiversity in Tyrrhenian Sea sediments as revealed by cultivation-dependent and -independent approaches. *Syst Appl Microbiol* 2010, 33: 222-231. 10.1016/j.syapm.2010.02.005

6. Finnegan L, Garcia-Melgares M, Gmerek T, Huddleston WR, Palmer A, Robertson A, Shapiro S, Unkles SE: A survey of culturable aerobic and anaerobic marine bacteria in de novo biofilm formation on natural substrates in St. Andrews Bay, Scotland. *Antonie Leeuwenhoek* 2011, 100: 399-404. 10.1007/s10482-011-9594-x

7. Fuhrman JA, Hagström A: Bacterial and archaeal community structure and its patterns. In *Microbial ecology of the oceans*. 2nd edition. Edited by: Kirchman DL. New York: Wiley-Liss; 2008.

8. Gärtner A, Blümel M, Wiese J, Imhoff JF: Isolation and characterization of bacteria from the Eastern Mediterranean deep sea. *Antonie Leeuwenhoek* 2011, 100: 421-435. 10.1007/s10482-011-9599-5

9. Gontang EA, Fenical W, Jensen PR: Phylogenetic diversity of Gram-positive bacteria cultured from marine sediments. *Appl Environ Microbiol* 2007, 73: 3272-3282. 10.1128/AEM.02811-06

10. Gregg WW, Conkright ME, Ginoux P, O'reilly JE, Casey NW: Ocean primary production and climate: global decadal changes. *Geophys Res Lett* 2003. 10.1029/2003GL016889

11. Hekinian R: Petrology of the Ninety East Ridge (Indian Ocean) compared to other aseismic ridges. *Contr Mineral Petrol*1974, 43: 125-147. 10.1007/BF00572715

12. Hensen C, Zabel M, Schulz HD: Benthic cycling of oxygen, nitrogen and phosphorus. In *Schulz HD*. Edited by: Zabel M. New York: Marine Geochemistry. Springer; 2006.

13. Ivanova EP, Romanenko LA, Chun J, Matte MH, Matte GR, Mikhailov VV, Svetashev VI, Huq A, Maugel T, Colwell RR: Idiomarina gen. nov., comprising novel indigenous deep-sea bacteria from the Pacific Ocean, including descriptions of two species, *Idiomarina abyssalis* sp. nov. and *Idiomarina zobellii* sp. nov. *Int J Syst Evol Microbiol* 2000, 50: 901-907. 10.1099/00207713-50-2-901

14. Jansen JHE, Van Weering TCE, Gieles R, Van Iperen J: Middle and late quaternary oceanography and climatology of the Zaire-Congo fan and the adjacent eastern Angola Basin. *Neth J Sea Res* 1984, 17: 201-249. 10.1016/0077-7579(84)90048-6

15. Jorgensen BB, Boetius A: Feast and famine – microbial life in the deep-sea bed. *Nat Rev Microbiol* 2007, 5: 770-781. 10.1038/nrmicro1745

16. Kaye JZ, Baross JA: High incidence of halotolerant bacteria in Pacific hydrothermal-vent and pelagic environments. *FEMS Microbiol Ecol* 2000, 32: 249-260. 10.1111/j.1574-6941.2000.tb00718.x

17. Kaye JZ, Baross JA: Synchronous effects of temperature, hydrostatic pressure, and salinity on growth, phospholipid profiles, and protein patterns of four *Halomonas* species isolated from deep-sea hydrothermal-vent and sea surface environments. *Appl Environ Microbiol* 2004, 70: 6220-6229. 10.1128/AEM.70.10.6220-6229.2004

18. Kaye JZ, Márquez MC, Ventosa A, Baross JA: *Halomonas neptunia* sp. nov., *Halomonas sulfidaeris* sp. nov., *Halomonas axialensis* sp. nov. and *Halomonas hydrothermalis* sp. nov.: halophilic bacteria isolated from deep-sea hydrothermal-vent environments. *Int J Syst Evol Microbiol* 2004, 54: 499-511. 10.1099/ijs.0.02799-0

19. Kaye JZ, Sylvan JB, Edwards KJ, Baross JA: *Halomonas* and *Marinobacter* ecotypes from hydrothermal vent, subseafloor and deep-sea environments. *FEMS Microbiol Ecol* 2011, 75: 123-133. 10.1111/j.1574-6941.2010.00984.x

20. Kobayashi T, Koide O, Mori K, Shimamura S, Matsuura T, Miura T, Takaki Y, Morono Y, Nunoura T, Imachi H, Inagaki F, Takai K, Horikoshi K: Phylogenetic and enzymatic diversity of deep subseafloor aerobic microorganisms in organics- and methane-rich sediments off Shimokita Peninsula. *Extremophiles* 2008, 12: 519-527. 10.1007/s00792-008-0157-7

21. Levin LA, Gooday AJ: The deep Atlantic Ocean. In *Ecosystems of the Deep Oceans*. Edited by: Tyler PA. Amsterdam: Elsevier; 2003:39-79.

22. Lu J, Nogi Y, Takami H: *Oceanobacillus iheyensis* gen. nov., sp. nov., a deep-sea extremely halotolerant and alkaliphilic species isolated from a depth of 1050 m on the Iheya Ridge. *FEMS Microbiol Lett* 2001, 205: 291-297. 10.1111/j.1574-6968.2001.tb10963.x

23. Maruyama A, Honda D, Yamamoto H, Kitamura K, Higashihara T: Phylogenetic analysis of psychrophilic bacteria isolated from the Japan Trench, including a description of the deep-sea species *Psychrobacter pacificensis* sp. nov. *Int J Syst Evol Microbiol* 2000, 50: 835-846. 10.1099/00207713-50-2-835

24. Okamoto T, Maruyama A, Imura S, Takeyama H, Naganuma T: Comparative phylogenetic analyses of *Halomonas variabilis*and related organisms based on 16S rRNA, gyrB and ectBC gene sequences. *Syst Appl Microbiol* 2004, 27: 323-333. 10.1078/0723-2020-00271

25. Orcutt BN, Sylvan JB, Knab NJ, Edwards KJ: Microbial ecology of the dark ocean above, at, and below the seafloor.*Microbiol Mol Biol Rev* 2011, 75(2):361-422. 10.1128/MMBR.00039-10

26. Pettit RK: Culturability and secondary metabolite diversity of extreme microbes: expanding contribution of deep sea and deep-sea vent microbes to natural product discovery. *Mar Biotechnol* 2011, 13: 1-11. 10.1007/s10126-010-9294-y

27. Polymenakou PN, Lampadariou N, Mandalakis M, Tselepides A: Phylogenetic diversity of sediment bacteria from the southern Cretan margin, Eastern Mediterranean Sea. *Syst Appl Microbiol* 2009, 32: 17-26. 10.1016/j.syapm.2008.09.006

28. Priest FG: Systematics and ecology of *Bacillus* . In *Bacillus subtilis and other Gram-positive bacteria: biochemistry, physiology and molecular genetics*. Edited by: Sonenshein AL, Losick R. Washington: AMS Press; 1993.

29. Salters VJM, Sachi-Kocher A: An ancient metasomatic source for the Walvis Ridge basalts. *Chem Geol* 2010, 273: 151-167. 10.1016/j.chemgeo.2010.02.010

30. Sass AM, Mckew BA, Sass H, Fichtel J, Timmis KN, Mcgenity TJ: Diversity of *Bacillus* -like organisms isolated from deep-sea hypersaline anoxic sediments. *Saline Syst* 2008, 4: 8-18. 10.1186/1746-1448-4-8

31. Schauer R, Bienhold C, Ramette A, Harder J: Bacterial diversity and biogeography in deep-sea surface sediments of the South Atlantic Ocean. *ISME J* 2010, 4: 159-170. 10.1038/ismej.2009.106

32. Simon-Colin C, Raguénès G, Cozien J, Guezennec JG: *Halomonas profundus* sp. nov., a new PHA-producing bacterium isolated from a deep-sea hydrothermal vent shrimp. *J Appl Microbiol* 2008, 104: 1425-1432. 10.1111/j.1365-2672.2007.03667.x

33. Smith CR, Leo FC, Bernardino AF, Sweetman AK, Arbizu PM: Abyssal food limitation, ecosystem structure and climate change. *Trends Ecol Evol* 2008, 23: 518-528. 10.1016/j.tree.2008.05.002

34. Stackebrandt E, Ebers J: Taxonomic parameters revisited: tarnished gold standards. *Microbiol Today* 2006, 33: 152-155.

35. Tamura K, Peterson D, Peterson N, Stecher G, Nei M, Kumar S: MEGA5: molecular evolutionary genetics analysis using maximum likelihood, evolutionary distance, and maximum parsimony. *Method Molecul Biol Evol* 2011, 28: 2731-2739. 10.1093/molbev/msr121

36. Toffin L, Webster G, Weightman AJ, Fry JC, Prieur D: Molecular monitoring of culturable bacteria from deep-sea sediment of the Nankai Trough, Leg 190 Ocean Drilling Program. *FEMS Microbiol Ecol* 2004, 48: 357-367. 10.1016/j.femsec.2004.02.009

37. Velmurugana N, Kalpanaa D, Chob JY, Leec GH, Parkc SH, Leea YS: Phylogenetic analysis of culturable marine bacteria in sediments from South Korean Yellow Sea. *Microbiol* 2011, 80: 261-272. 10.1134/S0026261711010188

38. Xu M, Wang P, Wang F, Xiao X: Microbial diversity at a deep-sea station of the Pacific nodule province. *Biodivers Conserv*2005, 14: 3363-3380. 10.1007/s10531-004-0544-z

39. Yanagibayashi M, Nogi Y, Li L, Kato C: Changes in the microbial community in Japan Trench sediment from a depth of 6292 m during cultivation without decompression. *FEMS Microbiol Lett* 1999, 170: 271-279. 10.1111/j.1574-6968.1999.tb13384.x

40. Zeng X, Xiao X, Wang P, Wang F: Screening and characterization of psychrotrophic, lipolytic bacteria from deep-sea sediments. *J Microbiol Biotechnol* 2004, 14(5):952-958.

41. Zengler K, Toledo G, Rappe M, Elkins J, Mathur EJ, Short JM, Keller M: Cultivating the uncultured. *P Natl Acad Sci USA* 2002, 99: 15681-15686. 10.1073/pnas.252630999

42. Zhuang W-Q, Tay J-H, Maszenan AM, Krumholz LR, Tay ST-L: Importance of Gram-positive naphthalene-degrading bacteria in oil-contaminated tropical marine sediments. *Lett Appl Microbiol* 2003, 36: 251-257. 10.1046/j.1472-765X.2003.01297.x

10
CHAPTER

CULTURE INDEPENDENT MOLECULAR ANALYSIS OF BACTERIAL COMMUNITIES IN THE MANGROVE SEDIMENT OF SUNDARBAN, INDIA

Abhrajyoti Ghosh[1,3], Nirmalya Dey[1,4], Amit Bera[1], Amit Tiwari[1,5], KB Sathyaniranjan[1], Kalyan Chakrabarti[2], Dhrubajyoti Chattopadhyay[1]

[1]Department of Biochemistry and Department of Biotechnology, University of Calcutta, 35, Ballygunge Circular Road, Kolkata- 700019, West Bengal, India

[2] Department of Agricultural Chemistry & Soil Science, Institute of Agricultural Science, University of Calcutta, 35, Ballygunge Circular Road. Kolkata- 700019, West Bengal, India

[3] Current address: Max Planck Research Group "Molecular Biology of Archaea", Max Planck Institute for Terrestrial Microbiology, Karlvon-Frisch-Strasse, D-35043 Marburg, Germany

[4] Current address: Division of Nephrology Department of Medicine, 7703 Floyd Curl Drive, University of Texas Health Science Center at San Antonio, San Antonio, Texas 78229 3900, USA

[5] Current address: Institute of Molecular Cancer Research, University of Zurich, 17 K 28, Winterthurerstrasse 190, Zurich-8057, Switzerland

ABSTRACT

Background

Sundarban is the world's largest coastal sediment comprising of mangrove forest which covers about one million hectares in the south-eastern parts of India and southern parts of Bangladesh. The microbial diversity in this sediment is largely unknown till date. In the present study an attempt has been made to understand the microbial diversity in this sediment using a cultivation-independent molecular approach.

Citation: Abhrajyoti Ghosh, Nirmalya Dey, Amit Bera, Amit Tiwari, KB Sathyaniranjan, Kalyan Chakrabarti, Dhrubajyoti Chattopadhyay, Culture independent molecular analysis of bacterial communities in the mangrove sediment of Sundarban, India, DOI: 10.1186/1746-1448-6-1.

Results

Two 16 S rRNA gene libraries were constructed and partial sequencing of the selected clones was carried out to identify bacterial strains present in the sediment. Phylogenetic analysis of partially sequenced 16 S rRNA gene sequences revealed the diversity of bacterial strains in the Sundarban sediment. At least 8 different bacterial phyla were detected. The major divisions of detected bacterial phyla were Proteobacteria (alpha, beta, gamma, and delta), Flexibacteria (CFB group), Actinobacteria, Acidobacteria, Chloroflexi, Firmicutes, Planctomycetes and Gammatimonadates.

Conclusion

The gammaproteobacteria were found to be the most abundant bacterial group in Sundarban sediment. Many clones showed similarity with previously reported bacterial lineages recovered from various marine sediments. The present study indicates a probable hydrocarbon and oil contamination in this sediment. In the present study, a number of clones were identified that have shown similarity with bacterial clones or isolates responsible for the maintenance of the S-cycle in the saline environment.

BACKGROUND

The majority (60-70%) of the world tropical and subtropical coastlines are covered with mangrove ecosystems. Mangroves are known to be highly productive ecosystems and have immense ecological values. They protect and stabilize the costal zones, nourish and nurture the coastal water with nutrients. They play important role as the feeding and breeding areas of many organisms including plants, animals and micro-organisms. The microbial community in the mangrove sediment is strongly influenced by bio-geographical, anthropological and ecological properties. These properties include food web in the ecosystem, nutrient cycling and the presence of organic and inorganic matters. During the past decade, the development of molecular techniques using nucleic acids has led to many new findings in the studies of microbial ecology [1]. As a basic approach to clarify the microbial communities, 16S rRNA genes are amplified by PCR from nucleic acids extracted from environmental samples, and then the PCR products are cloned and sequenced. This approach can avoid the limitation of the traditional culturing techniques for assessing the microbial diversity in the natural environments.

Both the sediment and soil probably represent some of the most complex microbial habitats on the Earth. There may be several thousand species of bacteria in 1 g of soil [2]. To study the genetic diversity and to analyse the members of mixed microbial populations are the two most important steps in microbial community studies. However, little research has been done on microbial diversity in marine sediments, and little information is currently available [3].

Mangrove ecosystems are in general nutrient-deficient, especially of nitrogen and phosphorus [4, 5, 6, 7, 8]. In spite of this, mangroves are highly productive. Microbial activity is responsible for major nutrient transformations within a mangrove ecosystem

[4, 9]. In tropical mangroves, bacteria and fungi constitute 91% of the total microbial biomass, whereas algae and protozoa represent only 7% and 2%, respectively [10]. Bacteria are responsible for most of the carbon flux in tropical mangrove sediments. They process most of the energy flow and nutrients, and act as a carbon sink. For example, in semiarid mangrove ecosystems on the Indus river in Pakistan, bacteria were attached to the sediment particles and processed most of the ecosystem nutrients [11].

Several studies have shown the uniqueness of mangrove sediments with respect to their microbial composition [3, 11, 12, 13, 14, 15]. Studies on microbial diversity in the mangrove sediments are important to understand the process of biogeochemical cycling and pollutants removal [16]. Sundarban is world's largest coastal wetland comprising of mangrove forest which covers about one million hectares in the delta of the rivers Ganga, Bramhaputra, and Meghna [17]. This mangrove region is shared between Bangladesh (~ 60%) and India (~ 40%). The area experiences a subtropical monsoon climate with the annual rainfall of about 1600-1800 mm and several cyclonic storms. The dynamics of this region is mainly maintained by sedimentations from all the three major rivers. Both the spatial and temporal influences have been demonstrated on the salinity in this region. The biodiversity of Sundarban includes numerous species of phytoplankton, zooplankton, micro-organisms, benthic invertebrates, molluscs, amphibians and mammals [17]. About 350 species of vascular plants, 250 species of fishes, and 300 species of birds are reported in Sundarban region [17]. Little work has been carried out on the microbial diversity in the Sundarban sediments.

This paper describes the culture independent microbial diversity analysis of the sediment sample from one of the famous islands of Sundarban, Netidhopani. Our molecular phylogenetic analysis reveals the occurrence of the bacterial 16S rRNA gene sequences that are unique and sequences that are previously reported in other mangrove sediments. Microbial community structure analysis can provide a better understanding about the microbial population and their interactions in a defined geographical region. Moreover such data are important with respect to our understanding of mangrove ecosystem processes and the role of micro-organisms in maintaining these processes [18]. The present results considerably extend our understanding on microbial diversity in Sundarban sediments. Furthermore, these results will open a new door towards understanding microbial diversity in the largest mangrove sediment in the world.

METHODS

Site Selection

Sediment samples were collected between 1st to 10th November' 2006 from one of the composite islands of Sundarban, Netidhopani (21°55›13» N, 88°44›46» E). This area is inundated with sea water about every twelve hours. Soil was collected after recessation when the land was exposed. This soil was highly saline with Ec_e 12.6 dSm^{-1}.

Soil Sampling, Analyses, and Site Climate

Soil samples were collected from the top 15 cm of the five different sites on the island

and brought to the laboratory in properly labelled, autoclaved, and sealed polythene bags on ice. All the soil analyses were carried out in the Department of Biotechnology, University of Calcutta. Microbiological and biochemical analyses were performed with the field moist soils. Physico-chemical analysis was carried out with air-dried soil samples. The soil pH was measured in 1:2.5 soil water suspensions. The Ec_e and ionic composition of soil saturation extract were measured following the method described in United States Department of Agriculture (USDA), 1954 [19]. The organic C (OC) and total N (TN) were measured by the methods proposed by Nelson & Sommers 1982 and Black, 1965 [4, 20] respectively. Textural composition (i.e., determination of sand, silt and clay) was determined by International Pipette Method as described previously [21]. The soil sample contained around 53.5% sand, 23.32% silt, 28% clay, 4.3% organic matter, and 0.395% total N (Average values of the triplicate analysis). The soil pH was found to be alkaline and it was 8.1. Prior to the total DNA isolation, all the soil samples (five samples were collected from different sites of the island) were mixed to homogeneity to make a single composite sample for DNA isolation.

Isolation of total soil DNA

Total soil (sediment) DNA was isolated using a modified CTAB-SDS based DNA extraction technique [22]. Soil sample of 5 g was mixed with 13.5 ml of the DNA extraction buffer [100 mM Tris-HCl (pH 8.0), 100 mM sodium-EDTA (pH 8.0), 100 mM sodium phosphate buffer (pH 8.0), 1.5 M NaCl, 1% CTAB] and 100 μl of proteinase K (10 mg/ml) in oakridge tube by horizontal shaking at 200 rpm for 30 min at 37°C. After the shaking treatment, 1.5 ml of 20% SDS was added, and then the sample was incubated at 65°C water bath for two hours with gentle end-over-end inversions every 15-20 min. The supernatant was collected after centrifugation at 6000 × g for 10 min at room temperature into 50 ml centrifuge tube. The soil pellet was extracted two more times by adding 4.5 ml of the extraction buffer and 0.5 ml of 20% SDS; vortexing for 10 s, incubating at 65°C for 10 min, and centrifuging as before. Supernatants from three cycles of extractions were combined and mixed with an equal volume of chloroform-isoamyl alcohol (24:1 v/v). The aqueous phase was recovered by centrifugation and total nucleic acid was precipitated with 0.6 volume of isopropanol at room temperature for 1 h. The pellet of crude nucleic acid was obtained by centrifugation at 16,000 × g for 20 min at room temperature, washed twice with cold 70% ethanol, and resuspended in sterile deionised water to give a final volume of 250 μl. RNase (10 mg/ml) treatment followed by phenol extraction and the re-precipitation was carried out prior to PCR amplification of 16S rRNA gene sequences.

PCR amplification and partial 16S rRNA gene library construction

Partial amplification of the 16S rRNA gene was performed with the thermal cycler ABI 2700 (ABI, Foster City, USA). The PCR of the 16S rRNA gene sequence from the total soil DNA was conducted in a final volume of 50 μl. The reaction mixture included 20-50 ng of isolated total soil DNA, 2 U taq polymerase (Recombinant, Cat. No. SKU# 10342-020, Invitrogen, Germany), 1× PCR buffer with 1.5 mM $MgCl_2$, 200 mM each dNTP, and 10 pmol of each primer (IDT, USA). The primers were chosen to amplify a

977-bp segment of 16S rRNA gene spanning V3-V9 region to construct the first library. In the first library (D16S_pMOS library) construction, forward primer used was 515F (5'-3') GTGCCAGCAGCCGCGGTAA and the reverse primer was 1492R (5'-3') TACGGYTACCTTGTTACGACTT [23]. This pair of primers was chosen to amplify both the bacterial and archaeal 16S rRNA gene sequences in the total soil DNA. Before amplification cycle DNA was denatured for 2 min at 94°C and after amplification an extension step (7 min at 72°C) was performed. The cycling parameters consisted of 28 cycles at: denaturation at 94°C for 30 sec, primer annealing at 45°C for 1 min, extension at 72°C for 1 min. The samples were held at 4°C until separated electrophoretically in a 2% agarose gel in 0.5 × Tris-Borate-EDTA buffers and visualized using ethidium bromide under ultraviolet illumination. All the amplified PCR products were agarose-gel-eluted using Qiagen gel elution kit.

Since the proteobacterial population has previously been reported to be quite abundant in other saline sediments [3, 24, 25, 26] a second library was also constructed using the proteobacteria specific universal SSU primer set [27], uni-forward (5'-3') TGCCAGCAGCCGCGGTA and uni-reverse (5'-3') GACGGGCGGTGTGTACAA, to screen for proteobacterial population in Sundarban sediments. The amplification cycle was as follows; initial denaturation for 5 min at 94°C, followed by 35 cycles at: 30 sec denaturation at 94°C, 1 min for primer annealing at 50°C, and 1 min of extension at 72°C. After the amplification a final extension step (7 min at 72°C) was performed. The samples were held at 4°C until analysis by agarose gel electrophoresis followed by elution using the Qiagen gel elution kit.

Two partial 16S rRNA gene libraries were constructed using gel eluted amplified PCR products (515F/1492R amplicon and uni-for/uni-rev amplicon) and pMOS-Blue vector (Pharmacia). The gel eluted PCR products were ligated to pMOS-Blue vector and then transformed into competent *Escherichia coli* XL1-Blue. The clones were screened for α-complementation by using X-Gal (5-bromo-4-chloro-3-indolyl-β-D-galactopyranoside) and IPTG (isopropyl-β-D thiogalactoside). All the positive clones were confirmed by PCR amplification and restriction digestion. All the positive clones were stored as glycerol stock at -70°C freezer.

Sequencing of the 16S rRNA gene Fragment

The sequencing of the partial 16S rRNA gene fragments in each of the recombinant plasmids was performed in ABI Prism 3100 automated DNA sequencer (Applied Biosystem, Foster City, California, USA) with the single primer 515F for D16S_pMOS library and uni-for for DUni_pMOS library, respectively. The sequencing reaction was performed using 5 pmoles 515F/uni-for primer and the Big Dye Terminator V3.1 sequencing kit as per manufacturer's protocol. The sequencing reaction conditions were as follows: 96°C for 10 sec, 50°C for 10 sec, and 60°C for 4 min for 25 cycles. After the sequencing PCR, the products (10 μl) were treated with 2 μl of 125 mM Na-EDTA, pH 8.0, and then precipitated using 2 μl of 3 M NaOAc (pH 4.6) and 50 μl absolute ethanol for 20 minutes at room temperature. The DNA was recovered by centrifugation (13,000 rpm for 30 min at 20°C), washed with 70% EtOH, dried, and resuspended in 15 μl Hi Di formamide (Applied Biosystems, Foster City, California, USA). Sequencing was performed in the ABI Prism 3100 Genetic Analyzer. Raw sequences were edited and assembled using the Auto Assembler program (V5.2). All the sequences were used to identify the bacteria with the help of the BLASTn program http://www.ncbi.nlm.nih.gov/BLAST, and all the sequences were submitted to GenBank. The multiple sequence alignment was performed using CLUSTAL-W software packagehttp://www.ebi.ac.uk/Tools/clustalw2/index.html.

Blast Search & Phylogenetic Analysis

The partial 16S rRNA gene sequences of the clones were compared with those available in the public databases. Identification to the species level was determined as a 16S rRNA gene sequence similarity of ≥ 97% with that of the prototype strain sequence in the GenBank. Sequence alignment and comparison was performed using the multiple sequence alignment program CLUSTALX (v 1.83) [28], with default parameters and the data converted to PHYLIP format. Minor modifications in the alignment were made using the BIOEDIT sequence editor. Rooted and unrooted phylogenetic trees were constructed using neighbor-joining (NJ) method and the TREEVIEW program for display of phylogenetic relationship [29]. Bootstrap analysis was performed as described by Felsenstein in 1985 [30] on 1000 random samples taken from the multiple alignments; analysis was done using the ClustalX programs.

Nucleotide sequence accession numbers

The 16S rRNA gene sequences reported in this study was submitted to the GenBank database under accession numbers EU939923-EU939972 (D16S_pMOS library) and EU999048-EU999127 (Duni_pMOS library).

RESULTS AND DISCUSSION

DNA extraction, Library construction and sequencing analysis

The total DNA was extracted from the sediment of Netidhopani, Sundarban, using modified CTAB-SDS based DNA isolation technique. Two partial 16S rRNA gene clone libraries were established from the PCR amplified partial 16S rRNA gene sequences using 515F/1492R and uni-for/uni-rev primer sets, respectively. The recombinant clones in the libraries were selected based on α-complementation (blue-white screening) technique and also confirmed by the re-PCR analysis and restriction enzyme digestion. Our sequencing analysis included 85 clones from D16S_pMOS library and 110 clones from DUni_pMOS library. All the sequenced clones were screened for sequences that repeat more than once in the library. Our final analysis included 50 clones from D16S_pMOS library and 80 clones from DUni_pMOS library, respectively (Table1) (Figures 1, 2, 3 and 4). *Methylophaga spp.* was found to be abundant in both the libraries. We also found four non bacterial chloroplastic DNA in recombinant clones from the two libraries. Although, the primer pair 515F & 1492R could amplify both bacterial and archaeal 16S rRNA gene sequences, we did not get any archaeal sequence in our library (D16S_pMOS). This was probably because of the limitation in our total

DNA extraction protocol and low primer specificities towards the archaeal 16S rRNA gene sequences. Our phylogenetic analysis revealed that 130 bacterial clones (50 clones from D16S_pMOS library and 80 clones from DUni_pMOS library) fell into 8 major phyla of the bacterial domain: Proteobacteria (Alpha-, Beta-, Gamma-, and Delta-), the Cytophaga-Flexibacter-Bacteroides (CFB) group, Actinobacteria, Chloroflexi, Firmicutes, Gemmatimonadetes, Acidobacteria group, and Planctomycetes (Table 1).

Table 1: Summary of the 16S rRNA gene sequences identified in the D16S_pMOS and DUni_pMOS clone libraries

Bacterial division	**Number of sequence types**	**Number of sequence types in D16S_pMOS library (%)**	**Number of sequence types in DUni_pMOS library (%)**	**Number of total clones (%)**	**Sequence similarity to the closest relatives[c](%)**
Proteobacteria	86	29 (58%)[a]	57 (71%)[a]	66.1	83-100
Alpha	6	1 (3.5%)[b]	5 (8.7%)[b]	4.6	87-96
Beta	7	4 (14%)[b]	3 (5.2%)[b]	5.4	92-99
Gamma	61	17 (59%)[b]	44 (77%)[b]	47	86-100
Delta	10	6 (21%)[b]	4 (7%)[b]	7.7	83-99
Unassigned proteobacteria	2	1 (2%)[b]	1 (1.25%)[b]	1.5	91-94
CFB group	1	1 (2%)[a]	0	0.8	93
Actinobacteria	2	0	2 (2.5%)[a]	1.53	87-98
Planctomycetes	4	2 (4%)[a]	2 (2.5%)[a]	3	86-92
Firmicutes	1	1 (2%)[a]	0	0.8	92
Chloroflexi	2	0	2 (2.5%)[a]	1.53	89-98
Gemmatimona-dates	2	1(2%)[a]	1 (1.25%)[a]	0.75	94-96
Acidobacteria	1	0	1 (1.25%)[a]	0.8	91
Marie eubacterium	1	1 (2%)[a]	0	0.8	89
Bacterial candidate division OP8	1	0	1 (1.25%)[a]	0.8	87
Uncultured	24	11 (22%)[a]	13 (16%)[a]	18.5	86-99
Unidentified	5	4 (8%)[a]	1 (1.25%)[a]	3.9	91-98
Total	130	50	80	--	83-100

"a" % of clones among all the 130 clones selected from D16S_pMOS and DUni_pMOS libraries

"b" % of clones among the proteobacterial clones under respective library

"c" Closest relatives as determined by the BLAST analysis

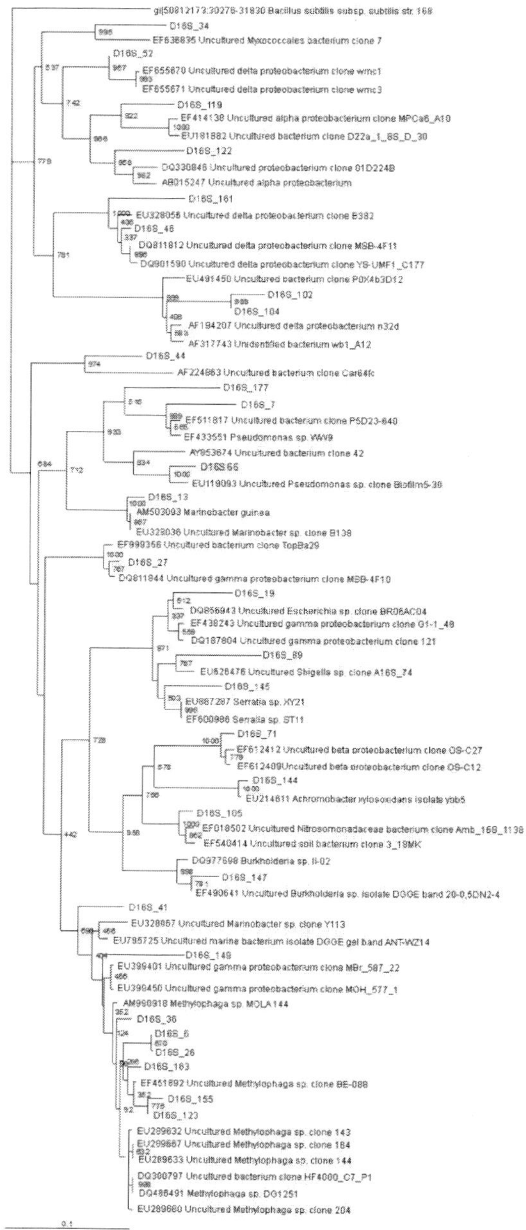

Figure 1: 16S rRNA gene tree showing positions of proteobacterial sequences in D16S_pMOS library including the reference sequences retrieved from GenBank. 16S rRNA gene sequence of *Bacillus subtilis* 168 is used to assign an out-group species.

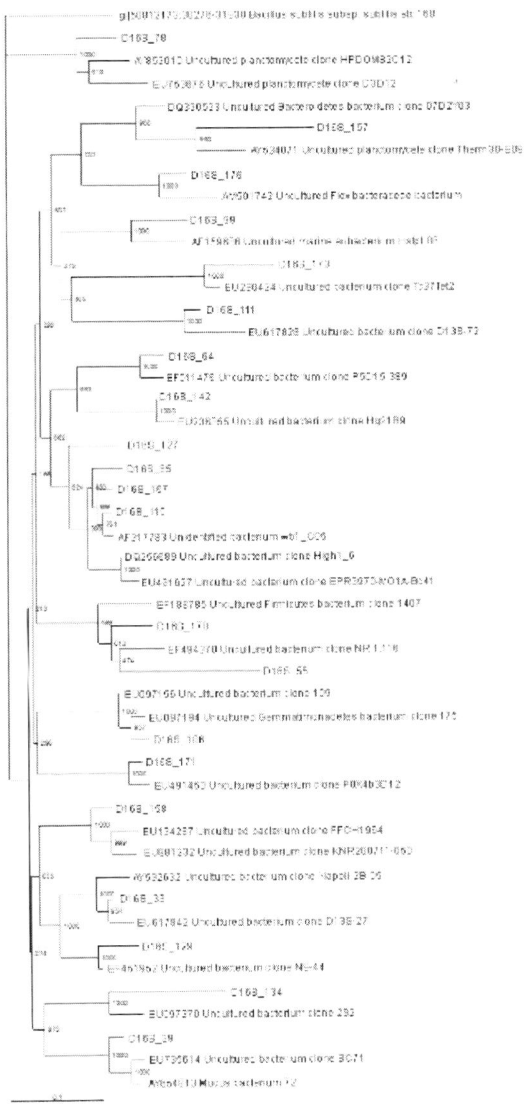

Figure 2: 16S rRNA gene tree showing positions of non-proteobacterial sequences (Flexibacteria, Planctomycetes, Firmicutes, Gemmatimonadates, unidentified/uncultured) in D16S_pMOS library including the reference sequences retrieved from GenBank. 16S rRNA gene sequence of *Bacillus subtilis* 168 is used to assign an out-group species.

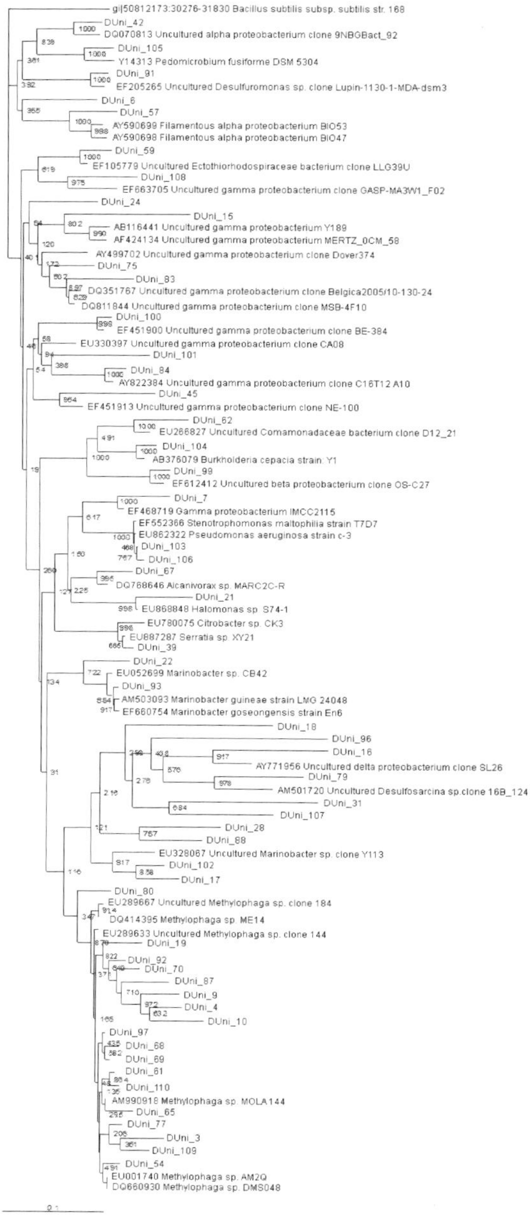

Figure 3: 16S rRNA gene tree showing positions of proteobacterial sequences in DUni_pMOS library including the reference sequences retrieved from GenBank. 16S rRNA gene sequence of *Bacillus subtilis* 168 is used to assign an out-group species.

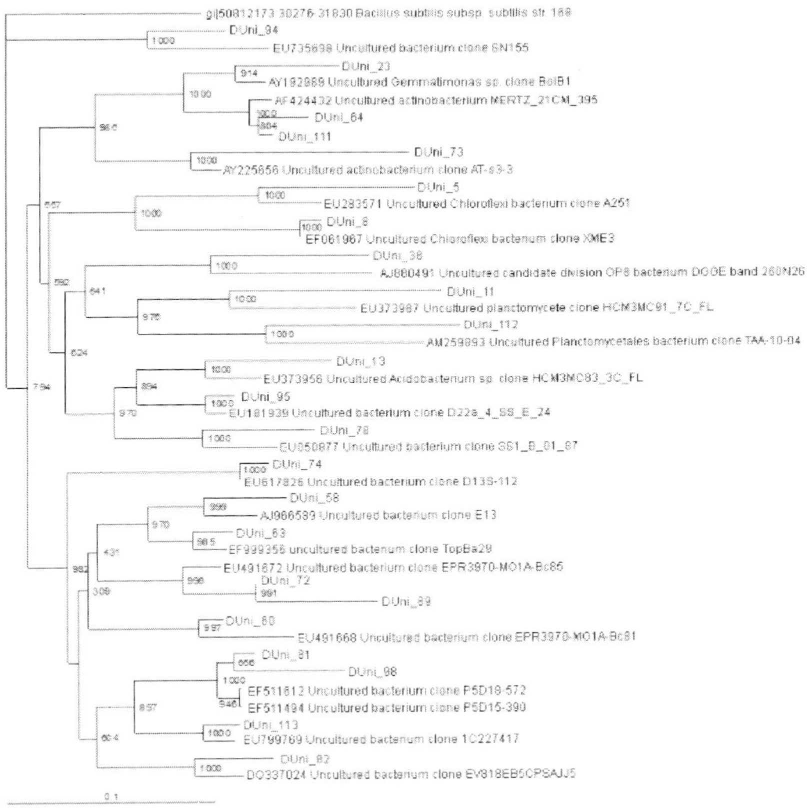

Figure 4: 16S rRNA gene tree showing positions of non-proteobacterial sequences (Actinobacteria, Planctomycetes, Chloroflexi, Acidobacteria, Gemmatimonadates, unidentified/uncultured) in D16S_pMOS library including the reference sequences retrieved from GenBank. 16S rRNA gene sequence of *Bacillus subtilis* 168 is used to assign an out-group species.

Proteobacteria

A total of 86 clones represented by 29 sequence types from D16S_pMOS library and 57 sequence types from DUni_pMOS library were identified as proteobacterial in origin by sequencing analysis (Table 1) (Figure 3). Most of the proteobacterial sequence types from both the libraries were similar to previously described isolates or metagenomic clones from coastal marine sediments or waters [GenBank description, sequences that were submitted to GenBank but yet to be published in scientific journals]. In D16S_pMOS library 29 proteobacterial sequence types accounted for 58% of the gene library and in DUni_pMOS library 57 proteobacterial sequence types accounted for 71% of the gene library (Figure 1).

Gammaproteobacteria

The gammaproteobacteria represented the most abundant proteobacterial subdivision (59% and 77% among the proteobacterial sequence types in D16S_pMOS and DUni_pMOS libraries, respectively) (Table 1). The most abundant sequence type in both the libraries showed similarity to *Methylophaga*, indicating a strong involvement of these bacterial species in the maintenance of the biogeochemical cycle in Sundarban sediment. A number of gammaproteobacterial sequence types showed similarity to organisms involved in the S-cycle (DUni_9, DUni_15, DUni_68, DUni_77, and DUni_91). A number of the gammaproteobacterial clones showed sequence similarity to the oil (D16S_41, DUni_17, DUni_102) and hydrocarbon (DUni_3, DUni_9, DUni_18, DUni_22, DUni_54, DUni_61, DUni_62, DUni_67, DUni_68, DUni_69, DUni_77, DUni_84, DUni_90, DUni_103, DUni_109, and DUni_110) degrading bacterial populations reported in different soil system [GenBank description, [31, 32, 33]]. Three sequence types (D16S_89, D16S_145 and DUni_83) have shown similarity to previously extracted sequences from the heavy metal contaminated soil or sediments [GenBank description, [34]]. Previous studies on Sundarban region revealed the contamination of hydrocarbon, petroleum and heavy metals in the soil [16, 35, 36, 37, 38]. The microbial composition further indicates the previous observations and a probable possibility of bio-conversion of those contaminating substances in this soil area.

Betaproteobacteria

14% and 5% of the proteobacterial sequence types were found to be betaproteobacterial from D16S_pMOS and DUni_pMOS libraries respectively (Table 1). In the D16S_pMOS library, the predominating clone (D16S_105) (three clones in the library) displayed a sequence similarity of 96% to its closest relative Amb_16S_1138, a betaproteobacterial clone previously recovered from trembling aspen [GenBank description]. In the DUni_pMOS library, among all the clones, DUni_99 showed 97% similarity to an uncultured betaproteobacterial clone OS-C27, recovered from the abandoned semiarid lead-Zn mine tailing site [GenBank description, [39]]; DUni_62 showed 92% similarity to D12-21, a betaproteobacterial clone recovered from a tar oil contaminated plume [GenBank description]; and finally the third clone DUni_104 showed 97% similarity to *Burkholderia cepacia* strain Y1 isolated from oil polluted soil.

Alphaproteobacteria

The six alphaproteobacterial sequence types detected in the two libraries comprised 3.5% and 8.7% of the total proteobacterial sequence types in D16S_pMOS and DUni_pMOS respectively (Table 1). In the D16S_pMOS library, the clone D16S_119 was the only sequence type and that showed 95% similarity to MPCa6_A10, an alphaproteobacterial clone recovered from wild and captive sponge *Microciona prolifera* in the Chesapeake Bay [GenBank description]. Five different alphaproteobacterial clones were obtained in DUni_pMOS library. The most predominating sequence type was DUni_6, which showed 87% similarity to MERTZ_OCM_210, an alphaproteobacterial clone reported within the Antarctic continental shelf sediment [GenBank description]. Among the other clones, sequence similarity revealed that they have shown identity to clones from the Seafloor Basalts from East Pacific Rise and the Juan de Fuca Ridge (DUni_42), industrial waste water treatment plants (DUni_56, DUni_57) [40] and *Pedomicrobium fusiform* DSM 5304 (DUni_105) [GenBank description, [41]].

Deltaproteobacteria

2.1% and 7% of the cloned sequence types have shown similarity to the deltaproteobacterial sequences in the database from D16S_pMOS and DUni_pMOS libraries, respectively (Table 1). In the D16S_pMOS library, all the deltaproteobacterial clones showed identity to organisms or clones recently described in Mangrove sediments. The sequence type D16S_52 was found to show 99% similarity with sulphate reducing bacterial strains [GenBank description]. In the DUni_pMOS library, clones DUni_79 and DUni_91 showed similarity with*Desulfosarcina spp.* and *Desulfuromonas spp.* respectively. These genera members have well-known metabolic features and involved in maintenance of S-cycle in soil. Probably the deltaproteobacterial strains in Sundarban sediments are largely involved in contributing towards the maintenance of sulphur cycle by sulphate reduction.

In addition to defined identities, two clones each from the D16S_pMOS and DUni_pMOS libraries showed only their proteobacterial identity. The clone D16S_122 showed similarity to uncultured proteobacterial clone 01D224B, which was described previously in the Guerrero Negro hyper saline microbial mat [GenBank description]. The clone DUni_55 showed identity to uncultured proteobacterial clone SIMO-1762, recovered from the salt marsh [GenBank description].

In both the libraries, many of the proteobacterial sequence types were found to show phylogenetic similarity with the strains (isolates) or clones recently described in marine sediments and waters, and were involved in S- or N-cycles, e.g., *Methylophaga spp.* DMS044, *Methylophaga spp.* DMS048, Uncultured gammaproteobacterium Y189 [42], Uncultured deltaproteobacterium wmc3 [GenBank description, EF655671], Uncultured *Nitrosomonadaceae* clone Amb_16S_1138 [GenBank description, EF018502], Uncultured*Desulfuromonas spp.* clone Lupin-1130-1-MDA-dsm3 [GenBank description, EF205265]. A large number of sequence types were also found to be phylogenetically similar to oil degradation associated microbes, e.g., Uncultured *Marinobacter spp.* Clone Y113 [GenBank description, EU328067], Uncultured gammaproteobacterium clone Y168 [GenBank description, EU328083]. There was also clone (DUni_67) identical to previously identified PAH degrading bacterial isolates [GenBank description]. Moreover in our libraries, we found many 16S rRNA gene sequences (D16S_6, D16S_66, D16S_89, D16S_106, D16S_123, D16S_134, D16S_155, D16S_163, DUni_4, DUni_19, DUni_45, DUni_65, and DUni_100) were similar to metagenomic clones or isolates reported in other studies from India [GenBank description, [7]]. In Sundarban sediment, compounds like polybrominated diphenyl ether (PBDE) and other hydrocarbons have been reported by different groups [35, 36, 38]. Moreover, in this sediment high concentration of heavy metal has been reported previously [37]. Furthermore, there are reports on the isolation of oil (petroleum) degrading bacterial strains from Sundarban sediment

[16]. All this analytical and microbiological evidences further support our findings of different bacterial species related to bacterial clones or strains previously reported in hydrocarbon, oil, and heavy metal contaminated soils and sediments.

Cytophaga-Flexibacteria-Bacteriodes

A single sequence type, representing a total of six clones and accounting for 2% of the D16S_pMOS library, was found to cluster with the CFB group (Table 1) (Figure 2). The representing clone D16S_176 showed similarity with the *Flexibacteraceae*, bacterium recently reported from Venice Lagoon anoxic sediments [GenBank description]. No clone related to the CFB group was detected in DUni_pMOS library.

Chloroflexi

Two clones in DUni_pMOS library showed similarity with the Chloroflexi (Table 1) (Figure 4). The clone DUni_5 showed sequence similarity with A251, an uncultured *Chloroflexi* [GenBank description]. Another clone DUni_8 showed similarity with XME3, an uncultured Chloroflexi, recently reported in mangrove sediment of Xiamen, China [GenBank description].

Planctomycetes

Two sequence types, representing 6 clones in D16S_pMOS library and two sequence types, representing 3 clones in DUni_pMOS library, were found to group within the Planctomycetes (Table 1) (Figures 2 and 4). In D16S_pMOS library, the clone D16S_78 showed similarity to D3D12, an uncultured planctomycete clone recovered from the fresh water stromatolites from the Ruidera Pools Natural Park, Spain [GenBank description]. The clone D16S_157 was found to be identical to Therm30-E09, an uncultured planctomycete clone reported in the sediment of the Eastern Mediterranean Sea [12]. In DUni_pMOS library, the clone DUni_112 showed similarity to TAA-10-04, an uncultured Planctomycete clone recovered as the phage associated bacterium and the clone DUni_11 was found to show similarity with HCM3MC91_7C_FL, the planctomycete clone recovered from Eastern Mediterranean Sea [GenBank description].

Other bacterial lineages

A total of six clones, represented by two sequence types and accounted for 2.5% of the DUni_pMOS library were found to group within the Actinobacteria (Table 1). Two representative sequence types, DUni_64 and DUni_73 showed similarities with uncultured actinobacteria clone MERTZ_21CM_395 [43] and uncultured actinobacterial clone AT-s3-3 [43], respectively (Figures 2 and 4).

Two clones each from the two libraries were grouped within the Gemmatimonadetes (Table 1), a recently discovered bacterial phylum [44]. The clone D16S_106 showed 96% similarity to Gemmatimonadetes bacterial clone 175, recovered from soil sample from radish rich area of Jaunpur, Uttarpradesh, India [GenBank description]. The other clone DUni_23 was found to be 94% identical to Gemmatimonadetes bacterial clone BolB1, reported recently as a member of phylogenetic division OP11 [45].

A single clone, DUni_38 was found to be identical to recently reported uncultured candidate division OP8 [GenBank description, [46]]. A single clone, D16S_178 was found to show similarity with the uncultured Firmicutes bacterium clone 1407, which was previously reported in the Altamira Cave [GenBank description]. Among the clones in the libraries, the clone DUni_13 showed similarity to HCM3MC83_3C-FL, an*Acidobacterium spp.* clone recently recovered from the sediment of Eastern Mediterranean Sea [GenBank description].

Other than known bacterial taxa, 22% and 16% of the clones from D16S_pMOS and DUni_pMOS libraries [26], were clustered within uncultured bacterial group (Table 1) (Figures 2 and 4). Most of the related uncultured bacterial clones from the blast search analysis revealed that they were reported from either marine sediments or from sea waters. In our library of clones, we also detected unidentified bacterial strains (four in D16S_pMOS and one in DUni_pMOS libraries). The only clone, D16S_99 showed identity to marine eubacterium HstpL86, previously described in the leaves of sea grass *Halophila stipulacea*.

CONCLUSIONS

In the present study, 16S rRNA gene clone library based analysis was performed on the world's largest mangrove ecosystem, Sundarban sediment, for the first time. Even no culture based analysis of the bacterial community is yet reported from this mangrove ecosystem. The present analysis revealed that the Sundarban sediment possesses diverse bacterial population. At least 8 major phyla of the bacterial domain were detected in this sediment. Previous studies on bacterial diversity analysis reported five to thirteen major lineages in sediments collected from a variety of coastal marine environments [3, 15, 24, 25, 42].

Sequencing analysis of the clones revealed the dominance of gammaproteobacterial sequences in both the libraries. Majority of the gammaproteobacterial clones resembled sequences recovered from oil and hydrocarbon rich marine sediments. This probably goes with the previous reports on Sundarban sediment where people have shown that in this sediment different hydrocarbons are present at high concentrations [35, 36, 38]. Moreover, a number of cultivable bacterial strains, which were capable of degrading petroleum, have been isolated from this sediment [16].

In the present report, a number of gammaproteobacterial clones were found to show similarity towards bacterial clones or isolates involved in sulfur cycling. Similar results were reported previously [42], while analysing coastal marine sediment beneath area of intensive shellfish aquaculture. Sulfur-oxidising bacterial strains are found to play an important role in detoxification of sulphide in marine sediments. Sulfur-reducing bacterial community instead is important in organic carbon oxidation in marine sediments and this observation is supported by the fact that sulphate is one of the main electron acceptors present in these environments. A number of gammaproteobacteria in the present study were found to show similarity to isolates or clones related to bioconversion of S-containing organic molecules (S-oxidisers). This interesting observation is supported by recent investigations, where it has been shown that the reduction of sulphate may be an important pathway of organic matter mineralization in organic rich deposits typical of mangrove forests. Furthermore, most of the identified deltaproteobacterial clones from the two libraries showed similarity to the sulfur and sulphate reducing bacteria recovered from a variety of marine sediments. Analysis of only 130 clones might not be enough to cover the whole picture of S-cycle but it could provide a little insight about what is happening in Sundarban sediment. In marine ecosystems, S-cycle has proved to be the important biogeochemical factor that dictates the flow of electrons along the biological systems under such an anaerobic condition. Identification of sulfur- oxidising and sulphur and sulphate reducing bacterial clones refer to the anaerobic condition in this sediment and a possible maintenance of the biogeochemical cycle in Sundarban sediment.

The evidence of the presence of hydrocarbons [35, 36, 38] in this sediment supports the finding of a comparatively lower number of alphaproteobacterial clones in Sundarban sediment. The special difference in the productivity of the water columns is probably enhances the reason for this observation in this sediment. Previous observations of Horner-devine *et.al.*, 2003 [47] showed the dependence of alphaproteobacteria richness on the productivity levels in aquatic ecosystem. Furthermore, identification of the clones associated with hydrocarbon/oil degradation probably confirms the reason for the lower abundance of alphaproteobacteria in Sundarban sediment.

In our study, epsilonproteobacteria are absent in both the libraries. It has been well documented that epsilonproteobacteria were absent or scarce in other clone libraries of coastal marine sediments [3, 15, 42]. Although, some reports have described the presence of epsilonbacteria in the library of clones made from marine sediments [24, 48].

Identification of Gemmatimonades in our clone libraries was interesting and probably the first report of recovery of this phylum from mangrove sediment.

In conclusion, the Sundarban surface sediment harboured a phylogenetically diverse population of organisms from bacterial domain. At least 8 major phyla have been recovered from Sundarban sediment. The proteobacteria, especially the gammaproteobacteria were found to be abundant in both the libraries. While some of the 16S rRNA gene sequence types detected were related to genera or taxa that were classically identified in Sundarban sediment and correlated to a defined functional arena; many were derived from uncultured/unidentified taxa. Previous studies have shown that the primer pair uni-for/uni-rev was good for proteobacteria. As in marine system, proteobacterial communities are major; this pair of primers was employed in the present study. Further studies are necessary to understand the bacterial diversity in more details. This is the first report describing the bacterial diversity in Sundarban sediment. We feel that the present study has obtained a fundamental insight into the major bacterial populations in Sundarban sediment. This study will definitely open a new era in understanding the microbial diversity in Sundarban.

DECLARATIONS

Acknowledgements

This study was supported by a grant from the Centre for Advance Studies (CAS) programme of the Department of Biochemistry, University Grants Commission, India. Support for DNA sequencing was obtained from the NCMP grant to University of Calcutta, India. A research fellowship for AG was provided by the B. C. Guha Endowment Fund, University of Calcutta, India.

Authors' contributions

AG designed and conducted the experiments in consultation with DJC, analyzed the data and drafted the manuscript. ND helped with the library constructions, screening and sequencing analysis. AB participated in sequencing analysis, sequence alignment and phylogenetic analysis. AT helped with the screening of the libraries and sequence alignment. SB helped with the screening of the libraries. KC helped with the sampling, soil analyses and interpretation. DJC was involved in the acquisition of funding, contributed to the concept, experimental design, analysis of the data and revising the manuscript. All authors read and approved the final manuscript.

REFERENCES

1. Amann RI, Ludwig W, Schleifer KH: Phylogenetic identification and in situ detection of individual microbial cells without cultivation. Microbiol Rev. 1995, 59: 143-169.
2. Torsvik V, Goksoyr J, Daae FL: High diversity in DNA of soil bacteria. Appl Environ Microbiol. 1990, 56: 782-787.
3. Gray JP, Herwig RP: Phylogenetic analysis of the bacterial communities in marine sediments. Appl Environ Microbiol. 1996, 62: 4049-4059.
4. Alongi DM, Christoffersen P, Tirendi F: The influence of forest type on microbial-nutrient relationships in tropical mangrove sediments. J Exp Mar Biol Ecol. 1993, 171: 201-223. 10.1016/0022-0981(93)90004-8.
5. Holguin G, Guzman MA, Bashan Y: Two new nitrogen-fixing bacteria from the rhizosphere of mangrove trees: their isolation, identification and in vitro interaction with rhizosphere *Staphylococcus* sp. FEMS Microbiol Ecol. 1992, 101: 207-216. 10.1111/j.1574-6968.1992.tb05777.x.
6. Holguin G, Vazquez P, Bashan Y: The role of sediment microorganisms in the productivity, conservation, and rehabilitation of mangrove ecosystems: an overview. Biol Fert Soils. 2001, 33: 265-278. 10.1007/s003740000319.
7. Sengupta A, Chaudhuri S: Ecology of heterotrophic dinitrogen fixation in the rhizosphere of mangrove plant community at the Ganges river estuary in India. Oecologia. 1991, 87: 560-564. 10.1007/BF00320420.

8. Vazquez P, Holguin G, Puente ME, Lopez-Cortes A, Bashan Y: Phosphate-solubilizing microorganisms associated with the rhizosphere of mangroves in a semiarid coastal lagoon. Biol Fertil Soils. 2000, 30: 460-468. 10.1007/s003740050024.

9. Holguin G, Bashan Y, Mendoza-Salgado RA, Amador E, Toledo G, Vazquez P, Amador A: La Microbiologia de los manglares. Bosques en la frontera entre el mar y la tierrra. Ciencia Desarrollo. 1999, 144: 26-35.

10. Alongi DM: Bacterial productivity and microbial biomass in tropical mangrove sediments. Microbiol Ecol. 1988, 15: 59-79. 10.1007/BF02012952.

11. Bano N, Nisa M-U, Khan N, Saleem M, Harrison PJ, Ahmed SI, Azam F: Significance of bacteria in the flux of organic matter in the tidal creeks of the mangrove ecosystem of the Indus river delta, Pakistan. Mar Ecol Prog Ser. 1997, 157: 1-12. 10.3354/meps157001.

12. Polymenakou PN, Bertilsson S, Tselepides A, Stephanou EG: Bacterial community composition in different sediments from the Eastern Mediterranean Sea: a comparison of four 16S ribosomal DNA clone libraries. Microb Ecol. 2005, 50: 447-462. 10.1007/s00248-005-0005-6.

13. Ravenschlag K, Sahm K, Pernthaler J, Amann R: High bacterial diversity in permanently cold marine sediments. Appl Environ Microbiol. 1999, 65: 3982-3989.

14. Todorov JR, Chistoserdov AY, Aller JY: Molecular analysis of microbial communities in mobile deltaic muds of Southeastern Papua New Guinea. FEMS Microbiol Ecol. 2000, 33: 147-155. 10.1111/j.1574-6941.2000.tb00737.x.

15. Urakawa H, Kita-Tsukamoto K, Ohwada K: Microbial diversity in marine sediments from Sagami Bay and Tokyo Bay, Japan, as determined by 16S rRNA gene analysis. Microbiology. 1999, 145: 3305-3315.

16. Roy S, Hens D, Biswas D, Biswas D, Kumar R: Survey of petroleum - degrading bacteria in coastal waters of Sunderban. Biosphere Reserve. World J Microbiol Biotech. 2002, 18: 575-581. 10.1023/A:1016362819746.

17. Gopal B, Chauhan M: Biodiversity and its conservation in the Sundarban Mangrove Ecosystem. Aquatic Sciences. 2006, 69: 338-354. 10.1007/s00027-006-0868-8.

18. Staley JT, Gosink JJ: Poles apart: biodiversity and biogeography of sea ice bacteria. Annu Rev Microbiol. 1999, 53: 189-215. 10.1146/annurev.micro.53.1.189.

19. Black CA: Methods of Soil Analysis. 1965, American Society of Agronomy, Wisconsin, USA

20. Nelson DW, Somers LE: Organic carbon. Methods in soil microbiology and biochemistry. Edited by: Alef K, Nannipieri P. 1975, Academic, London, 59-60.

21. Piper CS: Soil and Plant analysis. 1966, Maver Publishers, Bombay, India

22. Miller DN, Bryant JE, Madsen EL, Ghiorse WC: Evaluation and optimization of DNA extraction and purification procedures for soil and sediment samples. Appl Environ Microbiol. 1999, 65: 4715-4724.

23. Lane DJ: 16S/23S rRNA sequencing. Nucleic acid techniques in bacterial systematics. Edited by: Stackebrandt E, Goodfellow M. 1991, Wiley, New York, NY, 115-175.

24. Bowman JP, McCuaig RD: Biodiversity, community structural shifts, and biogeography of prokaryotes within Antarctic continental shelf sediment. Appl Environ Microbiol. 2003, 69: 2463-2483. 10.1128/AEM.69.5.2463-2483.2003.

25. Cifuentes A, Anton J, Benlloch S, Donnelly A, Herbert RA, Rodriguez-Valera F: Prokaryotic diversity in Zostera noltii-colonized marine sediments. Appl Environ Microbiol. 2000, 66: 1715-1719. 10.1128/AEM.66.4.1715-1719.2000.

26. Holmes AJ, Tujula NA, Holley M, Contos A, James JM, Rogers P, Gillings MR: Phylogenetic structure of unusual aquatic microbial formations in Nullarbor caves, Australia. Environ Microbiol. 2001, 3: 256-264. 10.1046/j.1462-2920.2001.00187.x.

27. Giovannoni SJ, Britschgi TB, Moyer CL, Field KG: Genetic diversity in Sargasso sea bacterioplancton. Nature. 1990, 344: 60-63. 10.1038/345060a0.

28. Thompson JD, Gibson TJ, Plewniak F, Jeanmougin F, Higgins DG: The ClustalX windows interface: flexible strategies for multiple sequence alignment aided by quality analysis tools. Nucleic Acids Res. 1997, 24: 4876-4882. 10.1093/nar/25.24.4876.

29. Page RD: TreeView: an application to display phylogenetic trees on personal computers. Comput Appl Biosci. 1996, 12: 357-358.

30. Falsenstein J: Condence limits on phylogenetics: an approach using bootstrap. Evolution. 1985, 39: 783-791. 10.2307/2408678.

31. Bissett A, Bowman J, Burke C: Bacterial diversity in organically-enriched fish farm sediments. FEMS Microbiol Ecol. 2006, 55: 48-56. 10.1111/j.1574-6941.2005.00012.x.

32. Neufeld JD, Schafer H, Cox MJ, Boden R, McDonald IR, Murrell JC: Stable-isotope probing implicates Methylophaga spp and novel Gammaproteobacteria in marine methanol and methylamine metabolism. Isme J. 2007, 1: 480-491. 10.1038/ismej.2007.65.

33. Schafer H: Isolation of Methylophaga spp. from marine dimethylsulfide-degrading enrichment cultures and identification of polypeptides induced during growth on dimethylsulfide. Appl Environ Microbiol. 2007, 73: 2580-2591. 10.1128/AEM.02074-06.

34. Gillan DC, Pernet P: Adherent bacteria in heavy metal contaminated marine sediments. Biofouling. 2007, 23: 1-13. 10.1080/08927010601108725.

35. Binelli A, Sarkar SK, Chatterjee M, Riva C, Parolini M, Bhattacharya BD, Bhattacharya AK, Satpathy KK: Concentration of polybrominated diphenyl ethers (PBDEs) in sediment cores of Sundarban mangrove wetland, northeastern part of Bay of Bengal (India). Mar Pollut Bull. 2007, 54: 1220-1229. 10.1016/j.marpolbul.2007.03.021.

36. Binelli A, Sarkar SK, Chatterjee M, Riva C, Parolini M, Bhattacharya B, Bhattacharya AK, Satpathy KK: A comparison of sediment quality guidelines for toxicity assessment in the Sunderban wetlands (Bay of Bengal, India). Chemosphere. 2008, 73: 1129-1137. 10.1016/j.chemosphere.2008.07.019.

37. Saha M, Cobelo-Garcia A, Sarkar SK, Prego R, Bhattacharya B: Distribution of metals in representative biota of sundarban mangrove wetland, northeast India. Bull Environ Contam Toxicol. 2006, 76: 656-662. 10.1007/s00128-006-0970-6.

38. Sahu LK, Lal S, Venkataramani S: Distributions of O_3, CO and hydrocarbons over the Bay of Bengal: A study to assess the role of transport from southern India and marine regions during September-October 2002. Atmospheric Environment. 2006, 40: 4633-4645. 10.1016/j.atmosenv.2006.02.037.

39. Mendez MO, Neilson JW, Maier RM: Characterization of a bacterial community in an abandoned semiarid lead-zinc mine tailing site. Appl Environ Microbiol. 2008, 74: 3899-3907. 10.1128/AEM.02883-07.

40. Levantesi C, Beimfohr C, Geurkink B, Rossetti S, Thelen K, Krooneman J, Snaidr J, Waarde van der J, Tandoi V: Filamentous Alphaproteobacteria associated with bulking in industrial wastewater treatment plants. Syst Appl Microbiol. 2004, 27: 716-727. 10.1078/0723202042369974.

41. Rainey FA, Ward-Rainey N, Gliesche CG, Stackebrandt E: Phylogenetic analysis and intrageneric structure of the genus Hyphomicrobium and the related genus Filomicrobium. Int J Syst Bacteriol. 1998, 48 (Pt 3): 635-639.

42. Asami H, Aida M, Watanabe K: Accelerated sulfur cycle in coastal marine sediment beneath areas of intensive shellfish aquaculture. Appl Environ Microbiol. 2005, 71: 2925-2933. 10.1128/AEM.71.6.2925-2933.2005.

43. Lopez-Garcia P, Duperron S, Philippot P, Foriel J, Susini J, Moreira D: Bacterial diversity in hydrothermal sediment and epsilonproteobacterial dominance in experimental microcolonizers at the Mid-Atlantic Ridge. Environ Microbiol. 2003, 5: 961-976. 10.1046/j.1462-2920.2003.00495.x.

44. Zhang H, Sekiguchi Y, Hanada S, Hugenholtz P, Kim H, Kamagata Y, Nakamura K: Gemmatimonas aurantiaca gen. nov., sp. nov., a gram-negative, aerobic, polyphosphate-accumulating micro-organism, the first cultured representative of the new bacterial phylum Gemmatimonadetes phyl. nov. Int J Syst Evol Microbiol. 2003, 53 (Pt 4): 1155-1163. 10.1099/ijs.0.02520-0.

45. Harris JK, Kelley ST, Pace NR: New perspective on uncultured bacterial phylogenetic division OP11. Appl Environ Microbiol. 2004, 70: 845-849. 10.1128/AEM.70.2.845-849.2004.

46. Wilms R, Köpke B, Sass H, Chang TS, Cypionka H, Engelen B: Deep biosphere-related bacteria within the subsurface of tidal flat sediments. Environ Microbiol. 2006, 8: 709-719. 10.1111/j.1462-2920.2005.00949.x.

47. Horner-Devine MC, Leibold MA, Smith VH, Bohannan BJM: Bacterial diversity patterns along a gradient of primary productivity. Ecol Lett. 2003, 6: 613-622. 10.1046/j.1461-0248.2003.00472.x.

48. Liang JB, Chen YQ, Lan CY, Tam NFY, Zan QJ, Huang LN: Recovery of novel bacterial diversity from mangrove sediment. Mar Biol. 2007, 150: 739-747. 10.1007/s00227-006-0377-2.

11

CHAPTER

DIVERSITY AND DISTRIBUTION OF ARCHAEA IN THE MANGROVE SEDIMENT OF SUNDARBANS

Anish Bhattacharyya[1] Niladri Shekhar Majumder[2] Pijush Basak[1] Shayantan Mukherji[3] Debojyoti Roy[1] Sudip Nag[1] Anwesha Haldar[4] Dhrubajyoti Chattopadhyay[1] Suparna Mitra[5] Maitree Bhattacharyya[1] and Abhrajyoti Ghosh[3]

[1]Department of Biochemistry, University of Calcutta, 35 Ballygunge Circular Road, Kolkata, West Bengal 700019, India

[2]Roche Diagnostics India Pvt. Ltd., Block 4C, Akash Tower, Near Ruby Hospital, 781 Anandapur, Kolkata 700107, India

[3]Department of Biochemistry, Bose Institute, P1/12, C. I. T. Road, Scheme VIIM, Kolkata, West Bengal 700054, India

[4]Department of Geography, University of Calcutta, 35 Ballygunge Circular Road, Kolkata, West Bengal 700019, India

[5]Norwich Medical School, University of East Anglia and Institute of Food Research, Norwich Research Park, Norwich, Norfolk NR4 7UA, UK

ABSTRACT

Mangroves are among the most diverse and productive coastal ecosystems in the tropical and subtropical regions. Environmental conditions particular to this biome make mangroves hotspots for microbial diversity, and the resident microbial communities play essential roles in maintenance of the ecosystem. Recently, there has been increasing interest to understand the composition and contribution of microorganisms in mangroves. In the present study, we have analyzed the diversity and distribution of archaea in the tropical mangrove sediments of Sundarbans using 16S rRNA gene amplicon sequencing. The extraction of DNA from sediment

Citation: Anish Bhattacharyya, Niladri Shekhar Majumder, Pijush Basak, et al., "Diversity and Distribution of Archaea in the Mangrove Sediment of Sundarbans," Archaea, vol. 2015, Article ID 968582, 14 pages, 2015. doi:10.1155/2015/968582.

samples and the direct application of 16S rRNA gene amplicon sequencing resulted in approximately 142 Mb of data from three distinct mangrove areas (Godkhali, Bonnie camp, and Dhulibhashani). The taxonomic analysis revealed the dominance of phyla Euryarchaeota and Thaumarchaeota (Marine Group I) within our dataset. The distribution of different archaeal taxa and respective statistical analysis (SIMPER, NMDS) revealed a clear community shift along the sampling stations. The sampling stations (Godkhali and Bonnie camp) with history of higher hydrocarbon/oil pollution showed different archaeal community pattern (dominated by haloarchaea) compared to station (Dhulibhashani) with nearly pristine environment (dominated by methanogens). It is indicated that sediment archaeal community patterns were influenced by environmental conditions.

INTRODUCTION

Archaea, representing the third domain of life, were originally anticipated to thrive under extreme environments, such as hydrothermal vents, hot water springs, salt brines, and extremely acidic and anoxic environments, where they contribute significantly towards the maintenance of the biogeochemical cycles [1–7]. However, with the advent of molecular techniques, it has become increasingly evident that archaea are much more widespread and metabolically diverse than originally postulated. A considerable portion of the microbial communities in a wide variety of "nonextreme" environments, for example, soil, ocean, and lakes, is constituted by archaea [8–13]. Despite the increasing interest to understand the ecophysiology of archaea, the lack of knowledge with respect to mesophilic and cold-adopted archaea is still enormous [9, 14, 15].

Mangrove wetlands are typical example of mesophilic and moderately halophilic environmental niches. They are commonly situated at the intertidal zones along the tropical and subtropical coasts and play a very important role in shaping the coastal ecology [16]. Mangrove forests are considered to be highly productive niche that support detritus-based food web [16, 17]. Particularly, in tropical mangroves, the high turnover rates for organic matters and nutrient cycling between the ocean and terrestrial habitats makes it the most productive ecosystem in the world [17]. The high primary productivity of mangroves implies a high demand for nutrients essential for plant growth and this appears to be achieved by a highly efficient system of nutrient trapping, uptake, and recycling in mangrove ecosystem [18]. The diverse microbial communities residing in the mangroves play important role in transformation of nutrients in the environment. While the importance of bacteria and fungi in mangrove biogeochemical cycles is well established, our knowledge of archaea in mangrove habitats remains extremely limited [16]. Sundarbans is the world's largest tidal halophytic mangrove ecosystem covering 20,400 square kilometers (7,900 sq mi) of area and has been recognized as a UNESCO World Heritage site. Situated in the delta of Ganges, Meghna, and Brahmaputra rivers on the Bay of Bengal, Sundarbans is shared between India and Bangladesh. This mangrove ecosystem is the home for diverse flora and fauna, including mangrove plant species like sundari (Heritiera fomes), goran (Ceriops decandra), geoa (Excoecaria agallocha), keora (Sonneratia apetala), and so forth, and the world's famous endangered royal Bengal tiger (Panthera tigris tigris). Microbial communities play an important role in generation of detritus in mangrove areas and Sundarbans is one such ecotype where microbes have been shown to be involved in biogeochemical cycling of the nutrients [18–20]. Despite global advancement in understanding the microbial diversity and role of microbes in different mangrove environments, little has been performed in Sundarbans [21, 22]. In a very recent study, a detailed description of the bacterial diversity has been presented in the backdrop of seasonal changes [23]. However, no data are yet available regarding the abundance and diversity of archaea in Sundarbans.

Hitherto, most of the research on mangrove environments has focused on understanding the diversity and functions of bacteria and fungi [23–26], and very little is known about archaeal assemblages in mangrove [16,27, 28]. In order to gain new insight into the archaeal community patterns and the influence of environmental conditions in the mangrove sediments and to build foundational information for future research, we have investigated the archaeal diversity in the sediments of Indian Sundarbans employing 454-pyrosequencing.

MATERIALS AND METHODS

Ethics Statement

No specific permits were required for the described field studies, which complied with all relevant regulation. The studied locations are not privately owned. Moreover, the study did not involve endangered or protected species. Indeed, the Indian Coastal Zone Management (ICZM) authority of the state of West Bengal, India, has approved this experimental exercise.

Study Area and Sediment Sampling

Samples were collected in triplicate from surface (2 cm) and subsurface (32 cm) sediments of the Sundarbans mangrove wetland, which is located on the northeastern coast of India (Figure 1). The samples were collected from three different stations; for example, Godkhali (station A; 22°06′32.570″N 88°46′22.220″E), Bonnie camp (station B; 21°49′53.581″N 88°36′44.860″E), and Dhulibhashani (station C; 21°37′40.837″N 88°33′47.762″E) spanning 90 km along the tidal gradient from the shoreline (Figure 1). In June 2013, the sediment samples were collected in triplicate from each station. A sediment corer (50 cm depth, 6 cm diameter) was used to collect the top 32 cm of sediments. The uppermost surface layer (0–2 cm) and deeper subsurface layer (30–32 cm) were sampled and immediately sieved by a 2 mm mesh in the field. The sieved fractions were stored in sealed sterile polypropylene containers and brought to the laboratory. Upon arrival, a portion of each sediment sample was flash-frozen at −80°C for polyaromatic hydrocarbon (PAH) analysis and DNA extraction. The remaining sediments were stored at 4°C for other analytical procedures such as nutrients and heavy metals estimation. The individual collection from each sampling station was used for physicochemical analysis and in case of DNA extraction for sequencing analysis; three individual collections were mixed to homogeneity to generate a representative composite sample.

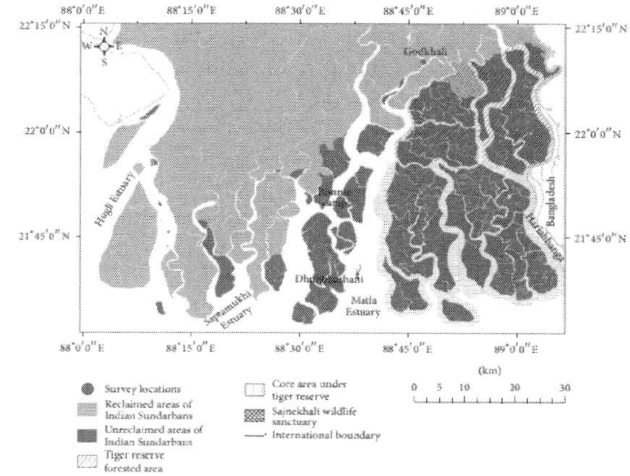

Figure 1: Geographical location of the sampling stations (Jharkhali-A, Sahidnagar-B, and Godkhali-C) in Indian Sundarbans. Coordinates of the sampling points and description of the stations are presented in Section 2.

Sediment Analyses and Site Climate

Microbiological and biochemical analyses were performed with the field moist sediments. Physical and chemical analyses were carried out with air-dried sediment samples. The sediment pH was measured in 1 : 2.5 sediment-water suspensions and found to be alkaline. The total organic carbon (TOC) was measured by the methods described previously [29, 30]. Briefly, TOC in a sample was determined by combusting the air-dried sediment sample catalytically in oxygen atmosphere into instrument chamber at 500°–900°C and the resulting carbon dioxide gas was detected by a nondispersive infrared (NDIR) detector in an Aurora TOC Analyzer that was calibrated to directly display the detected carbon dioxide mass. This mass was proportional to the TOC mass in the sediment sample and calculated as total mass of carbon per unit of sediment sample.

Conductivity and salinity were measured in situ with Hach Portable Meters (HQ40d). Measured salinity was expressed in parts per thousand (ppt) or gm Kg^{-1} as described previously [31]. Nutrients like inorganic nitrogen (ammonia, nitrite, and nitrate), soluble phosphate, and reactive silicate were measured after quantitative extraction in respective buffering conditions following standard methodologies [32]. Briefly, nitrite was measured after complexing with sulphanilamide followed by a coupling reaction with n(1-napthyl)-ethylenediamene dihydrochloride, which forms an azo dye upon coupling. The resulting azo dye was measured spectrophotometrically at 543 nm. The nitrate, in contrary, was quantitatively reduced to nitrite using cadmium (Cd) granules prior to measurement. The total nitrite was then measured spectrophotometrically as described earlier and further subtraction of the measured value of free nitrite in the sediment resulted in determination of nitrate in the sample. Ammonia was measured in a reaction with hypochlorite under alkaline condition, which results in formation of monochloramine. In a successive reaction with phenol and nitroprusside, monochloramine was converted into indophenol blue, which was measured spectrophotometrically at 630 nm. The soluble phosphate was measured using acidified molybdate reagent, which yields phosphomolybdate complex upon reaction with soluble phosphate. This complex was further reduced into molybdenum blue and measured spectrophotometrically at 880 nm. The reactive silicate was measured using the formation of yellow silicomolybdic acid in presence of molybdate under acidic condition.

Organic pollutants (polyaromatic hydrocarbons, PAH) were measured using a combined gas chromatography and mass spectrometry (GC/MS) method described previously [33, 34]. Heavy metals in the sediment samples were assessed using atomic absorption spectrophotometric technique (Agilent Technologies, CA, USA).

Sediment DNA Isolation

For 454-pyrosequencing, each of the sediment samples from a station (total n=3) and aliquots of homogenized sediment of 0.5 g were subjected to DNA extraction using the MoBio PowerSoil DNA Isolation Kit (MoBio Laboratories, Carlsbad, CA). After the extraction, DNA from all three samples from each sampling station was pooled together (approximately 200 ng of DNA from each extraction), and the pooled DNA was concentrated in a speed vacuum centrifuge (2,500 rpm, 30 min) to a final volume of 25 μL (approximately 24 ng μL^{-1}). A NanoDrop (Thermo Scientific, Wilmington, DE, USA) spectrophotometer was used to quantify the extracted pooled DNA and to measure other important parameters for DNA quality, such as the ratio of absorbance at 260/280 nm and 260/230 nm.

PCR Amplification of Archaeal 16S rRNA Gene

To analyze archaeal diversity, the V3–V5 region of archaeal 16S rRNA gene was amplified by PCR. The PCR reaction (25 μL) contained 5 μL of 5-fold Phusion GC buffer (Finnzymes, Vantaa, Finland), 200 μM of each of the four deoxynucleoside triphosphates, 1.5 mM $MgCl_2$, 4 μM of each primer (see Table S1 of the Supplementary Material available online at http://dx.doi.org/10.1155/2015/968582), 2.5% DMSO, 1 U of Phusion High Fidelity Hot Start DNA polymerase (Finnzymes), and 25 ng of pooled sediment DNA. The following thermal cycling scheme was used on a Veriti Thermal Cycler (Applied Biosystems, USA): initial denaturation at 98°C for 5 min, 25 cycles of denaturation at 98°C for 45 s, and annealing at 68°C for 45 s, followed by extension at 72°C for 30 s. The final extension was carried out at 72°C for 5 min. Negative controls were performed by using the reaction mixture without template. Primer sequences for amplification of the V3–V5 region [35] as well as 454 adaptors with the unique MIDs for each sample are listed in Table S1.

Library Preparation

PCR amplicons were evaluated by electrophoresis on a 1.5% agarose gel. The amplicon library was purified with Agencourt AMPure XP beads (Beckman Coulter

Inc., Canada) and quantified by fluorometry using the Quant-iT PicoGreen dsDNA Assay Kit (Invitrogen, Burlington, ON) according to the Roche 454 "Amplicon Library Preparation Method Manual" of the GS Junior Titanium Series (454 Life Sciences, USA). Pooled amplicons were diluted as recommended and amplified by emulsion PCR on a Thermal Cycler 9700 (Applied Biosystem) according to the Roche 454 "emPCR Amplification Method Manual Lib-L" (454 Life Sciences, USA).

Sequencing and Data Processing

Pyrosequencing was performed for 200 cycles on a Roche 454 GS Junior sequencing instrument according to the manufacturer's protocol (454 Life Sciences, USA). All reads were filtered using the standard read rejecting filters of the GS Junior sequencer, namely, key pass filters, dot and mixed filters, signal intensity filters, and primer filters (454 Sequencing System Software Manual, V 2.5.3, 454 Life Sciences, USA). Raw sequencing and processing data was carried out using a combination of mothur (software version 1.28.0) and Ribosomal Database Project (RDP-II). The raw data were subjected to initial quality trimming using the mothur software and all reads having an average quality value of <20 were discarded. Then we removed the number of chimeric sequences by using UCHIME algorithm. Then the chimeric free reads were further screened for the presence of ambiguous bases and any reads which have a length <200 were discarded and forward primer sequence was removed from the final dataset. The obtained processed reads were then demultiplexed to separate sample based on the 10 bp MID sequence and any reads which do not match the MID were discarded. The high quality reads were then aligned to archaeal 16S rRNA sequences and clustering was performed at 97% similarity; using the RDP pipeline, a representative sequence from each cluster was selected based on abundance using the sequence selection tool. The representative sequences obtained from each cluster were then classified using the naïve Bayesian Classifier (Ribosomal Data Project, release 10) at a bootstrap confidence of 80.

Taxonomic Annotation

After quality trimming all the sequence reads were aligned against the SILVA, which is a curated ribosomal RNA sequence database using BLASTN [36, 37]. The resulting output files of read sequences were imported and analyzed using the paired-end protocol of MEGAN [38] to obtain taxonomic profiles as described earlier [39]. When processing the BLAST output files by MEGAN we used parameter settings of "Min Score = 35," "Top Percent = 100," "Min Support = 5," and "Minimum Sequence Complexity Threshold = 0.44." Reads which did not have any match to the respective database were placed under "No hit" node. Any reads that were originally assigned to a taxon that did not meet our selected threshold criterion were pushed back using the lowest common ancestor (LCA) algorithm to higher nodes where the threshold was met. After importing the datasets in MEGAN we obtained a set of MEGAN-proprietary "rma files" for each data mapped onto the NCBI taxonomy based on our selected threshold for tree visualization purpose.

All six RMA files were normalized to the smallest data set size (without including reads classified as not having a taxonomic assignment) to allow intersample comparison of taxonomic abundances and to obtain comparative tree view for all samples. Comparative abundance of read counts at different levels of NCBI taxonomy (class, order, family, genus, and species) was exported from all sample comparison-file for later statistical analyses.

Statistical Analyses

Hierarchical cluster analyses have been performed using R 3.0.2 [40], with species abundance data where Pearson's correlation is used for clustering the probes/genes (rows) and Spearman rank correlation is used for clustering the sample datasets (cols), with complete linkage.

Composition (presence/absence) data were also analyzed to compare the community structure. SIMPER analyses were used to determine which taxa/function contributed most to the observed differences. Further data were analyzed using common multivariate ordination techniques: Nonmetric Multidimensional Scaling (NMDS) using Bray Curtis distance. To understand the influence of physicochemical parameters in different sampling stations, Principal Component Analysis (PCA) was performed. Additionally, we have performed correspondence analyses on genomic data as well as on PAH (polyaromatic hydrocarbons) data.

We have also plotted the genomic data using Voronoi diagram, which is a partitioning of a plane into regions based on distance to points in a specific subset of the plane. In this plot we have highlighted 6 major phyla.

Nucleotide Sequence Accession Numbers

All 454-GS Junior sequence data from this study were submitted to the NCBI Sequence Read Archive (SRA) under accession numbers SRR1632262 (Godkhali_surface), SRR1632263 (Godkhali_subsurface), SRR1632258 (Bonnie camp_surface), SRR1632259 (Bonnie camp_subsurface), SRR1632260 (Dhulibhashani_surface), and SRR1632261 (Dhulibhashani_subsurface).

RESULTS

Environmental Parameters

Physicochemical characteristics such as pH, dissolved oxygen, saturation, total carbon, total nitrogen, and salinity were estimated for surface and subsurface samples from all the three stations in the present study (Table1 and Table S2). The environmental parameters, for example, salinity, temperature, dissolved oxygen, pH, and saturation were measured in situ (Table 1). Temperature and salinity ranged from 29.7 to 32.8°C and from 21 to 22.8 psu, respectively. The lowest temperature and highest salinity were recorded in Dhulibhashani.

Table 1: Environmental parameters of sampling sites analyzed in this study

Sample	Site	Latitude °N	Longitude °E	Depth (cm)	T (°C)*	Salinity (psu)*	pH*	DO (mg/L)*
Surface	Godkhali	22°06′32.570″	88°46′22.220″	2	31.30	21.00	7.88	7.32
Subsurface	Godkhali	22°06′32.570″	88°46′22.220″	32	31.80	20.80	7.85	7.24
Surface	Bonnie camp	21°49′53.581″	88°36′44.860″	2	32.10	21.60	7.98	7.15
Subsurface	Bonnie camp	21°49′53.581″	88°36′44.860″	32	32.50	21.40	7.97	7.05
Surface	Dhulibhashani	21°37′40.837″	88°33′47.762″	2	29.70	22.80	7.96	7.15
Subsurface	Dhulibhashani	21°37′40.837″	88°33′47.762″	32	30.10	22.10	7.95	7.02

*Average of three independent measurements, $n = 3$.

Measurement of NO_2^-, NO_3^-, and NH_4^+ in the surface and subsurface samples from all the three sampling stations revealed a differential pattern for these nitrogen species (Table S2). Ammonia was found maximum in the sediment samples of Godkhali, while it was minimum in the sediment of Dhulibhashani ($P<0.05$). Nitrite was found negligible in all the stations confirming very high rate of nitrification, resulting in nitrate formation, which was evenly detected in all three sampling stations (Table S1). Other parameters such as phosphate, sulphate, and silicate were found evenly distributed in all the sampling stations possibly due to regular inundation.

Heavy metals have been shown to play a pivotal anthropogenic role in the marine environment. Previously, it has been shown that heavy metals like Fe, Zn, Cu, Mn, Hg, Pb, Ni, Cr, and Cd are present in the Sundarbans estuaries [41–44]. To ascertain the global ecological impact of different heavy metals in the present study, concentrations of heavy metals like Fe, Cd, Zn, Cu, Ni, and Pd were measured in surface and subsurface samples collected from all the three stations (Table S2). Heavy metal estimates for the Sundarbans sediment found in our study were fairly consistent with previously found results [45].

Archaeal Community Structure Revealed by 16S rRNA Gene Based Analysis

We analyzed archaeal 16S rRNA gene (SSU) amplicons prepared from six sediment samples originated from three sampling stations at two different depths. After quality filtering, denoising, and removal of potential chimeric sequences of pyrosequencing dataset, this resulted in recovery of 1,41,816 sequences with an average length ranges between 500 and 510 bases (506 bp average) (Table 2). Within our sequence dataset, we were able to assign 1,39,953 sequences to the domain archaea and to classify all of these sequences below domain level using BLASTN 2.2.25+ in SILVA database and MEGAN (Table 2). The classified sequences were affiliated to two archaeal phyla and five archaeal classes or similar phylogenetic group (Figure 2 and Figure S1). Euryarchaeota often was the most abundant phylum (36–60%) and Thermoplasmata was the predominant class across all samples (39–62%) (Figure S1). Besides Euryarchaeota, Thaumarchaeota (Marine group I) was found to be highly abundant in all the samples (Table S3). Within euryarchaeal sequences, a number of members of the class Halobacteria, for example, Halosarcina, Halorientalis, Halolamina, Halorhabdus, Halogranum,Haloferax, Halomarina, Halorussus, Haloplanus, and Halarchaeum, and of the class Methanomicrobia, for example, Methanosarcina, Methermicoccus, Methanocella, Methanococcoides, Methanosalsum, Methanolobus, and Methanogenium were detected (Figure 2). Besides, few sequences were also affiliated to classes Thermoplasmata and Methanobacteria. Unfortunately, within our datasets, we could not detect any representative of the archaeal phyla Crenarchaeota, Korarchaeota, Nanoarchaeota, and Nanohaloarchaeota due to lower coverage of the degenerate primers used in this study.

Table 2: Sample statistics

Samples	Total number of reads sequenced	Assigned against SILVA using BLASTN 2.2.25+ and MEGAN
Godkhali_surface	15,939	15,647
Godkhali_subsurface	23,437	23,091
Bonnie camp_surface	15,371	15,120
Bonnie camp_subsurface	15,788	15,654
Dhulibhashani_surface	33,771	33,395
Dhulibhashani_subsurface	37,510	37,046

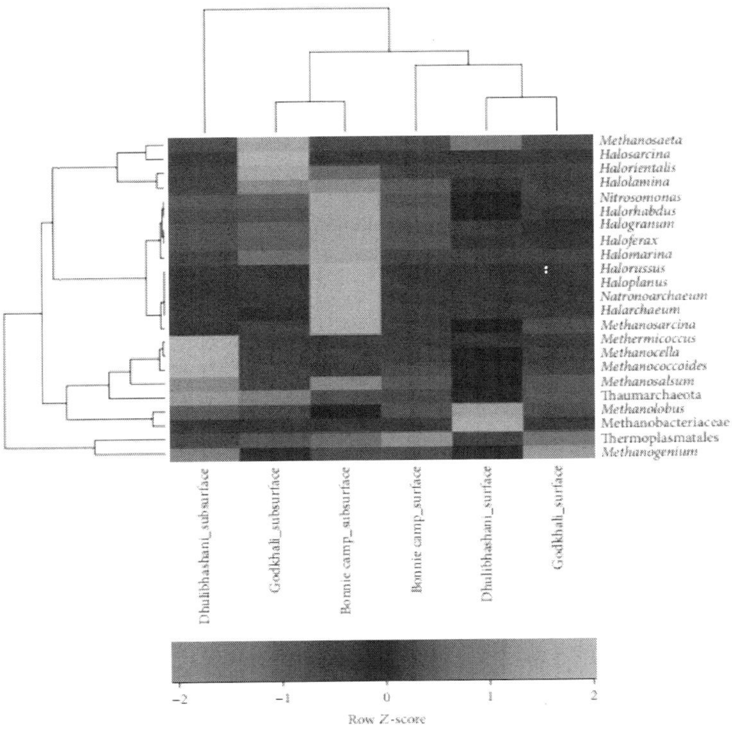

Figure 2: Hierarchical clustering (complete-linkage) heat map generated from taxonomic abundance profile (using Spearman's rank correlation) reflecting spatial distribution of archaeal class in the sediment of Sundarbans (using Pearson's correlation).

Diversity and Species Richness of Archaeal Communities

A diversity index is a mathematical measure of species diversity in a community. Diversity indices provide important information about community composition rather than only species richness (i.e., the number of species present); and simultaneously they also take the relative abundances of different species into account. We evaluated the archaeal diversity and evenness using the normalized dataset. Considering all the sampling sites, the Shannon-Weaver (H′) index varied from 0.67 to 1.08 and the Simpson diversity $(1 - \lambda')$ index varied from 0.29 to 0.52 (Table 3). Such variations of diversity indicated that the archaeal diversity is higher at the surface samples both in Bonnie camp and Godkhali. However, in Dhulibhashani, archaeal diversity was found to be almost similar both at the surface and at subsurface level with a marginally higher diversity in subsurface sample. The Pielou index (J′) was between 0.23 and 0.37 and the Margalef (d) index was between 1.52 and 3.69 (Table 3). Diversity and evenness are more informative for describing community composition than simple phylotype richness levels. Community diversity, as reflected by the Shannon index, was highest in Bonnie camp_subsurface and lowest in Godkhali_subsurface and is by definition generally correlated positively with the number of unique phylotypes and/or with greater community evenness.

Table 3: Univariate diversity indices

Sample	Total species (S)	Total individuals (N)	Species richness (Margalef): $d = (S-1)/\text{Log}(N)$	Pielou's evenness $J' = H'/\text{Log}(S)$	Shannon $H' = -\text{SUM}(P_i * \text{Log}(P_i))$	Simpson $(1 - \text{Lambda}') = 1 - \text{SUM}(N_i * (N_i - 1)/N * (N - 1))$
Godkhali_surface	11	100	2.171	0.3447	0.8264	0.5206
Godkhali_subsurface	17	100	3.474	0.2384	0.6753	0.2998
Bonnie camp_surface	8	100	1.52	0.3597	0.748	0.5104
Bonnie camp_subsurface	18	100	3.692	0.3763	1.088	0.4389
Dhulibhashani_surface	16	100	3.257	0.3454	0.9575	0.4761
Dhulibhashani_subsurface	13	100	2.606	0.357	0.9157	0.4858

Taxonomic Assignments and Statistics

In addition to surveying archaeal diversity at the surface and subsurface sediments of three sites in Sundarbans, pyrosequencing also allowed assessment of the relative abundance of the taxonomic levels of archaea detected. A total of 23 genera (genus/order/family) were commonly shared between six samples (Figure 2 and Figure S1). Interestingly, the dominant class/phylum showed some geographical characteristics. For example, Halobacteria were highly abundant in the subsurface layers of the representative samples from Godkhali and Bonnie camp. In contrary, Methanomicrobia were only dominant in the subsurface sediment of Dhulibhashani.

Cluster analysis showed that the archaeal community detected in the subsurface sediment of Dhulibhashani was the most dissimilar of all the sediment samples tested (Figure 2). Additionally, among the subsurface sediments, the most diverse archaeal community was detected in this sediment. Among others, there is a clear separation of two types of samples, surface and subsurface. The surface sediments have shown an overall similarity among Godkhali and Bonnie camp while Dhulibhashani surface showed a different community profile.

Further we have performed Nonmetric Multidimensional Scaling (NMDS) to compare the community composition using presence/absence data. NMDS employs an iterative algorithm that optimizes the position of samples into a low-dimensional space minimizing the difference between rank-order of multivariate ecological dissimilarity and distances in NMDS space (Figure 3). From the figure it is easily observed that despite the presence of very few abundant taxa in more than one sample, surface and subsurface communities from different sampling stations or locations (Godkhali and Bonnie camp) were more closely related to each other than to samples from the same location. Samples from Dhulibhashani show similar results to cluster analyses, showing much different community structure than the other locations for both surface and subsurface samples. To this end we believe that the presence or absence of certain taxonomic groups in the sediment of Dhulibhashani is probably due its proximity to the open ocean (Bay of Bengal) and the nutritional status of the sediment.

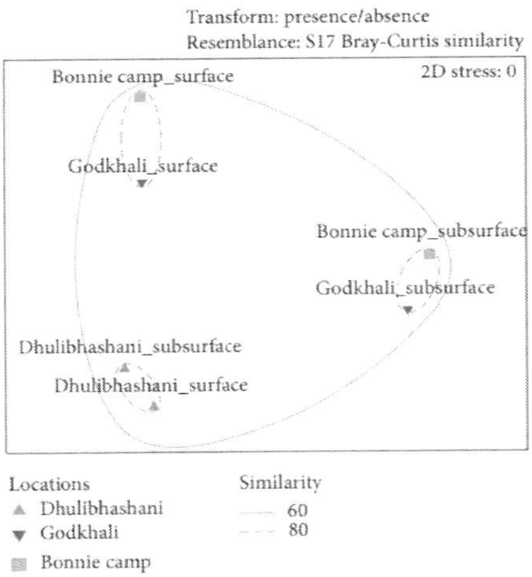

Figure 3: Archaeal community compositional structure in the sediments of Sundarbans indicated by Nonmetric Multidimensional Scaling (NMDS) using Bray-Curtis distance. Gene abundance data was transformed based on presence/absence before creating Bray-Curtis resemblance matrix for NMDS analysis (showing similarity based on community composition only).

Furthermore, our PCA analysis revealed that the physicochemical parameters influence sampling stations differentially. The first PCA axis showed high positive correlations with heavy metals Zn, Pb, and Cu (Figure4(a)). Samples from Bonnie camp showed high positive correlations with all of these heavy metals. In contrary, samples from Godkhali showed positive correlations with heavy metals Cd and Fe (Figure 4(a)). To this end, PCA analysis revealed that both of these sampling stations at Godkhali and Bonnie camp were at an elevated risk compared to Dhulibhashani, regarding heavy metal pollution. Notably, these two sampling stations also showed positive correlations with pH, DO, silicate, nitrite, ammonia, saturation, and total nitrogen (Figure4(a)). Dhulibhashani, on the other hand, showed positive correlations with TOC, nitrate, sulphate, and salinity (Figure 4(a)).

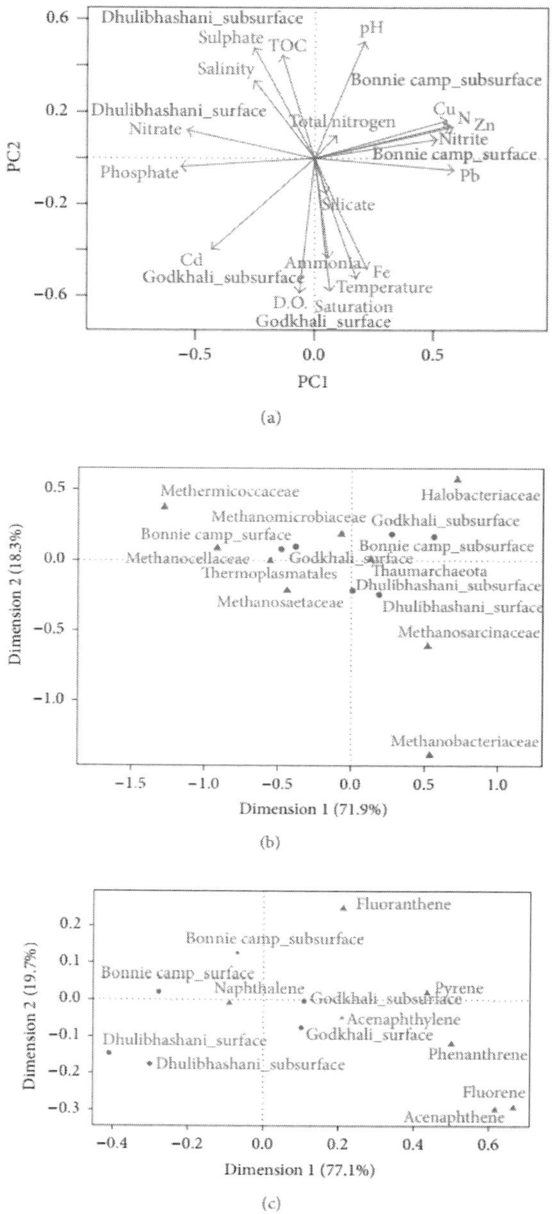

Figure 4: (a) Principal Component Analysis (PCA) of samples based on environmental parameters, nutrient, and heavy metal levels in the sediments samples. (b) Correspondence analysis (CA) based on distribution of major taxonomic lineages generated analyzing genomic data in the sediment samples. (c) Correspondence analysis (CA) based on distribution of polyaromatic hydrocarbons (PAH) in the sediment samples.

Further, we have performed correspondence analysis (CA) to compare the influence of abundant taxonomic groups and polyaromatic hydrocarbons (PAH) in these sampling sediments (Figures 4(b) and 4(c)). The correspondence analysis comparing abundant taxonomic groups in the sampling stations clearly indicated that the subsurface sediments from Godkhali and Bonnie camp were heavily dominated by Halobacteriaceae (Figure 4(b)), whereas surface archaeal communities in these sampling stations were dominated by Methanomicrobiaceae, Methermicoccaceae, and Methanocellaceae (Figure 4(b)). Both the surface and subsurface archaeal communities in Dhulibhashani were heavily dominated by Methanosarcinaceae and Methanobacteriaceae (Figure 4(b)). To summarize, methanogens were commonly encountered in the surface sediments of all three stations. However, while Halobacteriaceae dominated the subsurface archaeal population in Godkhali and Bonnie camp, methanogenic Methanosarcinaceae and Methanobacteriaceae dominated the subsurface sediment of Dhulibhashani.

Correspondence analysis, performed to understand the influence of PAHs in the sediments of the sampling stations, revealed that both the surface and subsurface sediments of Godkhali and Bonnie camp correlated with the detected PAH congeners (Figure 4(c)). In contrary, surface and subsurface sediments of Dhulibhashani showed little correlation to any of the identified PAHs. To this end, we believe that the pattern of correlation of PAHs in the sediments of Godkhali and Bonnie camp is in agreement with the detection of Halobacteriaceae, the most active hydrocarbon degrading archaeal representative, in the subsurface sediments of these stations. In general, hydrocarbons, by virtue of being hydrophobic in nature, are scarcely soluble in water and tend to adsorb readily onto organic matter, particularly sediments. In an environment like Sundarbans, surface sediments are in general more dynamic due to regular inundation. The hydrocarbons, therefore, remained mostly absorbed in the sediments and our data clearly shows that subsurface sediments contain more PAHs compared to surface sediments.

Finally, we have generated Voronoi plot (Figure 5), which describes the proportionate taxonomic composition pattern for each sample as a partition of a plane.

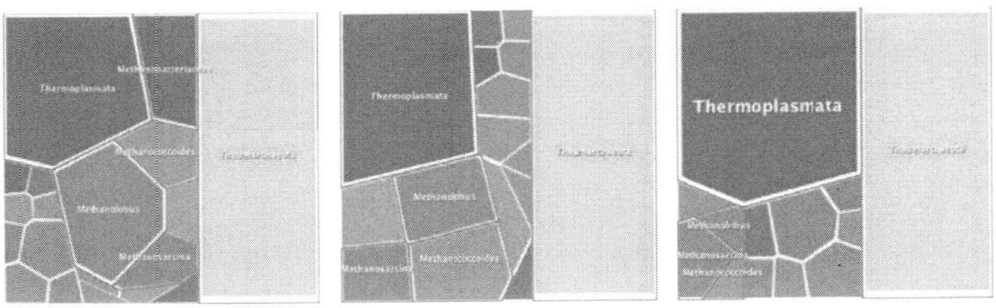

Dhulibhashani_surface Dhulibhashani_subsurface Bonnie camp_surface

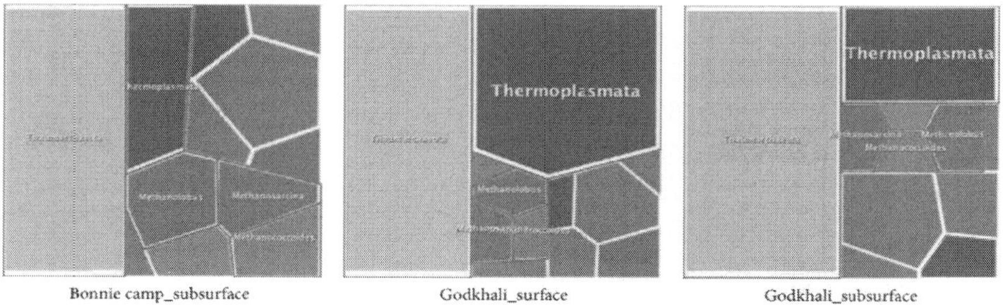

Figure 5: Voronoi diagram depicting comparative taxonomic profile composition pattern for each sample as a partition of a plane highlighting 6 major phyla.

Multivariate SIMPER (Similarity Percentage) Analysis

To reveal differences between three sampling stations, Godkhali, Bonnie camp, and Dhulibhashani, and to identify major taxonomic groups that contributed to the similarities or dissimilarities, SIMPER (Similarity Percentage) analysis was performed (Tables 4(a), 4(b), and 4(c)). At the regional scale, there were differences observed at order/family/genus level between three sampling stations. Overall, distribution of taxa was found similar in Godkhali and Bonnie camp with Thermoplasmatales being the major contributor towards the similarity (23.3%), and with only exception of Methanosarcina, the relative abundance of which was higher in Bonnie camp. In general, SIMPER analysis targeting relative abundance of taxa in different stations revealed higher relative abundance of class Halobacteria in Godkhali and Bonnie camp, while class Methanomicrobia was found relatively more abundant in Dhulibhashani (Tables 4(b) and 4(c)). Representatives of the class Methanomicrobia, such as Methanolobus, Methanococcoides, Methermicoccus, and Methanocella, were either higher or only present in Dhulibhashani. In contrary, representatives of class Halobacteria such as Halogranum,Halorientalis, Haloferax, and Halarchaeum were either higher or only present in Godkhali/Bonnie camp (Table4(a)). Putative ammonia-oxidizing archaea in the phylum Thaumarchaeota were highly abundant in all three sampling sites. To our surprise, the order Thermoplasmatales representing acidophilic Euryarchaeota were highly abundant in all three sampling stations despite the fact that the recorded pH fell between 7.85 and 7.98 in these sites.

Table 4: Results of SIMPER (Similarity Percentage) analyses indicating the contribution of specific taxa to observed differences in archaeal community structure among the stations. (a) shows differences between Godkhali and Bonnie camp; (b) shows differences between Dhulibhashani and Godkhali; and (c) shows differences between Dhulibhashani and Bonnie camp

(a) Godkhali versus Bonnie camp

Taxon[1]	Order/family/genus	Contribution[2] (%)	Average[3] % abundance in Godkhali	Average % abundance in Bonnie camp	Average dissimilarities
Thermoplasmata	Thermoplasmatales	23.30	4.85	4.64	5.23
Halobacteria	*Halogranum*	11.72	1.50	1.64	2.63
Thaumarchaeota	Thaumarchaeota	10.08	8.29	7.99	2.26
Methanomicrobia	*Methanolobus*	8.33	0.65	0.99	1.87
Methanomicrobia	Methanococcoides	6.46	0.40	0.82	1.45
Halobacteria	*Haloferax*	5.42	0.48	0.47	1.22
Halobacteria	*Natronomonas*	5.41	0.62	0.73	1.21
Methanomicrobia	*Methanosarcina*	5.00	0.55	1.03	1.12
Halobacteria	*Halorhabdus*	3.97	0.35	0.35	0.89
Halobacteria	*Halolamina*	3.49	0.38	0.30	0.78
Methanomicrobia	*Methanosalsum*	3.46	0.13	0.33	0.78
Halobacteria	*Halarchaeum*	3.17	0.08	0.30	0.71
Halobacteria	*Halorientalis*	2.17	0.19	0.16	0.49

(b) Dhulibhashani versus Godkhali

Taxon[1]	Order/family/genus	Contribution[2] (%)	Average[3] % abundance in Dhulibhashani	Average % abundance in Godkhali	Average dissimilarities
Methanomicrobia	*Methanolobus*	19.80	2.34	0.65	4.28
Thermoplasmata	Thermoplasmatales	18.75	4.59	4.85	4.05
Methanomicrobia	Methanococcoides	12.14	1.43	0.40	2.62
Thaumarchaeota	Thaumarchaeota	9.64	8.29	8.29	2.08
Halobacteria	*Halogranum*	8.80	0.75	1.50	1.90
Methanomicrobia	*Methermicoccus*	4.02	0.34	0.00	0.87
Methanomicrobia	*Methanosalsum*	3.84	0.45	0.13	0.83
Halobacteria	*Haloferax*	3.58	0.18	0.48	0.77
Halobacteria	*Natronomonas*	3.49	0.42	0.62	0.75
Methanomicrobia	*Methanocella*	2.47	0.21	0.00	0.53
Halobacteria	*Halorientalis*	2.20	0.00	0.19	0.47
Halobacteria	*Halolamina*	1.90	0.24	0.38	0.41

(c) Dhulibhashani versus Bonnie camp

Taxon[1]	Order/family/genus	Contribution[2] (%)	Average[3] % abundance in station A	Average % abundance in station B	Average dissimilarities
Thermoplasmata	Thermoplasmatales	20.76	4.59	4.64	5.12
Methanomicrobia	*Methanolobus*	14.23	2.34	0.99	3.51
Halobacteria	*Halogranum*	10.05	0.75	1.64	2.48
Methanomicrobia	Methanococcoides	8.65	1.43	0.82	2.13
Thaumarchaeota	Thaumarchaeota	6.72	8.29	7.99	1.66
Halobacteria	*Natronomonas*	4.73	0.42	0.73	1.17
Halobacteria	*Haloferax*	4.42	0.18	0.47	1.09
Methanomicrobia	*Methanosarcina*	3.96	0.70	1.03	0.98
Methanomicrobia	*Methermicoccus*	3.42	0.34	0.00	0.84
Methanomicrobia	*Methanosalsum*	3.40	0.45	0.33	0.84
Halobacteria	*Halorhabdus*	3.37	0.24	0.35	0.83
Halobacteria	*Halolamina*	2.92	0.24	0.30	0.72
Halobacteria	*Halarchaeum*	2.70	0.00	0.30	0.67
Methanomicrobia	*Methanocella*	2.10	0.21	0.00	0.52

[1]Phylum or class level for archaea.
[2]Contribution of each taxon to the overall dissimilarity between these two clusters.
[3]Average abundance of each taxon in the two clusters.

DISCUSSION

In marine environments, microorganisms play an important role in sustainable maintenance of the ecosystem. They are key mediators of global biogeochemical cycling of the essential elements in the marine environments. Recent advancement of molecular microbial ecology has emerged as multiple studies on the diversity and abundance of microorganisms in these habitats and their role in shaping the sustainable ecosystem. Most of these studies focused on marine bacteria and archaea, whereas little is known on the diversity and ecology of mangrove archaea [16, 17, 23, 24]. Interestingly, archaea have been found to be ubiquitous and abundant members of the microbiome in diverse marine environments including coastal waters [46, 47], marine sediments, estuaries [48–50], stratified basins [7, 51], mangrove sediments [52–54], and open ocean water columns [47, 55]. In marine habitat archaea play crucial roles in nitrification [54, 56, 57], sulfur metabolism [58], methane oxidation (ANME) [59], and methanogenesis [60].

The present study focused on accessing the archaeal community structure and diversity in the world's largest tropical mangrove sediments of Sundarbans using 16S rRNA gene-based 454-amplicon sequencing. To our knowledge, this is the first study on understanding archaeal diversity in Sundarbans ecosystem. The majority of sequences obtained were affiliated to the Euryarchaeota. Previous studies by Sapp et al. on marine sediment and Wemheuer et al. on German Bight found high abundance of members of Euryarchaeota [35]. We identified Thermoplasmatales as the most abundant euryarchaeal group in the investigated samples. Most sequences were affiliated to the MKCST-A3, CCA47, and VC2.1Arc6 groups. The MKCST-A3 group was first identified in the mangrove sediment of China and most of the members reported so far are uncultured archaeon [16]. The CCA47 group was originally identified by 16S rRNA gene analysis of oxygen-depleted marine environment and anoxic subsaline sediments. The number of sequences within this group was also affiliated to the Marine Group II (Euryarchaeota). Previous studies by DeLong and Karl have suggested that the members of Marine Group II are more abundant in temperate sea water than Marine Group I (Thaumarchaeota) [61]. In our case, however, overall abundance of Marine Group I (Thaumarchaeota) is higher compared to Marine Group II (Euryarchaeota) possibly emphasizing geographical differences between tropical and temperate marine sediment microbial community. Moreover, this might also be due to the tidal current that influences the microbial community structure within Sundarbans sediment. The regular inundation of the subtidal zones within Sundarbans whirls up microbial cells to the water column. An in-depth analysis of the archaeal diversity and abundance in the water column might be useful to know whether we could detect a large number of sequences affiliated to Marine Group II. Unfortunately, only limited studies have been performed targeting archaeal communities in the water column in tropical mangrove. Due to such knowledge gap, the habitat preference of Marine Group II cannot be addressed properly at this time. The Thermoplasmatales cluster VC2.1Arc6 was originally described within the microbial community obtained from in situ growth chamber placed on a deep-sea hydrothermal vent on the Mid-Atlantic Ridge [62]. Besides Thermoplasmatales, the number of euryarchaeal sequences was affiliated to Halobacteria. Members of this group can grow aerobically as well as anaerobically. The majority of the halobacterial sequences analyzed in the present study were affiliated to the Halogranum genus. Besides, a significant number of sequences affiliated to Natronomonas, Haloferax,Halorhabdus, and Halolamina were identified. Methanomicrobia and Methanobacteria were the other two abundant euryarchaeal classes in the investigated samples. Methanomicrobia and Methanobacteria are known for their putative importance in sulfate reduction and methanogenesis in anoxic marine sediments, such as mangroves. The members of these classes are methanogens and are involved in carbon-cycle through methanogenesis.

Beside euryarchaeota, another archaeal group found in all samples was the Marine Group I (MG1) or Thaumarchaeota. Marine Group I (MGI) Thaumarchaeota are one of the most abundant and cosmopolitan chemoautotrophs and prokaryotic picoplankton within the global deep sea water. They were originally identified by sequencing of environmental 16S rRNA genes derived from sea water. All organisms of this lineage thus far identified are chemolithoautotrophic ammonia-oxidizers and may play important roles in biogeochemical cycles, such as the nitrogen cycle and the carbon cycle. Within our sequence dataset, the number of ammonia oxidizing Thaumarchaeota was identified.

We observed an increased abundance of Halobacteriaceae in the investigated subsurface samples from Godkhali and Bonnie camp. In contrary, the number of Methanosarcinaceae and Methanobacteriaceae was higher in the samples derived from Dhulibhashani. This might be correlated with the high amounts of organic matter in resident algal blooms in Godkhali and Bonnie camp (Figures 4(b) and 4(c)). Furthermore, correspondence analysis revealed that detected PAHs were mostly correlated to the surface and subsurface sediments of Godkhali and Bonnie camp. Previous studies from our group have shown that in Godkhali and Bonnie camp, because of the presence of jetty for transportation, large amount of hydrocarbon-derived chemicals and oils are released in water and in turn increase algal blooms in surrounding areas (Personal Communication). To this end, we believe that abundance of Halobacteriaceae in Godkhali and Bonnie camp corroborates well to inflated detection PAHs in the sediment. In hypersaline environment, Halobacteria are the most active organisms capable of organic matter degradation. Thus the higher abundance of Halobacteria in Godkhali and Bonnie camp might indicate an involvement in marine organic matter degradation under high nutrient conditions found during algal blooms. Similar results were reported by Teeling et al., where they demonstrated that bacterial community structures were highly influenced by the presence of an algal bloom [63]. Furthermore, Wemheuer et al. have recently demonstrated that archaeal diversity was influenced by algal blooms in German Bight [35]. Taken together, present observation indicates that marine microbial communities are influenced by algal blooms or by environmental parameters correlated with bloom presence.

CONCLUSIONS

In summary, the spatial variations of the sedimentary archaeal diversity have been studied in the Sundarbans mangrove ecosystem for the first time by means of 16S rRNA 454-pyrosequencing. We found highly diverse archaeal communities in the sediment of Sundarbans. Our analyses have confirmed the influence of environment in shaping the archaeal community diversity within the sediment. However, due to the lack of pure cultures and large comparative investigations, robust conclusions related to the involvement of identified archaeal communities in biogeochemical cycle cannot be drawn. Further research including analyses of expression of functional genes and determination of the active archaeal population within the sediment might unravel the role of archaea in Sundarbans mangrove ecosystem.

ACKNOWLEDGMENTS

The authors would like to acknowledge the instrument facility provided by World Bank, ICZM Project, in the Department of Biochemistry, University of Calcutta, India. Anish Bhattacharyya, Debojyoti Roy, and Sudip Nag were supported with Research Fellowships from the ICZM Project, World Bank. Abhrajyoti Ghosh was supported by Ramanujan Fellowship from Department of Science and Technology, India (SR/S2/RJN-106/2012) and an extramural grant (no. 38(1391)/14/EMR-II) from the Council of Scientific & Industrial Research (CSIR), Government of India.

REFERENCES

1. S.-V. Albers, P. Forterre, D. Prangishvili, and C. Schleper, "The legacy of Carl Woese and Wolfram Zillig: from phylogeny to landmark discoveries," Nature Reviews Microbiology, vol. 11, no. 10, pp. 713–719, 2013. · ·
2. K. F. Jarrell, A. D. Walters, C. Bochiwal, J. M. Borgia, T. Dickinson, and J. P. J. Chong, "Major players on the microbial stage: why Archaea are important," Microbiology, vol. 157, no. 4, pp. 919–936, 2011. · ·
3. E. F. DeLong, "Archaea in coastal marine environments," Proceedings of the National Academy of Sciences of the United States of America, vol. 89, no. 12, pp. 5685–5689, 1992. · ·
4. K. Takai and K. Horikoshi, "Genetic diversity of archaea in deep-sea hydrothermal vent environments,"Genetics, vol. 152, no. 4, pp. 1285–1297, 1999. ·
5. J. L. Macalady, D. S. Jones, and E. H. Lyon, "Extremely acidic, pendulous cave wall biofilms from the Frasassi cave system, Italy," Environmental Microbiology, vol. 9, no. 6, pp. 1402–1414, 2007. · ·
6. A. Gittel, K. B. Sørensen, T. L. Skovhus, K. Ingvorsen, and A. Schramm, "Prokaryotic community structure and sulfate reducer activity in water from high-temperature oil reservoirs with and without nitrate treatment," Applied and Environmental Microbiology, vol. 75, no. 22, pp. 7086–7096, 2009. · ·
7. M. J. L. Coolen, B. Abbas, J. van Bleijswijk et al., "Putative ammonia-oxidizing Crenarchaeota in suboxic waters of the Black Sea: a basin-wide ecological study using 16S ribosomal and functional genes and membrane lipids," Environmental Microbiology, vol. 9, no. 4, pp. 1001–1016, 2007. · ·
8. J. Borneman and E. W. Triplett, "Molecular microbial diversity in soils from eastern Amazonia: evidence for unusual microorganisms and microbial population shifts associated with deforestation," Applied and Environmental Microbiology, vol. 63, no. 7, pp. 2647–2653, 1997. ·
9. D. H. Buckley, J. R. Graber, and T. M. Schmidt, "Phylogenetic analysis of nonthermophilic members of the kingdom Crenarchaeota and their diversity and abundance in soils," Applied and Environmental Microbiology, vol. 64, no. 11, pp. 4333–4339, 1998. ·

10. G. Jurgens, K. Lindström, and A. Saano, "Novel group within the kingdom Crenarchaeota from boreal forest soil," Applied and Environmental Microbiology, vol. 63, no. 2, pp. 803–805, 1997. ·
11. R. Massana, A. E. Murray, C. M. Preston, and E. F. DeLong, "Vertical distribution and phylogenetic characterization of marine planktonic Archaea in the Santa Barbara Channel," Applied and Environmental Microbiology, vol. 63, no. 1, pp. 50–56, 1997. ·
12. C. Schleper, W. Holben, and H.-P. Klenk, "Recovery of crenarchaeotal ribosomal DNA sequences from freshwater-lake sediments," Applied and Environmental Microbiology, vol. 63, no. 1, pp. 321–323, 1997. ·
13. E. F. DeLong and N. R. Pace, "Environmental diversity of bacteria and archaea," Systematic Biology, vol. 50, no. 4, pp. 470–478, 2001. · ·
14. K. Knittel, T. Lösekann, A. Boetius, R. Kort, and R. Amann, "Diversity and distribution of methanotrophic archaea at cold seeps," Applied and Environmental Microbiology, vol. 71, no. 1, pp. 467–479, 2005. · ·
15. R. Margesin and V. Miteva, "Diversity and ecology of psychrophilic microorganisms," Research in Microbiology, vol. 162, no. 3, pp. 346–361, 2011. · ·
16. B. Yan, K. Hong, and Z.-N. Yu, "Archaeal communities in mangrove soil characterized by 16S rRNA gene clones," Journal of Microbiology, vol. 44, no. 5, pp. 566–571, 2006. ·
17. T. J. Lyimo, A. Pol, M. S. M. Jetten, and H. J. M. Op den Camp, "Diversity of methanogenic archaea in a mangrove sediment and isolation of a new Methanococcoides strain: research Letter," FEMS Microbiology Letters, vol. 291, no. 2, pp. 247–253, 2009. · ·
18. G. Holguin, P. Vazquez, and Y. Bashan, "The role of sediment microorganisms in the productivity, conservation, and rehabilitation of mangrove ecosystems: an overview," Biology and Fertility of Soils, vol. 33, no. 4, pp. 265–278, 2001. · ·
19. S. Das, M. De, T. K. Jana, and T. K. De, "Environmental influence on cultivable microbial community in the sediment of Sundarban mangrove forest, India," African Journal of Microbiology Research, vol. 7, no. 38, pp. 4655–4665, 2013. ·
20. S. Das, P. S. Lyla, and S. A. Khan, "Marine microbial diversity and ecology: importance and future perspectives," Current Science, vol. 90, no. 10, pp. 1325–1335, 2006. ·
21. A. Ghosh, N. Dey, A. Bera et al., "Culture independent molecular analysis of bacterial communities in the mangrove sediment of Sundarban, India," Saline Systems, vol. 6, no. 1, article 1, 2010. · ·
22. A. Chakraborty, A. Bera, A. Mukherjee et al., "Changing bacterial profile of Sundarbans, the world heritage mangrove: impact of anthropogenic interventions," World Journal of Microbiology and Biotechnology, vol. 31, no. 4, pp. 593–610, 2015. · ·

23. P. Basak, N. S. Majumder, S. Nag et al., "Spatiotemporal analysis of bacterial diversity in sediments of Sundarbans using parallel 16S rRNA gene tag sequencing," Microbial Ecology, vol. 69, no. 3, pp. 500–511, 2014. · ·
24. X.-T. Jiang, X. Peng, G.-H. Deng et al., "Illumina Sequencing of 16S rRNA Tag Revealed Spatial Variations of Bacterial Communities in a Mangrove Wetland," Microbial Ecology, vol. 66, no. 1, pp. 96–104, 2013. · ·
25. F. L. D. D. Sebastianes, A. S. Romão-Dumaresq, P. T. Lacava et al., "Species diversity of culturable endophytic fungi from Brazilian mangrove forests," Current Genetics, vol. 59, no. 3, pp. 153–166, 2013. · ·
26. Y. Arfi, M. Buée, C. Marchand, A. Levasseur, and E. Record, "Multiple markers pyrosequencing reveals highly diverse and host-specific fungal communities on the mangrove trees Avicennia marinaandRhizophora stylosa," FEMS Microbiology Ecology, vol. 79, no. 2, pp. 433–444, 2012. · ·
27. T. J. Lyimo, A. Pol, M. S. M. Jetten, and H. J. M. Op Den Camp, "Diversity of methanogenic archaea in a mangrove sediment and isolation of a new Methanococcoides strain," FEMS Microbiology Letters, vol. 291, no. 2, pp. 247–253, 2009. · ·
28. A. C. C. Pires, D. F. R. Cleary, A. Almeida et al., "Denaturing gradient gel electrophoresis and barcoded pyrosequencing reveal unprecedented archaeal diversity in mangrove sediment and rhizosphere samples," Applied and Environmental Microbiology, vol. 78, no. 16, pp. 5520–5528, 2012. · ·
29. D. W. Nelson and L. E. Somers, Organic Carbon, Academic Press, London, UK, 1975.
30. C. A. Black, Methods of Soil Analysis, American Society of Agronomy, Madison, Wis, USA, 1965.
31. M. Knudsen, Hydrographical Tables, GEC GAD, Copenhagen, Denmark, 1901.
32. A. Knap, A. Michaels, A. Close, H. Ducklow, and A. Dickson, Eds., Protocols for the Joint Global Ocean Flux Study (JGOFS) Core Measurements, JGOFS Report No. 19, 1996.
33. I. Bragança, A. Plácido, P. Paíga, V. F. Domingues, and C. Delerue-Matos, "QuEChERS: a new sample preparation approach for the determination of ibuprofen and its metabolites in soils," Science of the Total Environment, vol. 433, pp. 281–289, 2012. · ·
34. L. Correia-Sá, V. C. Fernandes, M. Carvalho, C. Calhau, V. M. F. Domingues, and C. Delerue-Matos, "Optimization of QuEChERS method for the analysis of organochlorine pesticides in soils with diverse organic matter," Journal of Separation Science, vol. 35, no. 12, pp. 1521–1530, 2012. · ·
35. B. Wemheuer, F. Wemheuer, and R. Daniel, "RNA-based assessment of diversity and composition of active archaeal communities in the German Bight," Archaea, vol. 2012, Article ID 695826, 8 pages, 2012. · ·

36. E. Pruesse, C. Quast, K. Knittel et al., "SILVA: a comprehensive online resource for quality checked and aligned ribosomal RNA sequence data compatible with ARB," Nucleic Acids Research, vol. 35, no. 21, pp. 7188–7196, 2007. · ·

37. S. F. Altschul, W. Gish, W. Miller, E. W. Myers, and D. J. Lipman, "Basic local alignment search tool,"Journal of Molecular Biology, vol. 215, no. 3, pp. 403–410, 1990. · ·

38. D. H. Huson, S. Mitra, H.-J. Ruscheweyh, N. Weber, and S. C. Schuster, "Integrative analysis of environmental sequences using MEGAN4," Genome Research, vol. 21, no. 9, pp. 1552–1560, 2011. · ·

39. S. Mitra, M. Stärk, and D. H. Huson, "Analysis of 16S rRNA environmental sequences using MEGAN,"BMC Genomics, vol. 12, supplement 3, p. S17, 2011. ·

40. R Core Team, R: A Language and Environment for Statistical Computing, R Foundation for Statistical Computing, Vienna, Austria, 2014.

41. S. K. Sarkar, B. Bhattacharya, S. Debnath, G. Bandopadhaya, and S. Giri, "Heavy metals in biota from Sundarban Wetland Ecosystem, India: implications to monitoring and environmental assessment,"Aquatic Ecosystem Health & Management, vol. 5, no. 4, pp. 467–472, 2002. · ·

42. M. T. Rahman, M. S. Rahman, S. B. Quraishi, J. U. Ahmad, T. R. Choudhury, and M. A. Mottaleb, "Distribution of heavy metals in water and sediments in Passur River, Sundarban Mangrove Forest, Bangladesh," Journal of International Environmental Application & Science, vol. 6, no. 4, pp. 537–546, 2011.

43. M. Saha, S. K. Sarkar, and B. Bhattacharya, "Interspecific variation in heavy metal body concentrations in biota of Sunderban mangrove wetland, northeast India," Environment International, vol. 32, no. 2, pp. 203–207, 2006. · ·

44. K. Banerjee, B. Senthilkumar, R. Purvaja, and R. Ramesh, "Sedimentation and trace metal distribution in selected locations of Sundarbans mangroves and Hooghly estuary, Northeast coast of India,"Environmental Geochemistry and Health, vol. 34, no. 1, pp. 27–42, 2012. · ·

45. S. Massolo, A. Bignasca, S. K. Sarkar, M. Chatterjee, B. D. Bhattacharya, and A. Alam, "Geochemical fractionation of trace elements in sediments of Hugli River (Ganges) and Sundarban wetland (West Bengal, India)," Environmental Monitoring and Assessment, vol. 184, no. 12, pp. 7561–7577, 2012. · ·

46. T. J. Mincer, M. J. Church, L. T. Taylor, C. Preston, D. M. Karl, and E. F. DeLong, "Quantitative distribution of presumptive archaeal and bacterial nitrifiers in Monterey Bay and the North Pacific Subtropical Gyre," Environmental Microbiology, vol. 9, no. 5, pp. 1162–1175, 2007. · ·

47. J. M. Beman, B. N. Popp, and C. A. Francis, "Molecular and biogeochemical evidence for ammonia oxidation by marine Crenarchaeota in the Gulf of California," The ISME Journal, vol. 2, no. 4, pp. 429–441, 2008. · ·

48. A. C. Mosier and C. A. Francis, "Relative abundance and diversity of ammonia-oxidizing archaea and bacteria in the San Francisco Bay estuary," Environmental Microbiology, vol. 10, no. 11, pp. 3002–3016, 2008. · ·

49. A. E. Bernhard, Z. C. Landry, A. Blevins, J. R. de la Torre, A. E. Giblin, and D. A. Stahl, "Abundance of ammonia-oxidizing archaea and bacteria along an estuarine salinity gradient in relation to potential nitrification rates," Applied and Environmental Microbiology, vol. 76, no. 4, pp. 1285–1289, 2010. · ·

50. H. Urakawa, W. Martens-Habbena, and D. A. Stahl, "High abundance of ammonia-oxidizing archaea in coastal waters, determined using a modified DNA extraction method," Applied and Environmental Microbiology, vol. 76, no. 7, pp. 2129–2135, 2010. · ·

51. P. Lam, M. M. Jensen, G. Lavik et al., "Linking crenarchaeal and bacterial nitrification to anammox in the Black Sea," Proceedings of the National Academy of Sciences of the United States of America, vol. 104, no. 17, pp. 7104–7109, 2007. · ·

52. Y.-F. Wang, Y.-Y. Feng, X. Ma, and J.-D. Gu, "Seasonal dynamics of ammonia/ammonium-oxidizing prokaryotes in oxic and anoxic wetland sediments of subtropical coastal mangrove," Applied Microbiology and Biotechnology, vol. 97, no. 17, pp. 7919–7934, 2013. · ·

53. M. Li, Y.-G. Hong, H.-L. Cao, and J.-D. Gu, "Mangrove trees affect the community structure and distribution of anammox bacteria at an anthropogenic-polluted mangrove in the Pearl River Delta reflected by 16S rRNA and hydrazine oxidoreductase (HZO) encoding gene analyses," Ecotoxicology, vol. 20, no. 8, pp. 1780–1790, 2011. · ·

54. H. Cao, M. Li, Y. Hong, and J.-D. Gu, "Diversity and abundance of ammonia-oxidizing archaea and bacteria in polluted mangrove sediment," Systematic and Applied Microbiology, vol. 34, no. 7, pp. 513–523, 2011. · ·

55. M. J. Church, B. Wai, D. M. Karl, and E. F. DeLong, "Abundances of crenarchaeal amoA genes and transcripts in the Pacific Ocean," Environmental Microbiology, vol. 12, no. 3, pp. 679–688, 2010. · ·

56. D. A. Stahl and J. R. de la Torre, "Physiology and diversity of ammonia-oxidizing archaea," Annual Review of Microbiology, vol. 66, pp. 83–101, 2012. · ·

57. H. Cao, J.-C. Auguet, and J.-D. Gu, "Global ecological pattern of ammonia-oxidizing archaea," PLoS ONE, vol. 8, no. 2, Article ID e52853, 2013. · ·

58. Y. Zhang, Z. Zhao, C.-T. A. Chen, K. Tang, J. Su, and N. Jiao, "Sulfur metabolizing microbes dominate microbial communities in Andesite-hosted shallow-sea hydrothermal systems," PLoS ONE, vol. 7, no. 9, Article ID e44593, 2012. · ·

59. J. Milucka, T. G. Ferdelman, L. Polerecky et al., "Zero-valent sulphur is a key intermediate in marine methane oxidation," Nature, vol. 491, no. 7425, pp. 541–546, 2012. · ·

60. K. Knittel and A. Boetius, "Anaerobic oxidation of methane: progress with an unknown process," Annual Review of Microbiology, vol. 63, pp. 311–334, 2009.

61. E. F. DeLong and D. M. Karl, "Genomic perspectives in microbial oceanography," Nature, vol. 437, no. 7057, pp. 336–342, 2005. · ·

62. A.-L. Reysenbach, K. Longnecker, and J. Kirshtein, "Novel bacterial and

archaeal lineages from an in situ growth chamber deployed at a mid-atlantic ridge hydrothermal vent," Applied and Environmental Microbiology, vol. 66, no. 9, pp. 3798–3806, 2000. · ·

63. H. Teeling, B. M. Fuchs, D. Becher et al., "Substrate-controlled succession of marine bacterioplankton populations induced by a phytoplankton bloom," Science, vol. 336, no. 6081, pp. 608–611, 2012. · ·

12
CHAPTER

ABUNDANCE AND DIVERSITY OF AMMONIA-OXIDIZING ARCHAEA AND BACTERIA IN SEDIMENTS OF TROPHIC END MEMBERS OF THE LAURENTIAN GREAT LAKES, ERIE AND SUPERIOR

Annette Bollmann[1], George S. Bullerjahn[2], Robert Michael McKay[2]

[1] Department of Microbiology, Miami University, Oxford, Ohio, United States of America

[2] Department of Biological Sciences, Bowling Green State University, Bowling Green, Ohio, United States of America

ABSTRACT

Ammonia oxidation is the first step of nitrification carried out by ammonia-oxidizing Archaea (AOA) and Bacteria (AOB). Lake Superior and Erie are part of the Great Lakes system differing in trophic status with Lake Superior being oligotrophic and Lake Erie meso- to eutrophic. Sediment samples were collected from both lakes and used to characterize abundance and diversity of AOA and AOB based on the ammonia monooxygenase (amoA) gene. Diversity was accessed by a pyro-sequencing approach and the obtained sequences were used to determine the phylogeny and alpha and beta diversity of the AOA and AOB populations. In Lake Erie copy numbers of bacterial amoA genes were in the same order of magnitude or even higher than the copy numbers of the archaeal amoA genes, while in Lake Superior up to 4 orders of magnitude more archaeal than bacterial amoA copies were detected. The AOB detected in the samples from Lake Erie belonged to AOB that are frequently detected in freshwater. Differences were detected between the phylogenetic affiliations of the AOA from the two lakes. Most sequences detected in Lake Erie clustered in the Nitrososphaera cluster

Citation: Annette Bollmann, George S. Bullerjahn, Robert Michael McKay, Abundance and Diversity of Ammonia-Oxidizing Archaea and Bacteria in Sediments of Trophic End Members of the Laurentian Great Lakes, Erie and Superior, doi: 10.1371/journal.pone.0097068.

(Thaumarchaeal soil group I.1b) where as most of the sequences in Lake Superior were found in the Nitrosopumilus cluster (Thaumarchaeal marine group I.1a) and the Nitrosotaleacluster. Pearson correlations and canonical correspondence analysis (CCA) showed that the differences in abundance and diversity of AOA are very likely related to the sampling location and thereby to the different trophic states of the lakes.

INTRODUCTION

Ammonia oxidation is the first and rate-limiting step in nitrification, the oxidation of ammonia to nitrate via nitrite. Understanding this process and its controls is of high importance because it controls the availability of two major nitrogen compounds (ammonium and nitrate) in nature. The long-known Ammonia-oxidizing Bacteria (AOB) and the recently discovered Ammonia-oxidizing Archaea (AOA) use the oxidation of ammonia to nitrite as an energy-generating step[1], [2]. Since both groups use the same energy substrate it is important to understand the environmental conditions under which AOA or AOB dominate. Among the factors reported to influence the abundance and diversity of AOA and AOB are fertilizers (ammonium addition) [3], [4]; pH [5], [6]; salinity [7], [8] and oxygen [9]. For example AOA have much higher affinity for ammonium/ammonia than AOB [10]–[13] and are often detected in more oligotrophic environments like the open ocean or oligotrophic lakes [14]. In contrast AOB grow with higher rates in soils [1]–[4] and enrichment cultures [3], [4], [15].

AOB comprise a phylogenetically distinct group in the phylum *Beta-Proteobacteria* as well as a few marine strains in the *Gamma-Proteobacteria* [5], [6], . The betaproteobacterial AOB cluster in different groups based on environmental characteristics such as high- and low ammonium availability, salinity and pH [7], [8],[16], [17]. AOB found in freshwater systems generally belong to the *Nitrosomonas oligotropha*,*Nitrosomonas communis* and the *Nitrosospira* clusters [9], [18]–[21]. Recently the phylum *Thaumarchaeota* was described as a new deep-branching phylum in the archaeal domain [10]–[13], [22]–[24]. Besides groups of microorganisms with unknown physiology such as the groups pSL12 from hot springs, ALOHA from the open ocean and I.1c from acidic soils, AOA are a large group within the *Thaumarchaeota* [14], [25]. The AOA have been split into four groups: *Nitrosopumilus*(Thaumarchaeal marine group I.1a), *Nitrososphaera* (Thaumarchaeal soil group I.1b), *Nitrosotalea*(SAGMGC-1, formerly group I.1a associated) and *Nitrosocaldus* cluster (formerly, ThAOA group). Representatives of the *Nitrosopumilus* cluster have mainly been detected in aquatic marine and freshwater systems, *Nitrososphaera* cluster in soils and sediments, *Nitrosotalea* cluster in acidic soils and oligotrophic freshwater systems and *Nitrosocaldus* in extreme environments like hot springs [22], [26]. However, it has been shown that not all *amoA* encoding Thaumarchaeota are autotrophic ammonia oxidizers. Some exhibit a mixotrophic physiology, while others express the *amoA* gene but don't oxidize ammonium [27], [28].

Overall AOA communities in marine and soil environments are much better studied than the AOA in freshwater systems. Molecular surveys have been conducted to analyze AOA and/or AOB communities and the factors that control them in freshwater systems.

Trophic status and ammonium availability are among the factors that influence the abundance as well as the structure of the AOA and AOB communities [7], [14],[18]–[21], [29]–[38]. The Laurentian Great Lakes system is the largest system of freshwater lakes on earth and is located in the eastern part of North America forming part of the border between the United States and Canada. Lake Superior, the largest and deepest of the five lakes, is mainly surrounded by forest and coincident with low human population abundance in the watershed, is least affected by pollution. At the opposite end of a trophic continuum is Lake Erie, the shallowest of the Great Lakes. With high population abundance and a watershed allocated largely to agricultural and industrial activities, Lake Erie is heavily impacted by urban and agricultural [39] runoff from the areas surrounding the lake and ranges from mesotrophic to eutrophic. In stark contrast, Lake Superior, which serves as the headwaters for the Great Lakes system, has remained pristine and is characterized as oligotrophic [40]. This is supported by historical data showing mainly flat profiles of total dissolved solids as well as concentrations of major ions which serve as indicators of anthropogenic impacts on the system [40], [41].

Seemingly counter to the static trends in major ions is the observation that Lake Superior has exhibited a continuous, century-long five-fold increase in nitrate levels from 5 μmol/l to 26 μmol/l [42]. Nitrification rates in the water column of Lake Superior are lower than in other freshwater systems including measurements in other parts of the lakes, but higher than in the open ocean – another indication that Lake Superior is an oligotrophic system [43], [44]. Here we present a study investigating the abundance and diversity of Ammonia-oxidizing Archaea and Bacteria in the sediments of western Lake Superior and embayments of western Lake Erie using a deep sequencing approach (454 pyrosequencing with barcoded primers). We chose samples from these two different sections of the Great Lakes, because they represent trophic end members of the Great Lakes system with Lake Superior being very oligotrophic and Lake Erie being meso/eutrophic. The results will give an insight into the phylogeny and distribution of AOA and AOB in the Great lakes and on the impact of the trophic state of freshwater environments on both groups of organisms.

MATERIALS AND METHODS

Sediment sampling

Sediment samples were taken at Lake Superior and Erie stations (Fig. 1) in October 2010, maintained at 4°C and transported within 4 days after sampling to the laboratory at Miami University where they were frozen at −20°C upon arrival. The samples were taken in US territorial waters of the Great Lakes where no permissions are required as overseen by the International Joint Commission. No endangered species are involved. Two sampling techniques focusing on the top 0–10 cm of the sediment were utilized: ponar dredge (0-max. 10 cm) and sediment cores (0–5 cm). All Lake Erie samples (EC1300, EC1301, EC1302, EC1303) were taken with the ponar dredge (Table 1). At four process stations (CD, SteC, UWM, WM) in Lake Superior (Table 1), sediment cores were collected using an Ocean Instruments (San Diego, CA) MC-400 multi-corer. The other four Lake Superior samples were taken with the ponar dredge (Grab5, Grab6,

Grab9, Grab10) (Table 1). Subsamples were taken from the mixed sediment material for determination of mineral nitrogen (ammonium, nitrite, and nitrate), dry weight and molecular analysis in triplicate.

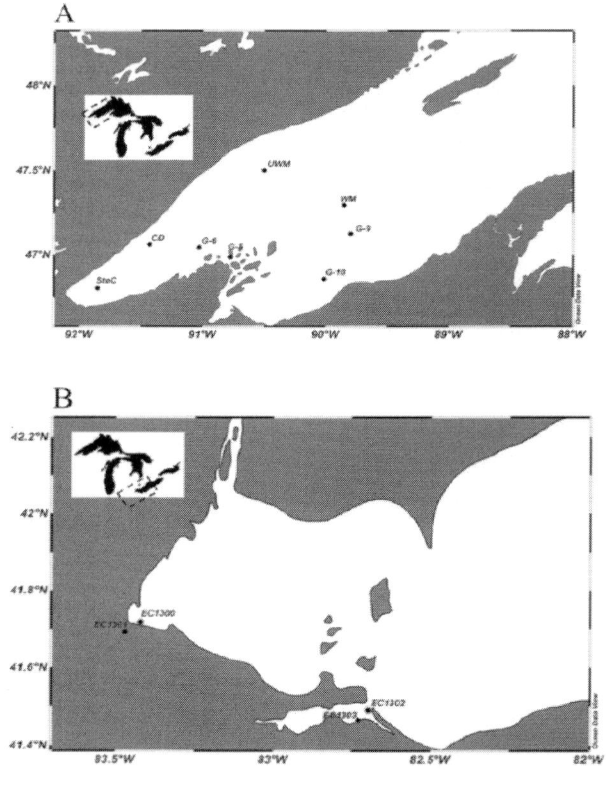

Figure 1: Sampling sites for Lake Superior (A) and Lake Erie (B) sediments.

Table 1: Environmental data from the sediment samples and the overlaying water

	Lake	Sampling Date	Depth [m]	Distance to shore [km]	Sample	NH_4^+ [2] [µg N/g dw]	NO_3^- [2] [µg N/g dw]	Chl *a* [1] [µg Chl a/l]	DO [mg/l] [3]
EC1300	Erie	10/13/10	9.6	Near shore	Ponar;	149.8±2.7	0.4±0.0	4.4±1.9	7.2
EC1301	Erie	10/13/10	8.3	River mouth	Ponar;	202.2±6.4	1.2±0.0	17.9±7.7	5.5
EC1302	Erie	10/13/10	12.6	Bay mouth	Ponar;	7.4±0.2	0.5±0.1	13.6±1.3	10.0
EC1303	Erie	10/13/10	7.2	Near shore	Ponar;	59.5±2.5	0.5±0.0	20.2±1.1	nd
Grab5	Superior	10/07/10	30	Near shore	Ponar	0.1±0.1	1.3±0.2	nd	nd
Grab6	Superior	10/07/10	30	Near shore	Ponar	1.3±0.9	2.6±0.5	nd	nd
Grab9	Superior	10/06/10	nd	28.2	Ponar	0.5±0.1	1.4±0.1	nd	nd
Grab10	Superior	10/06/10	nd	22.2	Ponar	23.0±0.9	3.4±0.3	nd	nd
CD	Superior	10/05/10	245	8.2	Multi-corer; 0–5 cm	8.8±4.2	5.3±0.2	1.2±0.1	12.4
SteC	Superior	10/07/10	41	10	Multi-corer; 0–5 cm	22.7±2.9	4.3±0.3	nd	11.4
UWM	Superior	10/06/10	171	31	Multi-corer; 0–5 cm	13.2±6.9	12.8±2.3	nd	12.3
WM	Superior	10/06/10	165	56.3	Multi-corer; 0–5 cm	0.9±0.2	5.1±0.4	1.1±0.1	12.5

[1] determined in surface water.
[2] in the sediment.
[3] dissolved oxygen concentration in the water directly above the sediment.
nd: not determined.
doi:10.1371/journal.pone.0097068.t001

Determination of environmental data

Mineral nitrogen was measured in 1 M KCl extracts. 3–4 g mixed sediment samples were mixed with 1 M KCl in the ratio 1:10, shaken for 1 h at 200 rpm and centrifuged at 7000 g for 10 min. The supernatant was stored at −20°C until further analysis. Mineral nitrogen (ammonium, nitrite and nitrate) was determined colorimetrically [45]–[48]. Dry weight was determined after drying the sediment samples for 24 h at 110°C.

Water column phytoplankton biomass was assessed using parallel approaches: Chlorophyll *a*, a proxy for biomass, was measured fluorometrically following acetone extraction of seston retained on 0.2 μm polycarbonate filters [49]. Oxygen concentrations were measured during water sampling by a Seabird (Bellevue, WA) model 911 dissolved oxygen sensor.

Molecular analysis: DNA isolation

DNA was isolated with the PowerSoil DNA isolation kit (MoBIO, Carlsbad, CA, USA) according to the manufacturer's recommendations. DNA concentration was measured spectrophotometrically with a NanoDrop 2000 (Thermo Fisher Scientific, Wilmington, DE). DNA isolation was conducted in triplicate per sample.

Quantitative PCR (qPCR)

DNA for the qPCR was diluted to concentrations between 1–10 ng/μl. qPCR to determine the abundance of archaeal and bacteria *amoA* genes was performed using AOA and AOB-specific *amoA* primers and the Bioline SensiMix SYBR No-ROX kit according to the manufacturer›s recommendations (Bioline, Taunton, MA). The number of gene copies in the samples was calculated using the standard curve method and the specificity of the primers was determined by agarose gel electrophoresis and melting curves.

Pyrosequencing of the archaeal and bacteria *amoA* genes

DNA was first amplified with AOA and AOB *amoA* primers under conditions presented. The PCR products were diluted 1:10 and used as template for a second PCR using the barcoded primers. Twelve bar-coded primers were generated, six for archaeal *amoA* and six for bacterial *amoA*, allowing sequencing of the *amoA* genes of six independent samples in 1/16 of a 454 sequencing run. The two-step PCR prevents amplification biases and increases reproducibility [52]. Per DNA isolation, three PCR reactions were conducted and at the end all PCR products (triplicate PCR runs × triplicate DNA isolations) per sample were mixed and used as one sample for pyrosequencing. The samples were purified with AMPureXP (Beckman-Coulter, Inc., Indianapolis, IN, USA). The samples were quantified with PicoGreen assay, diluted, pooled, and purified again with AMPureXP (Beckman-Coulter, Inc., Indianapolis, IN, USA). The concentration in the pooled samples was determined using KAPA qPCR (KAPA Biosystems, Woburn MA, USA). For the first library 0.5 copies per bead and for the second library 2 copies per bead were sequenced on the Roche GS FLX system at the Plant-Microbe Genomics Facility at The Ohio State University (Columbus, Ohio, USA).

Analysis of the pyrosequencing data

The data were processed into quality (.qual) and sequence (.fasta) files using GSRmBrowser version 2.5.3. QIIME was used for initial quality filtering of the sequences [53]. The overall sequences files were split based on the barcodes, quality filtered with the average quality score being 25, truncated at 400 bp length and exported as sequence (.fasta) files. The sequences were imported into ARB [54] and translated into protein sequences. The protein sequences were screened to exclude sequences with stop codons and frame shifts. The remaining sequences were exported as nucleotide sequences and further analyzed with QIIME [53]. First the sequences files were merged and the sequences were grouped based on identity into groups with 89% and 98% identity. Since there has been a discussion that the increase in diversity in pyrosequencing libraries could be due to sequencing errors [55], we eliminated all OTU›s with just single sequences per sample. The AOA and AOB libraries, respectively, were 100 times rarefied based on the number of sequences in the library with the lowest sequence number (AOA: 538 sequences at 85% and 525 sequences at 98% similarity; AOB: 221 sequences at 85% and 208 sequences at 98% similarity). The rarified libraries were used for the determination of the alpha diversity (number of OTU›s, chao1 and Shannon index) and the beta-diversity using two different measures (abundance weighted Jaccard distance and weighted Unifrac analysis), to integrate abundance and phylogenetic information in the beta-diversity analysis. The phylogenetic trees used for the Unifrac analysis were constructed with representative sequences in ARB using the Neighbor-joining method [54].

Representative AOA and AOB sequences based on 98% identity after excluding singletons were aligned to the ARB-AOA file published by Pester et al (2012) [22] and an AOB file and added with the ARB parsimony addition tool [54]. Phylogenetic trees for AOA and AOB were constructed in ARB with the added representative sequences and close related sequences using the Neighbor-joining method.

Statistical analysis

Correspondence analysis was conducted using CANOCO (http://www.canoco5.com) and statistical analysis (One-Way ANOVA and Pearson correlations) with SPSS (version 19). The sequences were deposited in the NCBI SRA database under the accession number PRJNA217461 with the individual accession numbers for each library: SRS474329, SRS474331-333, SRS474335-345, SRS474348.

RESULTS

We analyzed twelve sediment samples from the western basin of Lake Superior (8) and the western basin of Lake Erie (4) (Figure 1, Table 1). Ammonium concentration was by one to two orders of magnitude higher in the sediment samples from Lake Erie than in the samples from Lake Superior, whereas nitrate was by one to two orders of magnitude higher in the Lake Superior samples (Table 1). Chlorophyll *a* in the water column as measure of the trophic state was much higher in the samples from Lake Erie than in the Lake Superior samples (Table 1).

The abundance of AOA and AOB in the sediment was determined using AOA and AOB specific *amoA*primers (Figure 2). In the samples EC1300 and EC1301 from Lake Erie's Maumee Bay, AOA and AOB abundances were in the same order of magnitude. AOB were by one order of magnitude more abundant than AOA in the samples EC1302 and EC1303 from Sandusky Bay showing a dominance of the AOB in all samples from Lake Erie. By contrast, AOA were 2–4 orders of magnitude more abundant than AOB in Lake Superior. In summary, sediment samples from embayments of Lake Erie were dominated by AOB, whereas Lake Superior samples were dominated by AOA.

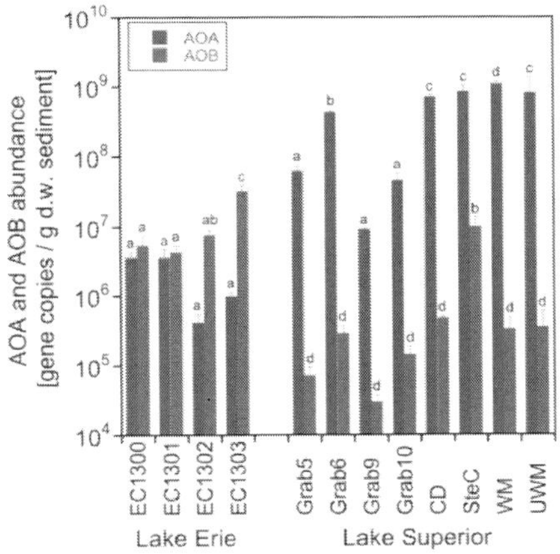

Figure 2: Abundance (amoA gene copy number) of AOA and AOB in the sediment of Lake Erie and Superior (mean ± SD, n = 3; different letters above the columns indicate significant differences between samples determined by one-way ANOVA followed by Tukey test; p,0.05).

The archaeal and bacterial *amoA* genes were sequenced using a pyrosequencing approach in two separate runs. AOB *amoA* genes were only sequenced from the four samples from Lake Erie, because the AOB abundance in Lake Superior was very low. Run 1 resulted in 11847 sequences after sequencing and 6668 sequences after quality control with QIIME and ARB whereas run 2 resulted in 20892 and 13563 sequences, respectively.

The number of OTU›s, the Chao1 index (richness) and the Shannon index (evenness) were calculated at cutoff values of 85% and 98% similarity using the software package QIIME [53]. The number of OTU›s and the Chao1 index for the AOA were at both cutoff values higher for the samples from Lake Erie (OTU›s cutoff 85%: 9–12 and cutoff 98%: >40) than for the samples from Lake Superior (OTU›s cutoff 89%: 2–7 and cutoff 98%: around 20) indicating a higher AOA diversity in Lake Erie than in Lake Superior. The evenness of the AOA in the samples from Lake Erie was higher than in Lake Superior, as shown by the higher Shannon index in the samples from Lake Erie compared to Lake Superior. The number of OTU›s for AOB in Lake Erie ranged from 7–12 (89% similarity) and 15–40 (98% similarity). The chao1 indices for the AOB were similar or a little higher than the OTU numbers for the AOB in Lake Erie.

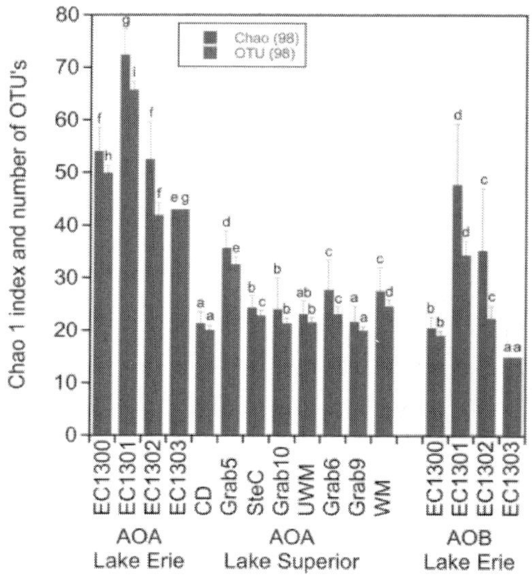

Figure 3: Alpha-diversity at 98% identity and singletons removed of the AOA and AOB *amoA* **sequence libraries.**

AOA and AOB abundances were correlated using Pearson correlation to environmental factors and alpha-diversity data (Table 2). The AOA abundance was positively correlated with nitrate concentration ($p<0.05$) while the AOB abundance was positively correlated with ammonium concentration in the sediment ($p<0.05$). Chlorophyll *a* concentration in the water column as an indicator of the trophic state of the lakes showed positive correlation with the AOB abundance ($p<0.05$) and negative correlation with the AOA abundance ($p<0.05$) indicating that AOA are more prevalent under low nutrient oligotrophic conditions and AOB under nutrient rich meso- to eutrophic conditions. Of note, the AOA abundance showed negative correlation with alpha diversity; high AOA abundances in the Lake Superior samples coincided with low species diversity and low AOA abundances in the Lake Erie samples with high species diversity.

Table 2: Pearson correlations of AOA and AOB abundances with environmental and alpha diversity data

	AOA abundance	AOB abundance	n =
Ammonium (sediment)	−0.344	0.734*	12
Nitrate (sediment)	0.906*	−0.445	12
Chlorophyll A (watercolumn)	−0.930*	0.912*	6
Chao1 index	−0.744*	−0.756	12/4
Number of OTU's	−0.741*	−0.790	12/4
Shannon index	−0.591*	−0.372	12/4

* $p<0.05$. (All data log-transformed.)
doi:10.1371/journal.pone.0097068.t002

Phylogenetic affiliation of AOA and AOB was determined by aligning representative sequences (98% similarity; 400 bp length) in ARB. AOA were classified using the classification published by Pester et al., 2012 (Figure 4). AOA sequences in Lake Superior and Lake Erie exhibited very different community compositions (Figure 4). The AOA communities in Lake Erie were dominated by a wide variety of *Nitrososphaera*-like sequences, except sample EC1302 from the mouth of Sandusky Bay that contained a high number of *Nitrosopumilus*-like sequences. All other Lake Erie samples contained only low numbers of*Nitrosopumilus* sequences. Sequences from the group *Nitrosopumilus* subcluster 1.1 were found in all samples from Lake Superior in high quantities (>50% of total sequences). In addition members of*Nitrosopumilus* subcluster 15 were detected in quantities higher than 10% in samples from Grab6, Grab10, and UWM and *Nitrosotalea* cluster 2 in samples from Grab9 and WM. Only in the samples Grab9 and WM were low numbers of *Nitrososphaera* sequences detected. A detailed phylogenetic tree showed that only two sequences in the *Nitrososphaera* cluster were more than 5% abundant (Sequence 59 and 88 in samples EC1300, EC1301, EC1303). All other sequences were detected in lower abundances (<5% of the total sequences). The sequences from Lake Erie clustering in *Nitrosopumilus* cluster 1.1 were different from the sequences found in Lake Superior. Sequences from the *Nitrosopumilus* cluster 5.1/5.2, a*Nitrosopumilus* cluster dominated by freshwater and ground water sequences, were only detected in Lake Erie (Figure 4).

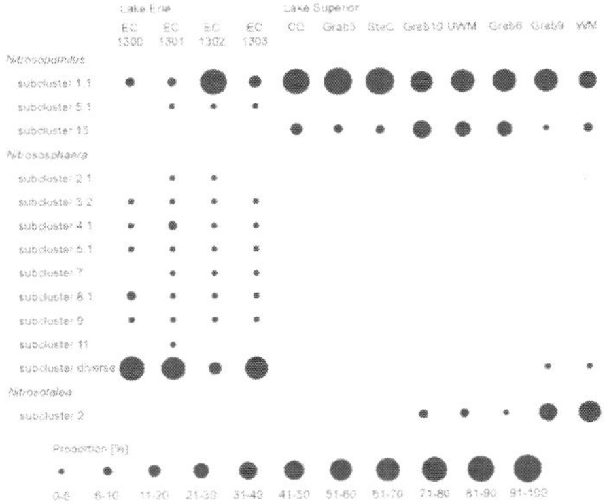

Figure 4: Relative frequency [%] of the AOA in the different phylogenetic groups.

The AOB community in Lake Erie was dominated by members of the *Nitrosomonas communis* cluster (*Nitrosomonas* cluster 8) and the *Nitrosospira* cluster (*Nitrosospira* cluster 0; 3A and uncultured) (Figure 5). Only low numbers (up to 10% of the total sequences) of sequences were found in the typical freshwater cluster *Nitrosomonas* cluster 6a. The Maumee Bay samples EC1300 and EC1301 contained*Nitrosomonas communis* and *Nitrosospira* cluster 3A sequences as well as low numbers from other*Nitrosospira* clusters. The AOB communities in the Sandusky Bay samples from EC1302 and EC1303 were less diverse, as only 2-3 different *Nitrosospira* clusters could be detected.

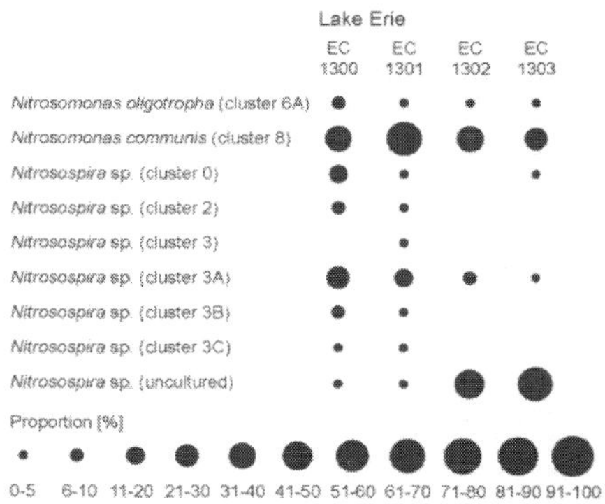

Figure 5: Relative frequency [%] of the AOB in the different phylogenetic groups.

The community composition of AOA was compared using Weighted Unifrac distance and Jaccard abundance [53], [56]. Weighted Unifrac distance analysis is based on phylogenetic relationship and abundance, while Jaccard abundance is based only on abundance. Overall both analyses showed similar relationships between the different communities (Figure 6). Samples from Lake Superior and samples from Lake Erie, respectively, clustered together with the exception of the sample EC1302 from Lake Erie. This sample clustered together with the Lake Erie samples when analyzed with Jaccard abundance but with Lake Superior samples when analyzed with Weighted Unifrac distance. Sample EC1302 exhibited a high number of sequences belonging to the *Nitrosopumilus* subcluster 1.1 while the other Lake Erie samples contained low numbers and the Lake Superior samples high numbers of sequences from that cluster (Figure 4;). The AOB communities of EC1300 and EC1301 as well as EC1302 and EC1303 clustered together based on Weighted Unifrac analysis.

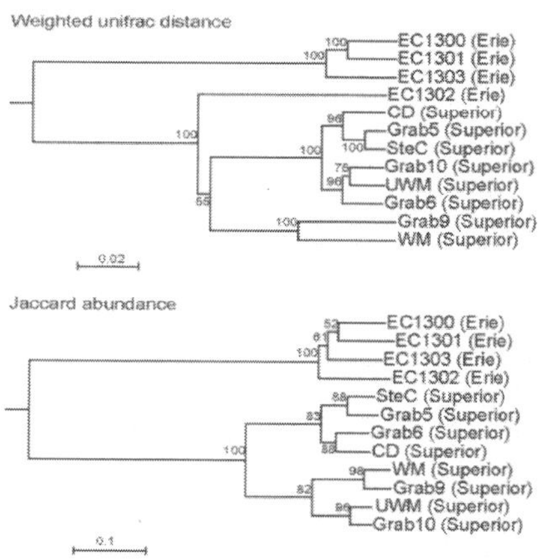

Figure 6: UPGMA clustering of Weighted Unifrac distances (A) and Jaccard abundances (B) of AOA in the sediment of Lake Erie and Superior at 98% similarity.

Canonical correspondence analysis (CCA) was used to determine relationships between species, communities and the environmental factors (Figure 7). The AOA communities from Lake Superior were positive related with AOA abundance and nitrate concentration and from Lake Erie to ammonium concentration (Figure 7). The AOA abundance explained 47% and the ammonium 18.3% of the variation. Ammonium concentration and AOB abundance were the most important factors explaining the AOB community composition.

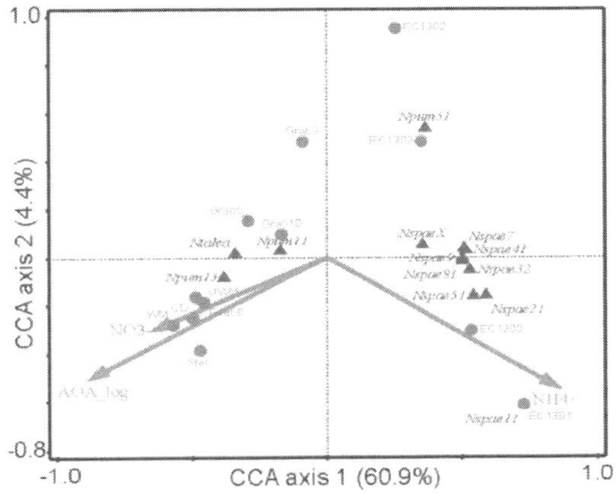

Figure 7: Canonical correspondence analysis (CCA) triplot (arrows: environmental variables; circles: samples; triangles: species) for quantitative data as presented in Fig. 3 of the AOA *amoA*-sequence libraries in Lake Erie and Superior (Eigenvalues: Axis 1: 0.5865, ...

DISCUSSION

Niche differentiation between AOA and AOB in freshwater environments

AOA were more abundant in the sediments of oligotrophic Lake Superior than AOB whereas the situation in two meso-eutrophic embayments of Lake Erie was vice versa (Figure 2). The AOA abundance was negatively, and the AOB abundance positively correlated with chlorophyll *a* concentration – an indirect measure of the trophic state of the lakes (Table 2). Our results are consistent with recent reports showing dominance by AOA over AOB in the water column of two oligotrophic lakes, Lake Superior [43] and Lake Lucerne [37]. Likewise, AOA are the dominant ammonia oxidizers in the open ocean where nutrients are scarce [57]-[59]. By contrast, the ratio of AOB to AOA increases with increasing ammonium concentrations in freshwater streams [33] and soils [3]. These results confirm the trend that AOA are found in more nutrient poor and AOB in nutrient rich environments.

The ammonium concentration was one of the major factors regulating abundance and distribution of the AOA (Table 2; Figure 7). In freshwater aquarium biofilters and in a wastewater treatment plant, the AOA abundance was negatively related to the ammonium concentration [60], [61]. AOA enrichment cultures and the pure culture *Nitrosopumilus maritimus* have much lower K_m values for ammonium/ammonia than AOB[10]–[12]. Overall these results show that the majority of AOA can be found in and is very likely adapted to environments with low ammonium concentrations and availability.

Based on the abundances of and ratios between AOA and AOB in the oligotrophic Lake Superior and meso/eutrophic Lake Erie and under the assumption that AOA and AOB use primarily ammonia oxidation for energy generation it is likely that AOA were mainly responsible for ammonia oxidation in Lake Superior and AOB in Lake Erie. Nitrification in the water column in Lake Superior has been attributed to the activity of AOA, which were found to be the dominant ammonia oxidizers in that environment [43]. This observation and the high abundance of AOA in the sediments of Lake Superior support our assumption that AOA are very likely the main ammonia oxidizers in the sediments of Lake Superior.

Niche differentiation between different groups of AOA

The results showed not only a niche differentiation between AOA and AOB, but also between different groups of AOA (Figure 4). In 3 out of 4 Lake Erie samples members of the *Nitrososphaera* soil/sediment cluster and *Nitrosopumilus* cluster 5 were the dominant AOA while the Lake Superior samples were dominated by members of *Nitrosopumilus* cluster 1 and 15 and the *Nitrosotalea* cluster (Figure 4). Also these differences could be due to the trophic states of the two lakes. Sequences from the *Nitrosopumilus* cluster 1.1 cluster together with sequences from the Qiantang River in China [34], the rhizosphere of the freshwater macrophyte *Littorella uniflora* [31], oligotrophic freshwater lakes [14], [62] and a drinking water distribution system [63]. The freshwater enrichment cultures AOA-AC2 and *Candidatus* Nitrosoarchaeum limnia belong to the same AOA subcluster[11], [15], [64].

High proportions of AOA *amoA* sequences from Lake Superior were detected in the cluster *Nitrosopumilus*cluster 15. Interestingly this cluster has not been dominated by freshwater strains. Most sequences were found in estuarine sediments such as Douro River estuary (Portugal) [65], Plum Island Sound [8], the Elkhorn Slough [50] and also in high altitude lakes of the Tibetan Plateau [66].

Finally many sequences from Lake Superior were detected in the *Nitrosotalea* cluster. Sequences from both*Nitrosotalea* clusters have been found frequently in oligotrophic freshwater [14], [29], [31], [32], groundwater [63], [67] and acidic soils [6], [68]. The only cultivated member of the *Nitrosotalea* cluster is*Candidatus* Nitrosotalea devanaterra, an obligate acidiphilic AOA [69]. However, the pH in Lake Superior waters average 7.2, indicating that acidophilic conditions were not the cause for the high abundance of members of the *Nitrosotalea* cluster in these sediments.

Overall many sequences in the *Nitrosopumilus* cluster 1.1 and the *Nitrosotalea* cluster have been detected in rather nutrient poor systems, an observation that is in accordance with our observations that *Nitrosopumilus*cluster 1.1 and the *Nitrosotalea* cluster are highly abundant in the sediment samples of the oligotrophic Lake Superior.

In Lake Erie a few sequences from *Nitrosopumilus* cluster 5 were detected. *Nitrosopumilus* cluster 5 was found in other freshwater environments such as Lake Taihu [70], groundwater [63], [67], freshwater sediments [31], [34], roots of macrophytes [32], aquarium filters [60], a wastewater treatment plant [28] and enriched from the sediments of two meso/eutrophic lakes in Ohio [15] and hot springs [71]. Overall, sequences clustering in this group were detected in meso-to-eutrophic environments rather than in oligotrophic environments. This observation indicates that freshwater *Nitrosopumilus* cluster 5 strains could be adapted to different trophic states in their environment than the AOA strains detected in Lake Superior.

In addition many AOA sequences in Lake Erie belonging to the *Nitrososphaera* cluster I.1b were related to sequences from soils samples. The sampling sites in Lake Erie were close to the mouth of two rivers with agricultural watersheds. The presence of these sequences could have different reasons: (1) *Nitrososphaera*-like AOA in the eutrophic sediments could coexist with the AOB due to spatial separation; (2) some*Nitrososphaera*-like AOA sequences/strains could have originated from agricultural runoff rather than exist as members of the active ammonia-oxidizing community in those sediments; or (3) AOA could have additional metabolic capabilities providing them with an advantage over AOB (i.e. mixotrophy or non-autotrophic growth) [27], [28]. Mixotrophic growth was observed in AOA from both large groups (*Nitrosopumilus* cluster I.1.a and *Nitrososphaera* cluster I.1.b) [72], [73] and AOA belonging to the*Nitrososphaera* cluster I.1.b found in a refinery wastewater sample did express *amoA* but did not oxidize ammonia [28].

An additional important difference between the samples from the two lakes is that the overall AOA abundance in Lake Superior is by several orders of magnitude higher than in Lake Erie while the overall diversity of the AOA is lower. Similar observations have been made in other oligotrophic freshwater environments [14], [32], [62] and peatland soil[74]. In all those environments were AOA much more abundant than AOB, while the AOA diversity was rather low.

AOB phylogeny

In Lake Erie the AOB are as or more abundant than the AOA (Figure 2). Members of the *Nitrosomonas communis* cluster were found in high abundance in all samples from Lake Erie whereas *Nitrosomonas oligotropha*-like AOB were only present in low abundances (Figure 5). *N. communis* is found typically in eutrophic-, and *N. oligotropha* in oligotrophic freshwater environments [20], [75]-[77] indicating that the dominant *Nitrosomonas* strains reflect the trophic state of the Lake Erie.

Pyrosequencing of AOA and AOB *amoA* genes

Molecular surveys of AOA and AOB use 16S rRNA or *amoA* genes as marker genes [50], [51], [78]. Up to now most studies used a cloning-sequencing approach rather than a pyrosequencing approach to access the*amoA* diversity. Only a few studies have already used pyrosequencing of the *amoA* gene to describe AOA and AOB diversity [22], [68], [79], [80]. One of the major problems with pyrosequencing is the quality control of the sequences to ensure that the diversity is not overestimated based on sequencing errors [55]. The study demonstrated inflated 16SrDNA diversity due to a non-stringent quality control of the sequences. However, the use of functional genes has the advantage that the genes can be analyzed on the nucleotide- and the protein-level to eliminate sequences with sequencing errors more effectively [81]. During the ARB step (conversion into proteins) in the quality control around 30–40% of the sequences were removed due to the presence of stop codons and frame shift in the sequences. The presence of these sequences in the libraries for analysis increased the overall diversity in the samples artificially (results not shown) demonstrating the need for rigorous quality control of pyrosequencing data using functional genes. One downside of eliminating the sequences with frame shifts and stop codons could be to eliminate not just sequences with sequence errors, but also sequences of non-functional gene copies. However, this is rather unlikely to be a large problem, because most of the sequences that were removed had the stop codons and frame shifts at the end of the sequence reads or were single sequences with mistakes resulting in the conclusion that these errors are very likely due to sequencing mistakes.

Conclusions and Outlook

AOA dominated the sediments in Lake Superior and AOB in Lake Erie. In addition differences in the community composition of the AOA were observed between Lake Superior and Lake Erie. Due to the environmental conditions and the abundances of AOA and AOB in both lakes, it can be concluded that trophic state of the environment and ammonium availability play a key role in the niche differentiation between AOA and AOB as well as between the different groups of AOA in Lake Erie and Superior. Based on these observations future experiments should include the enrichment of new freshwater AOA-strains, investigation of the niche differentiation between AOA and AOB, as well as between different groups of AOA, and investigation of the basic physiology in connection with the main environment from which the strains were obtained to get better insights into the physiological capacities of AOA and AOB in freshwater systems.

ACKNOWLEDGMENTS

We thank Mike Zianni and Anthony McCoy from the Plant-Microbe Genomics Facility at The Ohio State University (Columbus, OH, USA) for their support with primer design and pyrosequencing. Samples were acquired during surveys aboard RV *Blue Heron* (Lake Superior) and CCGS *Kelso* (Lake Erie). We thank the captains and crews of these vessels as well as Technical Operations personnel from Environment Canada and Ben Beall (BGSU) for their assistance with sampling and Derek Smith (BGSU) for providing the maps.

FUNDING STATEMENT

The work was funded by the National Science Foundation (NSF) Department of Environmental Biology (DEB) Grant No: 1120443 to AB and Department of Ocean

Sciences (OCE) Grant No: 0927277 to RMM and GSB. The funders had no role in study design, data collection and analysis, decision to publish, or preparation of the manuscript.

REFERENCES

1. Arp D, Sayavedra-Soto L, Hommes N (2002) Molecular biology and biochemistry of ammonia oxidation by *Nitrosomonas europaea*. Arch Microbiol 178: 250–255doi:10.1007/s00203-002-0452-0
2. Könneke M, Bernhard AE, la Torré de JR, Walker CB, Waterbury JB, et al. (2005) Isolation of an autotrophic ammonia-oxidizing marine archaeon. Nature 437: 543–546doi:10.1038/nature03911
3. Verhamme DT, Prosser JI, Nicol GW (2011) Ammonia concentration determines differential growth of ammonia-oxidising archaea and bacteria in soil microcosms. ISME J 5: 1067–1071doi:10.1038/ismej.2010.191
4. Pratscher J, Dumont MG, Conrad R (2011) Ammonia oxidation coupled to CO_2 fixation by archaea and bacteria in an agricultural soil. Proc Natl Acad Sci USA 108: 4170–4175doi:10.1073/pnas.1010981108
5. Nicol GW, Leininger S, Schleper C, Prosser JI (2008) The influence of soil pH on the diversity, abundance and transcriptional activity of ammonia oxidizing archaea and bacteria. Environ Microbiol 10: 2966–2978doi:10.1111/j.1462-2920.2008.01701.x
6. Gubry-Rangin C, Nicol GW, Prosser JI (2010) Archaea rather than bacteria control nitrification in two agricultural acidic soils. FEMS Microbiol Ecol 74: 566–574doi:10.1111/j.1574-6941.2010.00971.x
7. Mosier AC, Francis CA (2008) Relative abundance and diversity of ammonia-oxidizing archaea and bacteria in the San Francisco Bay estuary. Environ Microbiol 10: 3002–3016doi:10.1111/j.1462-2920.2008.01764.x
8. Bernhard AE, Landry ZC, Blevins A, la Torré de JR, Giblin AE, et al. (2010) Abundance of ammonia-oxidizing archaea and bacteria along an estuarine salinity gradient in relation to potential nitrification rates.Appl Environ Microbiol 76: 1285–1289doi:10.1128/AEM.02018-09
9. Coolen MJL, Abbas B, van Bleijswijk J, Hopmans EC, Kuypers MMM, et al. (2007) Putative ammonia-oxidizing Crenarchaeota in suboxic waters of the Black Sea: a basin-wide ecological study using 16S ribosomal and functional genes and membrane lipids. Environ Microbiol 9: 1001–1016doi:10.1111/j.1462-2920.2006.01227.x
10. Martens-Habbena W, Berube PM, Urakawa H, la Torré de JR, Stahl DA (2009) Ammonia oxidation kinetics determine niche separation of nitrifying Archaea and Bacteria. Nature 461: 976–979doi:10.1038/nature08465
11. Jung M-Y, Park S-J, Min D, Kim J-S, Rijpstra WIC, et al. (2011) Enrichment and characterization of an autotrophic ammonia-oxidizing archaeon of mesophilic crenarchaeal group I.1a from an agricultural soil.Appl Environ Microbiol 77: 8635–8647doi:10.1128/AEM.05787-11

12. Kim J-G, Jung M-Y, Park S-J, Rijpstra WIC, Sinninghe Damsté JS, et al. (2012) Cultivation of a highly enriched ammonia-oxidizing archaeon of thaumarchaeotal group I.1b from an agricultural soil. Environ Microbiol 14: 1528–1543doi:10.1111/j.1462-2920.2012.02740.x
13. Park B-J, Park S-J, Yoon D-N, Schouten S, Damste JSS, et al. (2010) Cultivation of autotrophic ammonia-oxidizing archaea from marine sediments in coculture with sulfur-oxidizing bacteria. Appl Environ Microbiol 76: 7575–7587doi:10.1128/AEM.01478-10
14. Auguet J-C, Casamayor EO (2013) Partitioning of *Thaumarchaeota* populations along environmental gradients in high mountain lakes. FEMS Microbiol Ecol 84: 154–164doi:10.1111/1574-6941.12047
15. French E, Kozlowski JA, Mukherjee M, Bullerjahn G, Bollmann A (2012) Ecophysiological characterization of ammonia-oxidizing archaea and bacteria from freshwater. Appl Environ Microbiol 78: 5773–5780doi:10.1128/AEM.00432-12
16. Kowalchuk G, Stephen J (2001) Ammonia-oxidizing bacteria: A model for molecular microbial ecology.Annu Rev Microbiol 55: 485–529doi:10.1146/annurev.micro.55.1.485
17. Koops H, Purkhold U, Pommerening-Roeser A, Timmermann G, Wagner M (2006) The lithoautotrophic ammonia-oxidizing bacteria. The Prokaryotes 5: 778–811doi:_10.1007/0-387-30745-1_36
18. Speksnijder A, Kowalchuk G, Roest K, Laanbroek HJ (1998) Recovery of a *Nitrosomonas*-like 16S rDNA sequence group from freshwater habitats. Syst Appl Microbiol 21: 321–330doi:10.1016/S0723-2020(98)80040-4
19. Coci M, Bodelier PLE, Laanbroek HJ (2008) Epiphyton as a niche for ammonia-oxidizing bacteria: Detailed comparison with benthic and pelagic compartments in shallow freshwater lakes. Appl Environ Microbiol 74: 1963–1971doi:10.1128/AEM.00694-07
20. Chen G-Y, Qiu S-L, Zhou Y-Y (2009) Diversity and abundance of ammonia-oxidizing bacteria in eutrophic and oligotrophic basins of a shallow Chinese lake (Lake Donghu). Res Microbiol 160: 173–178doi: 10.1016/j.resmic.2009.01.003
21. de Bie M, Speksnijder A, Kowalchuk G, Schuurman T, Zwart G, et al. (2001) Shifts in the dominant populations of ammonia-oxidizing beta-subclass Proteobacteria along the eutrophic Schelde estuary. Aquat Microb Ecol 23: 225–236doi:10.3354/ame023225
22. Pester M, Rattei T, Flechl S, Groengroeft A, Richter A, et al. (2012) *amoA*-based consensus phylogeny of ammonia-oxidizing archaea and deep sequencing of *amoA* genes from soils of four different geographic regions. Environ Microbiol 14: 525–539doi:10.1111/j.1462-2920.2011.02666.x
23. Brochier-Armanet C, Boussau B, Gribaldo S, Forterre P (2008) Mesophilic crenarchaeota: proposal for a third archaeal phylum, the *Thaumarchaeota*. Nat Rev Microbiol 6: 245–252doi:10.1038/nrmicro1852

24. Spang A, Hatzenpichler R, Brochier-Armanet C, Rattei T, Tischler P, et al. (2010) Distinct gene set in two different lineages of ammonia-oxidizing archaea supports the phylum *Thaumarchaeota*. Trends Microbiol 18: 331–340doi:10.1016/j.tim.2010.06.003

25. Pester M, Schleper C, Wagner M (2011) The *Thaumarchaeota*: an emerging view of their phylogeny and ecophysiology. Curr Opin Microbiol 14: 300–306doi:10.1016/j.mib.2011.04.007

26. Hatzenpichler R (2012) Diversity, physiology, and niche differentiation of ammonia-oxidizing archaea.Appl Environ Microbiol 78: 7501–7510doi:10.1128/AEM.01960-12

27. Tourna M, Stieglmeier M, Spang A, Koenneke M, Schintlmeister A, et al. (2011) *Nitrososphaera viennensis*, an ammonia oxidizing archaeon from soil. Proc Natl Acad Sci USA 108: 8420–8425doi:10.1073/pnas.1013488108

28. Mussmann M, Brito I, Pitcher A, Sinninghe Damsté JS, Hatzenpichler R, et al. (2011) Thaumarchaeotes abundant in refinery nitrifying sludges express *amoA* but are not obligate autotrophic ammonia oxidizers.Proc Natl Acad Sci USA 108: 16771–16776doi:10.1073/pnas.1106427108

29. Auguet J-C, Nomokonova N, Camarero L, Casamayor EO (2011) Seasonal Changes of Freshwater Ammonia-Oxidizing Archaeal Assemblages and Nitrogen Species in Oligotrophic Alpine Lakes. Appl Environ Microbiol 77: 1937–1945doi:10.1128/AEM.01213-10

30. Herfort L, Kim J-H, Coolen MJL, Abbas B, Schouten S, et al. (2009) Diversity of Archaea and detection of crenarchaeotal *amoA* genes in the rivers Rhine and Tet. Aquat Microb Ecol 55: 189–201doi:10.3354/ame01294

31. Herrmann M, Saunders AM, Schramm A (2008) Archaea dominate the ammonia-oxidizing community in the rhizosphere of the freshwater macrophyte *Littorella uniflora*. Appl Environ Microbiol 74: 3279–3283doi:10.1128/AEM.02802-07

32. Herrmann M, Saunders AM, Schramm A (2009) Effect of Lake Trophic Status and Rooted Macrophytes on Community Composition and Abundance of Ammonia-Oxidizing Prokaryotes in Freshwater Sediments.Appl Environ Microbiol 75: 3127–3136doi:10.1128/AEM.02806-08

33. Herrmann M, Scheibe A, Avrahami S, Kuesel K (2011) Ammonium Availability Affects the Ratio of Ammonia-Oxidizing Bacteria to Ammonia-Oxidizing Archaea in Simulated Creek Ecosystems. Appl Environ Microbiol 77: 1896–1899doi:10.1128/AEM.02879-10

34. Liu S, Shen L, Lou L, Tian G, Zheng P, et al. (2013) Spatial distribution and factors shaping the niche segregation of ammonia-oxidizing microorganisms in the Qiantang river, China. Appl Environ Microbiol 79: 4065–4071doi:10.1128/AEM.00543-13

35. Lliros M, Gich F, Plasencia A, Auguet J-C, Darchambeau F, et al. (2010) Vertical Distribution of Ammonia-Oxidizing Crenarchaeota and Methanogens in the Epipelagic Waters of Lake Kivu (Rwanda-Democratic Republic of the Congo). Appl Environ Microbiol 76: 6853–6863doi:10.1128/AEM.02864-09

36. Merbt SN, Auguet J-C, Casamayor EO, Marti E (2011) Biofilm recovery in a wastewater treatment plant-influenced stream and spatial segregation of ammonia-oxidizing microbial populations. Limnol Oceanogr 56: 1054–1064doi:10.4319/lo.2011.56.3.1054

37. Vissers EW, Blaga CI, Bodelier PLE, Muyzer G, Schleper C, et al. (2013) Seasonal and vertical distribution of putative ammonia-oxidizing thaumarchaeotal communities in an oligotrophic lake. FEMS Microbiol Ecol 83: 515–526doi:10.1111/1574-6941.12013

38. Vissers EW, Anselmetti FS, Bodelier PLE, Muyzer G, Schleper C, et al. (2013) Temporal and Spatial Coexistence of Archaeal and Bacterial *amoA* Genes and Gene Transcripts in Lake Lucerne. Archaea. doi:10.1155/2013/289478

39. Baker DBD, Richards RPR (2002) Phosphorus budgets and riverine phosphorus export in northwestern Ohio watersheds. J Environ Qual 31: 96–108doi:10.2134/jeq2002.0096

40. Chapra SC, Dove A, Warren GJ (2012) Long-term trends of Great Lakes major ion chemistry. J Gt Lakes Res 38: 550–560doi: 10.1016/j.jglr.2012.06.010

41. Beeton AM (1965) Eutrophication of the St. Lawrence Great Lakes. Limnol Oceanogr 10: 240–254

42. Sterner RW, Anagnostou E, Brovold S, Bullerjahn GS, Finlay JC, et al. (2007) Increasing stoichiometric imbalance in North America›s largest lake: Nitrification in Lake Superior. Geophys Res Lett 34doi:10.1029/2006GL028861

43. Small GE, Bullerjahn GS, Sterner RW, Beall BFN, Brovold S, et al. (2013) Rates and controls of nitrification in a large oligotrophic lake. Limnol Oceanogr 58: 276–286doi:10.4319/lo.2013.58.1.0276

44. Lavrentyev PJ, Gardner WS, Johnson JR (1997) Cascading trophic effects on aquatic nitrification: experimental evidence and potential implications. Aquat Microb Ecol 13: 161–175doi: 10.3354/ame013161

45. Bollmann A, French E, Laanbroek HJ (2011) Isolation, cultivation, and characterization of ammonia-oxidizing bacteria and archaea adapted to low ammonium concentrations. Meth Enzymol 486: 55–88doi:10.1016/B978-0-12-381294-0.00003-1

46. Kandeler E, Gerber H (1988) Short-Term Assay of Soil Urease Activity Using Colorimetric Determination of Ammonium. Biol Fert Soils 6: 68–72

47. Keeney DR, Nelson DW (1982) Nitrogen - inorganic forms. In: Page AL. Methods of Soil analysis - part 2. American Society of Agronomy, Madison, WI, USA. pp. 643–698.

48. Shand CA, Williams BL, Coutts G (2008) Determination of N-species in soil extracts using microplate techniques. Talanta 74: 648–654doi:10.1016/j.talanta.2007.06.039

49. Welschmeyer NA (1994) Fluorometric Analysis of Chlorophyll-a in the Presence of Chlorophyll-B and Pheopigments. Limnol Oceanogr 39: 1985–1992

50. Francis C, Roberts K, Beman J, Santoro A, Oakley B (2005) Ubiquity and diversity of ammonia-oxidizing archaea in water columns and sediments of the ocean. Proc Natl Acad Sci USA 102: 14683–14688doi:10.1073/pnas.0506625102
51. Rotthauwe J, Witzel K, Liesack W (1997) The ammonia monooxygenase structural gene *amoA* as a functional marker: Molecular fine-scale analysis of natural ammonia-oxidizing populations. Appl Environ Microbiol 63: 4704–4712
52. Berry D, Ben Mahfoudh K, Wagner M, Loy A (2011) Barcoded primers used in multiplex amplicon pyrosequencing bias amplification. Appl Environ Microbiol 77: 7846–7849doi:10.1128/AEM.05220-11
53. Caporaso JG, Kuczynski J, Stombaugh J, Bittinger K, Bushman FD, et al. (2010) QIIME allows analysis of high-throughput community sequencing data. Nat Methods 7: 335–336doi:10.1038/nmeth.f.303
54. Ludwig W, Strunk O, Westram R, Richter L, Meier H, et al. (2004) ARB: a software environment for sequence data. Nucleic Acids Res 32: 1363–1371doi:10.1093/nar/gkh293
55. Kunin V, Engelbrektson A, Ochman H, Hugenholtz P (2010) Wrinkles in the rare biosphere: pyrosequencing errors can lead to artificial inflation of diversity estimates. Environ Microbiol 12: 118–123doi:10.1111/j.1462-2920.2009.02051.x
56. Lozupone C, Hamady M, Knight R (2006) UniFrac—an online tool for comparing microbial community diversity in a phylogenetic context. BMC Bioinformatics 7: 371.doi:10.1186/1471-2105-7-371
57. Beman JM, Popp BN, Francis CA (2008) Molecular and biogeochemical evidence for ammonia oxidation by marine Crenarchaeota in the Gulf of California. ISME J 2: 429–441doi:10.1038/ismej.2007.118
58. Wuchter C, Abbas B, Coolen MJL, Herfort L, van Bleijswijk J, et al. (2006) Archaeal nitrification in the ocean. Proc Natl Acad Sci USA 103: 12317–12322doi:10.1073/pnas.0600756103
59. Mincer TJ, Church MJ, Taylor LT, Preston C, Kar DM, et al. (2007) Quantitative distribution of presumptive archaeal and bacterial nitrifiers in Monterey Bay and the North Pacific Subtropical Gyre.Environ Microbiol 9: 1162–1175doi:10.1111/j.1462-2920.2007.01239.x
60. Sauder LA, Engel K, Stearns JC, Masella AP, Pawliszyn R, et al. (2011) Aquarium nitrification revisited: Thaumarchaeota are the dominant ammonia oxidizers in Freshwater Aquarium Biofilters. PlOS One 6doi:10.1371/journal.pone.0023281
61. Sauder LA, Peterse F, Schouten S, Neufeld JD (2012) Low-ammonia niche of ammonia-oxidizing archaea in rotating biological contactors of a municipal wastewater treatment plant. Environ Microbiol 14: 2589–2600doi:10.1111/j.1462-2920.2012.02786.x
62. Auguet J-C, Triado-Margarit X, Nomokonova N, Camarero L, Casamayor EO (2012) Vertical segregation and phylogenetic characterization of ammonia-oxidizing Archaea in a deep oligotrophic lake.ISME J 6: 1786–1797doi:10.1038/ismej.2012.33

63. van der Wielen PWJJ, Voost S, van der Kooij D (2009) Ammonia-oxidizing bacteria and archaea in groundwater treatment and drinking water distribution systems. Appl Environ Microbiol 75: 4687–4695doi:10.1128/AEM.00387-09

64. Mosier AC, Lund MB, Francis CA (2012) Ecophysiology of an ammonia-oxidizing archaeon adapted to low-salinity habitats. Microb Ecol 64: 955–963doi:10.1007/s00248-012-0075-1

65. Magalhaes CM, Machado A, Bordalo AA (2009) Temporal variability in the abundance of ammonia-oxidizing bacteria vs. archaea in sandy sediments of the Douro River estuary, Portugal. Aquat Microb Ecol56: 13–23doi:10.3354/ame01313

66. Hu A, Yao T, Jiao N, Liu Y, Yang Z, et al. (2010) Community structures of ammonia-oxidising archaea and bacteria in high-altitude lakes on the Tibetan Plateau. Freshwater Biol 55: 2375–2390doi:10.1111/j.1365-2427.2010.02454.x

67. Reed DW, Smith JM, Francis CA, Fujita Y (2010) Responses of ammonia-oxidizing bacterial and archaeal populations to organic nitrogen amendments in low-nutrient groundwater. Appl Environ Microbiol76: 2517–2523doi:10.1128/AEM.02436-09

68. Gubry-Rangin C, Hai B, Quince C, Engel M, Thomson BC, et al. (2011) Niche specialization of terrestrial archaeal ammonia oxidizers. Proc Natl Acad Sci USA 108: 21206–21211doi:10.1073/pnas.1109000108

69. Lehtovirta-Morley LE, Stoecker K, Vilcinskas A, Prosser JI, Nicol GW (2011) Cultivation of an obligate acidophilic ammonia oxidizer from a nitrifying acid soil. Proc Natl Acad Sci USA 108: 15892–15897doi:10.1073/pnas.1107196108

70. Ye W, Liu X, Lin S, Tan J, Pan J, et al. (2009) The vertical distribution of bacterial and archaeal communities in the water and sediment of Lake Taihu. FEMS Microbiol Ecol 70: 263–276doi:10.1111/j.1574-6941.2009.00761.x

71. Lebedeva EV, Hatzenpichler R, Pelletier E, Schuster N, Hauzmayer S, et al. (2013) Enrichment and genome sequence of the group I.1a ammonia-oxidizing archaeon "*Ca.* Nitrosotenuis uzonensis" representing a clade globally distributed in thermal habitats. PLOS One 8: e80835.doi:10.1371/journal.pone.0080835

72. Tourna M, Freitag TE, Nicol GW, Prosser JI (2008) Growth, activity and temperature responses of ammonia-oxidizing archaea and bacteria in soil microcosms. Environ Microbiol 10: 1357–1364doi:10.1111/j.1462-2920.2007.01563.x

73. Stahl DA, la Torré de JR (2012) Physiology and diversity of ammonia-oxidizing archaea. Annu Rev Microbiol 66: 83–101doi:10.1146/annurev-micro-092611-150128

74. Herrmann M, Hädrich A, Küsel K (2012) Predominance of thaumarchaeal ammonia oxidizer abundance and transcriptional activity in an acidic fen. Environ Microbiol 14: 3013–3025doi:10.1111/j.1462-2920.2012.02882.x

75. Bollmann A, Laanbroek H (2001) Continuous culture enrichments of ammonia-oxidizing bacteria at low ammonium concentrations. FEMS Microbiol Ecol 37: 211–221doi: 10.1016/S0168-6496(01)00163-5

76. Koops H, Pommerening-Roser A (2001) Distribution and ecophysiology of the nitrifying bacteria emphasizing cultured species. FEMS Microbiol Ecol 37: 1–9doi: 10.1111/j.1574-6941.2001.tb00847.x
77. Stehr G, Zorner S, Bottcher B, Koops HP (1995) Exopolymers - an ecological characteristic of a floc-attached, Ammonia-oxidizing Bacterium. Microb Ecol 30: 115–126
78. Kowalchuk G, Stephen J, DeBoer W, Prosser J, Embley T, et al. (1997) Analysis of ammonia-oxidizing bacteria of the beta subdivision of the class *Proteobacteria* in coastal sand dunes by denaturing gradient gel electrophoresis and sequencing of PCR-amplified 16S ribosomal DNA fragments. Appl Environ Microbiol63: 1489–1497
79. Mao Y, Yannarell AC, Mackie RI (2011) Changes in N-transforming archaea and bacteria in soil during the establishment of bioenergy crops. PLOS One 6: e24750.doi:10.1371/journal.pone.0024750
80. Mao Y, Yannarell AC, Davis SC, Mackie RI (2013) Impact of different bioenergy crops on N-cycling bacterial and archaeal communities in soil. Environ Microbiol 15: 928–942doi:10.1111/j.1462-2920.2012.02844.x
81. Lüke C, Frenzel P (2011) Potential of *pmoA* amplicon pyrosequencing for methanotroph diversity studies. Appl Environ Microbiol 77: 6305–6309doi:10.1128/AEM.05355-11

13
CHAPTER

ISOLATION AND CHARACTERIZATION OF SUPERIOR RUMEN BACTERIA OF CATTLE (BOS TAURUS) AND POTENTIAL APPLICATION IN ANIMAL FEEDSTUFF

Krushna Chandra Das[1, 2], Wensheng Qin[2]

[1]National Research Centre on Mithun, Indian Council of Agricultural Research, Medziphema, India

[2]Biorefining Research Institute, Department of Biology, Lakehead University, Thunder Bay, Canada

ABSTRACT

Rumen of cattle harbors many microorganisms responsible for bioconversion of nutrients into a source of energy for the animals. In recent years many rumen microbes have been isolated and characterized by sequence analysis of 16S ribosomal RNA gene. Some of the microbes have also been recommended as feed additives for improving the overall growth or production of animals. Rumen bacteria which have potential application in animal feed stuffs were isolated and characterized in this experiment. Isolation was carried out from the rumen of cattle (Bos taurus) using techniques of serial dilutions and repeated tubing of the selectively enriched microbial cultures by using the specific media for rumen bacteria. All the isolates were then screened for in vitro gas production and cellulase enzyme activity and four superior isolates were selected and characterized. There were 18.00% to 23.00% increases in gas production on addition of these isolates to the rumen fluid of cattle and there was better cellulase enzyme activity. Two isolates were identified as Butyrivibrio fibrisolvens, one isolate as Streptococcus species and one isolate as Clostridium aminophilum. This indicated that, these isolates are superior and may have potential to be used as microbial feed additive in ruminants if fed in higher quantity.

Citation: Das, K. and Qin, W. (2012) Isolation and characterization of superior rumen bacteria of cattle (Bos taurus) and potential application in animal feedstuff. Open Journal of Animal Sciences, 2, 224-228. doi:10.4236/ojas.2012.24031.

INTRODUCTION

Rumen microbes are of great importance for livestock and they are able to utilize plant nutrients efficiently as source of energy. Microbes survive in the rumen under different constraints which may be either natural or feed associated as some of the feeds contain a significant amount of anti-nutritional factors [1]. Any contaminant which cannot survive these constraints is eliminated. Therefore, the microbes which harbor the rumen are unique and for better understanding of rumen environment, identification and molecular characterization of microbes are highly essential. Moreover, the rumen manipulation by improving the digestibility of plant nutrient is a priority area for improving the growth or production of livestock.

In recent years many rumen microbes have been isolated and characterized by sequence analysis of 16S ribosomal RNA gene. Prokaryote diversity in the rumen of yak (Bos grunniens) and Jinnan cattle (Bos Taurus) were estimated by 16S rDNA homology analysis [2]. Some of the microbes have also been recommended as feed additive for improving the overall growth or production of animals [3]. The tropical forage Leucaena leucocephala contains mimosine, a non-protein amino acid. When consumed by ruminants in Australia and some parts of India, DHP causes goitrogenic effects. The detoxification of the 3-hydroxy-4(1H)-pyridone (DHP) by Synergistes jonesii, isolated from Hawaiian cattle, is one of the examples in manipulating rumen fermentation with bacteria [4]. Similarly addition of Megasphaera elsdenii has also experimentally prevented acidosis in steers [5]. The enzymes produced by rumen bacteria may contribute to the breakdown of switch grass, a renewable biofuel energy source [6]. Cows in early lactation produced 5.9% more milk when enzyme mixture was added to the concentrate feed [7]. The average increase in Dry Matter Intake (DMI) was 1.0 ± 1.3 kg/day and the average increase in milk yield was 1.1 ± 1.5 kg/d. In our study some of the superior bacteria of cattle were isolated and characterized which has got potential to be used as feed additive.

MATERIALS AND METHODS

Isolation of bacteria was done from the rumen of cattle (Bos Taurus, Simmental breed) of Canada. The slaughter house of Thunder Bay (Thunder Bay Meat Processing Plant, Murille, Thunder Bay) Canada was selected for sample collection and samples were collected immediately after slaughtering of animals (5 to 6 no) in the slaughter house. About 3 to 4 samples of rumen content were collected from rumen of each animal in the slaughter house and then mixed together to form a composite sample which was brought to the laboratory in strict anaerobic conditions for further processing.

Isolation Procedure

The isolation was carried out using conventional techniques of serial dilutions and repeated tubing of the selectively enriched microbial cultures by using the specific media for rumen bacteria (Figure 1). Strict anaerobic conditions were maintained during the isolation procedure.

The growth media contained: 15 ml Mineral Solution I (KH_2PO_4 3.0 g; $(NH_4)_2SO_4$ 6.0 g; NaCl 6.0 g; $MgSO_4$ 0.6 g; $CaCl_2 \cdot 2H_2O$ 0.795 g per Litre), 15 ml Mineral Solution II (K_2HPO_4 3 g/litre), 0.25 g Yeast Extract, 1 g Tryptone, 0.1 ml Resazurine (0.1%), 0.2 ml Hemin (0.05%), 0.5 g Microcrystaline Cellulose, 0.1 g Cellobiose, 0.4 g Sodium Carbonate, 20 ml Clear Rumen Fluid, 50 ml Distilled Water and 50 mg Cysteine Hydrochloride. Fifty bacterial isolates were obtained and screened for superior isolates.

In Vitro Gas Production

The pure isolates were screened for in vitro gas production [8]. 100 ml calibrated glass syringes (haeberle Labortechnik, Germany) were used for this purpose and 24 hours gas production was recorded in triplicate for each isolate. The amount of gas produced during the incubation of isolates with rumen liquor is closely related to degradability and energetic value of isolates.

Rumen fluid was obtained from three adult Simmental breed of cattle. The rumen liquor was collected into the thermo flask that had been pre warmed to a temperature of 39°C.

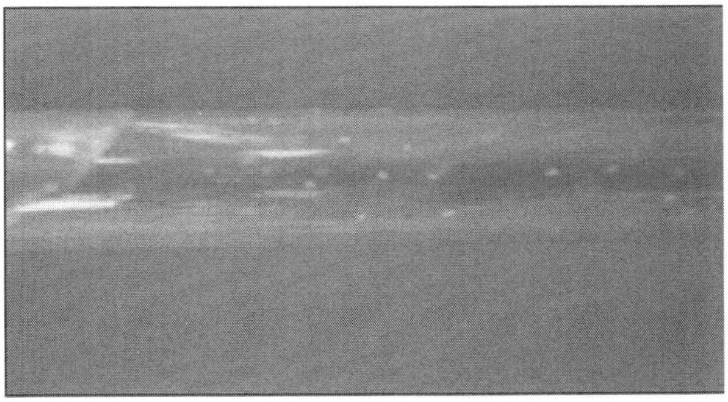

Figure 1: Bacteria colonies in roll tube.

Incubation procedure was done by using special type of calibrated transparent glass syringes with fitted silicon tube. Inoculums (30 ml) containing strained rumen liquor, buffer and mineral solution under continuous flushing with CO_2 was dispensed using 50 ml plastic calibrated syringe after which 2 ml of isolates were immediately pushed by using another 10 ml syringe. The glass syringe was tapped and pushed upward by the piston in order to completely eliminate air in the inoculums. The silicon tube in the syringe was then tightened by a metal clip so as to prevent escape of gas. Incubation was carried out at 39°C ± 1°C and the volume of gas production was measured after 24 hours of incubation. The average of the volume of gas produced from the blanks was deducted from the volume of gas produced per sample against the incubation time.

Carboxy Methyl Cellulase (CMC) Activity

The pure isolates were also screened for enzymatic activity. The CMC enzyme activity was determined by measuring the amount of reducing sugar released from carboxymethyl cellulose. The reaction mixture contained 1.0 ml phosphate buffer (0.1 M pH 6.8), 0.5 ml of 1% carboxymethyl cellulose solution prepared in 0.1 M phosphate buffer (pH 6.8), 0.5 ml enzyme and incubated for 60 minutes at 39°C. The reaction was halted and reducing sugars were determined by the addition of 3.0 ml of Dinitrosalicylic acid reagent [9]. Glucose was used as standard for determination of reducing sugars. The enzyme activity is expressed as International Unit (IU) which is micromole of glucose released per ml per hour.

Total Cellulase Activity by Filter Paper

Microplate-based filter paper assay [10] was conducted to measure total cellulose activity. Total cellulose activity is measured by the amount of glucose released from filter paper after two hrs incubation and is referred to as glucose equivalent. One FPU is defined as an average of one micromole of glucose equivalents released per min in the assay reaction.

DNA Isolation and PCR Amplification

Genomic DNA was extracted from the microbes of five isolates by using the standard kit (Nucleospin soil, manufactured by Macherey-Nagel GmbH & Co. KG). The DNA and its concentration were checked by agarose gel electrophoresis. Polymerase chain reaction (PCR) amplification of bacterial 16S rDNA was performed using the universal primer (forward: 5'-AGAGTTTGATCCTCAGAACGAACGCT-3' and reverse: 5'-TACGGCTACCTTGACGACTTCACCCC-3') of bacteria. The PCR reaction was set up in 20 μl volumes containing 1 μl template (100 ng/μl), 2 μl 10× buffer, 1.6 μl 25 mM $MgCl_2$, 1 μl of each primer (10 μM), 0.4 μl of 10 mM dNTP mix, 0.2 μl Taq DNA polymerase and 12.8 μl distilled water. The amplification conditions were standardized for universal primer. The amplification conditions were as follows: 3 min of initial denaturation at 95°C, followed by 30 cycles of denaturation at 94°C for 30 sec, annealing at 60°C for 30 sec, extension at 72°C for 2 min with the last cycle followed by a 10 min extension step at 72°C.

The PCR product was then visualized on an agarose gel (Figures 2 and 3). The bands were excised and DNA was purified from the gel slices using the standard kit (NORGEN Biotek Corporation, ON, Canada). The purified PCR product was sent for sequencing. The sequences were queried online in the NCBI BLAST database. The isolates were again confirmed by using the specific primer of bacteria.

Phylogenetic Analysis

The sequences of the isolates were compared with those available in the database. The reference sequences were downloaded from the nucleotide sequence databases and were used in further analysis for comparison. The obtained sequences were aligned using clustal V method of megAlign software (DNASTAR) and then phylogenetic tree was plotted.

RESULTS AND DISCUSSION

The isolation was carried out using conventional techniques of serial dilutions and repeated tubing of the selectively enriched microbial cultures by using the specific media for rumen bacteria. Strict anaerobic conditions were maintained during the isolation procedure.

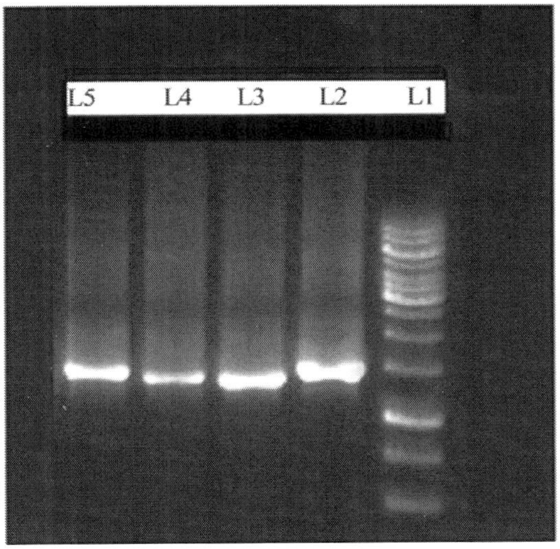

Figure 2: Agarose gel electrophoresis of PCR Products after amplification with Universal Primer (L1: 1 kb DNA Marker, L2-L5: isolates of BF12, BF14, SB13 and CA22).

In Vitro Gas Production

Four bacterial isolates i.e. BF12, BF14, SB13 and CA22 were observed in microscope and found to be cocci. On addition of these isolates to the rumen fluid of cattle, there was increase in gas production (Table 1). The gas production of isolates after 24 hours incubation varied from 8 to 10.5 ml/ml of isolate, highest in CA22 and lowest in BF14. It was observed that, there were 18.00% to 23.00% increases in gas production indicating release of energy on fibre-based mixed ration. In vitro gas production is the indicator of digestibility and nutritive value of feed stuffs [11,12]. Increased gas production in this experiment indicated that, these isolates may be termed as superior isolates and have potential to be used as microbial feed additives in ruminants (Table 1).

Enzymatic Analysis of Isolates

The isolates were subjected to enzymatic analysis like carboxymethyl cellulose (CMC), and total cellulose activity (filter paper activity). All the four isolates showed superior fibre degrading activity like CMC, and total cellulose activity, highest in CA22 and lowest in BF14 (Table 2). The in vitro gas productions were in same order as enzymatic analysis of the isolates.

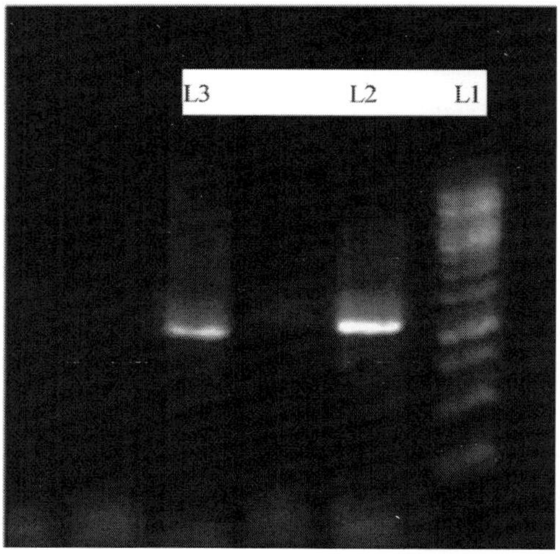

Figure 3: Agarose gel electrophoresis of PCR products amplification with specific primers of B. fibrisolvens. L1: 1 kb DNA Marker, L2-L3: B. fibrisolvens isolates (BF12, BF14).

Table 1: In vitro gas production of bacterial isolates

Isolate no	*In vitro* gas production	
	Increase in gas production (ml per 1 ml isolate)	Percentage of increase in gas production (%)
BF12	10.0	22.0
BF14	8.0	18.0
SB13	9.0	20.0
CA22	10.5	23.0

Identification of Isolates

For identification of isolates, DNA was isolated from all pure isolates and DNA was found to be sufficient for PCR reaction as quantified by spectrophotometer. The DNA was amplified by universal and specific primers of bacteria and amplification product of required band size was obtained (Figures 2 and 3). The bands were excised and purified PCR products were sent for sequencing. One isolate (BF12) was homologus to Butyrivibrio fibrisolvens (Genbank Accession number EU684229) where as other isolate (BF14) was homologus to Butyrivibrio fibrisolvens (Genbank accession number EF 427365). Isolate SB13 was homologus to Streptococcus species (Genbank Accession number FJ611790) where as isolate CA22 was homologus to Clostridium aminophilum (Genbank Accession number DQ278862). This indicated that, these isolates have potential to be used as microbial feed additive in ruminant animal for better growth and production. There is also ample scope in using the enzymes of these bacteria as microbial feed additives.

Phylogenetic Characterisation

The phylogenetic tree displayed three main groups of bacteria i.e. Butyrivibrio, Streptococcus and Clostridium under the phylum Firmicutes (Figure 4). The isolates BF12 was closely related to Butyrivibrio fibrisolvens isolated from sheep rumen (Accession no EU684229) and isolate no BF14 was closely related to Butyrivibrio fibrisolvens strain of 16S ribosomal RNA gene (Acession no EF427365). The isolate SB13 was closely related to Streptococcus equinus isolated from sheep rumen (Acession no AB563264) and the isolate CA22 was closely related to Clostridium aminophilum isolated from cow rumen (Accession no DQ278862). This shows that, three bacteria isolated in the present experiment are inhabitant of rumen.

Table 2: Enzymatic analysis of bacterial isolates

Isolate no	Enzyme analysis	
	CMC activity (MIU/ml)	Total cellulase activity (FPU/ml)
BF12	4.47	21.0
BF14	3.20	18.5
SB13	3.39	20.0
CA22	5.18	23.0

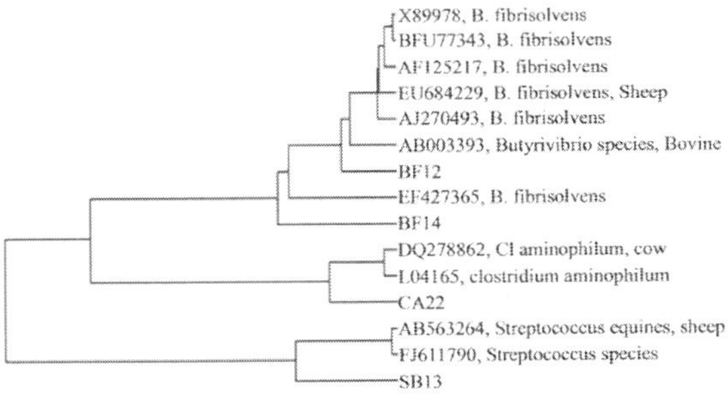

Figure 4: Phylogenetic characterization of isolates.

Potential of Rumen Bacteria as Microbial Feed Additives

Butyrivibrio species is the major cellulose and hemicellulose degrading bacteria in the rumen [13] and rapidly degrade xylan necessary for improving the overall efficiency of livestock production [14]. This may also be used as probiotics for reducing the aberrant crypt foci in the colon and rectum of mice [15]. Butyrivibrio species isolated in this experiment contained higher source of cellulolytic enzyme and in vitro gas production was increased on addition of these isolates to the rumen fluid of cattle.

Streptococcus species are the predominant cellulolytic micro-organisms that are associated with the possession of complex cellulose enzyme systems [16,17]. Some strains of streptococcus species are responsible for starch hydrolysis in the rumen [18]. Pulse dosing alone or pulse dosing with daily feeding of Streptococcus faecium M74 resulted in improved performance of incoming feedlot cattle [19]. Streptococcus species identified in this experiment are superior because of higher enzymatic activity and in vitro gas production. It has also been reported that, Streptococcus equinus are phylogenetically similar to Streptococcus bovis [20].

Clostridium strain IY-2 isolated from gut content of giant Panda possess cellulase activity towards untreated natural materials such as bamboo fibres, cellulose, rice straw and saw dust and comprises three major cellulases [21]. Clostridial isolate from swine were effective in degrading cell wall components from grasses [22]. As per the result of other worker, Clostridium aminophilum produces ammonia and is responsible for digestion of protein in the feed stuffs [23]. Though effectiveness depends on the strain of microbes, positive results were observed in this experiment.

CONCLUSION

Based on the findings of the result, it can be concluded that three species of microbes i.e. Butyrivibrio fibrisolvens, Steptococcus species and Clostridium aminophilus are superior and may have potential to be used as microbial feed additive as a source of microbial enzymes if fed in higher quantity. However, animal experiments are required to confirm the effectiveness of these three species of microorganism.

ACKNOWLEDGEMENTS

We are thankful to lab personnel Dr. Wenju Wan, Mehdi Dashtban, Miranda Maki, Bruce Rosa, Xi Chen and Yuanyuan Hong during the research work at Biorefining Research Initiative and Department of Biology, Lakehead University, Thunder Bay, Canada for providing help and cooperation for successful completion of this research work. This work was supported by NSERC-RCD funding of Canada to W. Q. and Indian Overseas fellowship of DBT to K. C. D.

REFERENCES

1. Kamra, D.N. (2005) Rumen microbial ecosystem. Current Science, 89, 124-135.
2. An, D.D., Dong, X.Z. and Dong, Z.Y. (2005) Prokaryote diversity in the rumen of yak (Bos grunniens) and Jinnan cattle (Bos taurus) estimated by 16S rDNA homology analyses. Anaerobe, 11, 207-215. doi:10.1016/j.anaerobe.2005.02.001
3. Mamen, D., Vadivel, V., Pugalenthi, M. and Parimelazhagan, T. (2010) Evaluation of fibrolytic activity of two different anaerobic rumen fungal isolates for their

utilization as microbial feed additive. Animal Nutrition and Feed Technology, 10, 37-49.

4. Allison, M.J., Mayberry, W.R., McSweeney, C.S. and Stahl, D.A. (1992) Synergistes jonesii, gen. nov., sp.nov: A rumen bacterium that degrades toxic pyridinediols. Systematic and Applied Microbiology, 15, 522-529. doi:10.1016/S0723-2020(11)80111-6

5. Robinson, J.A., Smolenski, W.J., Greening, R.C., Ogilvie, M.L., Bell, R.L., Barsuhn, K. and Peters, J.P. (1992) Prevention of acute acidosis and enhancement of feed intake in the bovine by Megasphaera elsdenii 407A. Journal of Animal Science, 70, 310 (Abstract).

6. Anonymous (2011) Cow rumen enzymes for better biofuels. Science Daily, 2011.

7. Yang, W.Z., Beauchemin, K.A. and Rode, L.M. (2000) A comparison of methods of adding fibrolytic enzymes to lactating cow diets. Journal of Dairy Science, 83, 2512-2520.doi:10.3168/jds.S0022-0302(00)75143-5

8. Menke, H.H. and Steingass, H. (1988) Estimation of the energetic feed value obtained from chemical analysis and in vitro gas production using rumen fluid. Animal Research and Development, 28, 7-55.

9. Miller, G.L. (1959) Use of dinitrosalicylic acid reagent for determination of reducing sugar. Analytical Chemistry, 31, 426-428. doi:10.1021/ac60147a030

10. Xiao, Z., Storms, R. and Tsang, A. (2004) Microplatebased filter paper assay to measure total cellulose activity. Biotechnol Bioeng, 88, 832-837. doi:10.1002/bit.20286

11. Akinfemi, A., Adesanya, A.O. and Aya, V.E. (2009) Use of an in vitro gas production technique to evaluate some Nigerian feedstuffs. American-Eurasian Journal of Scientific Research, 4, 240-245.

12. Das, K.C., Hundal, J., Mahapatra, P.S., Subudhi, P.K. and Sharma, K. (2010) Chemical composition and in vitro gas production of fodder tree leaves and shrubs. Indian Veterinary Journal, 87, 899-901.

13. Krause, D.O., McSweeney, C.S. and Forster, R.J. (1999) Molecular ecological methods to study fibrolytic ruminal bacteria: Phylogeny, competition, and persistence. In: Bell, C.R., Brylinsky, M. and Johnson-Green, P., Eds., Microbial Biosystems: New Frontiers Proceedings of the 8th International Symposium on Microbial Ecology, Atlantic Canada Society for Microbial Ecology, Halifax, 1999.

14. Cotta, M.A. and Zeltwanger, R.L. (1995) Degradation and utilization of xylan by the ruminal bacteria Butyrivibrio fibrisolvens and Selenomonas ruminantium. Applied and Environmental Microbiology, 61, 4396-4402.

15. Ohkawara, S., Furuya, H., Nagashima, K., Asanuma, N. and Hino, T. (2005) Oral administration of Butyrivibrio fibrisolvens, a butyrate-producing bacterium, decreases the formation of aberrant crypt foci in the colon and rectum of mice. Journal of Nutrition, 135, 2878-2883.

16. Schwarz, W.H. (2001) The cellulosome and cellulose degradation by anaerobic bacteria. Applied Microbiology & Biotechnology, 56, 634-649. doi:10.1007/s002530100710

17. Oyeleke, S.B. and Okusanmi, T.A. (2008) Isolation and characterization of cellulose hydrolyzing microorganism from the rumen of ruminants. African Journal of Biotechnology, 7, 1503-1504.

18. Dunican, L.K. and Seeley, H.W. (1962) Starch hydrolysis by Streptococcus equinus. Journal of Bacteriology, 83, 264-269. http://www.ncbi.nlm.nih.gov/pubmed/13888473

19. Lee, R.W. and Botts, R.L. (1988) Evaluation of a single oral dosing and continuous feeding of Streptococcus faecium M74 (Syntabac) on the performance of incoming feedlot cattle. Journal of Animal Science, 66, 460.

20. Saitou, N. and Nei, M. (1987) The neighbor-joining method: A new method for reconstructing phylogenetic trees. Molecular Biology and Evolution, 4, 406-425.

21. Yu, I.K. (1996) Secretion of Clostridium cellulase by E. coli. United States Patent 5496725.

22. Varel, V.H. and Pond, W.G. (1992) Characteristics of a new cellulolytic Clostridium sp. isolated from pig intestinal tract. Applied and Environmental Microbiology, 58, 1645-1649.

23. Atiwood, G.T., Reilly, K. and Patel, B.K.C. (1996) Clostridium proteoclasticum sp. nov., a novel proteolytic bacterium from the bovine rumen. International Journal of Systematic Bacteriology, 46, 753-758. doi:/10.1099/00207713-46-3-753

14
CHAPTER

CHARACTERIZATION OF TOMATO-ASSOCIATED RHIZOBACTERIA RECOVERED FROM VARIOUS TOMATO-GROWING SITES IN TUNISIA

Nada Ouhaibi-Ben Abdeljalil[1,2], Jessica Vallance[3,4], Jonathan Gerbore[5], Emilie Bruez[3,4], Guilherme Martins[6], Patrice Rey[3,4] **and Mejda Daami-Remadi**[2]

[1]Higher Agronomic Institute of Chott-Mariem, Sousse University, 4042-Chott Mariem, Tunisia

[2]UR13AGR09-Integrated Horticultural Production in the Tunisian Centre East, Regional Centre of Research on Horticulture and Organic Agriculture, University of Sousse, 4042, Chott-Mariem, Tunisia

[3]INRA, UMR1065 Santé et Agroécologie du Vignoble (SAVE), ISVV, F-33140 Villenave d'Ornon, France

[4]Université de Bordeaux, Bordeaux Sciences Agro, ISVV, UMR1065 SAVE, F-33140 Villenave d'Ornon, France

[5]BIOVITIS, 15400 Saint Etienne de Chomeil, France

[6]USC Oenologie-INRA, Université Bordeaux Segalen, Bordeaux Sciences Agro, ISVV, Villenave d'Ornon, France

ABSTRACT

In the present study, a total of 200 rhizobacterial isolates were obtained from rhizosphere of healthy tomato plants grown in fields with a history of severe soilborne diseases and mainly crown and root rots. Screened their capacity to suppress in vitro growth of Sclerotinia sclerotiorum and Rhizoctonia solani, 69 and 57 isolates out of the 200 tested were shown able to inhibit significantly the mycelial growth of target pathogens by 11-62% relative to control. The 25 most effective isolates, leading to suppression of both fungi by more than 45% over control, were selected and subjected to morphological, biochemical, molecular, and metabolic

Citation: Abdeljalil NOB, Vallance J, Gerbore J, Bruez E, Martins G, et al. (2016) Characterization of Tomato-associated Rhizobacteria Recovered from Various Tomato-growing Sites in Tunisia. J Plant Pathol Microbiol 7: 351. doi:10.4172/2157-7471.1000351.

characterizations. This collection of tomato-associated rhizobacteria exhibited a great morphological and biochemical diversity. Sequencing of 16S rRNA and rpoB genes led to the identification of four genera namely Bacillus, Chryseobacterium, Enterobacter, and Klebsiella. The most frequent species were B. amyloliquefaciens, B. thuringiensis, B. megaterium, B. subtilis, E. cloacae, C. jejuense, and K. pneumoniae. Screening for their plant growth-promoting properties, 20 isolates were shown able to produce siderophore, 18 had solubilized phosphate, and 19 were capable to synthesize indole-3- acetic acid (IAA). PCR amplification of lipopeptide biosynthetic genes revealed the presence of genes encoding fengycin A and bacillomycin D biosynthesis in 18 and 16 isolates, respectively. Metabolic characterization performed using Biolog™ Ecoplates indicated that tomato-associated rhizobacteria displayed a large metabolic activity and they were able to use a wide range of carbon sources with the increase of the incubation duration. Based on their metabolic profiles, these rhizobacterial isolates were grouped into eight major clusters generated at the different sampling times (24, 48 and 120 h of incubation). Average well-color development (AWCD) values were found to be positively correlated with the Shannon diversity index.

INTRODUCTION

Tomato (*Solanum lycopersicum* L.) is one of the most widely grown vegetable crop worldwide and is ranked second after potato [1]. In Tunisia, tomato is a strategic crop but unfortunately threatened by various bioagressors and mainly soilborne fungi due to the practice of monocropping and the lack of crop rotation due to the limited producing areas [2-5]. Moreover, the widely grown tomato cultivars were susceptible to soilborne diseases and especially to Rhizoctonia Root Rot caused by *Rhizoctonia solani* and Sclerotinia Stem Rot induced by*Sclerotinia sclerotiorum*. These pathogens are responsible for damping-off and/or collar, stem and roots rots and eventually plant death leading to serious yield losses both under greenhouse and open field growing systems [6,7]. They survive in soil either as sclerotia or as mycelium in crop debris [8]. Currently, few fungicides are effective in controlling these serious pathogens but even when allowed, chemical tools still involve technical, environmental and toxicological risks [9]. Moreover, it is very difficult to protect tomato plants from crown and root rot diseases using fungicides only particularly when the crown zone was covered with dense plant foliage. Such issues have encouraged research efforts on new, effective, and eco-friendly alternative methods such as biocontrol [10].

Several microorganisms have been reported as effective biocontrol agents (BCA) for suppression of plant diseases caused by soilborne pathogens [11-13]. A particular interest was attributed to root-associated bacteria which displayed interesting antagonistic potential towards Rhizoctonia spp. [14]. Plant growth promoting rhizobacteria (PGPR) produce various types of antifungal metabolites, including antibiotics and siderophores, capable of reducing or suppressing infection by pathogenic fungi in several pathosystems [15,16]. Furthermore, rhizobacteria have received particular attention because of their excellent root colonizing ability and their ability to produce a wide range of antifungal metabolites and auxins such as indole-3-acetic acid or IAA) [17-19].

Due to the limited knowledge in Tunisia on the nature of native tomato-associated rhizobacteria and their ability to suppress root rot diseases in tomato, the main objectives of the present study were i) to isolate bacterial isolates from tomato rhizosphere and to assess their ability to suppress the *in vitro* growth of *S. sclerotiorum* and *R. solani*, ii) to identify and characterize the most effective ones in suppressing both pathogens, and iii) to assess their potential to produce antifungal and plant growth-promoting compounds.

MATERIALS AND METHODS

Soil sampling and processing

Tomato-associated rhizobacteria were isolated from the rhizospheric soils of apparently healthy and vigorous tomato plants grown in various fields and greenhouses with a history of severe soilborne diseases. Surveys concerned different tomato-growing sites namely Kairouan, Sidi Bouzid, Cap-Bon, Sahline and Chott-Mariem regions (Table 1). Soil samples were removed from the top layer (15-20 cm deep) using a small shovel. Each soil sample was placed in a separate plastic bag before being brought to the laboratory and processed.

Table 1: Origin and number of soil samples used for isolation of tomato-associated rhibacteria

Geographical location	Growing region	Crop type and number of samples		Cropping system
		Protected	Open field	
Cap Bon	Korba Takelsa	- 5	5 -	Conventional
Centre	Chebika (Kairouan) Sidi Ali Ben Salem	- -	10 5	Conventional
Centre-Ouest	Regueb (Sidi Bouzid)	-	5	Conventional
Sahel Centre-East	Chott-Mariem Sahline Téboulba Ouardanine Békalta Ouled chamekh	5 5 5 5 5 5	- - - - - -	Organic Conventional Conventional Conventional Conventional Conventional
Total	-	**35**	**25**	

Bacterial isolation and culture

A subsample of 1 g was taken from each air-dried soil sample and homogenized in 10 mL SDW. The obtained soil suspension was shaken at 150 rpm at 25°C for 24 h. The resulting sludge was filtered and serially diluted four times (10^{-1} to 10^{-4}), and 0.1 ml of each diluted suspension was injected in a separate Petri dish containing 10 ml of molten Levan Peptone Glucose Agar (LPGA) and stirred well before medium solidification.

After 2 to 3 days of incubation at 28°C, distinct individual developing bacterial colonies were selected based on of their macro morphological diversity and picked separately onto Nutrient agar (NA) medium. Purification was performed by streaking out each distinct type of colony on NA in a separate plate and grown during 48 h at 28°C before being used for characterization bioassays. Stock cultures were stored at -20°C in Nutrient Broth (NB) medium supplemented with 15% glycerol.

Assessment of antifungal potential of collected rhizobacteria

R. solani and *S. sclerotiorum* isolates used for the assessment of the antifungal potential of the collected rhizobacterial isolates was originally recovered from naturally infected tomato plants exhibiting typical Rhizoctonia Root Rot and Sclerotinia Stem Rot symptoms, respectively. Fungal cultures were gratefully provided by the Plant Pathology Laboratory at the Regional Centre of Research on Horticulture and Organic Agriculture of Chott-Mariem, Tunisia. They were grown onto Potato Dextrose Agar (PDA) medium supplemented with streptomycin sulfate (300 mg/L w/v) and incubated at 25°C during 5 days before use.

A total of 200 rhizobacterial isolates was collected from the different sampling sites. They were evaluated for their antifungal activity against both target pathogens using the dual culture technique. Agar plugs (5 mm in diameter), removed from 5-day old cultures, were plated at one side of a Petri plate containing PDA. From a 48 h -old culture, 10 μL of bacterial cell suspension (108 CFU/mL) were injected into a well (5 mm in diameter) performed at the opposite side using a sterile cork borer. Plates challenged with pathogen plugs only were used as control. Each individual treatment was replicated thrice. The whole experiment was repeated twice.

Plates were incubated at 28°C for 5 days. The diameter of target pathogen colony was measured and the percentage of growth inhibition was calculated using the following formula [20]:

$$GI(\%) = \frac{(C\text{-}T)\times 100}{C}$$

where I: Percent growth inhibition; C: Colony diameter of the target pathogen in control plates; T: Colony diameter of the target pathogen dual cultured with rhizobacterial isolates.

A growth inhibition score (GIS), estimated based on GI values, was attributed to each rhizobacterial isolate tested using an arbitrary 0-5 scale where 0 = no inhibition, 1= GI 1- 15%; 2 = GI 16- 35%; 3 = GI 36-45%; 4 = GI 46-65%, and 5 = GI $\geq$ 66%.

The most effective 25 isolates in suppressing growth of both pathogens by more than 45% (GIS estimated at 4 and more) will be selected for further investigations.

Characterization of selected rhizobacterial isolates

To explore the diversity of tomato-associated rhizobacteria, the selected 25 bacterial isolates were subjected to various characterizations.

Morphological and biochemical characterization

Morphology of bacterial colonies was characterized based on various traits such shape, elevation, surface, margin, color, and pigmentation. This macro-morphological characterization was performed, following Patel et al. [21] protocol, on bacterial cultures previously grown for 48 h on NA medium and incubated at 28°C.

The biochemical traits considered in the present study were the Gram staining, levan activity, catalase activity, oxidase activity, and ability to produce fluorescent pigment on King's B medium (KB) [22]. Results of these tests were scored as positive or negative.

Characterization for plant growth-promoting traits

Siderophore production: Bacterial isolates were grown on NA medium for 48 h at 28°C. Siderophore production was assessed by spot inoculation of bacterial isolates onto the Chrome Azurol S (CAS) agar medium. Control plates were spot treated with SDW only. Plates were incubated at 28°C for 5 days. Development of yellow-orange halos around the growing colonies indicated positive siderophore production as indicated by Shobha and Kumudini [23]. The halo diameter was measured and compared to the untreated control.

Phosphate solubilization: Phosphate solubilization ability of isolates was assessed qualitatively using a modified Pikovskaya's agar medium containing tricalcium phosphate according to Venkadesaperumal et al. [24] method. Briefly, each bacterial culture was spot inoculated in the centre of a Petri plate and incubated at 28°C for 7 days. Control plates were treated with SDW only. Clear halo zones formed around colonies indicate positive Phosphate solubilization activity.

Indole-3-Acetic Acid (IAA) production: Production of the phytohormone indole-3-acetic acid (IAA) by the selected isolates was assessed using Ghodsalavi et al. [25] protocol. Briefly, each bacterium was grown onto NB and incubated at 28°C for 48 h in an incubatorshaker at 170 rpm. Then, 50 μL of each bacterial liquid culture were transferred to NB containing 50 μg/mL of L-tryptophane. After 48 h of incubation, the bacterial suspensions were centrifuged at 10000 rpm for 10 min. About 1 mL of supernatant was mixed with 4 mM of Salkowski reagent (1 mL 0.5 M FeC13 and 50 mL of 35% HC1O4) [26]. Reagent mixed with SDW was used as control. After 30 min of incubation, the inoculated reagents that turned red indicate that the tested isolate is an IAA-producing agent.

Characterization of the hypersensitive reaction: The tomato-associated rhizobacteria isolated were also screened for their hypersensitive reaction (HR) on tobacco leaves. Each isolate was suspended in SDW and the concentration of the cell suspension was adjusted to 10^8cells/mL using a heamocytometer. The bacterial suspension was injected into tobacco leaves. Leaves injected with *Pseudomonas syringae*pv. *tomato*, a phytopathogenic bacteria, were used as positive controls and others injected with SDW only were considered as negative control. Bacteria with positive reaction i.e. causing the development of necrotic area 24 h after injection were ranked as pathogenic and therefore excluded from any further investigation on their antagonistic potential as suggested by Nawangsih et al. [27].

Metabolic characterization: Metabolic characterization of the 25 bacterial isolates was performed using the Biolog® system. Among the 31 substrates present in triplicate, 10 contained carbon and nitrogen (including six amino-acids), 2 composed of carbon and phosphorus and 19 with only carbon. Bacterial cells were recuperated after centrifugation at 3000 rpm during 30 min at 15°C. The pellet cells were resuspended in 15 mL of 0.85% NaCl solution. About 150 µL of the obtained bacterial cell suspension were dropped in each well of the Biolog® EcoPlates as indicated in Renault et al. work [28]. The plates were incubated at 28°C in darkness. The rate of substrate use was indicated by the reduction of the tetrazolium, a redox indicator dye turning from colorless to purple. The color development was determined at 590 nm with Microlog-System Release 4.0 software absorbance after 24 h, 48 h and 120 h incubation.

Microbial response in each microplate, expressed through average well-color development (AWCD), was determined according to Gomez et al. [29] using the following formula:

$$AWCD = \Sigma ODi/31$$

where OD_i is the optical density in each well, corrected by subtracting the water blank well (inoculated but without any carbon source). The reading for each well was then corrected by subtracting the values of the water blank for a given replicate, and standardized by dividing by AWCD for that replicate as indicated in Garland and Mills [30].

If OD_{ijt} represents the corrected optic density for well for replicate j at time t, then the AWCD for replicate j at time t is given as follows:

$$AWCDjt = \frac{1}{31}\sum_{i=1}^{n31} OD_{ijt}$$

The standardized OD values are determined as follows:

$$\overline{ODijt} = \frac{ODijt}{AWCDjt}.$$

The Shannon diversity index (H), which was commonly used to characterize species diversity within a microbial community [8,30], was calculated to assess the diversity of the 25 rhizobacterial isolates using the following Chojniak et al. [31] formula:

$$H = -\sum (P_i \times \ln P_i)$$

where P_i is the proportional optical density value of each well.

The relationships among different samples on the basis of the transformed data were determined using Principal Component Analysis (PCA) according to Grove et al. [32]. This technique allowed projection of the original data onto new axis (principal components) thus, reflecting any intrinsic pattern in the multidimensional data group [30].

Molecular identification and detection of antibiotic producing genes

DNA was extracted from pure cultures grown in NB using the Invisorb Spin Plant Mini Kit (Invitek) according to manufacturer's instructions. The DNA extracts were then quantified with a nanodrop (ND-1000, Thermoscientific, Labtech) and homogenized at a concentration of 10 ng/μL.

Bacterial isolates were first identified by 16S rDNA sequence analysis, amplified with the universal primers 799f/1492r. For distinction of isolates belonging to *Bacillus* genus, two other genes were sequenced i.e. *rpo*B and *gyr*B. Also, all isolates were screened for 5 antibiotic-producing genes i.e. 4 fengycins (A, B, D, and E), and one gene encoding bacillomycin D biosynthesis (Table 2).

Table 2: Pairs of primers used for identification of tomato-associated rhizobacteria and detection of antibiotic biosynthesis genes

Primer name	Sequence	Amplicon size (pb)	Annealing Temperature (°C)
799f	AACMGGATTAGATACCCKG	400	52
1492r	GTTACCTTGTTACGACTT	400	52
rpoB-f	GACGATCATYTWGGAAACCG	500	52
rpoB-r	GGNGTYTCRATYGGACACT	500	52
gyrB-f	AAAACAACCRATTCATGAAG	500	52
gyrB-r	TCGCTTCACTATTYCCAAGT	500	52
FENA1F	GACAGTGCTGCCTGATGAAA	757	54
FENA1R	GTCGGTGCATGAAATGTACG		
FENB1F	CAGCCGCTGTCAACAAGATA	950	54
FENB1R	ACACGACATTGCGATTGGTA		
FEND1F	TTTGGCAGCAGGAGAAGTTT	964	53
FEND1R	GCTGTCCGTTCTGCTTTTTC		
FENE1F	GCCAAAAAGAAACGAGCAAG	756	53
FENE1R	GTCGGAGCTAACGCTGAAAC		
BACC1F	GAAGGACACGGCAGAGAGTC	814	60
BACC1R	CGCTGATGACTGTTCATGCT		

The 8 genes were amplified under the same PCR conditions. Briefly, each PCR mixture contained 2 μL of extracted DNA (10 ng/μL), 1 μL of MgCl2 (50 mM), 0.6 μL of dNTP (10 mM), 0.6 μL of each primer (10 ng/μL), 3 μL Bovine Serum Albumine (10 μg/μL) (BioLabs), 0.1 μL of Taq polymerase (SilverStar DNA polymerase, Eurogentec), 3 μL of PCR buffer (10X) and 19.7 μL of SDW in a total volume of 30 μL.

PCR amplification reactions were performed in a Mastercycler Gradient thermocycler (Eppendorf) using De Clerck and De Vos method [33]. The following cycling conditions were used: initial denaturation of 5 min at 94°C followed by 30 cycles of denaturation (1 min at 94°C), annealing of primers (45 s at 52°C), elongation (1 min at 72°C) and a final elongation for 8 min at 72°C. Amplified fragments were purified and sequenced by GATC Biotech (Germany). For species level identification, sequences were compared with the Genbank database using the BLAST program (http://blast.ncbi.nlm. nih.gov/Blast.cgi) [34]. Then, sequences were deposited in the Genbank database for getting the accession numbers.

Statistical analysis

Mean colony diameters of each target pathogen depending on bacterial treatments tested (200 isolates and untreated control) were analyzed using one-way analysis of variance according to a completely randomized design. Means were separated using the Duncan's Multiple Range test at ($P \leq 0.05$). Based on growth inhibition records relative to control, isolates were ranked into different classes as previously described and the number of isolates within each class was determined. Data analyses were performed using SPSS version 16.0 and each individual treatment was replicated thrice.

RESULTS

Assessment of the antifungal potential of tomato-associated rhizobacteria towards *Rhizoctonia solani* and *Sclerotinia sclerotiorum*

All the 200 bacterial isolates collected were screened for their antifungal potential against two tomato fungal pathogens namely *S. sclerotiorum*and *R. solani* using the dual culture method. Their effects on pathogen colony diameter were compared to the untreated control. ANOVA analysis, performed for each target pathogen, revealed that mean colony diameter, noted after 5 days of incubation at 25°C, varied significantly (at $P \leq 0.05$) depending on bacterial treatments tested (collected isolates and control). In fact, 69 isolates out of the 200 tested had significantly inhibited by 11.11 to 62.22% the mycelial growth of S. *sclerotiorum* relative to the untreated control (data not shown). Based on growth inhibition scores (GISs) exhibited toward *S. sclerotiorum*, 155 isolates were ranked within class 1 indicating that their relative percentages of growth inhibition were less than 15%. However, interestingly, 27 isolates led to more than 45% decrease in pathogen growth and were, thus, ranked within class 4 (Figure 1).

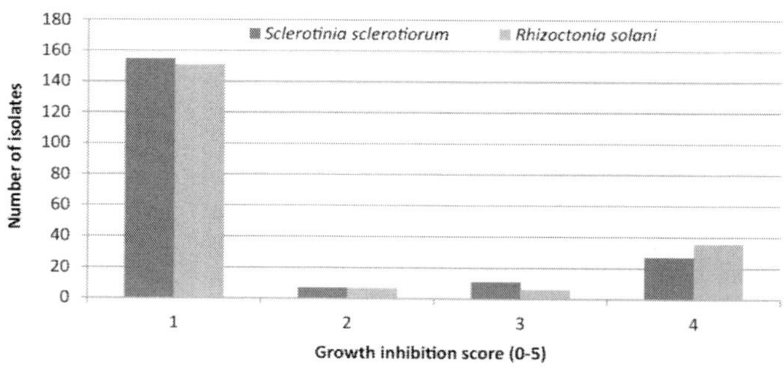

Figure 1: Number of tomato-associated rhizobacterial isolates and their repartition into four growth inhibition classes depending on their relative ability to suppress the *in vitro* growth of *Sclerotinia sclerotiorum* and*Rhizoctonia* solani using the dual culture method. Growth

inhibition score was estimated based on growth inhibition (GI) percentage calculated using Prashar et al. [20] formula using an arbitrary 0-5 scale where 0 = no inhibition, 1= GI 1- 15%; 2 = GI 16- 35%; 3 = GI 36-45%; 4 = GI 46-65%, and 5 = GI ≥ 66%.

As for *R. solani,* mycelial growth was significantly lowered by 11.85 to 61.48%, compared to the untreated control, when dual cultured with 57 isolates. Data shown in Figure 1 indicated that when assessed toward *R. solani,* 151 isolates had inhibited by less than 15% pathogen growth versus control and were thereafter ranked within class 1 but, interestingly, more than 45% decrease in pathogen radial growth was achieved using 36 isolates.

The most effective 25 rhizobacterial isolates against both targeted pathogens and raked with class 4 were selected for further identification and characterizations.

Morphological and biochemical characterization of tomato-associated rhizobacteria

Based on their macro-morphological traits on NA medium, the selected isolates exhibited a great diversity color in their color (white, transparent, cream, light yellow), shape and texture as shown in Table 3.

Table 3: Molecular, morphological and biochemical characterization of tomato-associated rhizobacteria and assessment their ability to produce lipopeptide antibiotics and plant growth-promoting compounds. Positive reaction (+); Negative reaction (–); Fen A : Fengycin A, Bac : Bacillomycin D, IAA: Indole-3- acetic acid, P.solubilization: Phosphate solubilization. All isolates were negative for hypersensitive reaction (HR) on tobacco leaves, and they were negative for the detection of Fengycin B, D and E. *: Average diameter of orange halo of the siderophore production by rhizobacteria. PGPR: Plant growth promoting rhizobacteria

Molecular identification			Morphological characterization	Biochemical characterization				Detection of antibiotic biosynthesis genes		PGPR traits		
Isolate	Similarity (%)	Accession No.	Colony morphology	Gram	Oxidase	Catalase	Levan	Fen A	Bac D	IAA	P. Solubilization	Siderophore production (halo diameter in mm)*
Bacillus megaterium B1	99	KU168423	White, flat, opaque	-	+	+	-	-	+	+	-	48.3
B. thuriengiensis B2	99	KU158884	White, flat, opaque	+	+	+	-	-	+	+	+	31.7
Enterobacter cloacae B3	99	KT923049	White, flat, opaque	-	-	+	-	+	-	+	+	55.0
E.cloacae B4	100	KT923050	Transparent, flat, Translucide	-	-	+	+/-	+	+	+	+	50.0
B. megaterium B5	100	KT923054	Transparent, flat,Translucide	-	+/-	+	+/-	+	-	+	+	45.0
B. subtillis B6	100	KT921427	White, convex, translucide	+	-	+	-	+	+	-	+	0.0
B. amyloliquefaciens B7	99	KT921428	White, flat, opaque	+	-	+	-	+	+	+	-	83.3
B. subtillis B8	99	KU158885	White, flat, opaque	+	+	+	+/-	+	+	+	+	35.0
B. amyloliquefaciens B9	100	KU158887	Transparent, convex, opaque	+	+	-	-	+	+	+	+	36.7
B. subtillis B10	99	KT921327	Light yellow, flat, opaque	+	+	+	-	+	+	+	+	90.0
Chryseobacterium jejuense B11	99	KU158886	White, convex, translucide	+	-	-	-	-	+	+	+	43.3

Klebsiella pneumoniae B12	99	KT921328	Transparent, convex, translucide	-	-	+	-	+	-	+	+	90.0
B. amyloliquefaciens B13	99	KT951658	Transparent, convex, translucide	+	+	+/-	-	+	-	+	+	36.7
B. subtillis B14	99	KU161090	White, flat, opaque	+	+	+/-	-	+	-	-	-	73.3
B. amyloliquefaciens B15	99	KT923051	White, plate, opaque	+	+	+/-	-	-	+	-	-	28.3
E. cloacae B16	100	KT921429	Cream, flat, opaque	-	-	+	-	+	+	+	+	90.0
B. subtillis B17	100	KT923055	White, flat, opaque	+	-	+	-	+	+	+	+	0.0
B. amyloliquefaciens B18	100	KT923052	White, plate, opaque	+	+	-	-	+	+	+	+	0.0
B. subtillis B19	99	KT921430	Cream flat, opaque	+	+	+/-	-	+	-	-	-	75.0
B. subtillis B20	99	KT921431	White, plate, opaque	+	+	+	-	+	+	+	+	66.7
B. amyloliquefaciens B21	100	KT923047	Transparent, plate, translucide	+	+	+	-	+	+	+	+	0.0
B. amyloliquefaciens B22	100	KT923053	White, plate, opaque	+	+	+	-	+	+	+	+	0.0
B. thuriengiensis B23	99	KT923056	White, plate, opaque	+	+	+	-	+	+	+	+	13.3
B. megaterium B24	100	KT923048	Cream, flat, opaque	-	+	+	-	-	+	-	-	75.0
B. subtillis B25	99	KU161091	Transparent, flat, Translucide	+	+/-	+/-	-	-	+	-	-	10.0

The results of biochemical characterizations indicated also a large diversity in their biochemical traits. In fact, among the collected rhizobacteria, 18 isolates out of the 25 tested were Gram positive and the remaining ones were Gram negative. The oxidase activity (oxidase +) was detected in 15 isolates and 2 were oxidase +/-. The majority of isolates (22/25) were Levane – and 3 were Levan +/-. Catalase positive activity (catalase +) was detected in 17 isolates while 5 exhibited catalase +/- and the remaining isolates gave negative reaction. It should be also mentioned that all the isolates tested did not produce fluorescent pigments in King's medium B; this result indicated the absence of*Pseudomonas fluorescens* within the bacterial collection subjected to this chemical characterization (Table 3).

Molecular identification of tomato-associated rhizobacteria

In the present study, molecular identification of 25 isolates was performed using the 16S rRNA and *rpo*B gene sequences. Based on 16S rRNA gene sequencing, the results revealed that the collected isolates belonged to different genera; the sequence analysis of 16S rRNA (alignment on 700 pb) obtained for *Bacillus* sp. was similar to each other and did not allow evident distinction at the species level although the fragments revealed more than 98% similarity. In fact, as shown in Table 3, among the 25 isolates tested, 20 were identified as belonging to the genus Bacillus, three as being *Enterobacter cloacae*, one as *Chryseobacterium jejuense* and one as *Klebsiella pneumoniae*.

Unlike the level of sequence similarity noted between Bacillus isolates using the 16S rRNA gene sequencing, rpoB gene sequences showed more variation and were used to identify isolates at the species level. In fact, the rpoB amplicons were of approximately 400 bp in length and the sequence similarity of the collected isolates and the bacterial sequences available at the Gene Bank ranged from 99 to 100%. Based on comparison of the partial sequences using BLAST, 3 isolates (B1, B5 and B24) were identified as *B. megaterium*, 9 isolates (B2, B6, B8, B10, B14, B17, B19, B20, and B25) were affiliated to *B. subtilis*, 7 isolates (B7, B9, B13, B15, B18, B21 and B22) were attributed to *B. amyloliquefaciens*, and one isolate (B23) was determined as being *B. thuriengiensis*.

Sequences of the isolates, used in the present study, were submitted to the GenBank and their relative accession numbers are cited in Table 3.

Detection of antibiotic biosynthesis genes

To determine whether the collected tomato-associated rhizobacteria have the potential to produce different types of antimicrobial lipopeptides, PCR amplifications were performed to detect genes responsible for fengycin (A, B, D and E) and bacillomycin D biosynthesis. Among the 5 primer pairs tested, 2 genes encoding fengycin A and bacillomycin D biosynthesis were detected. Among the 25 isolates screened using FENA1F/FENA1R primer pairs, 18 yielded the indicative presence of fenA gene involved in fengycin A biosynthesis. Screened using BACC1F/BACC1R primer pairs, 16 isolates were found to be bacillomycin D-producing agents (Table 3 and Figure 2).

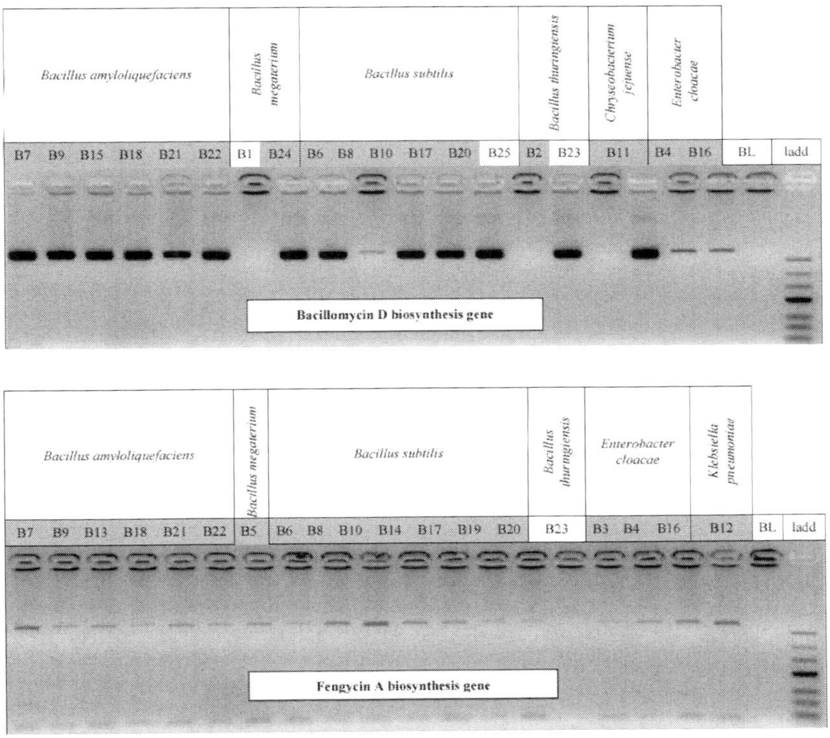

Figure 2: Results of PCR amplification of Bacillomycin D and Fengycin A biosynthesis genes in the 25 tomato-associated rhizobacterial isolates recovered from Tunisian rhizospheric soils. Lane Bl: negative control; lane ladd: molecular weight markers (Ladder).

After sequencing the antibiotic biosynthesis genes, the results showed that the gene encoding the enzymes of fengycin A production does not distinguish several different haplotypes within the tested collection of isolates and the obtained sequences were aligned on 757 pb. However, the gene encoding bacillomycin D biosynthesis allowed the discrimination of 3 different haplotypes (13 isolates attributed to the same haplotype and 2 having a distinct haplotype, namely B15 (6 mutations) and B22 (10 mutations) (Figure 2).

Data given in Table 3 also indicated that among the rhizobacterial isolates screened for peptide synthetase biosynthetic genes of fengycin A and bacillomycin D, 20 were able to produce at least one of these antibiotics and 15 isolates were shown able to produce both fengycin A and bacillomycin D.

Screening of plant growth-promoting traits

In vitro assays were performed to select bacterial isolates exhibiting interesting plant growth-promoting characteristics. Isolates were screened for siderophore production, IAA production and phosphate solubilization. Data given in Table 3 revealed that 20 out of the 25 isolates naturally associated with tomato plants were found able to produce siderophore on CAS medium agar plates. *B. subtilis* B10 (KT921327), *E. cloacae*B16 (KT921429) and *K. pneumoniae* B12 (KT921328) isolates produced the largest halo on CAS medium agar, reaching 90 mm in diameter, followed by *B. amyloliquefaciens* B7 (KT921428), *B. megaterium* B24 (KT923048), and *C. jejuense* B11 (KU158886) forming halos of 83.33, 75, and 43.33 mm in diameter, respectively.

When screened for their phosphate solubilization ability on PVK agar, 18 isolates out of 25 tested were shown capable to solubilize phosphate as given in Table 3. Also, most of the tested bacterial isolates (19/25) had produced the plant growth-promoting hormone i.e. indole-3-acetic acid (IAA) (Table 3).

It can be concluded through whole analysis of PGPR traits data given in Table 3 that all screened collected bacterial isolates, originally recovered from rhizosphere of apparently healthy and vigorous tomato plants, showed positive response to at least one PGPR trait (siderophore, IAA production or phosphate solubilization) and that interestingly, 13 and 6 strains showed positive result for three and two traits, respectively.

Hypersensitive reaction on tobacco

Data of this test indicated that no necrotic lesions were observed in all rhizobacteria-inoculated tobacco leaf areas assuming that these 25 bacterial isolates were not pathogenic and were therefore eligible to be used as biocontrol agents once their disease-suppression ability demonstrated in the future.

Metabolic characterization of rhizobacteria collection

The 25 tomato-associated rhizobacterial isolates were assessed for their ability to utilize carbon sources using Biolog™ Ecoplates system based on 31 carbon sources.

The utilization rate was indicated by the reduction of tetrazolium violet, a redox indicator dye that changes from colorless to purple. At the first overall look, all isolates displayed very similar substrate richness indicating a good catabolic activity. Data of the relative metabolic profile of the tested isolates revealed a change in the degree of AWCD with the increase of the incubation duration. Therefore, to more differentiate isolates based on their metabolic profiles and ability to use of the provided carbon sources, a principal components analysis (PCA) was performed.

PCA results given in Figure 3 showed that all isolates were grouped into eight major clusters at each incubation duration (i.e. 24, 48 and 120 h of incubation), indicating that they had different patterns and capacity to metabolize carbon sources. In fact, depending on their metabolic capacity and incubation durations, the bacterial collection tested showed distinctive patterns in their use of carbon source. Figure 3 showed that after 24 h of incubation, the 2 main axes (the 2 first principal components), Dim1 and Dim2, accounted for 65.71% of the total variation. After 48 h of incubation, the two principal components, Dim1 and Dim2, covered 67.32% of the total data variance. After 120 h of incubation, the first and the second principal component (Dim1 and Dim2) explained 51.06 and 9.41% of data variance, respectively, and gave a consistent description of the global database.

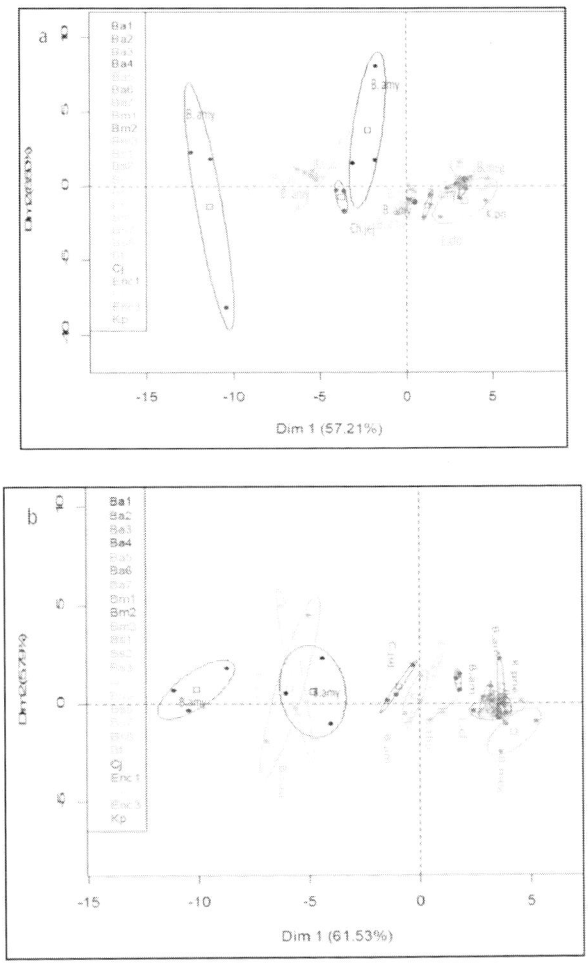

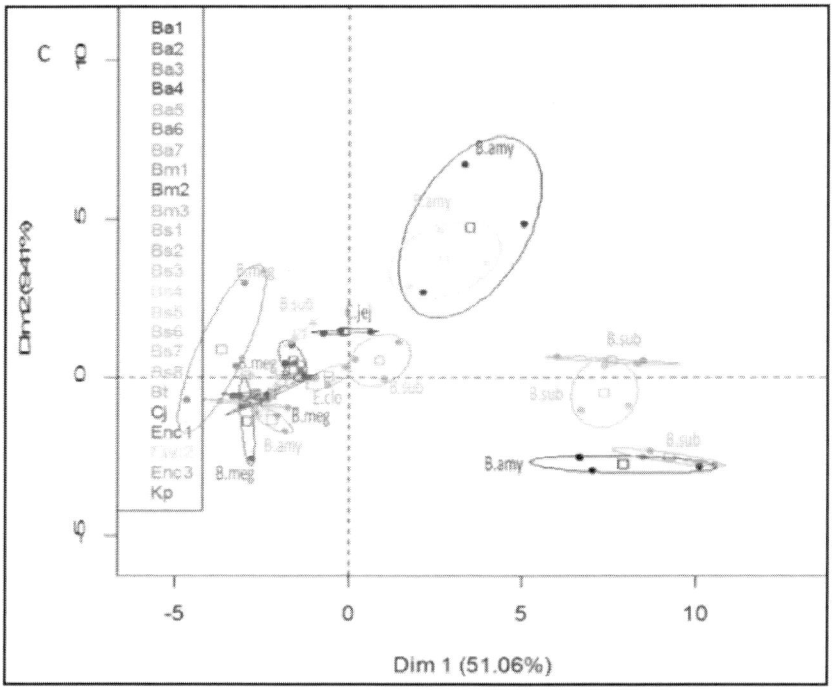

Figure 3: Distribution of the 25 tomato-associated rhizobacterial isolates on the principal plan designed by the global Principal Components (ACP) based on their metabolic profiles using Biolog system based on 31 carbon sources. a: After 24 h of incubation; b: After 48 h of incubation and c: After 120 h of incubation. For each sample, triplicates are represented by small scores and averaged by a larger one. Percentage of total variation accounted by each PC axis is indicated in parenthesis. Ba1: B7; Ba2 :B9; Ba3: B13; Ba4:B15; Ba5: B18; Ba6: B21; Ba7: B22; Bm1: B1; Bm2: B5; Bm3: B24; Bs1: B6; Bs2: B8; Bs3: B10; Bs4: B14; Bs5: B17; Bs6: B19; Bs7: B20; Bs8: B25; Bt: B2; Bt: B23; Cj: B11; Enc1: B3; Enc2: B4; Enc3: B16, and Kp: B12.

The mean scores for each of the 25 isolates tested were calculated and presented to discriminate among microbial populations. Principal components' percentage of the data variance, whatever the duration of incubation considered, clearly separated *B. amyloliquefaciens*, *B. subtilis* and *C. jejuense* (negative correlation) species from *B. megaterium, E. cloacae* and *K. pneumoniae* (positive correlation). Thus, these isolates exhibited variable capacity to metabolize carbon sources and there was a systematic change in the carbon source use patterns associated with the different genera.

The Shannon diversity index was calculated for each isolate. Data shown in Figure 4 indicated that for all the isolates tested, AWCD values, representing the overall average activity of these bacteria, and their Shannon indexes (calculated for each isolate) were positively correlated ($R^2 = 0.759$).

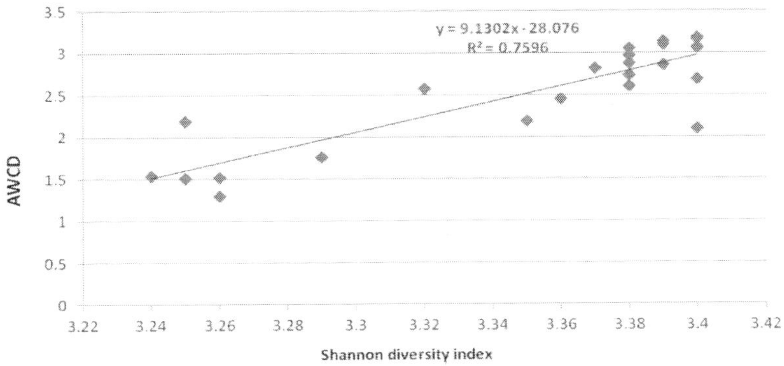

Figure 4: Correlation between average well color development (AWCD) and Shannon diversity index calculated from substrates degraded in Biolog® EcoPlates by the 25 tested tomato-rhizobacterial isolates (performed for means of three incubation durations).

The average AWCD values recorded at the three incubation durations (24, 48 and 120 h of incubation) ranged between 0.18 and 3.19. The most important values were noted for isolates *E. cloacae* B4 (KT923050) and *C. jejuense* B11 (KU158886). However, average values of Shannon indexes calculated at three reading times varied between 3.2 and 3.40. The highest values were recorded with *E. cloacae* B4 (KT923050) and *B. subtilis* B8 (KU158885) isolates. These results indicate that these native tomato-associated rhizobacteria displayed a large metabolic activity and they were able to use a wide range of carbon sources.

DISCUSSION

Bacterial community in rhizosphere area is known to play a significant effect in protecting plant roots from soilborne pathogens and in improving plant growth [9,35,36]. In the study, bacterial isolates were isolated from soil samples collected around healthy tomato plants grown in different tomato-producing regions of Tunisia. Among a total of 200 isolates collected, 25 were selected based on their ability to suppress by more than 45% *R. solani* and *S. sclerotiorum* growth relative to controls. They were identified and subjected to various characterizations in order to elucidate their main traits involved in fungal biocontrol and eventual plant growth promotion.

For the development of reliable biocontrol agents, it is important to elucidate their mechanisms of action and their nutritional and physiological requirements for optimum growth [37,38]. Thus, deep characterization of a given biocontrol agent is an essential step for improving the level and the reliability of its biocontrol activity [39].

In the present study, the selected 25 isolates were morphologically characterized based on colony color, shape, margin, elevation and surface where a great morphological diversity was recorded within the bacterial collection tested. Additionally, performed biochemical characterizations revealed the presence of great diversity within tomato-associated rhizobacteria based on Gram staining, levan activity, catalase activity, and oxidase activity.

Identification of biocontrol agents may be performed using biological and molecular methods. The 16S rRNA sequencing is widely used as an alternative method to identify microbial microorganisms. In the present study, the 25 isolates were subjected to 16S rRNA sequences analysis for phylogenetic comparisons. The obtained results revealed that the collected isolates belonged to different genera but, in some cases, the sequences obtained were very similar to each other and did not allow distinction of isolates at the species level especially within Bacillus group. Similarly, Chen and Tsen [40] failed in discriminating between species of Bacillus using 16S rRNA sequences analysis only. Also, González et al. [41] reported that analysis based on pair wise alignment of 16S rDNA sequences solely showed limited variations and sequence similarities between isolates and strain types were frequently encountered. In the same way, Hutsebaut et al. [42] and 43 Maughan and van der Auwera [43] indicated that the 16S rRNA gene sequences of species within the Bacillus group were highly conserved to enable easy and evident discrimination among species and subspecies. In the present study, analysis of 16S rDNA sequences showed limited variation where sequence similarities between isolates ranged between 96.5 and 99%. To overcome this constraint, protein-coding genes are reported to lead to greater genetic variation and were widely used as molecular tool for the classification of closely related species [44-46]. Some genes like gyrA and rpoB have been proposed to provide conclusive evidence on the evolutionary relationships of related taxa [47,48]. Unlike the high level of 16S rRNA gene sequence similarity noted within Bacillus group, sequence of the rpoB gene exhibited greater diversity and allowed identification of distinct species [41,49]. The sequencing of the 16S rRNA and rpoB genes of the 25 selected isolates revealed a species diversity where the tomato-associated isolates were classified into seven bacterial species namely *B. subtilis* (8 isolates), *B. amyloliquefaciens* (7 isolates), *B. megaterium* (3 isolates), *E. cloacae* (3 isolates), *B. thuringiensis* (2 isolates), *C. jejuense* (1 isolate), and *K. pneumoniae* (1 isolate).

Antibiotic production by bacterial antagonists is an essential component in the biological control of fungal phytopathogens and is characterized genetically through the detection of various antibiotic producing genes as shown in various studies [50-52]. In recent years, cyclic lipopeptide antibiotics have been reported to inhibit a wide range of plant pathogenic fungi [53,54]. In the present study, it was possible to associate the presence of antibiotic in the culture suspension of some strains with their corresponding biosynthetic genes. PCR primers used for the detection of multi-modular enzymes, the peptide synthetases involved in the biosynthesis of antifungal lipopeptides, showed the presence of fengycin A- and bacillomycin D-producing agents within this collection of rhizobacteria. The production of antibiotics by Bacillus spp. and their roles in suppression of plant pathogens are mentioned in many reviews [9,55]. *B. subtilis* and *B. thuringiensis* are previously reported to be subtillin, bacitracin, bacillin, subtenolin, and bacillomycin producers [56-58]. The production of cyclic lipopeptides is largely detected within*Bacillus* species but Mandal et al. [59] also demonstrated that *E. cloacae* can also produce different lipopeptide antibiotics like kurstakin, iturin, surfactin, and fengycin.

Chryseobacterium species were also commonly found in soil and water and their ability to suppress some soilborne diseases was reported [60-62]. Their mechanisms of action as biocontrol agents vary depending on species and isolates. In fact, *C.*

wanjuense KJ9C8, effective in controlling Phytophthora blight of pepper, produced protease and Hydrogen cyanide (HCN) but not antibiotics [63], and according to Kim et al. [62], ability to produce antibiotics has not been reported in *Chryseobacterium* species. However, in the present study, *C. jejuense* B11 (KU158886) naturally associated to tomato was found to be a bacillomycin D-producing agent. This is, to our knowledge, the first report of presence of bacillomycin D encoding gene in *C. jejuense*. Interestingly, our *K. pneumoniae* B12 (KT921328) isolate was also able to produce fengycin A.

This collection of rhizobacterial isolates was also screened for PGPR properties and particularly siderophore and IAA production and phosphate solublization. In fact, several microorganisms were found to be able to make insoluble soil phosphorous available to plants by solubilizing mineral phosphates through the production of organic acids or phosphatases. In the present study, 18 isolates out of the 25 tested, belonging to *Bacillus, Enteroabacter, Klebsiella* and *Chrysobacterium*, were shown able to solubilize inorganic forms of phosphorous. Bacillus strains were reported to have this capacity [64,65]. The phosphate solubilization ability recorded in some tomato-associated rhizobacteria in the present study is in line with other investigations performed using *Bacillus* [66], *Enterobacter* and *Klebsiella* species [64,67-69].

Siderophore production is another plant growth promoting (PGP) trait that may affect plant growth by binding to the available iron form ($Fe3+$) in the rhizosphere [70]. Through this process, iron becomes unavailable to plant pathogens; thus plants are protected from biotic stress. In the current investigation, 20 rhizobacterial isolates out the 25 tested were shown able to produce siderophore. Rhizobacteria with siderophore-production ability were reported in various other previous studies as PGP agents [25,71].

IAA is one of the most important phytohormone involved in the regulation of plant growth. IAA production by PGPR can vary between and within species [72]. In the present study, among the 25 isolates tested, belonging to four genera, 19 were found to be IAA producers. Production of phytohormones (e.g. ethylene) by PGPR may affect plant growth and development, and can improve fruit quality as well [73,74].

Overall, based on PGP traits, *Bacillus* and *Pseudomonas* species are well known by their strong phosphorus solubilization activity and by their IAA- and siderophore- production abilities compared to the other rhizobacterial species [75]. Native *Bacillus* spp. isolates, used in the current study, showed PGP traits but also, interestingly, some *Enterobacter, Chryseobacterium* and *Klebsiella* isolates cumulated the three traits as they were found capable to produce IAA and siderophore and to solubilize inorganic phosphate. Previous investigations indicated that members of the genus *Chryseobacterium* are an important bacterial group associated with plants [76,77] and exhibiting plant-growth promoting activities [78]. Reports concerning the role of *Klebsiella* as PGPR are relatively rare [79]. Ahemad and Khan [80] demonstrated that *Klebsiella*sp. strain PS19 is able to solubilize inorganic phosphate and produce IAA and siderophores like DHBA (2,3-dihydroxybenzoic acid). Similarly,*K. oxytoca* Rs-5, isolated from a saline cotton field in Xinjiang Province, China, has also the ability to relieve salt stress, promote plant growth and produce IAA [81].

In previous studies, rhizobacteria were reported to have at least on PGP trait. For example, 216 bacterial strains, isolated from rice rhizospheric soils in Northern Thailand, have solubilized inorganic phosphate but only 18.05% of them are able to produce IAA where the best IAA producer was identified as *Klebsiella* SN 1.1 [82]. Also, according to Yasmin et al. [83], two isolates namely *Klebsiella* UPMSP9 and Erwinia UPMSP10 produce high IAA with L-tryptophan, solubilize phosphate, produce N in culture, resist to antibiotics but fail to produce siderophore. *K. oxytoca* is able to produce IAA, to solubilize phosphate and displays nitrogenase activity but six *K. pneumoniae* strains, recovered from wheat (var. Lokwan) rhizosphere produce only IAA [79].

Ability to use a sole carbon source, through the Biolog™ EcoPlate test, was widely explored as a reliable method for characterization of microbial communities based on their metabolic profiles [84]. Biolog system based physiological profiling generated a large number of characters and produced robust data and more realistic representation of relationships among strains of a given microbial collection. Data from this study showed that isolates, with similar profiles and being closest, were grouped together. Dawson et al. [85] reported that the metabolic profiling analysis allowed samples to be represented on scatter plots of two or more axes. This analysis is considered to be a highly effective method for differentiating among and within species groups through cluster analysis. In the current study, variable metabolic profiles were found within the rhizobacteria collection tested as indicated by the variations recorded in AWCD values and Shannon indexes. After direct incubation of rhizobacteria isolates in Biolog EcoPlates, results revealed a great diversity as determined through the well color change to violet, permitting colorimetric determination of respiration increase occurring when bacterial cells are oxidizing a carbon source. Liao and Xie [86] indicated that color development seems to mainly reflect species metabolic activity and the ability of a given bacterial community to utilize substrates. Campbell et al. [87] showed that, although the use of environmentally relevant substrates to be beneficial, these substrates would only be useful if their use resulted in the differentiation between samples. Indeed, it can be concluded from the current the study the metabolic profiles may be a useful tool for classifying the soil microbial communities on ecological basis using their functional characteristics (single-use of carbon sources).

CONCLUSION

The present study was undertaken in order to isolate and identify tomato-associated rhizobacteria from various tomato-producing sites and growing systems and to perform morphological, biochemical, metabolic and molecular characterization for the 25 most effective ones in suppressing growth of *S. sclerotiorum* and *R. solani* by more than 45%.

The current investigation provided strong evidence that tomato rhizosphere was populated by a numerous and a diverse array of rhizobacteria exhibiting a great morphological and biochemical diversity. These 25 isolates were classified into seven species, based on sequencing of 16S rRNA and rpoB genes, which were *B. amyloliquefaciens, B. megaterium, B. subtilis, E. cloacae, C. jejuense*, and *K. pneumoniae*. These isolates exhibited a large metabolic activity and were shown able to metabolize a wide range of carbon sources with the increase of the incubation duration. Furthermore, based on their metabolic profiles obtained using Biolog system, the selected isolates were grouped into eight major clusters at each incubation duration

(24, 48 and 120 h of incubation) and their AWCD was found be correlated to Shannon diversity index.

Screening of their ability to synthesize plant growth-promoting and antifungal compounds revealed that the majority of isolates were able to produce siderophore, to solubilize phosphate and to synthesize IAA and that interestingly, some isolates cumulated the three plant growth promoting traits. Moreover, when screened for their potential to produce lipopeptide antibiotics, PCR amplification of genes encoding fengycin A and bacillomycin D biosynthesis revealed the presence of these genes in 18 and 16 isolates, respectively. Thus, this study clearly demonstrated that tomato rhizospheric soils harbor a diversity of beneficial bacteria which may be promising biocontrol agents due to their interesting metabolic activity, their capacity to produce lipopeptide antibiotics and their interesting antifungal potential displayed toward both target fungi. Moreover, their ability to produce plant growth-promoting compounds indicated that they may exhibit a bio-fertilizing action. Future investigations will be focused on their ability to suppress root and crown rots caused by *R. solani* and to enhance growth of tomato plants.

ACKNOWLEDGMENTS

This work was funded by the Ministry of Higher Education Scientific Research in Tunisia through the budget assigned to UR13AGR09-Integrated Horticultural Production the Tunisian Centre-East, the Regional Centre of Research on Horticulture and Organic Agriculture (CRRHAB) of Chott Mariem, and INRA Bordeaux-France through the budget allocated to SAVE UMR. Authors thank the research teams of UMR SAVE / INRA Bordeaux-France and CFBP / INRA Anger-Nante team for their hospitality, their guidance and support.

REFERENCES

1. Pastor N, Carlier E, Andrés J, Rosas SB, Rovera M (2012) Characterization of rhizosphere bacteria for control of phytopathogenic fungi of tomato. J Environ Manage 95 Suppl: S332-337.
2. Hibar K, Daami-Remadi M, Hamada W, El Mahjoub M (2006) Biofungicides as an alternative for tomato *Fusarium* crown and root rot control. Tunisian J Plant Prot 1: 19-29.
3. Jabnoun-KhiareddineH,Daami-RemadiM,AyedF,ElMahjoubM(2009)Biological control of tomato Verticillium wilt by using indigenous*Trichoderma* spp. In: Daami-Remadi M (Ed) Tunisian Plant Science and Biotechnology I. Afr J Plant Sci Biotech 3: 26-36.
4. Jabnoun-Khiareddine H, Abdallah RAB, El-Mohamedy RSR, Abdel-Kareem F, Gueddes-Chahed M, et al. (2016) Comparative efficacy of potassium salts against soil-borne and air-borne fungi and their ability to suppress tomato wilt and fruit rots. J Microb Biochem Technol 8: 45-55.
5. Abdallah RAB, Tlili MS, Nefzi A, Jabnoun-Khiareddine H, Daami-Remadi M (2016) Biocontrol of Fusarium wilt and growth promotion of tomato plants using

endophytic bacteria isolated from *Nicotiana glauca* organs. Biol Control 97: 80-88.

6. Jiskani MM, Pathan MA, Wagan KH, Imran M, Abro H (2007) Studies on the control of tomato damping of disease caused by *Rhizoctonia solani* Kühn. Pak J Bot 39: 2749-2754.
7. Arora NK, Khare E, Oh H, Kang SC, Maheshwari DK (2008) Diverse mechanisms adopted by fluorescent *Pseudomonas* PGC2 during the inhibition of *Rhizoctonia solani* and *Phytophthora capsici*. World J Microbiol Biotechnol 24: 581-585.
8. Zachow C, Grosch R, Berg G (2011) Impact of biotic and a-biotic parameters on structure and function of microbial communities living on sclerotia of the soil-borne pathogenic fungus *Rhizoctonia solani*. Appl Soil Ecol 48: 193-200.
9. Singh JS, Pandey VC, Singh DP (2011) Efficient soil microorganisms: A new dimension for sustainable agriculture and environmental development. Agric Ecosys Environ 140: 339-353.
10. De Curtis F, Lima G, Vitullo D, De Cicco V (2010) Biocontrol of *Rhizoctonia solani* and *Sclerotium rolfsii* on tomato by delivering antagonistic bacteria through a drip irrigation system. Crop Prot 29: 663-670.
11. Kita N, Ohya T, Uekusa H, Nomura K, Manago M, et al. (2005) Biological control of damping-off of tomato seedlings and cucumber Phomopsis root rot by *Bacillus subtilis* RB14-C. Japan Agric Res Quart 39: 109-114.
12. Cazorla FM, Romero D, Perez-Garcia A, Lugtenberg BJJ, de Vicente A, et al. (2007) Isolation and characterization of antagonistic *Bacillus subtilis* strains from the avocado rhizoplane displaying biocontrol activity. J Appl Microbiol 103: 1950-1959.
13. Taheri P, Tarighi S (2012) The role of pathogenesis-related proteins in the tomato-*Rhizoctonia solani* interaction. Journal of Botany pp: 6.
14. Blazier SR, Conway KE (2004) Characterization of *Rhizoctonia solani* isolates associated with patch diseases on turfgrass. Proc Okla Acad 4 Sci 8: 41-45
15. Yang J, Kloepper JW, Ryu CM (2009) Rhizosphere bacteria help plants tolerate abiotic stress. Trends Plant Sci 14: 1-4.
16. Joshi P, Bhatt AB (2011) Diversity and function of plant growth promoting rhizobacteria associated with wheat rhizosphere in North Himalayan region. Int J Environ Sci 1: 1135-1143.
17. Arias AA, Ongena M, Halimi B, Lara Y, Brans A, et al. (2009) *Bacillus amyloliquefaciens* GA1 as a source of potent antibiotics and other secondary metabolites for biocontrol of plant pathogens. Microb Cell Fact 8: 63-70.
18. Lugtenberg B, Kamilova F (2009) Plant-growth-promoting rhizobacteria. Annu Rev Microbiol 63: 541-556.
19. Ahemad M, Kibret M (2014) Mechanisms and applications of plant growth promoting rhizobacteria: Current perspective. Journal of King Saud University - Science 26: 1-20.

20. Prashar P, Kapoor N, Sachdeva S (2013) Isolation and characterization of *Bacillus* sp. with in-vitro antagonistic activity against *Fusarium oxysporum* from rhizosphere of tomato. J Agric Sci Technol 15: 1501-1512.

21. Patel M, Sheth U, Hajela P, Joshi B, Animasaun DA (2014) Isolation and diversity analysis of rhizobacteria from sugarcane and its biocontrol potential against *Rhizoctonia solani* a common plant pathogen. Int J Curr Microbiol App Sci 3: 69-76.

22. Agrawal KDP, Agrawal Sh (2013) Characterization of Bacillus sp. strains isolated from rhizosphere of tomato plants (*Lycopersicon esculentum*) for their use as potential plant growth promoting rhizobacteria. Int J Curr Microbiol App Sci 2: 406-417.

23. Shobha G, Kumudini BS (2012) Antagonistic effect of the newly isolated PGPR *Bacillus* spp. on *Fusarium oxysporum*. Int J Appl Sci Eng Res 1: 463-474.

24. Venkadesaperumal G, Amaresan N, Kumar K (2015) Plant growth promoting capability and genetic diversity of bacteria isolated from mud volcano and lime cave of Andaman and Nicobar Islands. Braz J Microbiol 45: 1271-1281.

25. Ghodsalavi B, Ahmadzadeh M, Soleimani M, Madloo PB, Taghizad-Farid R (2013) Isolation and characterization of rhizobacteria and their effects on root extracts of Valeriana officinalis. Aust J Crop Sci 7: 338-344.

26. Mishra VK, Kumar A (2015) Biosynthesis of indole-3-acetic acid by plant growth promoting rhizobacteria, *Klebsiella pneumonia*, *Bacillus amyloliquefaciens* and *Bacillus subtilis*. Afr J Microbiol Res 9: 1139-1149.

27. Nawangsih AA, Aditya R, Tjahjono B, Negishi H, Suyama K (2012) Bio-efficacy and characterization of plant growth promoting bacteria to control the bacterial wilt disease of peanut in Indonesia. ISSAAS Journal 18: 185-192.

28. Renault D, Vallance J, Déniel F, Wery N, Godon JJ, et al. (2012) Diversity of bacterial communities that colonize the filter units used for controlling plant pathogens in soilless cultures. Microb Ecol 63: 170-187.

29. Gomez E, Garland J, Conti M (2004) Reproducibility in the response of soil bacterial community-level physiological profiles from a land use intensification gradient. Appl Soil Ecol 26: 21-30.

30. Garland JL1, Mills AL (1991) Classification and characterization of heterotrophic microbial communities on the basis of patterns of community-level sole-carbon-source utilization. Appl Environ Microbiol 57: 2351-2359.

31. Chojniak J, Wasilkowski D, Plaza G, Mrozik A, Brigmon R (2015) Application of biolog microarrays techniques for characterization of functional diversity of microbial community in phenolic-contaminated water. Int J Environ Res 9: 785-794.

32. Grove JA, Kautola H, Javadpour S, Young MM, Anderson WA (2004) Assessment of changes in the microorganism community in a biofilter. Biochem Eng J 18: 111-114.

33. De Clerck E, De Vos P (2004) Genotypic diversity among *Bacillus licheniformis* strains from various sources. FEMS Microbiol Lett 231: 91-98.

34. Tamura K, Dudley J, Nei M, Kumar S (2007) MEGA4: Molecular Evolutionary Genetics Analysis (MEGA) software version 4.0. Mol Biol Evol 24: 1596-1599.

35. Harish S, Kavino M, Kumar N, Balasubramanian P, Samiyappan R (2009) Induction of defense-related proteins by mixtures of plant growth promoting endophytic bacteria against *Banana Bunchy Top Virus*. Biol Control 51: 16-25.

36. Yang MM, Mavrodi DV, Mavrodi OV, Bonsall RF, Parejko JA, et al. (2011) Biological control of take-all by fluorescent *Pseudomonas* spp. from Chinese wheat fields. Phytopathology 101: 1481-1491.

37. Maleki M, Mostafaee S, Mokhtarnejad L, Farzaneh M (2010) Characterization of Pseudomonas fluorescens strain CV6 isolated from cucumber rhizosphere in varamin as a potential biocontrol agent. Aust J Crop Sci 4: 676-683.

38. Naveed M, Mubeen S, Khan S, Ahmed I, Khalid N, et al. (2014) Identification and characterization of rhizospheric microbial diversity by 16S ribosomal RNA gene sequencing. Braz J Microbiol 45: 985-993.

39. Kamalaknnan A, Mohan L, Harish S, Radjacommare R, Amutha G, et al. (2004) Biological agents induce disease resistance in Phyllanthus niruri Linn against damping-off disease caused by *Rhizoctonia solani*. Phytopathol Medit 43: 187-194.

40. Chen ML, Tsen HY (2002) Discrimination of *Bacillus cereus* and *Bacillus thuringiensis* with 16S rRNA and gyrB gene based PCR primers and sequencing of their annealing sites. J Appl Microbiol 92: 912-919.

41. González MJ, Gorgoroso F, Reginensi SM, Olivera JA, Bermúdez J (2013) Polyphasic identification of closely related *Bacillus subtilis* and*Bacillus amyloliquefaciens* isolated from dairy farms and milk powder. J Microbiol Biotechnol Food Sci 2: 2326-2331.

42. Hutsebaut D, Vandroemme J, Heyrman J, Dawyndt P, Vandenabeele P, et al. (2006) Raman microspectroscopy as an identification tool within the phylogenetically homogeneous '*Bacillus subtilis*' group. Syst Appl Microbiol 29: 650-660.

43. Maughan H, Van der Auwera G (2011) Bacillus taxonomy in the genomic era finds phenotypes to be essential though often misleading. Infect Genet Evol 11: 789-797.

44. Bavykin SG, Lysov YP, Zakhariev V, Kelly JJ, Jackman J, et al. (2004) Use of 16S rRNA, 23S rRNA, and gyrB gene sequence analysis to determine phylogenetic relationships of Bacillus cereus group microorganisms. J Clin Microbiol 42: 3711-3730.

45. Konstantinidis KT, Tiedje JM (2005) Towards a genome-based taxonomy for prokaryotes. J Bacteriol 187: 6258-6264.

46. Zeigler DR (2005) Application of a recN sequence similarity analysis to the

identification of species within the bacterial genus *Geobacillus*. Int J Syst Evol Microbiol 55: 1171-1179.

47. Ko KS, Kim JM, Kim JW, Jung BY, Kim W, et al. (2003) Identification of *Bacillus anthracis* by rpoB sequence analysis and multiplex PCR. J Clin Microbiol 41: 2908-2914.

48. Wang K, Kang L, Anand A, Lazarovits G, Mysore KS (2007) Monitoring in planta bacterial infection at both cellular and whole-plant levels using the green fluorescent protein variant GFPuv. New Phytol 174: 212-223.

49. Meintanis C, Chalkou KI, Kormas KA, Lymperopoulou DS, Katsifas EA, et al. (2008) Application of rpoB sequence similarity analysis, REP-PCR and BOX-PCR for the differentiation of species within the genus *Geobacillus*. Lett Appl Microbiol 46: 395-401.

50. Haas D, Keel C (2003) Regulation of antibiotic production in root-colonizing *Pseudomonas* spp. and relevance for biological control of plant disease. Annu Rev Phytopathol 41: 117-153.

51. Stein T (2005) *Bacillus subtilis* antibiotics: structures, syntheses and specific functions. Mol Microbiol 56: 845-857.

52. Mavrodi DV, Blankenfeldt W, Thomashow LS (2006) Phenazine compounds in fluorescent *Pseudomonas* spp. biosynthesis and regulation. Annu Rev Phytopathol 44: 417-445.

53. Nagorska K, Bikowski M, Obuchowski, M (2007) Multicelluar behaviour and production of a wide variety of toxic substances support usage of *Bacillus subtilis* as a powerful biocontrol agent. Acta Biochim Pol 54: 495-508.

54. Ongena M, Jacques P (2008) Bacillus lipopeptides: versatile weapons for plant disease biocontrol. Trends Microbiol 16: 115-125.

55. Yuan J, Raza W, Shen Q, Huang Q (2012) Antifungal activity of *Bacillus amyloliquefaciens* NJN-6 volatile compounds against *Fusarium oxysporum* f. sp. *cubense*. Appl Environ Microbiol 78: 5942-5944.

56. Montealegre JR, Reys R, Perez LM, Herrera R, Silva P, et al. (2003) Selection of bioantagonistic bacteria to be used in biological control of*Rhizoctonia solani* in tomato. Elect J Biotechnol 6: 115-127.

57. From C, Pukall R, Schumann P, Hormazábal V, Granum PE (2005) Toxin-producing ability among *Bacillus* spp. outside the Bacillus cereus group. Appl Environ Microbiol 71: 1178-1183.

58. Killani AS, Abaidoo RC, Akintokun AK, Abiala MA (2011) Antagonistic effect of indigenous *Bacillus subtilis* on root-/soil-borne fungal pathogens of cowpea. Researcher 3: 11-18.

59. Mandal SM, Sharma S, Pinnaka AK, Kumari A, Korpole S (2013) Isolation and characterization of diverse antimicrobial lipopeptides produced by Citrobacter and Enterobacter. BMC Microbiol 13: 152.

60. Krause MS, Madden LV, Hoitink HA (2001) Effect of potting mix microbial carrying capacity on biological control of rhizoctonia damping-off of radish and

rhizoctonia crown and root rot of poinsettia. Phytopathology 91: 1116-1123.

61. McSpadden Gardener BB, Weller DM (2001) Changes in populations of rhizosphere bacteria associated with take-all disease of wheat. Appl Environ Microbiol 67: 4414-4425.

62. Kim HS, Sang MK, Jung HW, Jeun YC, Myung IS, et al. (2012) Identification and characterization of *Chryseobacterium wanjuense* strain KJ9C8 as a biocontrol agent of Phytophthora blight of pepper. Crop Prot 32: 129-137.

63. Park MS, Jung SR, Lee KH, Lee MS, Do JO, et al. (2006) *Chryseobacterium soldanellicola* sp. nov. and *Chryseobacterium taeanense* sp. nov., isolated from roots of sand-dune plants. Int J Syst Evol Microbiol 56: 433-438.

64. Kumar A, Kumar A, Devi S, Patil S, Payal Ch, et al. (2012) Isolation, screening and characterization of bacteria from rhizospheric soils for different plant growth promotion (PGP) activities: an in vitro study. Recent Res Sci Technol 4: 1-5.

65. Ajilogba CF, Babalola OO, Ahmad F (2013) Antagonistic effects of *Bacillus* species in biocontrol of tomato *Fusarium* wilt. Ethno Med 7: 205-216.

66. Calvo P, Ormeño-Orrillo E, Martínez-Romero E, Zúñiga D (2010) Characterization of Bacillus isolates of potato rhizosphere from andean soils of Peru and their potential PGPR characteristics. Braz J Microbiol 41: 899-906.

67. Chatli AS, Beri V, Sidhu BS (2008) Isolation and characterization of phosphate solubilizing microorganisms from the cold desert habitat of Salix alba Linn. in trans Himalayan region of Himachal Pradesh. Indian J Microbiol 48: 267-273.

68. Khan AA, Jilani G, Akhtar MS, Naqvi SMS, Rasheed M (2009) Phosphorus solubilizing bacteria: Occurrence, mechanisms and their role in crop production. J Agric Biol Sci 1: 48-58.

69. Singh AV, Chandra R, Goel R (2013) Phosphate solubilization by *Chryseobacterium* sp. And their combined effect with N and P fertilizers on plant growth promotion. Arch Agron Soil Sci 59: 641-651.

70. Siddiqui ZA (2005) PGPR: prospective biocontrol agents of plant pathogens. Pages 111-142. In: PGPR: Biocontrol and Biofertilization, Springer, Netherlands.

71. Raval AA, Desai PB (2012) Rhizobacteria from rhizosphere of sunflower (*Helianthus annuus* l.) and their effect on plant growth. Res J Biol Sci 1: 58-61.

72. Mirza MS, Ahmad W, Latif F, Haurat J, Bally R, et al. (2001) Isolation, partial characterization, and the effect of plant growth-promoting bacteria (PGPB) on micro-propagated sugarcane in vitro. Plant and Soil 237: 47-54.

73. Violante HGM, Portugal VO (2007) Alteration of tomato fruit quality by root inoculation with plant growth-promoting rhizobacteria (PGPR):*Bacillus subtilis* BEB-13bs. Scientia Horticulturae 113: 103-106.

74. Lucas-Garcia JA, Probanza A, Ramos B, Ruiz-Palomino M, Gutierrez Manero FJ (2004) Effect of inoculation of *Bacillus licheniformis* on tomato and pepper. Agronomie 24: 169-176.

75. Gechemba OR, Budambula NLM, Makonde HM, Julius M, Matiru VN (2015) Potentially beneficial rhizobacteria associated with banana plants in Juja, Kenya. J Bio Env Sci 7: 181-188.

76. Anderson M, Habiger J (2012) Characterization and identification of productivity-associated rhizobacteria in wheat. Appl Environ Microbiol 78: 4434-4446.

77. Brown SD, Utturkar SM, Klingeman DM, Johnson CM, Martin SL, et al. (2012) Twenty-one genome sequences from *Pseudomonas* species and 19 genome sequences from diverse bacteria isolated from the rhizosphere and endosphere of Populus deltoides. J Bacteriol 194: 5991-5993.

78. Dardanelli MS, Manyani H, González-Barroso S, Rodríguez-Carvajal MA, Gil-Serrano AM, et al. (2010) Effect of the presence of the plant growth promoting rhizobacterium (PGPR) *Chryseobacterium balustinun* Aur9 and salt stress in the pattern of flavoids exuded by soybean roots. Plant Soil 328: 483-493.

79. Sachdev DP, Chaudhari HG, Kasture VM, Dhavale DP, Chopade BA (2009) Isolation and characterization of indole acetic acid (IAA) producing *Klebsiella pneumoniae* strains from rhizosphere of wheat (*Triticum aestivum*) and their effect on plant growth. Indian J Exp Biol 47: 993-1000.

80. Ahemad M, Khan MS (2011) Effects of insecticides on plant-growth-promoting activities of phosphate solubilizing rhizobacterium *Klebsiella*sp. strain PS19. Pestic Biochem Phys 100: 51-56.

81. Yue HT, Mo WP, Zheng YY, Li CH, Li H (2007) The salt stress relief and growth promotion effect of Rs-5 on cotton. Plant Soil 297: 139-145

82. Chaiharn M, Lumyong S (2011) Screening and optimization of indole-3-acetic acid production and phosphate solubilization from rhizobacteria aimed at improving plant growth. Curr Microbiol 62: 173-181.

83. Yasmin F, Othman R, Sijam K, Saad MS (2009) Characterization of beneficial properties of plant growth-promoting rhizobacteria isolated from sweet potato rhizosphere. Afr J Microbiol Res 3: 815-821.

84. Di Salvo L, De Salamone IG (2012) Evaluation of soil-microbial communities by their CLPP. Standardization of a laboratory technique to replace commercial available microplates. Ecologia Austral 22: 129-136.

85. Dawson SL, Fry JC, Dancer BN (2002) A comparative evaluation of five typing techniques for determining the diversity of fluorescent pseudomonads. J Microbiol Meth 50: 9-22.

86. Liao M, Xie XM (2007) Effect of heavy metals on substrate utilization pattern, biomass,and activity of microbial communities in a reclaimed mining wasteland of red soil area. Ecotoxicol Environ Safety 66: 217-223

87. Campbell CD, Grayston SJ, Hirst DJ (1997) Use of rhizosphere carbon sources in sole carbon source tests to discriminate soil microbial communities. J Microbiol Meth 30: 33-41.

15
CHAPTER

454-PYROSEQUENCING ANALYSIS OF BACTERIAL COMMUNITIES FROM AUTOTROPHIC NITROGEN REMOVAL BIOREACTORS UTILIZING UNIVERSAL PRIMERS: EFFECT OF ANNEALING TEMPERATURE

Alejandro Gonzalez-Martinez[1] Alejandro Rodriguez-Sanchez[2] Belén Rodelas[2] Ben A. Abbas[3] Maria Victoria Martinez-Toledo[2] Mark C. M. van Loosdrecht[3] F. Osorio[1] and Jesus Gonzalez-Lopez[2]

[1]Department of Civil Engineering, University of Granada, Campus de Fuentenueva, s/n, 18071 Granada, Spain

[2]Institute of Water Research, University of Granada, C/Ramón y Cajal 4, 18071 Granada, Spain

[3]Department of Biotechnology, Technical University of Delft, Julianalaan 67, 2628 BC Delft, Netherlands

ABSTRACT

Identification **of anaerobic ammonium oxidizing (anammox) bacteria by molecular tools aimed at the evaluation of bacterial diversity in autotrophic nitrogen removal systems is limited by the difficulty to design universal primers for the Bacteria domain able to amplify the anammox 16S rRNA genes. A metagenomic analysis (pyrosequencing) of total bacterial diversity including anammox population in five autotrophic nitrogen removal technologies, two bench-scale models (MBR and Low Temperature CANON) and three full-scale bioreactors (anammox, CANON, and DEMON), was successfully carried out by optimization of primer selection and PCR conditions (annealing temperature). The universal primer 530F was identified as the best candidate for total bacteria and anammox bacteria diversity coverage. Salt-adjusted optimum annealing temperature of primer 530F was calculated (47°C) and hence a range of annealing temperatures of 44–49°C was tested. Pyrosequencing data showed that annealing temperature of 45°C yielded the best results in terms of species richness and diversity for all bioreactors analyzed.**

Citation: Alejandro Gonzalez-Martinez, Alejandro Rodriguez-Sanchez, Belén Rodelas, et al., "454-Pyrosequencing Analysis of Bacterial Communities from Autotrophic Nitrogen Removal Bioreactors Utilizing Universal Primers: Effect of Annealing Temperature," BioMed Research International, vol. 2015, Article ID 892013, 12 pages, 2015. doi:10.1155/2015/892013.

INTRODUCTION

Anaerobic ammonium oxidizing (anammox) bacteria belong to the Candidatus Brocadiales order first described in 1999 [1]. Their ability to perform anaerobic ammonium oxidation has attracted the attention of many researchers due to the change it made for the understanding of the nitrogen cycle. They have been found as important bacteria for the ecology of nitrogen in oceanic environments [2–4] and have been proposed to play an important role in the nitrogen cycle at global scale [2]. Anammox bacteria have been identified in many natural and engineered environments such as marine sediments, agricultural soils, or wastewater treatment plants [5–7]. In addition, anammox bacteria are the basis for promising technologies aimed at removing nitrogen from wastewater. The autotrophic, anaerobic ammonium oxidation has been utilized for the set-up of several technologies for nitrogen removal, such as partial nitritation/anammox, DEMON, OLAND, or CANON [8]. Autotrophic nitrogen removal technologies account for several advantages over traditional total nitrification-denitrification processes, such as the lesser bioreactor volume required, lower biomass production, or saving costs in aeration and carbon source requirements [9–11]. Thus, anammox bacteria have become of relevance in both natural and engineered systems, and more research about the ecology of the ecosystems where they develop is expected in the following years.

PCR-based molecular biology techniques such as qPCR quantification and high-throughput pyrosequencing are powerful tools for the estimation of the abundance and diversity of microorganisms in natural and engineered ecosystems [12, 13]. Most frequently, anammox-specific primers have been applied for PCR-based evaluation of the occurrence and diversity of these organisms [12, 14–18]. The reason under this common practice is that universal primers targeting the whole Bacteria domain tend to amplify poorly the 16S rRNA gene of anammox due to their not very high identity ($<87.1\%$) [19, 20]. In this sense, when universal primers are used PCR amplification of some anammox phylotypes might be missing, with the consequent loss of fundamental information on the microbial diversity of the system. Nevertheless, the use of anammox-specific primers is not sufficient for a complete understanding of microbial ecosystems, given that other bacteria are not taken into account.

In this research, several universal primers and annealing temperature conditions for PCR amplification were tested in order to achieve the best combination possible for its use on the metagenomic analysis of ecosystems, such as autotrophic nitrogen removal bioreactors. After an in silico testing, a universal primer was selected for the best coverage of 16S rRNA genes of the domain Bacteria including anammox (Planctomycetes). Using this primer, bacterial diversity in samples from five different lab-scale and full-scale autotrophic nitrogen removal bioreactors (see Table S1 in Supplementary Materials available online athttp://dx.doi.org/10.1155/2015/892013) were analyzed by high-throughput pyrosequencing, using six different annealing temperatures. Based on the results obtained, a robust method for the evaluation of the bacterial diversity in ecosystems where anammox bacteria are of importance is proposed.

MATERIALS AND METHODS

Bioreactors

Five different autotrophic nitrogen removal bioreactors which represent the main existing anammox technologies were sampled in the analysis (Table S1). Two of them were bench-scale models and the other three were full-scale plant bioreactors. The bench-scale models analyzed in the study were named Lab MBR and Low Temperature CANON. Lab MBR is a membrane bioreactor (MBR) anammox system and Low Temperature CANON is a CANON system operated at 15°C. Both bioreactors were built in Netherlands in 2010 and 2009, respectively, and were fed synthetic wastewater.

The full-scale bioreactors sampled in the study were located at three cities in Netherlands, Apeldoorn, Olburgen, and Rotterdam, and were named A, N, and R, respectively. Bioreactor A is a DEMON system built in 2010 which treats reject water from anaerobic digester. Bioreactor N is a CANON process treating sewage from a potato processing factory built in 2009. Bioreactor R is the anammox reactor in a two-step anammox plant constructed in 2002 and treats reject water from anaerobic digester.

Sampling

Five samples (200 mL) were taken from five evenly distributed points within each bioreactor volume. The procedures for sampling and pretreatment for DNA extraction followed those described by Ni et al. [16]. Biomass was separated from the collected wastewater by centrifugation at 3500 rpm for 10 minutes at room temperature in a Kokusan H-103N series apparatus (Kokusan Enshinki Co., Ltd., Tokyo, Japan). Pelleted biomass was stored at −20°C before DNA extraction.

In Silico Primer Evaluation

The primer pair 530F (5′-GTGCCAGCMGCNGCGG)-1100R (5′-GGGTTNCGN TCGTTG), described by Dowd et al. [21], was selected after an in silico analysis of several universal primers targeting the 16S rRNA gene and commonly used in earlier literature. The primers were tested by correlating the accession numbers of the matching sequences, using the Probe Match function of the Ribosomal Data Project (RDP) (http://rdp.cme.msu.edu/probematch/search.jsp) (see Supplementary Material). Moreover, to double check the performance of the selected primer set over the Planctomycetes phylum, an extra in silico analysis study using the SiLVA Test Probe tool was done (http://www.arb-silva.de/search/testprobe).

In Silico Calculation of Optimum Annealing Temperature

Calculation of optimum annealing temperature for primer 530F was done following the expression for the optimum salt-adjusted annealing temperature of a primer developed by Rychlik et al. [22–25]. For calculation purposes, combined concentration of Na^+ and K^+ was taken as 50 mM.

DNA Extraction, PCR Amplification, and Pyrosequencing

300 mg of pelleted biomass for each centrifuged sample was used for DNA extraction using the Fast DNA SPIN Kit for Soil and the Fast-Prep24 apparatus (MP Biomedicals, Solon, OH, USA) following the instructions given by the manufacturer. DNA extracts from samples collected in the same bioreactor were merged into a pool. The DNA pool of each bioreactor was divided into 6 subsamples with equal volume for further pyrosequencing analysis.

The primer pair 530F-1100R [21] was used to amplify 500 bp of the 16S rRNA gene of Bacteria, encompassing the V4-V5-V6 hypervariable regions. Research and Testing Laboratory (Lubbock, Texas, USA) proceeded with pyrosequencing following the procedure described by Dowd et al., 2008 [21], using the Roche 454 GS-FLX+ apparatus. Amplification of the six subsamples within the same bioreactor DNA pool was assayed under the same PCR conditions but at different annealing temperatures (44 to 49°C), yielding a total of 30 different pyrosequencing datasets in the five different technologies. In this sense the PCR conditions for pyrosequencing were the following: preheating step at 94°C for 3 minutes; 32 cycles at 94°C for 30 seconds, 44–49°C for 40 seconds, and 72°C for 1 minute; elongation at 72°C for 5 minutes.

Pyrosequencing Postrun Analysis

Elimination of poor-quality end reads from pyrosequencing raw data was done by quality trimming based on quality scores. USEARCH [26] was then used to generate seed sequences to which quality trimmed reads were clustered within a 4% divergence threshold. This procedure eliminates sequences that fail to encounter similar enough reads. Chimera detection was developed using the de novo method implemented in UCHIIME [27] over clustered quality trimmed sequences collected during the previous step. Denoising was then conducted for elimination of bad sequences and correction of base pair errors. Following denoising, a quality control screening was conducted. Quality criteria taken were the following: (1) failed sequence reads, (2) sequences with low quality tags, and (3) sequences that are shorter than half the expected amplicon length or 250 bp, whichever the shortest. Sequences that could not meet the defined quality criteria were eliminated. Reads that passed the quality screening control were then clustered for phylogenetic identification into 0% divergence using USEARCH [26]. The Kraken BLAST software (http://ccb.jhu.edu/software/kraken/) [28] was utilized to provide a seed sequence for each phylogenetic identification cluster from a high-quality database derived from the NCBI GenBank database. Based on BLASTN+ identity, sequences were affiliated to distinct taxonomic levels as follows: (1) at OTU level if divergence was less than 3%, (2) at genus level if divergence was in the range 3–5%, (3) at family level if divergence was 5–10%, (4) at order level if divergence was 10–15%, (5) at class level if divergence was 15–20%, and (6) at phylum level if divergence was 20–23%. Sequences that did not encounter queried sequences with less than 23% divergence were discarded.

Rarefaction Curves

Rarefaction curves for each sample were calculated using the aRarefactWin software developed by S. Holland (University of Georgia, Athens; http://strata.uga.edu/software/). For the purpose of microbial community analysis, rarefaction curves of samples belonging to the same bioreactor were interpolated to the lowest reads count among all of them and extrapolated to the highest reads count among all of them. Extrapolation of rarefaction curves was done following the mathematical model described in Colwell et al. [29].

Heat Maps

For each bioreactor, a heat map was generated defining the differences of community structure on the basis of the annealing temperatures selected for PCR. The heat maps were based on the relative abundance of OTUs > 1% relative abundance.

Cluster Analysis (CA)

Three different types of cluster analysis (CA) were performed over the samples: "analysis A" including all OTUs and using a non-phylogeny-dependent method; "analysis B" including only OTUs with >1% relative abundance and using a non-phylogeny-dependent method; and "analysis C" including only OTUs with >1% relative abundance and using a phylogeny-dependent method. Non-phylogeny-dependent A and B analysis were based on Bray-Curtis dissimilarity and carried out with the Vegan package v.2.0 implemented on the statistical software R-Project v.2.15.1 [30]. The software Fast UniFrac [31] was utilized for the phylogeny-dependent C analysis, following instructions given by the developers in the software tutorial (http://unifrac.colorado.edu/). For this purpose, reference trees were constructed for each bioreactor, using the MEGA 6.0 software [32]. Relative abundances of each of the OTUs were used as weight for the analysis.

Principal Coordinates Analysis (PCoA)

Principal coordinates analysis (PCoA) of samples coming from the same bioreactor was done based on a phylogeny-dependent method. OTUs with relative abundance >1% were selected for the study. For each bioreactor, a reference tree with selected OTUs was constructed using MEGA 6.0 software. Utilizing reference trees, phylogeny-based PCoA was conducted using Fast UniFrac software [31], following the instructions given in the software tutorial (http://unifrac.colorado.edu/).

Hill Diversity Indices

Hill diversity indices of order 1 (Shannon index) and order 2 (Simpson index) were calculated for each sample using the Vegan package v.2.0 implemented on the statistical software R-Project v.2.15.1 [30].

RESULTS AND DISCUSSION

In Silico Primer Evaluation

The design of a primer that covers all species within the domain Bacteria is impossible [33]. Therefore, when a primer is chosen for a bacterial community analysis, one has to accept that no total coverage could be found. For this reason, potential coverage of a primer is of primary importance in order to decide the best option for the analysis of microbial ecology of a natural ecosystem. Results of the in silico search for coverage of bacterial species of nine widely used universal primers within the RDP database are displayed in Table 1. Only results for the phylum Planctomycetes and the complete Bacteria domain are shown. Results for all other phyla can be seen in Table S2. The results of the in silico analysis showed that primer 530F offered the best coverage of species for both the Bacteria domain and the phylum Planctomycetes. Primer 530F was able to discern 76.01% of the total bacterial species known to date. The closest follower was primer 519R, with 8% less total coverage. With regard to Planctomycetes, primer 530F covered 75.58% of species belonging to this group, while the second best (910R) covered 53.70%.

Table 1: In silico evaluation of primer coverage of species in the Bacteria domain andPlancto-mycetes specific

Phylum	27F	519R	530F	787R	910R	1064R	1392R	1492R
Planctomycetes	17.82%	23.89%	75.58%	6.17%	53.70%	39.88%	33.25%	6.59%
Total	**11.58%**	**68.47%**	**76.01%**	**59.45%**	**56.53%**	**54.52%**	**22.08%**	**4.08%**

Coverage of all phylotypes of anammox bacteria has been reported as very difficult if a universal primer is utilized [19, 20]. In silico analysis of the coverage of all Candidatus Brocadiales microorganisms by primer 530F within the SiLVA database is summarized in Table S3. We found that primer 530F targeted the 16S rRNA genes of all the anammox Candidatus species known to date yielding a minimum of 90% perfect matches to sequences filed in the database. Thus, it can be stated that primer 530F is the best option for the metagenomic analysis of autotrophic nitrogen removal bioreactors.

For the purposes of pyrosequencing, reverse primer 1100R was taken, following the literature that has utilized primer 530F for pyrosequencing analysis [21].

In Silico Optimum Annealing Temperature Calculation

Following the expression given by Rychlik et al. [22] for the salt-adjusted optimum melting temperature of primer 530F with combined N^+ and K^+ concentration of 50 mM, as utilized by other authors [23–25], optimum melting temperature for primer 530F was 47°C. For the testing of different annealing temperatures an interval around optimum annealing temperature calculated was covered, ranging from 44°C to 49°C.

Rarefaction Curves

For each bioreactor, original rarefaction curves, interpolated ones to lowest number of reads within samples, and extrapolated ones to the highest number of reads within samples were generated and are shown in Figure S1. It can be seen that annealing temperatures of 44°C offer lower species richness than all the other annealing temperatures tested.

Diversity and Relative Abundance of Bacterial Species

Heat maps were generated for each bioreactor taking into account only the OTUs with >0.1% relative abundance as shown in Figures 1 and 2. Annealing temperature of 44°C showed a pattern that differed from those generated at all other temperatures in all bioreactors. Interestingly, at 44°C bacteria belonging to theCandidatus Brocadiales order (Candidatus Brocadia anammoxidans, Brocadia fulgida, and Brocadia sp.) showed a much lower relative abundance than that at all the other annealing temperatures tested for every bioreactor. Particularly, for the Lab MBR bioreactor, which stands as a highly enriched (>90%) [34] culture of Brocadiales, the relative abundance found was 1.27% when 44°C was the annealing temperature, while at the other temperatures assayed it fell in the 75–86% range.

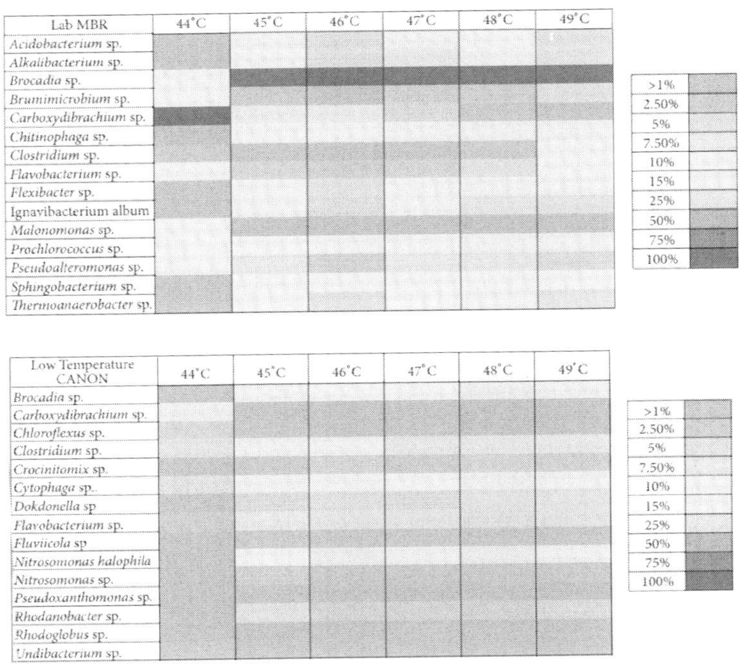

Figure 1: Heat maps of OTUs > 1% relative abundance in lab-scale bioreactors LAB MBR and Low Temperature CANON.

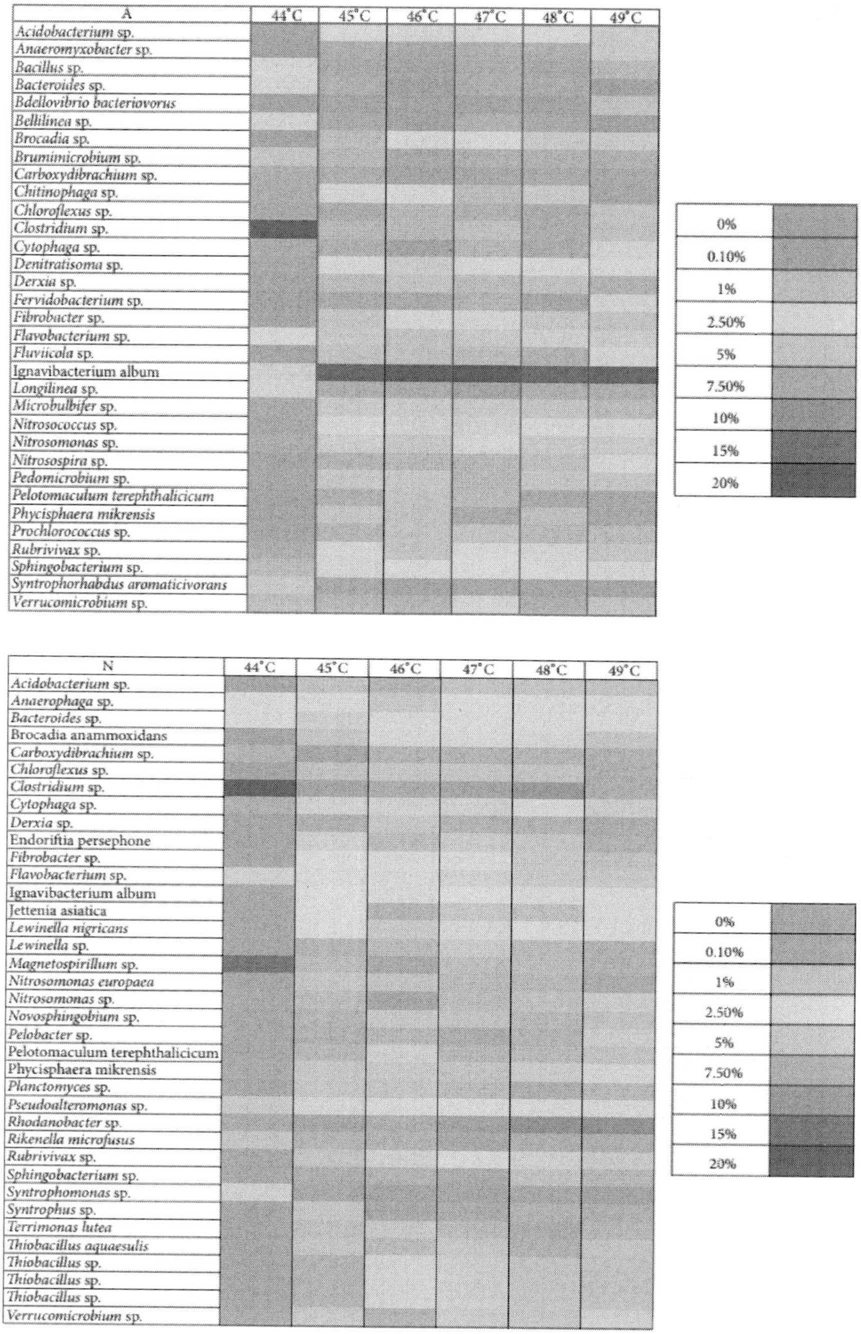
A
44°C
45°C
46°C
47°C
48°C
49°C
Acidobacterium sp.
Anaeromyxobacter sp.
Bacillus sp.
Bacteroides sp.
Bdellovibrio bacteriovorus
Bellilinea sp.
Brocadia sp.
Brumimicrobium sp.
Carboxydibrachium sp.
Chitinophaga sp.
Chloroflexus sp.
Clostridium sp.
Cytophaga sp.
Denitratisoma sp.
Derxia sp.
Fervidobacterium sp.
Fibrobacter sp.
Flavobacterium sp.
Fluviicola sp.
Ignavibacterium album
Longilinea sp.
Microbulbifer sp.
Nitrosococcus sp.
Nitrosomonas sp.
Nitrosospira sp.
Pedomicrobium sp.
Pelotomaculum terephthalicicum
Phycisphaera mikrensis
Prochlorococcus sp.
Rubrivivax sp.
Sphingobacterium sp.
Syntrophorhabdus aromaticivorans
Verrucomicrobium sp.
0%
0.10%
1%
2.50%
5%
7.50%
10%
15%
20%
N
44°C
45°C
46°C
47°C
48°C
49°C
Acidobacterium sp.
Anaerophaga sp.
Bacteroides sp.
Brocadia anammoxidans
Carboxydibrachium sp.
Chloroflexus sp.
Clostridium sp.
Cytophaga sp.
Derxia sp.
Endoriftia persephone
Fibrobacter sp.
Flavobacterium sp.
Ignavibacterium album
Jettenia asiatica
Lewinella nigricans
Lewinella sp.
Magnetospirillum sp.
Nitrosomonas europaea
Nitrosomonas sp.
Novosphingobium sp.
Pelobacter sp.
Pelotomaculum terephthalicicum
Phycisphaera mikrensis
Planctomyces sp.
Pseudoalteromonas sp.
Rhodanobacter sp.
Rikenella microfusus
Rubrivivax sp.
Sphingobacterium sp.
Syntrophomonas sp.
Syntrophus sp.
Terrimonas lutea
Thiobacillus aquaesulis
Thiobacillus sp.
Thiobacillus sp.
Thiobacillus sp.
Verrucomicrobium sp.
0%
0.10%
1%
2.50%
5%
7.50%
10%
15%
20%

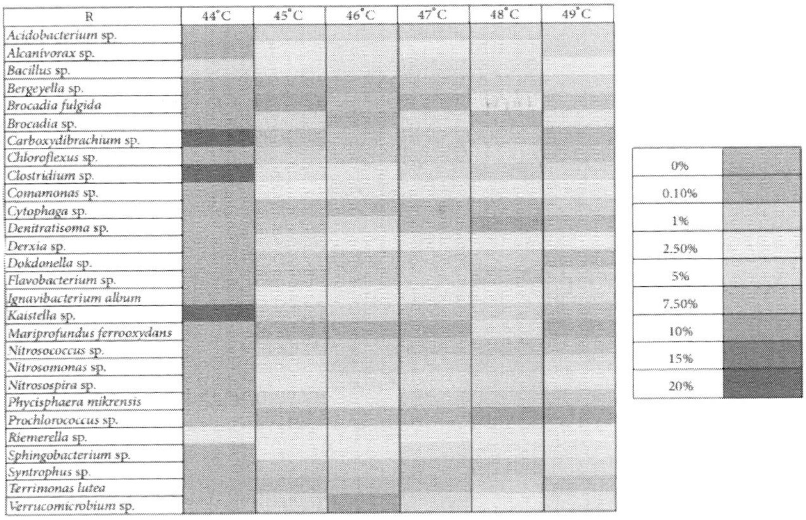

Figure 2: Heat maps of OTUs > 1% relative abundance in full scale bioreactors Appeldorn (A), Olburgen (N), and Rotterdam (R).

Amplification of partial 16S rRNA genes of bacteria other than anammox was also influenced by the annealing temperature selected. In the case of the Lab MBR, the relative abundance of copies of Carboxydibrachiumsp.was of 84.5% at 44°C, much higher compared to the other annealing temperatures tested. The inability to capture any Candidatus Brocadiales phylotypes in all bioreactors, with one of those being a high enrichment of these bacteria, when annealing was performed at 44°C is related to a consistent error occurring during the PCR procedure of primer 530F. In this sense, it is possible that annealing at 44°C hinds the relative abundance ofCandidatus Brocadiales bacteria under other bacterial species such as Carboxydibrachium sp. in the different pyrosequencing samples from the five bioreactors analyzed.

The results described here demonstrated that an annealing temperature of 44°C was inappropriate to analyze the bacterial community structure of the bioreactors sampled in the study, due to a severe underestimation of the relative abundance of Brocadiales.

Results of bacterial diversity obtained for bioreactors Lab MBR and Low Temperature CANON were further analyzed to check the performance of primer 530F for the evaluation of anammox diversity. Diversity of anammox bacteria in bioreactors Lab MBR and Low Temperature CANON was studied before by other authors utilizing different molecular biology techniques, such as FISH or qPCR [35, 36]. In these studies, it was concluded that Candidatus Brocadia fulgida was the only anammox bacteria in both Lab MBR and Low Temperature CANON bioreactors. In the present study, Candidatus Brocadia sp. was the only anammox identified in Lab MBR bioreactor, while in the Low Temperature CANON bioreactor, pyrosequencing with primer 530F retrieved partial 16S rRNA genes affiliated to four different anammox phylotypes: CandidatusBrocadia anammoxidans, Candidatus

Brocadia fulgida, Candidatus Brocadia sp., and Candidatus Jettenia asiatica, all of them represented in low relative abundance with the exception of Candidatus Brocadia sp. Sequences for the OTUs identified as Candidatus Brocadia sp. in both Lab MBR and Low Temperature CANON bioreactors (305 and 289 nucleotides, resp.) shared 99% similarity in 100% query cover with fourCandidatus Brocadia fulgida sequences found in the BLASTN database (JQ864319.1, JQ864321.1, JQ864322.1, Zheng & Zhang, unpublished, and JX243455.1 [36]). Thus, it can be said that universal primer 530F can express in a consistent way the diversity of anammox bacteria found in autotrophic nitrogen removal bioreactors.

Failure to find Candidatus Brocadiales sequences at annealing temperature of 44°C can be related to a bias of PCR procedure. The relative abundance of Brocadia sp. and Carboxydibrachium sp. along with their ratio for the bioreactor Lab MBR can be seen in Table 2. The Carboxydibrachium sp./Brocadia sp. ratio abruptly increases when dropping from 45°C to 44°C, while it stabilizes at temperatures of 45°C and higher. In accordance with our results, some studies have found that a small percentage of OTUs in chicken caecal samples suffered from distortions in relative abundance as their bacterial community structure was studied through RT-PCR at different annealing temperatures, even though the entire microbial community structures were not subjected to major changes [37]. Also, changes in amplification of Vibrio vulnificus by RT-PCR at different annealing temperatures have been found [38]. Mismatches of primers with the targeted region of the DNA templates are thought to be the cause of differences in the estimation of the relative abundance of bacterial species in environmental samples. A higher number of mismatches in the targeted region lead to a higher bias of the relative abundance of certain OTUs. Among other factors, the annealing temperature is thought to cause mismatch of primers during PCR processes [33]. In this way, low temperatures with respect to optimum have been related to the proliferation of amplicons with mismatches [39, 40]. On the other hand, it has been shown that high annealing temperatures with respect to optimum tend to increase the ratio of amplification of one-mismatch to no-mismatch sequences [41, 42]. Regardless of the diversity of opinions, knowledge of the factors that intervene in the differences of relative abundance of OTUs at different annealing temperatures is still incomplete; hence, primer mismatches may not be the only cause driving these changes [37].

Table 2: Relative abundances of Brocadia sp., Carboxydibrachium sp. and the ratio of relative abundances Carboxydibrachium sp. to Brocadia sp. for the Lab MBR reactor

	Lab MBR					
	44°C	45°C	46°C	47°C	48°C	49°C
Species						
Brocadia sp. (%)	1.04	75.15	86.76	79.51	77.68	75.77
Carboxydibrachium sp. (%)	84.50	0.81	0.77	0.44	0.26	0.13
C. sp./*B.* sp.	81.34	0.01	0.01	0.01	0.00	0.00

CA and PCoA

CA a) and b) are shown in Figure S2. When the similarity of all OTUs identified was compared, there were no relevant differences between results generated by PCR at the different annealing temperatures assayed (CA a)). However, CA b), which analyzes only OTUs with >1% relative abundance, clearly shows significant differences between the results generated at 44°C compared to the rest of annealing temperatures tested. This is in accordance with the heat maps generated for the bioreactors sampled (Figure 1). This implies that Bray-Curtis dissimilarities between annealing temperatures are less pronounced when samples are studied to their full sampling depths.

Phylogeny-dependent CA c) for each bioreactor is shown in Figure S3. Similarity between samples varies depending on the bioreactor analyzed. Samples from lab-scale bioreactors showed a higher phylogeny-based similarity for all the annealing temperatures tested. No significant differences at 80% similarity were found in bioreactor Low Temperature CANON, and three clusters were formed at 90% similarity. For bioreactor Lab MBR there were no significant differences at 70% similarity, but three clusters were generated at 80% and 90% similarity. Full-scale bioreactors showed lower phylogeny-based similarity, with significant differences at 70% for bioreactors N and R and sequences grouped in five or six clusters at 90% similarity in all cases. This leads to the assumption that differences in bacterial communities obtained at different annealing temperatures are more pronounced in more complex ecosystems.

Phylogeny-dependent PCoA 2D views are offered in Figure S4. PCoA was done to double check the results obtained by CA of samples. Phylogeny-dependent PCoA showed that 44°C tended to discriminate from the other annealing temperatures in terms of microbial community structure explanation. On the other hand, samples annealed at 45°C and 48°C seemed to be related in terms of microbial community structure. As also proven by cluster analysis, the differences between various annealing temperatures were found more pronounced as the complexity of the bacterial communities increased.

Some authors have reported that differences in annealing temperature do not significantly affect the microbial community structure of chicken caecal samples [37]. In our case, differences in microbial community structure at both full sampling depth and at >1% OTUs were confirmed.

Species Richness and Hill Diversity Indices

Values for individuals computed and original, interpolated and extrapolated species richness are summarized in Table 3. Rating for performance of the different annealing temperatures based on species richness is shown in Table 4. Differences in species richness are accounted at the original sequencing depth, interpolated to the lowest number of reads per bioreactor, and extrapolated to the highest number of reads per bioreactor. Even though species richness comparison cannot be performed at different sequencing depths, methods for interpolation and extrapolation of rarefaction curves can precisely estimate species richness diversity of samples with different number of reads, making comparison between them possible [29].

Table 3: Values for individuals computed, original species richness S(est), interpolated species richness S(int), and extrapolated species richness S(ext)

	Temperature	Individuals (computed)	S(est)	S(int)	S(ext)
Lab MBR (T2)	44°C	11646	42	34.8	42.99
	45°C	16895	148	85.6	148.4
	46°C	17336	117	67	117
	47°C	3162	78	76.8	85.67
	48°C	5883	100	79.2	108.9
	49°C	3024	105	105	117.3
Low Temperature CANON (TC)	44°C	11438	61	30.1	38.65
	45°C	14362	179	100.3	179
	46°C	9545	155	94.4	163.1
	47°C	4145	119	99.5	131.6
	48°C	5068	133	99.3	147.9
	49°C	2394	87	87	95.53
Apeldoorn DEMON (A)	44°C	9226	78	29.3	78.76
	45°C	9924	161.7	45.1	161.7
	46°C	654	73	55.6	79.11
	47°C	2192	141	66.5	156.9
	48°C	8206	258	68	265.2
	49°C	321	60	59.9	68.55
Olburgen CANON (N)	44°C	17193	87	41.9	87
	45°C	2893	198	134.2	218.3
	46°C	947	141	140.9	160.7
	47°C	3180	205	135.7	221.9
	48°C	2462	197	139.2	217.1
	49°C	1905	174	133.8	192.6
Rotterdam 2-stage anammox (R)	44°C	12977	81	43.2	81
	45°C	5183	211	147.3	228
	46°C	3879	188	143.5	205.2
	47°C	4120	198	151.6	212.3
	48°C	1679	143	143	159.6
	49°C	1791	145	142.3	159.7

Table 4: Ratings for the comparison of species richness for samples from different bioreactors at each annealing temperature. Higher mean ratio is related to higher performance of the annealing temperature

Temperature	Mean ratio individuals	Mean ratio S(int)	Mean ratio S(ext)
44°C	89.36%	32.89%	31.00%
45°C	72.36%	88.05%	91.86%
46°C	41.79%	86.87%	72.43%
47°C	24.22%	93.29%	76.70%
48°C	37.09%	93.51%	84.76%
49°C	12.49%	92.73%	63.01%

Differences in species richness and number of reads obtained at each annealing temperature can be appreciated. As a main trend, 44°C is the temperature that offered the highest number of reads, followed by 45°C and 46°C. It has been suggested that annealing temperatures lower than the optimum increase the number of PCR products generated [43]. The lowest number of reads was obtained at 49°C. 48°C was the temperature generating a higher interpolated species richness, while at 44°C a poorer performance was observed, and 44°C was also the worst among all annealing temperatures regarding original species richness. Extrapolated species richness of samples was higher at 45°C, followed by 48°C.

In conclusion, 45°C is the annealing temperature showing the best performance regarding the expression of species richness of bacterial communities in the bioreactors studied, while annealing at 49°C offers a worse performance taking into account the low number of reads consistently obtained.

Hill diversity indices of first order (Shannon index) and second order (Simpson index) have been defended as the most robust method for comparison of bacterial assemblages from natural ecosystems [31]. Hill diversities of first order and second order for each sample are shown in Table 5. Ratings for the comparison of each annealing temperature based on the indices can be seen in Table 6. Both indices showed the same quality pattern, with 45°C being the highest value followed closely by 49°C and 48°C. Once again, 45°C was the best annealing temperature for capturing diversity of bacterial communities inside the bioreactors analyzed.

Table 5: Values for the Shannon index and the Simpson index of all samples analyzed in the study

	Temperature	Shannon index	Simpson index
Lab MBR (T2)	44°C	1.559179	0.4866806
	45°C	1.657458	0.5265754
	46°C	0.9898754	0.2964455
	47°C	1.398619	0.4398892
	48°C	1.536843	0.480206
	49°C	1.665004	0.4945622
Low Temperature CANON (TC)	44°C	2.02004	0.7847361
	45°C	3.018484	0.9061173
	46°C	2.85764	0.8814217
	47°C	2.86951	0.8790168
	48°C	2.860865	0.8741682
	49°C	2.871034	0.8860556
Apeldoorn DEMON (A)	44°C	2.297351	0.6694541
	45°C	3.561685	0.9399574
	46°C	3.286833	0.928345
	47°C	3.51128	0.9295173
	48°C	3.651016	0.944623
	49°C	3.366082	0.9407517
Olburgen CANON (N)	44°C	2.959167	0.8714149
	45°C	3.979032	0.9632393
	46°C	4.039	0.9667856
	47°C	4.021377	0.9636953
	48°C	4.015465	0.9615441
	49°C	3.965338	0.9629424
Rotterdam 2-stage anammox (R)	44°C	2.497644	0.8499261
	45°C	3.910746	0.9644197
	46°C	3.87392	0.9619344
	47°C	3.903737	0.9636846
	48°C	3.865236	0.9630658
	49°C	3.902146	0.9643079

Table 6: Rating for the comparison of species diversity for samples from different bioreactors at each annealing temperature. Higher mean ratio is related to higher performance of the annealing temperature

Temperature	Mean ratio Shannon index	Mean ratio Simpson index
44°C	72.12%	85.63%
45°C	99.12%	99.83%
46°C	88.64%	90.32%
47°C	94.92%	95.71%
48°C	97.07%	97.40%
49°C	97.14%	98.24%

No significant differences in species richness and evenness of chicken caecal samples studied through RT-PCR have been reported [37]. Nevertheless, other authors have shown that differences in annealing temperature lead to different bonding of primer and targeted region of genetic templates, having an impact over ecological parameters of environmental samples. This is caused by enhanced amplification of certain strains, which deviates relative abundance of these species, or by nonspecific bonding, which increases the microbial diversity recorded on the samples [39, 44]. In our case, small differences in annealing temperature changed species richness, effective number of counts, and Hill diversities within the samples. Samples processed at 44°C accounted for the highest number of reads but surprisingly also contained the lowest species richness and diversity. This shows that annealing temperatures of 44°C reduced the estimation of the diversity of the system. Reduced diversity of samples analyzed at 44°C may rest on the fact that the primer is subjected to much more stringent bonding than that at the other annealing temperatures, therefore generating many sequences of a low number of species. All other annealing temperatures can capture the diversity of the ecosystem studied in a consistent fashion. 46°C offers a lower diversity quality values in comparison with the others.

In terms of species richness and diversity, 44°C offers poor results compared to the others, but it can be said that annealing temperatures ranging from 45 to 49°C offer good results. Nevertheless, 45°C stands as the optimal annealing temperature of all those tested due to superior species richness and diversity indices values. Interestingly, it yields better results in species richness and diversity than the in silico calculated salt-adjusted optimum for the primer. As suggested by Hecker and Roux, 1996 [43], annealing temperatures above and down the optimum annealing temperature theoretically calculated for a given primer increase number of reads obtained and specificity of PCR products. In this case, annealing temperature at 45°C gives better results, in terms of species richness and diversity, than the optimum salt-adjusted annealing temperature for the primer utilized.

CONCLUSIONS

The superior coverage of primer 530F with respect to other popular universal primers was indicated through in silico testing. The ability of primer 530F to target the majority of known anammox phylotypes was also demonstrated by in silico testing. Therefore, primer 530F stands as the best universal primer available for the metagenomic analysis of microbial communities where anammox bacteria are expected to develop important ecological functions.

The 30 pyrosequencing analysis showed that the annealing temperature produces a severe effect over the microbial community structure discerned through pyrosequencing. Strong bias in the identification of anammox species in particular at annealing temperature of 44°C was observed in all the pyrosequencing samples for each of the five autotrophic nitrogen removal technologies. The six different annealing temperatures analyzed in the study showed different microbial community structure compositions as proved by phylogeny-based and non-phylogeny-based CA and PCoA. Differences among annealing temperatures could also be observed with respect to the number of individuals sequenced and species richness with 45°C being the best in these terms. In this sense, annealing temperatures of 45°C demonstrate good coverage of total bacteria and anammox species, high number of reads, the highest species richness, and the highest diversity indices values. Therefore, results show that autotrophic nitrogen removal bioreactor bacterial community analyses including anammox bacteria can be done using the universal primer 530F with annealing temperature of 45°C. Under these conditions, this procedure is a good approach for the bacterial diversity study in autotrophic nitrogen removal technologies using pyrosequencing methods.

AUTHORS' CONTRIBUTION

Alejandro Gonzalez-Martinez was responsible for the bioreactors field study and in silico testing of the different primers. Alejandro Gonzalez-Martinez was mainly responsible for molecular biology experimental procedures, and Ben A. Abbas was also responsible for this. Alejandro Rodriguez-Sanchez was mainly responsible for the statistical analysis of samples and microbial ecology analysis, while Maria Victoria Martinez-Toledo and Belén Rodelas also were responsible for microbial ecology analysis. Alejandro Gonzalez-Martinez, Alejandro Rodriguez-Sanchez, Belén Rodelas, and Jesus Gonzalez-Lopez were responsible for writing the paper. Alejandro Gonzalez-Martinez and Alejandro Rodriguez-Sanchez were responsible for the conformation of all tables and figures. Alejandro Gonzalez-Martinez, Belén Rodelas, Maria Victoria Martinez-Toledo, Mark C. M. van Loosdrecht, F. Osorio, and Jesus Gonzalez-Lopez were responsible for revision of the paper.

ACKNOWLEDGEMENT

The authors received a financial support from a collaborator company but it has not participated in the work development.

REFERENCES

1. M. Strous, J. A. Fuerst, E. H. M. Kramer et al., "Missing lithotroph identified as new planctomycete,"Nature, vol. 400, no. 6743, pp. 446–449, 1999. · ·
2. T. Dalsgaard, B. Thamdrup, and D. E. Canfield, "Anaerobic ammonium oxidation (anammox) in the marine environment," Research in Microbiology, vol. 156, no. 4, pp. 457–464, 2005. · ·

3. M. M. M. Kuypers, A. O. Silekers, G. Lavik et al., "Anaerobic ammonium oxidation by anammox bacteria in the Black Sea," Nature, vol. 422, no. 6932, pp. 608–611, 2003. · ·

4. B. Thamdrup, T. Dalsgaard, M. M. Jensen, O. Ulloa, L. Farías, and R. Escribano, "Anaerobic ammonium oxidation in the oxygen-deficient waters off Northern Chile," Limnology and Oceanography, vol. 51, no. 5, pp. 2145–2156, 2006. · ·

5. S. Ganesh, D. J. Parris, E. F. Delong, and F. J. Stewart, "Metagenomic analysis of size-fractionated picoplankton in a marine oxygen minimum zone," ISME Journal, vol. 8, no. 1, pp. 187–211, 2014. · ·

6. L.-D. Shen, S. Liu, L.-P. Lou et al., "Broad distribution of diverse anaerobic ammonium-oxidizing bacteria in Chinese agricultural soils," Applied and Environmental Microbiology, vol. 79, no. 19, pp. 6167–6172, 2013. · ·

7. S. M. Kotay, B. L. Mansell, M. Hogsett, H. Pei, and R. Goel, "Anaerobic ammonia oxidation (ANAMMOX) for side-stream treatment of anaerobic digester filtrate process performance and microbiology," Biotechnology and Bioengineering, vol. 110, no. 4, pp. 1180–1192, 2013. · ·

8. A. González-Martínez, J. M. Poyatos, E. Hontoria, J. Gonzalez-Lopez, and F. Osorio, "Treatment of effluents polluted by nitrogen with new biological technologies based on autotrophic nitrification-denitrification processes," Recent Patents on Biotechnology, vol. 5, no. 2, pp. 74–84, 2011. · ·

9. M. S. M. Jetten, M. Wagner, J. Fuerst, M. C. M. van Loosdrecht, G. Kuenen, and M. Strous, "Microbiology and application of the anaerobic ammonium oxidation ('anammox') process," Current Opinion in Biotechnology, vol. 12, no. 3, pp. 283–288, 2001. · ·

10. S. Sri Shalini and K. Joseph, "Nitrogen management in landfill leachate: application of SHARON, ANAMMOX and combined SHARON-ANAMMOX process," Waste Management, vol. 32, no. 12, pp. 2385–2400, 2012. · ·

11. T. Liu, D. Li, H. Zeng et al., "Biodiversity and quantification of functional bacteria in completely autotrophic nitrogen-removal over nitrite (CANON) process," Bioresource Technology, vol. 118, pp. 399–406, 2012. · ·

12. Z.-X. Quan, S.-K. Rhee, J.-E. Zuo et al., "Diversity of ammonium-oxidizing bacteria in a granular sludge anaerobic ammonium-oxidizing (anammox) reactor," Environmental Microbiology, vol. 10, no. 11, pp. 3130–3139, 2008. · ·

13. Z. Hu, D. R. Speth, K.-J. Francoijs, Z.-X. Quan, and M. S. M. Jetten, "Metagenome analysis of a complex community reveals the metabolic blueprint of anammox bacterium 'Candidatus Jettenia asiatica',"Frontiers in Microbiology, vol. 3, article 366, 2012. · ·

14. Q. Yang, Z. Jia, R. Liu, and J. Chen, "Molecular diversity and anammox activity of novel planctomycete-like bacteria in the wastewater treatment system of a full-scale alcohol manufacturing plant," Process Biochemistry, vol. 42, no. 2, pp. 180–187, 2007. · ·

15. X.-R. Li, B. Du, H.-X. Fu et al., "The bacterial diversity in an anaerobic

ammonium-oxidizing (anammox) reactor community," Systematic and Applied Microbiology, vol. 32, no. 4, pp. 278–289, 2009. · ·

16. B.-J. Ni, B.-L. Hu, F. Fang et al., "Microbial and physicochemical characteristics of compact anaerobic ammonium-oxidizing granules in an upflow anaerobic sludge blanket reactor," Applied and Environmental Microbiology, vol. 76, no. 8, pp. 2652–2656, 2010. · ·
17. P. Han, Y.-T. Huang, J.-G. Lin, and J.-D. Gu, "A comparison of two 16S rRNA gene-based PCR primer sets in unraveling anammox bacteria from different environmental samples," Applied Microbiology and Biotechnology, vol. 97, no. 24, pp. 10521–10529, 2013. · ·
18. B. Xie, Z. Lv, C. Hu, X. Yang, and X. Li, "Nitrogen removal through different pathways in an aged refuse bioreactor treating mature landfill leachate," Applied Microbiology and Biotechnology, vol. 97, no. 20, pp. 9225–9234, 2013. · ·
19. M. S. M. Jetten, L. van Niftrik, M. Strous, B. Kartal, J. T. Keltjens, and H. J. M. Op Den Camp, "Biochemistry and molecular biology of anammox bacteria," Critical Reviews in Biochemistry and Molecular Biology, vol. 44, no. 2-3, pp. 65–84, 2009. ·
20. P. Junier, V. Molina, C. Dorador et al., "Phylogenetic and functional marker genes to study ammonia-oxidizing microorganisms (AOM) in the environment," Applied Microbiology and Biotechnology, vol. 85, no. 3, pp. 425–440, 2010. · ·
21. S. E. Dowd, T. R. Callaway, R. D. Wolcott et al., "Evaluation of the bacterial diversity in the feces of cattle using 16S rDNA bacterial tag-encoded FLX amplicon pyrosequencing (bTEFAP)," BMC Microbiology, vol. 8, article 125, 2008. · ·
22. W. Rychlik, W. J. Spencer, and R. E. Rhoads, "Optimization of the annealing temperature for DNA amplification in vitro," Nucleic Acids Research, vol. 18, no. 21, pp. 6409–6412, 1990. · ·
23. X. Chen, P. Liu, and H.-H. Chou, "Whole-genome thermodynamic analysis reduces siRNA off-target effects," PLoS ONE, vol. 8, no. 3, Article ID e58326, 2013. · ·
24. D. Dobrijevic, G. Di Liberto, K. Tanaka et al., "High-throughput system for the presentation of secreted and surface-exposed proteins from Gram-positive bacteria in functional metagenomic studies," PLoS ONE, vol. 8, no. 6, Article ID e65956, 2013. · ·
25. S. Nakano, Y. Morizane, N. Makisaka et al., "Development of a highly sensitive immuno-PCR assay for the measurement of α-galactosidase A protein levels in serum and plasma," PLoS ONE, vol. 8, no. 11, Article ID e78588, 2013. · ·
26. R. C. Edgar, "Search and clustering orders of magnitude faster than BLAST," Bioinformatics, vol. 26, no. 19, pp. 2460–2461, 2010. · ·
27. R. C. Edgar, B. J. Haas, J. C. Clemente, C. Quince, and R. Knight, "UCHIME improves sensitivity and speed of chimera detection," Bioinformatics, vol. 27, no. 16, pp. 2194–2200, 2011. · ·

28. D. E. Wood and S. L. Salzberg, "Kraken: ultrafast metagenomic sequence classification using exact alignments," Genome Biology, vol. 15, no. 3, article R46, 2014. · ·

29. R. K. Colwell, A. Chao, N. J. Gotelli et al., "Models and estimators linking individual-based and sample-based rarefaction, extrapolation and comparison of assemblages," Journal of Plant Ecology, vol. 5, no. 1, pp. 3–21, 2012. · ·

30. R Development Core Team, R: A Language and Environment for Statistical Computing, R Foundation for Statistical Computing, Vienna, Austria, 2008, http://www.R-project.org.

31. B. Haegeman, J. Hamelin, J. Moriarty, P. Neal, J. Dushoff, and J. S. Weitz, "Robust estimation of microbial diversity in theory and in practice," The ISME Journal, vol. 7, no. 6, pp. 1092–1101, 2013. · ·

32. K. Malhotra, L. Foltz, W. C. Mahoney, and P. A. Schueler, "Interaction and effect of annealing temperature on primers used in differential display RT-PCR," Nucleic Acids Research, vol. 26, no. 3, pp. 854–856, 1998. · ·

33. D. Bru, F. Martin-Laurent, and L. Philippot, "Quantification of the detrimental effect of a single primer-template mismatch by real-time PCR using the 16S rRNA gene as an example," Applied and Environmental Microbiology, vol. 74, no. 5, pp. 1660–1663, 2008. · ·

34. W. R. L. van der Star, A. I. Miclea, U. G. J. M. van Dongen, G. Muyzer, C. Picioreanu, and M. C. M. van Loosdrecht, "The membrane bioreactor: a novel tool to grow anammox bacteria as free cells,"Biotechnology and Bioengineering, vol. 101, no. 2, pp. 286–294, 2008. · ·

35. T. Lotti, Monitoraggio del proceso anammox: aspetti fisiologici per unutilizzo in piena scala, Università degli Studi della Basilicata, Potenza, Italy, 2011.

36. Z. Hu, T. Lotti, M. de Kreuk et al., "Nitrogen removal by a nitritation-anammox bioreactor at low temperature," Applied and Environmental Microbiology, vol. 79, no. 8, pp. 2807–2812, 2013. · ·

37. M. J. Sergeant, C. Constantinidou, T. Cogan, C. W. Penn, and M. J. Pallen, "High-throughput sequencing of 16s rRNA gene amplicons: Effects of extraction procedure, primer length and annealing temperature,"PLoS ONE, vol. 7, no. 5, Article ID e38094, 2012. · ·

38. S. Wang and R. E. Levin, "Thermal factors influencing detection of Vibrio vulnificus using real-time PCR," Journal of Microbiological Methods, vol. 69, no. 2, pp. 358–363, 2007. · ·

39. J. T. Hsu, S. Das, and S. Mohapatra, "Polymerase chain reaction engineering," Biotechnology and Bioengineering, vol. 55, no. 2, pp. 359–366, 1997. ·

40. M. J. Claesson, Q. Wang, O. O›Sullivan et al., "Comparison of two next-generation sequencing technologies for resolving highly complex microbiota composition using tandem variable 16S rRNA gene regions," Nucleic Acids Research, vol. 38, article e200, 2010. · ·

41. K. Ishii and M. Fukui, "Optimization of annealing temperature to reduce bias caused by a primer mismatch in multitemplate PCR," Applied and Environmental Microbiology, vol. 67, no. 8, pp. 3753–3755, 2001. ·

42. R. Sipos, A. J. Székely, M. Palatinszky, S. Révész, K. Márialigeti, and M. Nikolausz, "Effect of primer mismatch, annealing temperature and PCR cycle number on 16S rRNA gene-targetting bacterial community analysis," FEMS Microbiology Ecology, vol. 60, no. 2, pp. 341–350, 2007. · ·

43. K. H. Hecker and K. H. Roux, "High and low annealing temperatures increase both specificity and yield in touchdown and stepdown PCR," BioTechniques, vol. 20, no. 3, pp. 478–485, 1996. ·

44. K. Malhotra, L. Foltz, W. C. Mahoney, and P. A. Schueler, "Interaction and effect of annealing temperature on primers used in differential display RT-PCR," Nucleic Acids Research, vol. 26, no. 3, pp. 854–856, 1998. · ·

16
CHAPTER

ENRICHMENT OF A MICROBIAL COMMUNITY PERFORMING ANAEROBIC OXIDATION OF METHANE IN A CONTINUOUS HIGH-PRESSURE BIOREACTOR

Yu Zhang[1,2], Loïs Maignien[2], Xianxian Zhao[1], Fengping Wang[1] and Nico Boon[2]

[1] State Key Laboratory of Microbial Metabolism and School of Life Science & Biotechnology, State Key Laboratory of Ocean Engineering, Shanghai Jiao Tong University, Dongchuan Rd. 800, Shanghai, 200240, P. R. China

[2] Laboratory of Microbial Ecology and Technology (LabMET), Ghent University, Coupure Links 653, Gent, 9000, Belgium

ABSTRACT

Background

Anaerobic oxidation of methane coupled to sulphate reduction (SR-AOM) prevents more than 90% of the oceanic methane emission to the atmosphere. In a previous study, we demonstrated that the high methane pressure (1, 4.5, and 8 MPa) stimulated *in vitro* SR-AOM activity. However, the information on the effect of high-pressure on the microbial community structure and architecture was still lacking.

Results

In this study we analysed the long-term enrichment (286 days) of this microbial community, which was mediating SR-AOM in a continuous high-pressure bioreactor. 99.7% of the total biovolume represented cells in the form of small aggregates (diameter less then 15 μm). An increase of the total biovolume was observed (2.5 times). After 286 days, the ANME-2 (anaerobic methanotrophic archaea subgroup 2) and SRB (sulphate reducing bacteria) increased with a factor 12.5 and 8.4, respectively.

Citation: Yu Zhang, Loïs Maignien, Xianxian Zhao, Fengping Wang and Nico Boon, Enrichment of a microbial community performing anaerobic oxidation of methane in a continuous high-pressure bioreactor, DOI: 10.1186/1471-2180-11-137.

Conclusion

This paper reports a net biomass growth of communities involved in SR-AOM, incubated at high-pressure.

BACKGROUND

Anaerobic oxidation of methane coupled to sulphate reduction (SR-AOM) is a major process determining deep-sea geochemistry and cold-seep ecosystems. First of all, it controls the atmospheric methane efflux from the ocean floor, consuming more than 90% of the methane produced in marine sediments [1]. Moreover, it fuels the deep sea ecosystem by channelling thermal generated and biogenetic methane into organic matter and carbonate. Finally, SR-AOM shapes the sea floor landscape by contributing to bicarbonate and alkalinity production, resulting in massive carbonate precipitation [2]. The overall SR-AOM reaction is:

$$CH_4 + SO_4^{2-} = HCO_3^- + HS^- + H_2O \quad \Delta G^{\emptyset} = -16.5 kJ/mol$$

Two groups of microorganisms are the key players in SR-AOM process: anaerobic methanotrophic archaea (ANME) with three groups (ANME-1, ANME-2 and ANME-3) and sulphate reducing bacteria (SRB) [3, 4, 5, 6]. All ANME groups discovered so far are related clades of methanogens, while their SRB partner was always found in the same environment with or without forming spatial closely related consortia [7]. However, neither ANME nor SRB from SR-AOM active spots has been obtained in pure culture yet. The main difficulty lies on the extremely long doubling time (several months) and low growth yield (0.05 g dry weight/g carbon oxidized) of ANME and SRB from *in vitro* incubations [8, 9, 10].

To stimulate the *in vitro* SR-AOM activity and to enrich the SR-AOM community, different types of bioreactors, which can be operated at ambient/high pressure in continuous/batch mode, have been developed by different research groups [10, 11, 12, 13, 14]. Due to the extremely low affinity for methane (K_m of 37 mM) and the low methane solubility at ambient pressure, high-pressure bioreactors have the advantage of permitting a higher SR-AOM activity [11, 15]. Nevertheless, it is still unknown if the high-pressure bioreactor also confers advantage on biomass enrichment, and if it has an effect on selective enrichment of certain groups of ANME. Moreover, the information is lacking on the community architecture inside the high-pressure bioreactor, meaning if the microbes live as single cells or form consortia.

Through high-pressure incubation, we have obtained an enrichment originating from a Mud Volcano from the Gulf of Cadiz, performing anaerobic oxidation of methane. The SR-AOM activities at different incubation conditions have been described previously [11]. In this study, the community structure and architecture of this enrichment were investigated. The potential growth of ANME and SRB under high pressure has been evaluated.

RESULTS AND DISCUSSION

Community architectural distribution

To access the community architectural distribution, a DAPI (4', 6-diamidino-2-phenylindole) staining was applied on the samples S1 (before high-pressure incubation) and S2 (after 286 days high-pressure incubation). Based on DAPI staining cell counts, both single cells and aggregates were commonly observed in S1 and S2. The aggregates had different sizes ranging from 2 to 15 μm in diameter (Ø). In both S1 and S2 single cells were 1-2 orders more abundant than the aggregates (Figure 1A). Among all the aggregates, the ones with diameter from 2 to 5 μm were the most abundant ones (73.35 ± 2.63% in S1 and 73.28 ± 1.75% in S2). Few spherical aggregates bigger than 15 μm were observed in S1 or S2 (less then 4 × 10^4 aggregates/ml slurry). For some aggregates we observed that it was dividing into two smaller spherical aggregates in both S1 and S2 (data not shown). This was also reported in another enrichment from a semi-continuous bioreactor operated under 1.4 MPa methane pressure [9]. It is an indication that these large aggregates may have reached a «critical size» during growth, which then may disintegrate into smaller aggregates for further growth.

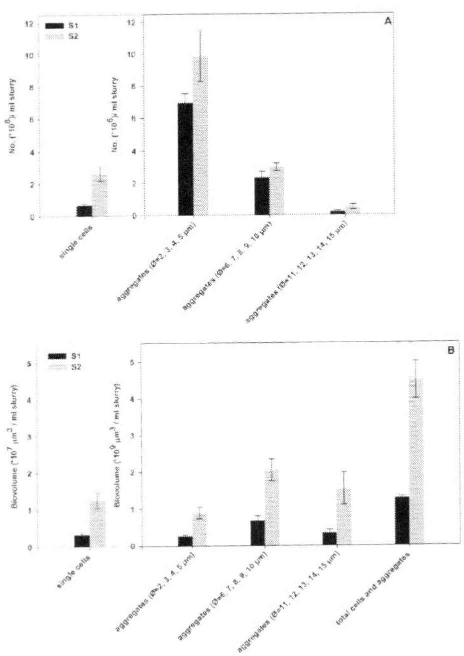

Figure 1: Numbers of cells and aggregates (A) and the biovolume of cells and aggregates (B) in S1 and S2. The average value and standard error were calculated from 4 individual staining for each sample. For each staining 50 fields of were counted for calculation. Note that the y axe scale is different for single cells.

Cell aggregates accounted for the major part of the biovolume (Figure 1B). The middle size aggregates (Ø = 6, 7, 8, 9, 10 μm) contributed for about half of the total biovolume (52.73 ± 9.04% in S1 and 47.02 ± 8.67% in S2). Although the big size aggregates (Ø = 11, 12, 13, 14, 15 μm) had very low concentrations (2.22 ± 0.74 $*10^5$/ml slurry in S1 and 4.93 ± 1.56 $*10^5$/ml slurry as shown in Figure 1A), they also contributed for large part of the biovolume (26.67 ± 7.83% in S1 and 33.34 ± 8.54% in S2).

Enrichment of total biomass

The total biovolume concentration increased from (1.28 ± 0.06)$*10^9$ μm^3/ml slurry in S1 to (4.49 ± 0.51)$*10^9 \mu m^3$/ml slurry in S2 (Figure 1B). Since the reactor volume was fixed and the biomass washing out during reactor operation was negligible [11], the total biomass inside the reactor increased 2.5 times within 286 days. This reactor system was the first system that was able to accumulate total biomass while maintaining high SR-AOM activity--0.5 mmol sulfide production per day while the reactor was operated at batch mode under 8 MPa methane pressure [11]. In the systems previously reported by other authors, either only specific groups but not the total biomass was quantified [16] or there was major loss of biomass due to sampling and decay [9, 10].

The biovolume data was converted into cell dry weight for a comparison with VSS (Volatile Suspended Solids) data. Taken the same assumption as described by Nauhaus et al. [9], there was about 0.2 g cell dry weight/ml biovolume in the sediment sample with SR-AOM activity. Therefore the biomass concentration in the high-pressure bioreactor increased from 0.3 (g cell dry weight/l slurry) in S1 to 0.9 (g cell dry weight/l slurry) in S2. However, this value was one order lower compared to the 8 g/l of VSS (based on weight difference between drying sample at 105°C and at 650°C) as reported by Zhang et al. [11]. One possibility is that the assumption 0.2 g cell dry weight/ml biovolume was based on analysis of two strains of small marine microorganism [9, 17], which could be not representative of the cells enriched in the reactor. Another possibility would be the extracellular polymeric substances (EPS) contributed large part of VSS. For example, for granular microbial aggregates enriched in an OLAND (oxygen-limited autotrophic nitrification-denitrification) reactor, as much as 50-80% of the space occupied by bacteria was constituted of EPS [18]. For the deep-sea sediment, the presence of EPS has been reported both from *in situ* sediment and *in vitro* enrichments at different locations [9, 19]. However whether the production of EPS was stimulated during high-pressure incubations and what was the mechanism behind still needs to be further investigated.

Community structure

To identify the cells and aggregates observed under microscope, catalyzed reporter deposition fluorescence *in situ* hybridization (CARD-FISH) with probes on ANME-1, 2, 3 and SRB (Table 1) was applied on S1 and S2. Based on CARD-FISH counts, ANME-2 and SRB were the most abundant ones compared to other types of ANME, especially in the form of aggregates. Among the free-living cells, only less than 10% belonged to

ANME-2 or SRB (Table 2). The number of ANME-2 aggregates accounted for 37.1 ± 6.2% of the total aggregates in S1 and 47.2 ± 8.2% in S2, while SRB accounted for 32.0 ± 6.2% of the total aggregates in S1 and 37.6 ± 5.0% in S2. However, it has to be taken into account that the CARD-FISH in this study was performed with single probe hybridization. Aggregates with ANME-2 are most probably also containing SRB as well, because they tend to live closely and form consortia [7, 9]. No ANME-1 was detected in S1 and S2. About 2% of ANME-3 was detected in the aggregates (Table 2).

Table 1: Primers and probes used in this study

Name (labelling)	Sequence (5' to 3')	Positions	Specificity	References
PCR primers				
Arch-21f	TTC CGG TTG ATC CYG CCG GA	21-40	Archaea	[28]
Arch-958r	YCC GGC GTT GAM TCC AAT T	958-976	Archaea	[28]
27f	AGA GTT TGA TCC TGG CTC AG	27-46	Eubacteria	[29]
1492r	GGT TAC CTT GTT ACG ACT T	1492-1510	Eubacteria	[30]
CARD-FISH probes				
ANME1-350	AGT TTT CGC GCC TGA TGC	350-367	ANME-1 archaea	[4]
EelMS932	AGC TCC ACC CGT TGT AGT	932-949	ANME-2 archaea	[4]

ANME3-1249	TCG GAG TAG GGA CCC ATT	1250-1267	ANME-3 archaea	[31]
ANME3-1249H3	GTC CCA ATC ATT GTA GCC GGC	1229-1249	Helper probe for ANME3-1249	[32]
ANME3-1249H5	TTA TGA GAT TAC CAT CTC CTT	1268-1288	Helper probe for ANME3-1249	[32]
DSS658	TCC ACT TCC CTC TCC CAT	658-685	*Desulfosarcina* spp., *Desulfofaba* spp., *Desulfococcus*spp., *Desulfofrigus* spp.	[33]

Table 2: Community composition based on CARD-FISH analysis

Samples	% of cell count [1]	% of aggregate count [1]	% of biovolume[1]
S1			
ANME-1	Below detection limit[2]	Below detection limit[2]	Below detection limit[2]
ANME-2	8.2 ± 3.0	37.1 ± 6.2	13.4 ± 4.2
ANME-3	0.1 ± 0.1	2.1 ± 1.4	1.5 ± 1.5
SRB	2.9 ± 1.5	32.0 ± 6.2	22.7 ± 5.3
S2			
ANME-1	Below detection limit[3]	Below detection limit[3]	Below detection limit[3]
ANME-2	2.5 ± 2.0	47.2 ± 8.2	50.4 ± 15.9
ANME-3	0.1 ± 0.1	0.8 ± 0.7	2.4 ± 1.8
SRB	0.8 ± 0.4	37.6 ± 5.0	60.6 ± 5.5

[1] The average value and standard error were calculated based on 50 fields of on each hybridization. No ANME-1 cell or aggregate was observed based on our method. [2] Detection limit of 4×10^4 cells/ml slurry. [3] Detection limit of 9×10^4 cells/ml slurry

The CARD-FISH result showed that a large part of biomass in S1 and S2, especially single cells, did not belong to ANME or SRB. There was growth of other unknown microbes within a mixed community of ANME/SRB. Therefore a clone library analysis was performed on S2 to approach to the complete archaeal and bacterial communities. Archaeal community had extremely low diversity, where ANME-2a and MBG-D (marine benthic group D) were the only two groups of archaea detected. ANME-2a was the dominant, which accounted for 88% of the archaeal community (Figure 2). No 16S rRNA gene from ANME-3 was detected. The absence of ANME-3 in the archaeal clone library was contradictory to CARD-FISH result. The size of the clone library was not large enough to detect the rare ANME-3 or the hybridization experiment may have led to mis-hybridization, thus giving false positive signal. Dissimilar from archaeal community, the bacterial community was highly diverse (Figure 3). Gammaproteobacteria (43%) were the most dominant followed by the Deltaproteobacteria (17%), which includes the SRB. Among total bacteria population in S2, 8% was belonging to SEEP-SRB1a subgroup of *Deltaproteobacteria*, which were found to be specifically associated with ANME-2a in other enrichments mediating SR-AOM process [20]. Most of the Gammaproteobacteria found in the community were closely related to *Methylophaga* sp. and *Methylobacter* sp., which are known to use reduced one-carbon compounds, such as methane, methanol or dimethylsulphide [21]. The presence of such bacteria in our anaerobic reactor is intriguing since methane and sulphate were the only electron donor and acceptor supplied. The presence and even production of sulphide (sulphide concentration increased up to 0.5 mM everyday in the reactor) was an indication of anaerobic condition inside the reactor. However we cannot exclude the possibility of a limited amount of dissolved oxygen in the reactor influent, which could explain the presence of aerobic. Further tests need to show if these Gammaproteobacteria are playing an important active role in the reactor.

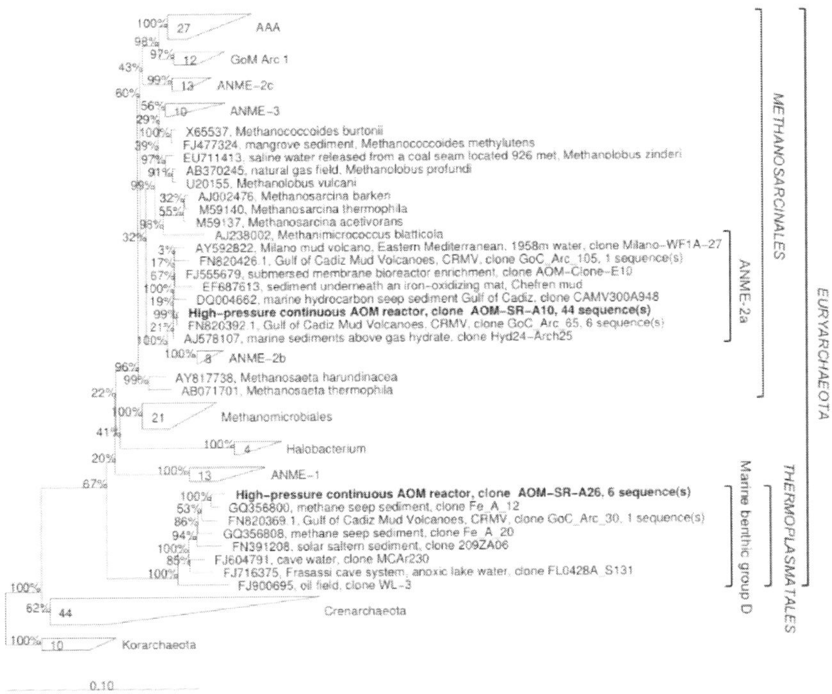

Figure 2: Phylogenetic tree showing the affiliations of archaeal 16S rRNA gene sequences detected from S2 to selected reference sequences.

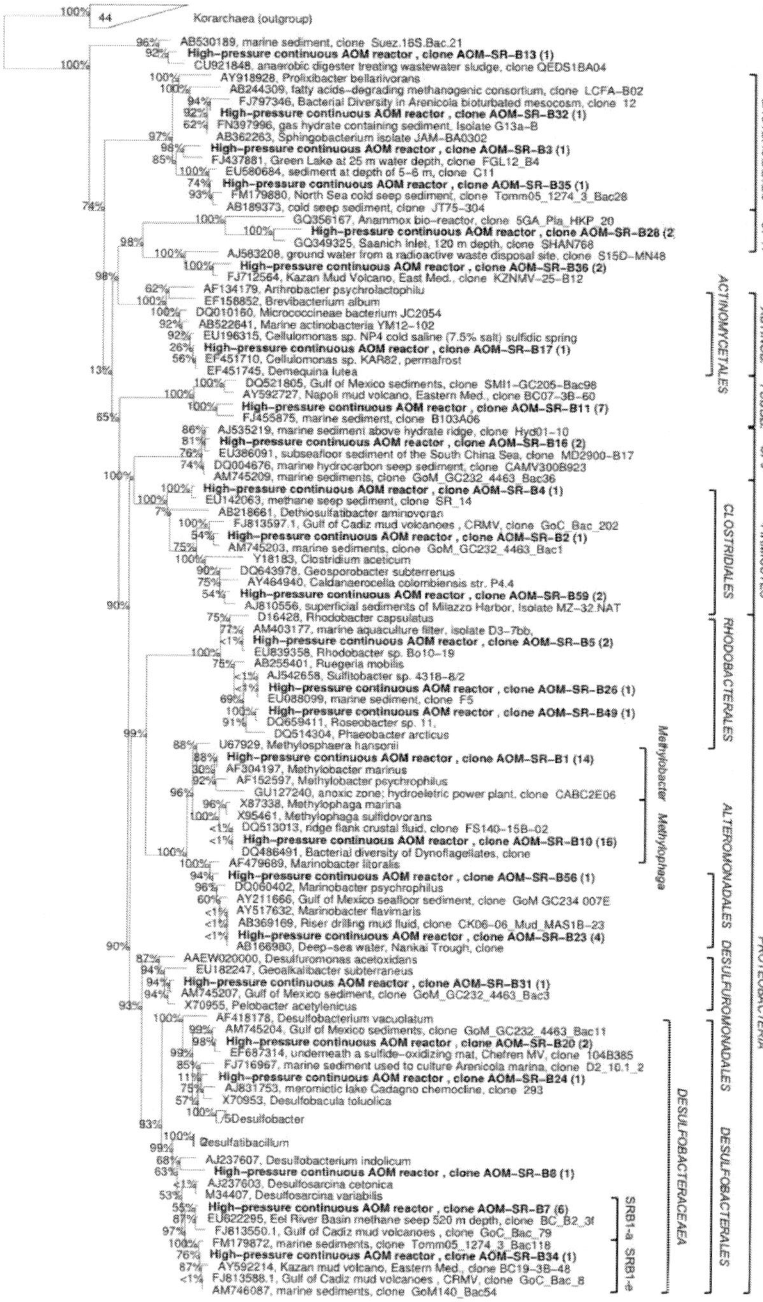

Figure 3: Phylogenetic tree showing the affiliations of bacterial 16S rRNA gene sequences detected from S2 to selected reference sequences.

Enrichment of ANME-2 and SRB

CARD-FISH results showed that percentages of ANME-2 and SRB biovolume increased from 13.4 ± 4.2% and 22.7 ± 5.3% in S1 to 50.4 ± 15.9% and 60.6 ± 5.5% in S2 (Table 2). By combining with the total biovolume data from DAPI staining (Figure 1B), the biovolume of ANME-2 in S1 was:

$(1.28*10^9\ \mu m^3/ml\ slurry) * 13.4\% = 1.7*10^8\ \mu m^3/ml\ slurry$

The biovolume of ANME-2 in S2 was:

$(4.49*10^9\ \mu m^3/ml\ slurry) * 50.4\% = 2.3*10^9\ \mu m^3/ml\ slurry$

Therefore after 286 days incubation, the ANME-2 population increased for 12.5 times. Following the same method of calculation, the SRB population increased for 8.4 times after 286 days incubation in this high-pressure bioreactor. The populations of ANME-2 and SRB both increased faster than the total biomass, which indicated that ANME-2 and SRB were selectively enriched in the system. This selective enrichment of ANME-2 and SRB was another proof that the incubation condition inside this high-pressure bioreactor was favourable for SR-AOM community.

To our knowledge, this is the first report on the enrichment of SR-AOM community under high methane pressure, although potential growth of ANME-1, ANME-2 and SRB has been reported in other engineered systems at ambient or low methane pressures (Table 3). The different inocula showed different doubling times. When ANME-1 and ANME-2c were incubated in continuous flow bioreactors under ambient methane partial pressure, ANME-1 had doubling time of 1.1 months while ANME-2c had doubling time of 1.4 months [16]. High methane partial pressure appeared to have advantage on stimulating the growth of ANME. In the experiment of Krüger *et al.* [22], the methane-dependent uptake of $^{15}N\text{-}NH_4$ by AOM community dominated by ANME-1 was higher at 1.5 MPa methane pressure than at ambient methane pressure. If we assume the ANME-2a cells in our system were following a logarithmic growth curve, a doubling time of 2.5 months can be estimated based on ANME-2 biovolume in S1 and S2, which is shorter than the result (3.8 months of doubling time of ANME-2a from an ambient pressure bioreactor) obtained by Meulepas *et al.* [10]. The increase of energy gained from SR-AOM process by increasing methane pressure may favour the biomass growth [8, 22]. Continuous flow also stimulated growth: ANME-2a/2c had longer doubling time in a fed-batch bioreactor (7.5 months) than in continuous flow bioreactors (1.4-3.8 months) (Table 3).

Table 3: Comparison of doubling times of ANME in different enrichment systems

Sediment origin	ANME group	Methane pressure	Operational mode	Doubling time (months)	Reference
Monterey Bay	ANME-1	Ambient	Continuous flow	1.1	[16]
Gulf of Mexico	ANME-1	1.5 MPa	Batch	2-3.4	[22]

Eckernforde Bay	ANME-2a	Ambient	Continuous flow	3.8	[10]
Monterey Bay	ANME-2c	Ambient	Continuous flow	1.4	[16]
Hydrate Ridge	ANME-2a/2c and SRB consortia	1.4 MPa	Fed-batch	7.5	[9]

CONCLUSIONS

After 286 days incubation in a simulated cold seep environment under high methane pressure, ANME-2 and SRB in the sediment from Captain Arutyunov Mud Volcano were enriched. Based on biovolume calculation, the populations of ANME-2 and SRB increased for 12.5 times and 8.4 times. Within total biomass volume, 99.7% was accounted from aggregates. Therefore the incubation condition apparently favoured the cells to form aggregates, especially in small size ($2<\text{Ø}\leq 5$ μm), rather than to live as single cells. No aggregate bigger than 15 μm in diameter was observed; they apparently divided after reaching a critical size. Based on the 16S rRNA gene clone library, the archaeal diversity was low, and contained only ANME-2 (88%) and MBG-D (12%). In contrast, the bacterial community was highly diverse.

METHODS

Incubation condition

In a previous study, the sediment sample originally from Captain Arutyunov Mud Volcano (Gulf of Cadiz, North East Atlantic) was diluted 12 times with artificial sea water medium and incubated in a continuous high-pressure bioreactor at 15°C [11]. This bioreactor system was a simulator for cold seep ecosystems, where sulphate and high-pressure methane were supplied. Because the high apparent affinity for methane (37 mM) in SR-AOM reaction and low dissolubility of methane in seawater (1.3 mM at 15°C at ambient pressure), it is necessary to supply high pressure methane to obtain high concentration of dissolved methane which can be directly used by microorganisms for high *in vitro* SR-AOM activity [11]. During this research, the reactor was operated in a fed-batch mode or a continuous mode. When it was in fed-batch mode, the methane pressures were switched between 1, 4.5 and 8 MPa. When it was in continuous mode, the methane pressure was either 1 or 8 MPa and the flow rate was 0.1 ml/min (HRT 100 hours). The SR-AOM activities under different operational conditions have been described previously [11]. To take a slurry sample, the incubation vessel was open under a nitrogen atmosphere and manually stirred to make the slurry sample homogeneous. The slurry samples before (S1) and after (S2) 286 days incubation were fixed in 4% formaldehyde and stored at 4°C for cell staining. Additional slurry from S2 was stored at -20°C for DNA extraction and clone library analysis.

Cell and aggregates quantification

To assess the number and the size of cells and aggregates, DAPI (4', 6-diamidino-2-phenylindole) staining was performed on S1 (after 2000 times dilution) and S2 (after 5600 times dilution). Subsequently, the samples were filtrated onto a circular GTTP polycarbonate filter (0.2 μm, Millipore, Germany) with a diameter of 2.5 cm. The number of cells (or aggregates) was quantified under a microscope (Zeiss, Carl Zeiss Microimaging GmbH, Germany) at 1,000 times magnification. The diameter of a single cell was assumed as 0.45 μm, which is the average size of ANME and SRB cells [3, 9]. The diameters of the aggregates were measured according to a reference scale bar built in the eyepiece of the microscope. The biovolume was calculated assuming that both cells and aggregates have spherical shapes. For each sample, 4 individual staining were applied. For each staining 50 fields of were counted for calculation.

Cell and aggregates identification

In order to evaluate which type of ANME and SRB were present and enriched in the reactor, catalyzed reporter deposition fluorescence *in situ* hybridization (CARD-FISH) was applied on S1 and S2. The slurry samples were embedded onto GTTP filters. The filters were incubated in methanol with 0.15% H_2O_2 for 30 min at room temperature before washed with water and ethanol and dried. For each sample, 2 filters were prepared. One was incubated in lysozyme solution (10 mg/ml in 0.05 M EDTA, pH 8.0; 0.1 M Tris-HCl, pH 8.0) for 15 min at 37°C to achieve permeablilization of bacterial cells, and another one was incubated in Proteinase K solution (15 μg/ml in MilliQ water) for 3 min at room temperature to achieve permeabilization of achaeal cells. Afterwards the filters were cut into 4 pieces. Each piece was for hybridization with one probe (Table 1). The hybridization was performed according to the protocol previously described [23]. After hybridization, the filter was stained with DAPI to target all cells present on the filter.

During CARD-FISH, a few steps of washing the filter may cause the loss of cells and aggregates. It was assumed that all types of cells or aggregates were washed out in the same ratio. Therefore the percentage of ANME or SRB among the total cells did not change after washing. For each hybridization, cells and aggregates in 50 fields of were analyzed under microscope. For each field, both probe staining and DAPI staining were counted to quantify the concentration of ANME-1 (or ANME-2, ANME-3 and SRB) among total biomass.

For a more detailed investigation on the microbial community, the archaeal and bacterial 16S rRNA gene clone libraries were performed on S2 according to protocol previously described [24, 25] with the primers listed in Table 1. For archaeal library, 56 clones were obtained while 50 clones were randomly picked for sequencing. For bacterial library, 110 clones were obtained while 100 clones were picked for sequencing. The sequences were compared with their best match in NCBI to classify their phylogenetic group (Additional file 1, Table S1). To calculate the percentage of each phylogenetic group into total archaeal/bacterial community, the number of clones within one phylogenetic group was divided by the number of sequenced clones within archaeal/bacterial library. All the sequences described in the paper have been deposited in the databases of GenBank, under accession numbers HQ405602 to HQ405741. The archaeal hpylogenetic tree has been constructed by Maximum Likelihood based on 212 sequences using the online RaxML tool athttp://phylobench.vital-it.ch/raxml-bb/index.php[26]. The bacterial phylogenetic tree has been constructed using the online maximum likelihood tool at http://www.atgc-montpellier.fr/phyml/[27].

DECLARATIONS

Acknowledgements

This work was supported by a PhD grant from the Bijzonder Onderzoeks Fonds of Ghent University (BOF08/DOC/016, BOF09/GOA/005), a research fund from the National High-Tech Program of China (2007AA091904), a research fund from the China State Key Laboratory of Ocean Engineering (AE010802) and a research fund from the European Science Foundation MicroSYSTEMS supported by Fonds voor Wetenschappelijk Onderzoek (506G.0656.05). We thank Siegfried E. Vlaeminck and Beatriz Guimarães for their critical re on this manuscript.

Authors' contributions

YZ carried out the incubation and DAPI staining, participated in CARD-FISH and drafted the manuscript. LM carried out the CARD-FISH and participated on the sequence analysis. XZ and FW carried the clone libraries and sequence analysis. NB conceived of the study, and participated in its design and coordination and helped to draft the manuscript. All authors read and approved the final manuscript.

REFERENCES

1. Reeburgh WS: Oceanic methane biogeochemistry. Chem Rev. 2007, 107 (2): 486-513. 10.1021/cr050362v.
2. Stadnitskaia A, Muyzer G, Abbas B, Coolen MJL, Hopmans EC, Baas M, van Weering TCE, Ivanov MK, Poludetkina E, Damste JSS: Biomarker and 16S rDNA evidence for anaerobic oxidation of methane and related carbonate precipitation in deep-sea mud volcanoes of the Sorokin Trough, Black Sea. Marine Geology. 2005, 217 (1-2): 67-96. 10.1016/j.margeo.2005.02.023.
3. Knittel K, Losekann T, Boetius A, Kort R, Amann R: Diversity and distribution of methanotrophic archaea at cold seeps. Applied and Environmental Microbiology. 2005, 71 (1): 467-479. 10.1128/AEM.71.1.467-479.2005.
4. Boetius A, Ravenschlag K, Schubert CJ, Rickert D, Widdel F, Gieseke A, Amann R, Jorgensen BB, Witte U, Pfannkuche O: A marine microbial consortium apparently mediating anaerobic oxidation of methane. Nature. 2000, 407 (6804): 623-626. 10.1038/35036572.

5. Hinrichs KU, Hayes JM, Sylva SP, Brewer PG, DeLong EF: Methane-consuming archaebacteria in marine sediments. Nature. 1999, 398 (6730): 802-805. 10.1038/19751.

6. Orphan VJ, Hinrichs KU, Ussler W, Paull CK, Taylor LT, Sylva SP, Hayes JM, Delong EF: Comparative analysis of methane-oxidizing archaea and sulfate-reducing bacteria in anoxic marine sediments. Applied and Environmental Microbiology. 2001, 67 (4): 1922-1934. 10.1128/AEM.67.4.1922-1934.2001.

7. Knittel K, Boetius A: Anaerobic Oxidation of Methane: Progress with an Unknown Process. Annual Re of Microbiology. 2009, 63: 311-334. 10.1146/annurev.micro.61.080706.093130.

8. Thauer RK, Shima S: Methane as fuel for anaerobic microorganisms. Incredible Anaerobes: From Physiology to Genomics to Fuels. Edited by: Wiegel J, Maier RJ, Adams MWW. 2008, Oxford: Blackwell Publishing, 1125: 158-170. 10.1196/annals.1419.000.

9. Nauhaus K, Albrecht M, Elvert M, Boetius A, Widdel F: In vitro cell growth of marine archaeal-bacterial consortia during anaerobic oxidation of methane with sulfate. Environmental Microbiology. 2007, 9 (1): 187-196. 10.1111/j.1462-2920.2006.01127.x.

10. Meulepas RJW, Jagersma CG, Gieteling J, Buisman CJN, Stams AJM, Lens PNL: Enrichment of Anaerobic Methanotrophs in Sulfate-Reducing Membrane Bioreactors. Biotechnol Bioeng. 2009, 104 (3): 458-470. 10.1002/bit.22412.

11. Zhang Y, Henriet J-P, Bursens J, Boon N: Stimulation of in vitro anaerobic oxidation of methane rate in a continuous high-pressure bioreactor. Bioresource Technology. 2010, 101 (9): 3132-3138. 10.1016/j.biortech.2009.11.103.

12. Girguis PR, Orphan VJ, Hallam SJ, DeLong EF: Growth and Methane Oxidation Rates of Anaerobic Methanotrophic Archaea in a Continuous-Flow Bioreactor. Applied and Environmental Microbiology. 2003, 69 (9): 5472-5482. 10.1128/AEM.69.9.5472-5482.2003.

13. Nauhaus K, Boetius A, Kruger M, Widdel F: In vitro demonstration of anaerobic oxidation of methane coupled to sulphate reduction in sediment from a marine gas hydrate area. Environ Microbiol. 2002, 4 (5): 296-305. 10.1046/j.1462-2920.2002.00299.x.

14. Deusner C, Meyer V, Ferdelman TG: High-Pressure Systems for Gas-Phase Free Continuous Incubation of Enriched Marine Microbial Communities Performing Anaerobic Oxidation of Methane. Biotechnol Bioeng. 2009, 105 (3): 524-533. 10.1002/bit.22553.

15. Yamamoto S, Alcauskas JB, Crozier TE: Solubility of methane in distilled water and seawater. J Chem Eng Data. 1976, 21 (1): 78-80. 10.1021/je60068a029.

16. Girguis PR, Cozen AE, DeLong EF: Growth and population dynamics of anaerobic methane-oxidizing archaea and sulfate-reducing bacteria in a continuous-flow bioreactor. Appl Environ Microbiol. 2005, 71 (7): 3725-3733. 10.1128/AEM.71.7.3725-3733.2005.

17. Robertson BR, Button DK, Koch AL: Determination of the biomasses of small bacteria at low concentrations in a mixture of species with forward light scatter measurements by flow cytometry. Applied and Environmental Microbiology. 1998, 64 (10): 3900-3909.

18. Vlaeminck SE, Terada A, Smets BF, De Clippeleir H, Schaubroeck T, Bolca S, Demeestere L, Mast J, Boon N, Carballa M, et al: Aggregate Size and Architecture Determine Microbial Activity Balance for One-Stage Partial Nitritation and Anammox. Applied and Environmental Microbiology. 2010, 76 (3): 900-909. 10.1128/AEM.02337-09.

19. Jagersma GC, Meulepas RJW, Heikamp-de Jong I, Gieteling J, Klimiuk A, Schouten S, Damste JSS, Lens PNL, Stams AJM: Microbial diversity and community structure of a highly active anaerobic methane-oxidizing sulfate-reducing enrichment. Environmental Microbiology. 2009, 11 (12): 3223-3232. 10.1111/j.1462-2920.2009.02036.x.

20. Schreiber L, Holler T, Knittel K, Meyerdierks A, Amann R: Identification of the dominant sulfate-reducing bacterial partner of anaerobic methanotrophs of the ANME-2 clade. Environmental Microbiology. 2010, 12 (8): 2327-2340. 10.1111/j.1462-2920.2010.02275.x.

21. Lidstrom M: Aerobic Methylotrophic Prokaryotes. The Prokaryotes. Edited by: Dworkin M, Falkow S, Rosenberg E, Schleifer K-H, Stackebrandt E. 2006, New York: Springer, 2: 618-634.

22. Kruger M, Wolters H, Gehre M, Joye SB, Richnow H-H: Tracing the slow growth of anaerobic methane-oxidizing communities by (15)N-labelling techniques. FEMS Microbiol Ecol. 2008, 63 (3): 401-411. 10.1111/j.1574-6941.2007.00431.x.

23. Pernthaler A, Pernthaler J, Amann R: Fluorescence in situ hybridization and catalyzed reporter deposition for the identification of marine bacteria. Applied and Environmental Microbiology. 2002, 68 (6): 3094-3101. 10.1128/AEM.68.6.3094-3101.2002.

24. Jiang LJ, Zheng YP, Peng XT, Zhou HY, Zhang CL, Xiao X, Wang FP: Vertical distribution and diversity of sulfate-reducing prokaryotes in the Pearl River estuarine sediments, Southern China. FEMS Microbiol Ecol. 2009, 70 (2): 249-262. 10.1111/j.1574-6941.2009.00758.x.

25. Wang SF, Xiao X, Jiang LJ, Peng XT, Zhou HY, Meng J, Wang FP: Diversity and Abundance of Ammonia-Oxidizing Archaea in Hydrothermal Vent Chimneys of the Juan de Fuca Ridge. Applied and Environmental Microbiology. 2009, 75 (12): 4216-4220. 10.1128/AEM.01761-08.

26. Stamatakis A, Hoover P, Rougemont J: A Rapid Bootstrap Algorithm for the RAxML Web Servers. Syst Biol. 2008, 57 (5): 758-771. 10.1080/10635150802429642.

27. Guindon S, Gascuel O: A simple, fast, and accurate algorithm to estimate large phylogenies by maximum likelihood. Syst Biol. 2003, 52 (5): 696-704. 10.1080/10635150390235520.

28. Delong EF: Archaea in coastal marine environments. Proc Natl Acad Sci USA. 1992, 89 (12): 5685-5689. 10.1073/pnas.89.12.5685.
29. Lane DJ: 16S/23S rRNA sequencing. Nucleic Acid Techniques in Bacterial Systematics. Edited by: Stackebrandt E. 1991, Goodfellow M: John Wiley & Sons, 142-175.
30. Reysenbach AL, Wickham GS, Pace NR: Phylogenetic analysis of the hyperthermophilic pink filament community in Octopus Spring, Yellowstone National Park. Appl Environ Microbiol. 1994, 60 (6): 2113-2119.
31. Niemann H, Losekann T, de Beer D, Elvert M, Nadalig T, Knittel K, Amann R, Sauter EJ, Schluter M, Klages M, et al: Novel microbial communities of the Haakon Mosby mud volcano and their role as a methane sink. Nature. 2006, 443 (7113): 854-858. 10.1038/nature05227.
32. Losekann T, Knittel K, Nadalig T, Fuchs B, Niemann H, Boetius A, Amann R: Diversity and abundance of aerobic and anaerobic methane oxidizers at the Haakon Mosby mud volcano, Barents Sea. Appl Environ Microbiol. 2007, 73 (10): 3348-3362. 10.1128/AEM.00016-07.
33. Manz W, Eisenbrecher M, Neu TR, Szewzyk U: Abundance and spatial organization of Gram-negative sulfate-reducing bacteria in activated sludge investigated by in situ probing with specific 16S rRNA targeted oligonucleotides. FEMS Microbiol Ecol. 1998, 25 (1): 43-61. 10.1111/j.1574-6941.1998.tb00459.x.

17

CHAPTER

MOLECULAR ASSESSMENT OF MICROBIAL SPECIES INVOLVED IN THE BIODEGRADATION OF CRUDE OIL IN SALINE NIGER DELTA SEDIMENTS USING BIOREACTORS

Chioma Blaise Chikere, Chinedu Christopher Obieze and Phillip Okerentugba

Department of Microbiology, University of Port Harcourt, Rivers State, Nigeria

ABSTRACT

Purpose: At elevated salinities conventional microbiological processes are not very effective, therefore clean up of contaminants using bioremediation strategy will involve the use of halophilic and halo-tolerant bacterial species. This research therefore aimed at isolating and identifying potential halophilic and halo-tolerant bacterial species capable of hydrocarbon degradation during bioreactor based treatment with exogenous nutrients. Methods: The diversity of indigenous bacterial species with potential to degrade hydrocarbons was investigated using both culture-dependent and independent techniques. Bioremediation of hydrocarbon contaminated saline sediments was carried out using seven 2.5 liter bioslurry bioreactors operated over a 64-day period. Physicochemical parameters monitored were pH, nitrate, phosphate, total petroleum hydrocarbon (TPH), polycyclic aromatic hydrocarbon (PAH), temperature, salinity, and total organic carbon (%TOC). Results: The baseline TPH, PAH and pH of the sediments were 19 ppm, 3.1 ppm and 7.0 respectively. The baseline salinity of the sediment was 10% thus the sediment was adjudged moderately saline. TPH ranged from 97 ppm-105 ppm on day zero and decreased to an average of 5.62 ppm on day 64, while PAH ranged from 56 ppm-61 ppm on day zero and decreased to an average of 4.02 ppm on day 64. The bacterial species identified as potential

Citation: Chikere CB, Obieze CC, Okerentugba P (2015) Molecular Assessment of Microbial Species Involved in the Biodegradation of Crude Oil in Saline Niger Delta Sediments Using Bioreactors. J Bioremed Biodeg 6: 307. doi:10.4172/2155- 6199.1000307.

hydrocarbon degraders includes Halomonas lutea, Achromobacter spp, Aquitalea magnusonii, Bacillus sp, Sphingobacterium sp, Shewanella sp, Brevundimonas naejangsanensis, Pseudomonas pseudoalcaligenes, Pseudomonas aeruginosa, unidentified bacterium BH23 and Gordonia sp. The genus Pseudomonas formed majority of the isolates successfully sequenced and exhibited similarity values ranging from 91% to 100% with sequences deposited in GenBank. A combination of both molecular and culture based technique allowed the identification to species level of twelve isolates. One isolate could not be identified while the remaining isolates were identified to their generic level. Treatment BCD recorded highest total culturable heterotrophic bacteria (TCHB) count (7.1 × 108 cfu/g) and total culturable hydrocarbon utilizing bacteria (TCHUB) count (6.7 × 108 cfu/g). There was a significant difference at P<0.05 in TCHUB bacteria counts between the unamended bioreactor slurries and those amended with organic and inorganic nutrients. There were also significant differences in TCHUB counts when the bioaugumented slurry was compared with those amended with NPK, Urea and cow dung using one way ANOVA and Tukey's multiple comparison tests. Conclusion: This study revealed potentially novel bacterial species and previously described hydrocarbon degrading bacterial species that can be characterized further to determine their role in hydrocarbon degradation as well as their salt tolerance level prior to application in bioremediation of saline environments.

INTRODUCTION

Petroleum is a complex mixture of different hydrocarbons including aliphatic, cycloalkanes, mono and polyaromatics, asphalthenes and resins with majority of these compounds toxic and carcinogenic [1]. The increasing worldwide demand for energy from petroleum and an increase in prospecting for crude oil in the marine environment would mean a higher risk of accidental oil discharge into the marine environment [2,3]. The Niger Delta region of Nigeria has witnessed an intermittent discharge of crude oil into its environment since the inception of crude oil exploration and exploitation, this has led to the destruction of its farmlands, aquaculture, rivers and creeks with hydrocarbon compounds [4].

Recently they have been an increasing agitation to cleanup, reclaim and restore all oil polluted environments within the Niger Delta [5]. Bioremediation is said to be the best approach for environmental cleanup because it is a cost effective and an ecofriendly strategy.

However in moderate to highly saline environments like the marine environment, application of bioremediation is very challenging due to the detrimental effect of salt on microbial life. The salty nature of the environment could even be compounded by produced water, a byproduct or waste associated with oil and gas production which contains high levels of salt (1000-250,000 mg/L), oil and grease, toxic chemicals, heavy metals, and naturally occurring radioactive materials [6,7]. Therefore in other to effectively carryout bioremediation in such environments a different bioremediation approach has to be applied. This approach would involve the use of halophilic and halo-tolerant microbial species capable of effectively degrading hydrocarbons under high salt conditions. Halophiles are classified into three groups according to their optimal salt concentration for growth: slightly halophilic (1-3% w/v), moderately halophilic (3-15% w/v), and extremely halophilic (15-32% w/v) [7-9]. Many hydrocarbon degrading organisms are known [10] and in the marine environment a number of specialist hydrocarbon degrading taxa are also known [1,11]. Aromatic hydrocarbon degraders include Cycloclasticus spp. which utilize biphenyl, naphthalene, anthracene, phenanthrene, toluene, and benzoate [12], and Neptunomonas which can degrade naphthalene, 2-methylnaphthalene and phenanthrene as sole carbon sources, but are unable to degrade 2,6-dimethylnaphthalene, 1-methylnaphthalene, biphenyloracenaphthene [13]. This highlights the drawback caused by the chemical complexity of crude oil which limits the capacity of single microbial specie to efficiently and effectively degrade all crude oil components. A combined effort of mixed bacterial consortia however usually improves hydrocarbon bioremediation. Marine saturated hydrocarbon degrading specialists include Alcanivorax [14,15], Planococcus [16], Oleiphilus [17], Oleispira [18], Thalassolituus [19].

With crude oil production shifting offshore in the Niger Delta region of Nigeria, it is expected that pollution of the marine environment with hydrocarbons will increase. Therefore an understanding of the impact of oil on indigenous microbial communities and the isolation and identification of key oil degrading indigenous microorganisms are prerequisite for directing the management and clean up of oil impacted marine environments. The use of artificial microbial consortia cannot substitute for highly complex and dynamic indigenous microbial population essential for complete and efficient hydrocarbon degradation [11]. Thus the objective of this study was to isolate and identify the bacterial species that are associated with hydrocarbon degradation in saline sediments of the Niger Delta which could be used as model organisms for pollution detection and as specialist organisms for oil spill clean-up.

METHODOLOGY

Site description

The examined sediments contaminated with crude oil originated from Bodo creek, in Bodo community of Gokana Local Government Area, in the lower Niger Delta Basin of Nigeria. Bodo creek is characterized by brackish water system and has undergone varying degrees of crude oil pollution due to exploration, exploitation and artisanal refining activities.

Sample collection

Sediment and water samples were collected in July 2013. Sediment was collected using Eckman grab, while water was collected using sterile 10 L Jerry can. All samples were transported to the Environmental Microbiology laboratory, University of Port Harcourt within 6 h for analysis.

Physico-chemical analysis

The parameters analysed were the pH, nitrate, phosphate, temperature and total organic carbon (%TOC). These parameters were analysed as described in ref. [20]. Salinity was determined using refractometer (S/mill-E Range 0-100%) after extracting water from sediment by centrifuging at 1000 rpm for 15 minutes. Gas chromatographic analyses were carried out as stated below.

Total petroleum hydrocarbons (TPH): Dried sediment samples were powder sieved and cold-extracted in conical flask for a total of two hours in each case using 100% dichloromethane. The solvent from the resultant solution was removed by means of rotary evaporator under vacuum (pressure not greater than 200 mbar) and finally by a flow nitrogen at not more than 30°C to yield the extracted organic matter (EOM). The semi volatile compounds were introduced into the GC/MS by injecting the sample extract into a gas chromatograph (GC) equipped with a narrow-bore fused silica capillary column. The GC column was temperature-programmed to separate the analytes, which were then detected with a mass spectrometer (MS) connected to the gas chromatograph. Analyte eluted from the capillary column were introduced into the mass spectrometer via a jet separator. Identification of target analytes was accomplished by comparing their mass spectra with the electron impact spectra of authentic standards. Quantitation was accomplished by comparing the response of a major (quantitation) ion relative to an internal standard using an appropriate calibration curve for the intended application.

Chromatograhic condition: The GC/MS system comprised of Agilent 6890GC, (Agilent technologies, Wilmington, USA) with 5975B MSD and MSD chemstation (version D. 03.00). Helium gas was used as the carrier gas at a constant flow rate of 1 mL/min at a pressure of 75 kpa. The injector temperature was set at 250°C. The program used was; 2 min hold time, ramp to 240°C at 7°C/min and a final ramp to 285°C at 12°C with an 8 min hold time. Column—30-m × 0.25-mm ID × 0.25 μm film thickness silicone-coated fused-silica capillary column.

MSD condition: Solvent delay: 4 min, Mode-Scan at 3.54, Solvent delay: 3 min, Quard temp: 150°C, Source temp: 230°C, Transfer line temp: 280°C, Sampling: 2, Low mass: 45.0 amu, High mass: 450 amn, Threshold: 150.

Polycyclic aromatic hydrocarbons (PAH): Solvents: pentane (Sigma-Aldrich, Germany) methylene chloride (99.50% Sigma-Aldrich Germany) acetone (BDH, chemicals, poole England), sodium sulphate anhydrous (May and Baker, Dagenham, England), silica gel 60/120 mesh (Hopkins and Williams, Essex England) distilled water. The samples were dried, crushed and sieved using 0.5 mm sieve. 2.0 ± 0.1 g of samples was weighed into a clean extraction container. 10 ml of extraction solvent (pentane) was added into the sample and mixed thoroughly and allowed to settle. The mixture was carefully filtered into clean solvent rinsed in extraction bottles using filter paper fitted into Buchner funnels. The extracts were concentrated to 2 ml and then transferred for clean up/separation. 1 cm of moderately packed glass wool was placed at the bottom of 10 mm long chromatographic column. Slurry of 2 g agitated silica gel in 10 ml methylene chloride was prepared and placed into the chromatographic column. To the top of the column was added 0.5 cm of sodium sulphate. The column was rinsed with addition of 10 ml methylene chloride. The column was pre-eluted with 20 ml pentane; this was allowed to flow through the column at the rate of about 2 min until the liquid in the column was just above the sulphate layer. Immediately 1 ml of the extracted sample was transferred into the column, the extraction column was rinsed with 1 ml pentane and added to the column as well.

The stop-cock of the column was opened and the eluent was collected with a 10 ml graduated cylinder. Just prior to exposure of the sodium sulphate layer to air. Accurately measured volume of 8-10 ml of the eluent was collected and labeled Aliphatic. Following recovery of the aliphatic fraction and just prior to exposure of the sodium sulphate layer, the column was eluted with 1:1 mixture of acetone and methylene chloride in 1-2 ml increments. Another accurately measured 8-10 ml of the eluent was collected and labeled ''Aromatics''. The aromatic fraction was concentrated to 1 ml for PAHs analysis using gas chromatography.

Bioreactor design and operation

Bioremediation of hydrocarbon-contaminated sediments from Bodo creek was carried out using seven 2.5 L bioslurry bioreactors operated over a 64-day period. Two reactors served as controls (unamended) and (heat-killed), four out of the remaining five as nutrient amended bioreactors while one was amended with 108 cfu/g consortium of indigenous hydrocarbon utilizing bacteria. The bioreactors were designated as BPD, BCD, BUR, BNPK, BAUG, BUNa and BHK (Poultry dropping, Cow dung, Urea, NPK, consortium of indigenous hydrocarbon utilizing bacteria, unamended control and heat killed control respectively). Each of the 7 bioreactors received 1 kg (wet weight) of sediments, 20 ml of crude oil and 20 mg of anthracene. For the controls, the unamended treatment was spiked with hydrocarbons without nutrient addition to determine whether the indigenous bacteria have the natural ability to degrade petroleum hydrocarbons, whereas the heat-killed treatment (killed by autoclaving sediments at 121°C for 15 min at 15 psi on 2 consecutive days) served to measure the role of abiotic factors in the loss of petroleum hydrocarbons. The bioreactors were continuously stirred (by 2 impellers) at 150 rpm throughout the 64-day experimental period. Filtered air was supplied to the bioreactors from the air compressor through hoses running in and out of them. The reactors were sealed with Teflon to prevent the ingress of atmospheric air and egress of the slurry. Throughout the 64-days of experimentation the reactors were operated at room temperature (30°C).

Microbiological analysis of samples

Enumeration of total culturable heterotrophic bacteria (TCHB): 1 g (wet weight) of sediment was homogenized in 0.85% of normal saline. Decimal dilutions (tenfold) of the suspensions was plated out in duplicate on plate count Agar (PCA) modified with 10% NaCl and incubated at 30°C for 24 h for the THB counts [15].

Enumeration of total culturable hydrocarbon utilizing bacteria (TCHUB): For hydrocarbon utilizing bacteria counts, appropriate dilutions of sediment suspensions (1 g wet weight of sediment homogenized in 0.85% of normal saline) was plated out in duplicate on Bushnell-Haas agar (Sigma-Aldrich, USA) modified with 10% NaCl. Hydrocarbons were supplied through the vapour phase transfer to putative hydrocarbon utilizers by placing sterile Whatmann No. 1 filter papers impregnated with 5 ml Okono medium crude oil in the lids of the inverted Petri plates. Plates were then incubated at 30°C for 7 days [15].

Biochemical and phenotypic analysis

Individual cultures were identified by morphological and biochemical techniques using Colour Atlas of Medical Bacteriology [21].

Molecular characterizations

Bacterial DNA extraction was done with Zymo Research Fungal/Bacteria DNA kit™. The DNA samples were quantified using NanoDrop ND-2000 spectrophotometer. A portion of the 16S rRNA gene sequence extracted from each bacterial species was subjected to polymerase chain reaction using the primer set pA8f-GC (59-CGCCCG-CCG-CGC-GCG-GCG-GCG-GGC-GGGGCG-GGG-GCAGGG-GAG-AGT-TTG-ATC-CTGGCT-CAG-39) and KPRUN518r (59-ATTACCGCGGCTG CTGG-39) [22]. Amplification of the template DNA was performed using 2 µl volume of the extracted bacterial DNA with Eppendorf thermal cycler. The 50 µl PCR mixture contained 5 µl of deoxy nucleoside triphosphates (dNTPs) mixture (2.5 µM) (Promega, USA), 5 µl of 5X Green Go Taq Flexi buffer (Promega, USA), 3.5 µl of 25 mM MgC12 (Promega, USA), 2 µl each of the primer set pA8f-GC (59-CGC-CCG-CCG-CGC-GCG-GCG-GCG-GGCGGGGCG-GGG-GCA-GGG-GAG-AGT-TTG-ATC-CTGGCTCAG-39) and KPRUN518r (59-ATTACCGCGGCTG CTGG-39), 0.25µl of 5U/µl hot start Go Taq DNA polymerase, 2.5µl of 20 mg/ml of bovine serum albumin and 27.75 µl of sterile water. A reaction tube without template DNA served as negative control.

Amplified DNAs were examined by electrophoresis in 1.4% agarose gel with 2 µl aliquots of PCR products in 1X Tris-Acetate-EDTA buffer using a UV-transilluminator. PCR products of the bacterial DNA were sequenced using GATC Sanger sequencer. (GATC Biotech AG European Custom Sequencing Centre Koln). Sequence identification was performed using the BLAST-N facility of the National Center for Biotechnology Information (NCBI) (http://www.ncbi.nlm.nih. gov/BLAST/). Each 16S rRNA sequence was subjected to a basic local alignment search tool (BLAST) analysis on the GenBank database and matching hits, with e-values closest to 0.0 indicating a statistically plausible match, was selected.

Okono medium crude oil degradation screening

Pure isolates obtained from Bushnell-Haas agar were inoculated into Bushnell-Haas broth amended with Okono medium crude oil (9 ml broth culture-1 ml crude oil) as sole carbon source and incubated at room temperature for 9 days. Biodegradation was determined by monitoring the change in turbidity of liquid culture (OD_{600nm}) using a spectrophotometer (Optima SP-300 spec. Japan).

Statistical analysis of data

Statistical analysis was performed on the data generated from the microbial counts, using one way ANOVA and Tukey's Multiple Comparison Test. The software Graph-Pad Prism (GraphPad Software, CA, USA) for Windows version 5.04 was used for the analysis.

RESULTS

Baseline analysis of sediment sample

Gas Chromatographic analysis of TPH detected the presence of C10-C40 carbons as shown in supplementary material. The PAHs detected includes Acenaphthene, Acenaphthylene, Anthracene, Benzo(a)pyrene, Benzanthracene, Chrysene, Fluoranthene, Fluorene, Naphthalene, Phenanthrene and Pyrene. The values for the total heterotrophic bacteria count and hydrocarbon utilizing bacteria count and other physicochemical parameters measured are shown in Table 1.

TPH and PAH analysis of bioreactor treatments

The TPH and PAH value for all bioreactor treatment on day 1 was 105 and 61 ppm respectively. On day 64 PAH content decreased to an average of 4.02 ppm while TPH also decreased to an average of 5.62 ppm. Figures 1 and 2 represents the TPH and PAH degradation profile for each bioreactor treatment.

Table 1: Baseline characteristics of sediment sample

Parameter	Concentration
Total petroleum hydrocarbon (TPH)	19 ppm
Polycyclic aromatic hydrocarbons (PAH)	3.1 ppm
pH	7.0
Nitrate	88 mg/kg
Salinity	10%
Phosphate	20 mg/kg
Temperature	27.6
Total organic carbon (TOC)	3.3%
Total heterotrophic bacterial count (THB)	6.3×10^5
Hydrocarbon utilizing Bacterial count (HUB)	4.6×10^5

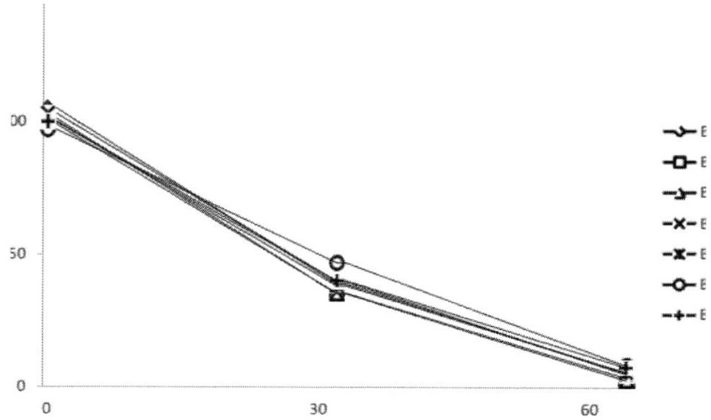

Figure 1: Total Petroleum Hydrocarbon degradation profile for each bioreactor treatment. Key: BUNa (Un-amended control); BHKD (Heat killed contol); BNPK (NPK); BAUG (Bioaugumentation); BURE (Urea); BPOUT (Poultry dropping); BCD (Cow dung). No statistically significant difference was recorded between the various bioreactor treatments for TPH.

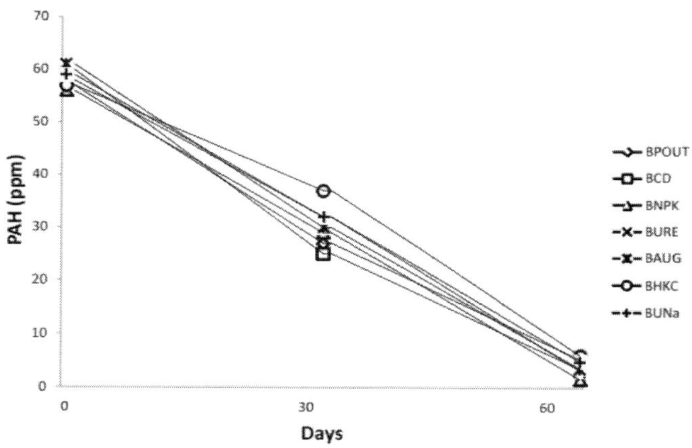

Figure 2: Polycyclic Aromatic Hydrocarbon degradation profile for each bioreactor treatment. Key: BUNa (Un-amended control); BHKD (Heat killed contol); BNPK (NPK); BAUG (Bioaugumentation); BURE (Urea); BPOUT (Poultry dropping); BCD (Cow dung). BHKC showed slight difference in PAH degradation profile when compared to other treatments, though not statistically significant.

Bacterial enumeration from bioreactor treatments Treatment

BCD recorded highest heterotrophic bacteria count (7.1×10^8 cfu/g) and hydrocarbon utilizing bacteria count (6.7×10^8 cfu/g) respectively. The heat killed control BHKC showed no growth for TCHB and TCHUB throughout the 64 days of experimentation. There was no significant difference in TCHB between the un-amended bioreactor slurry and those amended with organic and inorganic nutrients, but there were statistically significant differences for TCHUB counts when BUNa was compared with other treatments. There were also significant differences in TCHUB counts when the bioaugumented slurry was compared with those amended with NPK, Urea and cow dung. Figures 3 and 4 represent microbial counts across all bioreactor treatments and controls during the 64 days of bioremediation.

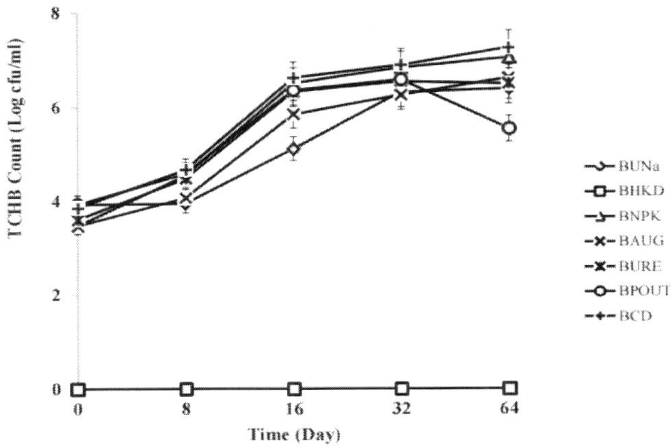

Figure 3: Total culturable heterotrophic bacteria counts. Key: BUNa (Un-amended control); BHKD (Heat killed contol); BNPK (NPK); BAUG (Bioaugumentation); BURE (Urea); BPOUT (Poultry dropping); BCD (Cow dung). BHKC showed no bacterial growth for TCHB throughout the 64 days of experimentation. There was no significant difference in TCHB between the other bioreactor treatments.

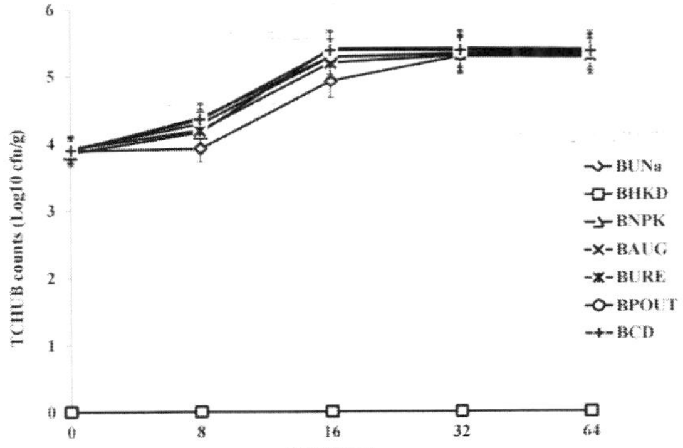

Figure 4: Total Culturable Hydrocarbon Utilizing Bacteria counts. Key: BUNa (Un-amended control); BHKD (Heat killed contol); BNPK (NPK); BAUG (Bioaugumentation); BURE (Urea); BPOUT (Poultry dropping); BCD (Cow dung). BHKD recorded no bacterial growth throughout the 64 day period of experimentation. BUNa showed statistically significant difference when compared to other amended treatments. There were significant differences in TCHUB counts when the bioaugumented slurry was compared with those amended with NPK, Urea and cow dung.

Characteristics of bacterial isolates

After performing a preliminary phenotypic characterization that included the colony morphology, gram reaction and relevant biochemical analysis to eliminate duplicate strains, 25 isolates recovered from the different bioreactor treatments were selected. In an attempt to identify as many bacterial isolates as possible, DNA sequencing results were combined with biochemical, macroscopic and microscopic morphological data as shown in Table 2. Amplification of bacterial DNA samples with primer pair pA8f-GC (5′-CGCCCG-CCG-CGC-GCG-GCG-GCG-GGC-GGGGCG-GGG-GCAGGG-GAG-AGT-TTG-ATC-CTGGCT-CAG-3′) and KPRUN518r (5′-ATTACCGCGGCTG CTGG-3′) resulted in fragments of approximately 550bp. PCR products were checked on 1% agarose gel in 1x TAE buffer. All PCR products obtained with primer pair pA8fGC-KPRUN518r were sequenced with primer KPRUN518r and were identified using the BLASTn program.

Comparative sequence analysis of the 16S rRNA of the selected isolates was performed. The sequences were aligned to the most similar 16S rRNA sequences in the data base. Table 2 shows the 16S rRNA closest match and accession numbers for the different isolates. The genus Pseudomonas formed majority of the isolates successfully sequenced and exhibited similarity values with those deposited in GenBank ranging from 91% to 100%. One isolate (A59) could not be identified as there was no sequence match available in GenBank database. The remaining isolates were identified as organisms related to Gordonia sp (98% sequence similarity), Brevundimonas naejangsanensis (99% sequence similarity), Shewanella sp. (99% sequence similarity), Achromobacter sp. (96% sequence similarity), Sphingobacterium sp. (97% sequence similarity), Bacillus sp. (98% sequence similarity), Aquitalea magnusonii (96% sequence similarity), Achromobacter sp. (100% sequence similarity) and Chromohalobacter salexigens (94%) sequence similarity).

Biodegradation screening

All bacterial isolates recovered using Bushnell Haas Agar (BHA) were further subjected to biodegradation screening for their ability to utilize Okono medium crude oil as their sole source of carbon and energy. The results obtained showed that all the isolates were significant for crude oil degradability assay evidenced by biomass increase, increase in turbidity, emulsification and a concomitant visual gradual decrease in the oil layer after 9 days of incubation when compared with isolates on day zero and controls.

DISCUSSION

In the last two decades, research in the marine environment has increased because of the prospects of isolating novel microbial species within the marine/moderately saline environment that possess enzymes adapted to crude oil degradation or that may be of general importance to the biotechnology industry [23]. In the Niger Delta region of Nigeria where crude oil exploration and exploitation activity is shifting offshore, there is a paucity of research in this regard. The examined sediments contaminated with crude oil originated from a moderately saline creek (Bodo) in the Niger Delta region of Nigeria. Total heterotrophic bacterial counts, hydrocarbon utilizing bacterial counts, as well as gas chromatography and other physicochemical analysis were carried out on the sediment in order to investigate the role of the autochthonous bacterial population in crude oil degradation.

There was a general increase in microbial population in the amended sediment slurries with attendant reduction in TPH and PAH. The hydrocarbon losses recorded in the bio-stimulated sediment slurries can be attributed to microbial activities which resulted in consumption of nitrogen and phosphorus added in the form of urea, NPK fertilizer, poultry droppings, and inorganic sources of nitrogen and phosphorus. Several researchers have reported that nutrient amendment with both organic and inorganic fertilizers significantly affected crude oil biodegradation [11,15,24]. Statistical analysis of the various bio-stimulated sediment slurries did not show significant differences for TPH and PAH. The significant differences in TCHUB counts recorded when the bio-augmented slurry was compared with those amended with NPK, Urea and cow dung is also consistent with previous research.

Table 2: Molecular and Biochemical identification and 16S sequence matches for bacteria isolates recovered from sediments

Isolate	Tentative identity	GenBank accession number	GenBank Closest cultured organism	Accession number/% Identity
A2	*Gordonia* sp. BH1	KR261407	*Gordonia* sp	EF197794 98%
A3	*Brevundimonas* sp. BH 2	KR261408	*Brevundimonas naejangsanensis*	KF548535 99%
A4	*Pseudomonas* sp. BH3	KR261409	*Pseudomonas* sp.	KJ7166166 97%
A6	*Pseudomonas* sp. BH4	KR261410	*Pseudomonas* sp.	KJ643426 100%
A7	*Pseudomonas* sp. BH5	KR261411	*Pseudomonas pseudoalcaligenes*	AB276372 99%
A11	*Pseudomonas aeruginosa*	KR261412	*Pseudomonas aeruginosa*	GU269267 99%
A13	*Shewanella* sp. BH7	KR261413	*Shewanella* sp.	KC211018 99%
A14	*Achromobacter* sp. BH8	KR261414	*Achromobacter* sp.	KJ716674 96%
A21	*Pseudomonas* sp. BH9	KR261415	*Pseudomonas* sp.	GQ250188 91%
A25	*Sphingobacterium* sp. BH10	KR261416	*Sphingobacterium* sp.	KC904977 97%

A31	*Pseudomonas* sp. BH11	KR261417	*Pseudomonas* sp.	JN626196 98%
A32	*Pseudomonas aeruginosa*	KR261418	*Pseudomonas aeruginosa*	KJ671465 97%
A37	*Bacillus* sp. BH13	KR261419	*Bacillus* sp.	KF889051 98%
A41	*Pseudomonas aeruginosa*	KR261420	*Pseudomonas aeruginosa*	AB920753 93%
A42	*Pseudomonas aeruginosa*	KR261421	*Pseudomonas aeruginosa*	AF193514 100%
A43	*Pseudomonas aeruginosa*	KR261422	*Pseudomonas aeruginosa*	KJ093485 99%
A45	*Pseudomonas* sp. BH17	KR261423	*Pseudomonas* sp.	KC147692 100%
A47	*Aquitalea* sp. BH18	KR261424	*Aquitalea magnusonii*	EU661705 96%
A52	*Pseudomonas aeruginosa*	KR261425	*Pseudomonas aeruginosa*	GU269267 100%
A55	*Achromobacter* sp. BH20	KR261426	*Achromobacter* sp.	JN196541 100%
A56	*Pseudomonas aeruginosa*	KR261427	*Pseudomonas aeruginosa*	KF7469577 99%
A58	*Halomonas* sp. BH22	KR261428	*Chromohalobacter salexigens*	KJ676975 94%
A59	Unidentified bacterium BH23	KR261429	Unavailable	Unavailable
A60	*Pseudomonas* sp. BH24	KR261430	*Pseudomonas* sp.	KC147692 100%
A64	*Pseudomonas aeruginosa*	KR261431	*Pseudomonas aeruginosa*	KF677012 100%

The reduction in TPH and PAH concentration in the contained system and the ability to culture the various species of bacteria recovered during the period of bioremediation using okono medium crude oil as the sole carbon source implies that the isolates are potential hydrocarbon degraders and could play important roles in bioremediation of pollutants. The study carried out suggests the proliferation of potential halophilic and halotolerant bacterial species capable of Okono medium crude oil degradation in a contained system. The recovery of Halomonas is consistent with previous research carried out in saline habitats. The Genus is known to contain no non halophilic species [25] and constitute the most commonly cultured inhabitants in saline environments. Halomonas sp. and Marinobacter sp. have been recovered from hydrocarbon contaminated saline soil collected from several regions in Iran [26]. Halomonas have also been previously suggested for use as model organism for the study of degradation of aromatic compounds at high salt concentrations, as this group of microorganisms have few nutritional requirements, can withstand a wide range of salinities, degrade a wide range of aromatic compounds and are easily cultured [23].

Pseudomonas and Bacillus, which are commonly isolated in physiologically normal environments, have species recognized as halotolerant and with potential for hydrocarbon degradation [25]. These organisms are often referred to as generalist hydrocarbon degraders, yet they seem to constitute a significant proportion of hydrocarbon degraders and play significant roles in hydrocarbon degradation in the marine environment. The Genus Pseudomonas formed the majority of the species recovered during the period of bioremediation; this could be as a result of their effective competitive capabilities in a wide range of environments [25] with saline environment not being an exception. Pseudomonas, Acinetobacter and Alcanivorax have been previously recovered from Niger-Delta marine sediments [15]. Previous scientific investigations have constantly reported Pseudomonas to be effectively involved in hydrocarbon degradation [4,15,27,28]. They are very versatile and have been reported to produce biosurfactants which aid in hydrocarbon emulsification and in reduction of hydrophobicity [29].

Halo-tolerant Achromobacter xylosoxidans capable of degrading multiple PAH have been isolated from a crude oil polluted saline site [30]. Bacillus sp has been reported as part of the diverse halophilic hydrocarbon utilizing bacteria from Kuwaiti coast of the Arabian Gulf [31]. The significance of the presence of the genus Brevundimonas and Aquitalea as part of the diversity of bacteria recovered would require further screening. Though it was able to grow on Bushnell Haas agar with crude oil as sole carbon source, there is paucity of information on its role in hydrocarbon degradation and thus would require further screening for the presence hydrocarbon degradative genes alongside other isolates recovered. The unidentified bacterium (strain BH23) can be said to be potentially novel species and will be investigated further for its role in hydrocarbon degradation.

The incubation of marine sediment in the presence of phenanthrene and bromodeoxyuridine revealed Shewanella as part of the diverse microorganisms capable of PAH degradation [32]. In another study of the influence of crude oil on changes in bacterial communities in Arctic sea-ice, the dominant phylotypes recovered were Pseudomonas spp., Shewanella sp and Marinobacter spp [33].

Three strains of Gordonia sp have been reported to be capable of degrading pyrene, benzo[a] pyrene, anthracene, naphthalene, phenanthrene, and fluoranthene as the sole source of carbon and energy [34]. Sphingobacterium multivorum have previously been reported as part of bacteria consortia capable of utilizing used lubricant [35].

CONCLUSION

This study revealed potentially novel bacterial species and previously described hydrocarbon degrading bacterial species that can be characterized further to determine their role in hydrocarbon degradation as well as their salt tolerance level prior to application in bioremediation of saline environments.

ACKNOWLEDGEMENT

We acknowledge the International Foundation of Science (IFS) Stockholm, Sweden, for funding this research through grant W4263-2 given to Dr. (Mrs.) Chioma Blaise

Chikere as well as staff members of Molecular Biology and Biotechnology Division, Nigerian Institute of Medical research (NIMR) Yaba, Lagos Nigeria.

REFERENCES

1. Philp JC, Whitely AS, Ciric L, Bailey MJ (2005) Monitoring bioremediation. In: Atlas RM, Philp J, editors. Bioremediation: applied microbial solutions for real-world environmental cleanup. (ASM) Press Washington DC, pp. 237-292.
2. Jernelov A (2010) The threats from oil spills: now, then, and in the future. Ambio 39: 353-366.
3. Kostka JE, Prakash O, Overholt WA, Green SJ, Freyer G (2011) Hydrocarbon-degrading bacteria and the bacterial community response in Gulf of Mexico beach sands impacted by the deep water horizon oil spill. Appl Environ Microbiol 77: 7962-7974
4. Eze VC, Okpokwasili GC (2010) Microbial and other related changes in a Niger Delta River Sediment receiving Industrial effluents. Cont J Microbiol4: 15-21.
5. United Nations Environmental Programme (2011) Environmental Assessment of Ogoniland, pp. 1-262.
6. Cuadros-Orellana S, Pohlschroder M, Durrant LR (2006) Isolation and characterization of halophilic archaea able to grow in aromaticcom-pounds. Int Biodeter Biodegrad 57: 151-154.
7. Fathepure BZ (2014) Recent studies in microbial degradation of petroleum hydrocarbons in hypersaline environments. Front microbiol.
8. Kushner DJ (1978) Life in high salt and solute concentrations: halophilic bacteria. In: Kushner DJ (ed) Microbial Life in Extreme Environments, London: Academic Press Ltd 317-368.
9. Oren A (2013) Life at high salt concentrations. In: Rosenberg E, Delong EF, Lory S, Stackebrandt E, Thompson F (eds) The Prokaryotes-Prokaryotic Communities and Ecophysiology. (Berlin; Heidelberg: Springer-Verlag), 421-440.
10. Prince RC (2005) The microbiology of marine oil spill bioremediation. In: Ollivier B, Magot B (eds) Petroleum microbiology. (ASM) Press Washington DC, pp. 317-335
11. Singh AK, Sherry A, Gray ND, Jones DM, Bowler BF, et al. (2014) Kinetic parameters for nutrient enhanced crude oil biodegradation in intertidal marine sediments. Front microbiol.
12. Dyksterhouse SE, Gray JP, Herwig RP, Lara JC, Staley JT (1995) Cycloclasticus pugetii gen. nov., sp. Nov., an aromatic hydrocarbon degrading-bacterium from marine sediments. Int J Syst Bacteriol45: 116-123.
13. Hedlund BP, Geiselbrecht AD, Bair TJ, Staley JT (1999) Polycyclic aromatic hydrocarbon degradation by a new marine bacterium, Neptunomonas naphthovorans gen. nov., sp. nov. Appl Environ Microbiol 65: 251-259.

14. Yakimov MM, Golyshin PN, Lang S, Moore ER, Abraham WR, et al. (1998) Alcanivorax borkumemsisgen. nov., sp. nov., a new hydrocarbon-degrading and surfactant producing marine bacterium. Int J Syst Bacteriol48: 339-348.

15. Chikere CB, Chikere BS, Okpokwasili GC (2012) Biorector-based bioremediation of hydrocarbon.-polluted Niger Delta marine sediment, Nigeria. 3 biotech 2: 53-66.

16. Engelhardt MA, Daly K, Swannell RPJ, Head IM (2001) Isolation and characterization of a novel hydrocarbon-degrading, Gram-positive bacterium, isolated from intertidal beach sediment, and description of *Planococcus alkanoclasticus* sp nov. J Appl Microbiol90: 237-247.

17. Golyshin PN, Chemikova TN, Abraham WR, Lunsdorf H, Timmis KN, et al. (2002) *Oleiphilaceae* fam nov., to include *Oleiphilus messinensis* gen. nov., sp. nov., a novel marine bacterium that obligatory utilizes hydrocarbons. Int J SystEvol Microbiol52: 901-911

18. Yakimov MM, Giullano L, Gentile G, Crisafi E, Chemikova TN, et al. (2003) *Oleispira antarctica* gen. nov., sp. nov., a novel hydrocarbon- oclastic marine bacterium isolated from Antarctic coastal seawater. Int J Syst Evol Microbiol53: 779-785.

19. Yakimov MM, Giuliano L, Denaro R, Grisafi E, Chernikova TN, et al. (2004) *Thalassolituus oleivorans* gen. nov., sp nov., a novel marine bacterium that obligately utilizes hydrocarbons. Int J Syst Evol Microbiol54: 141-148.

20. American Public Health Association (APHA) (2008) Standard Methods for the Examination of Water and Wastewater. 21st ed. Washington, D.C.

21. Luis M, Maria TP, Janet S, Grace T, Ellana MP (2013)Colour Atlas of Medical Bacteriology. ASM Press Washington DC.

22. Surridge AKJ (2007) Denaturing gradient gel electrophoresis characterization of microbial communities in polycyclic aromatic hydrocarbon and polychlorinated biphenyl contaminated soil. PhD Thesis, Department of Microbiology and Plant Pathology, University of Pretoria, South Africa.

23. Garcia MT, Ventosa A, Mellado E (2005) Catabolic versatility of aromatic compound-degrading halophilic bacteria. FEMS microbiol ecol 54: 97-109.

24. Ibiene AA, Orji FA, Ezidi CO, Ngwobia CL (2011) Bioremediation of hydrocarbon contaminated soil in the Niger Delta using spent mushroom compost and other organic wastes. Niger J Agric Food Environ. 7: 1-7.

25. Varnam AH, Evans MG (2000) Non-alkaline, highly saline environments. In: Varnam AH (ed) Environmental microbiology. ASM Press Washington DC, pp. 144-151.

26. Dastgheib SM, Amoozegar MA, Khajeh K, Shavandi M, Ventosa A (2012) Biodegradation of polycyclic aromatic hydrocarbons by a halophilic microbial consortium. Appl Microbiol Biotechnol 95: 789-798.

27. Said B, Goni-Urriza MS, El-Bour M, Dellali M, Aissa P, et al. (2007) Characterization of aerobic polycyclic aromatic hydrocarbon- degrading bacteria from Bizerte lagoon sediments Tunisia. J Appl Microbiol 107: 987-997.

28. Chikere CB, Azubuike CC (2013) Catechol 2, 3-dioxygenase screening in putative hydrocarbon utilizing bacteria. Int Res J Microbiol 4: 1-6.

29. Okore C, Mbanefo O, Onyekwere B, Onyewenjo S, Abba-Father C (2013) Isolation and Characterization of Biosurfactants Producing Bacteria from Oil Polluted Soil. J Nat Sci Res 3: 2225-0921.

30. Ghevariya C, Bhatt J, Dave B (2011) Enhanced chrysene degradation by halotolerant Achromobacter xylosoxidans using response surface methodology. Bioresour Technol 102: 9668-74

31. Al-Awadhi H, Sulaiman R, Mahmoud H, Radwan S (2007) Alkaliphilic and halophilic hydrocarbon utilizing bacteria from Kuwaiti coasts of Arabian Gulf. Appl Microbiol Biotechnol 77: 183-186.

32. McGenity TJ, Folwell, BD, McKew BA, Sanni GO (2012) Marine crude-oil biodegradation: a central role for interspecies interactions. Aquat Biosyst8: 10.

33. Gerdes B, Brinkmeyer R, Dieckmann G, Helmke E (2005) Influence of crude oil on changes of bacterial communities in Arctic sea-ice. FEMS Microbiol Ecol 53: 129-139

34. Hu FC, Li XY, Su ZC, Wang XJ, Zhang HW, et al. (2011) Identification and degradation capability of three pyrene-degrading *Gordonia* sp. strains. J Appl Ecol 22: 1857-62.

35. Saimmai A, Keauwrueng J, Maneerat S (2012) Used lubricating oil degradation and biosurfactant production by SC-9 consortia obtained from oil-contaminated soil. Ann Microbiol.

INDEX

M

N

O

P

Q

S

T

U

V

W